Mechanical Measurement & Control for Mechanical Engineering

AF520146

Arvind Kaushal,

BE. (Mechanical Engineering), ME. (Advance Production System),

Formerly Faculty of Mechanical Engineering, I.G.Govt. Engineering College, Sagar, MP.

Rekha Arvind Kaushal,

BE. (Information Technology), M.TECH. (Computer Science & Engineering),

Presently Faculty of Computer Science & Engineering,S.R. Govt. Polytechnic College, Sagar, MP.

Ajay Vardhan,

BE. (Mechanical Engineering), ME. (Heat Power & Thermal Engineering),

Presently Assistant Professor, Department of Mechanical Engineering,

Rajiv Gandhi Proudyogiki Vishwavidyalaya , Bhopal , MP.

Published by

BONFRING®

Intellectual Integrity

Mechanical Measurement & Control for Mechanical Engineering

9 789387 862326 >

ISBN 978-93-87862-32-6

Authors

Arvind Kaushal

Rekha Arvind Kaushal

Ajay Vardhan

Bonfring

309, 2nd Floor, 5th Street Extension, Gandhipuram,

Coimbatore-641 012.

Tamilnadu, India.

E-mail: info@bonfring.org

Website: www.bonfring.org

Phone: 0422 4213231

This Book is Dedicated To

My Parents, My Wife, for their Love & Supports

Preface

In present scenario, the article manufacturing is completely depending upon the manufacturing process or production process but the associated methodology can be implemented to achieve better performance with safe operation. Since safe operation indicate that there is a necessity of measurement & process controlling. Today we are glad to proposed the book which is entitled as "**Mechanical Measurement & Control for Mechanical Engineering**" This text book not only provides a basic foundation for engineers in industry, scientists, designers, managers, research personnel but also creates a guiding path & valuable reference for the students who will be become a parts of mechanical manufacturing in future. The overall study material has been especially designed and prepared for undergraduate and post-graduate students & equally applicable for mechanical engineering & instrumentation as well as production engineering.

Pedagogical Features of Book

(i) Each chapter begins with basics of respective title with easy & lucid language.

(ii) Each chapter includes suitable solve numerical problems & Practice problems at the end of the chapter

(iii) New advanced material is added to improve its usefulness.

(iv) The inclusion of well supported excellent theory with suitable plot and graph helps for easy understanding.

Course Outlines

The overall course coverage is based on the syllabus of Rajiv Gandhi proudyogiki Vishwavidyalaya (MP) which is a leading reputated technical university of Madhya Pradesh but it is also covers the syllabus of other technical universities. This book is divided into six parts/chapters, **Chapters 1** includes an introduction of measurement or basics concept of measurement, Calibration, Errors, data presenting technique, Measurement System Behavior etc. **Chapter 2** deals with Statistics, Uncertainty Analysis, Design stage uncertainty analysis, data acquisition, data reduction, Error propagation etc. **Chapter 3** deals with the Temperature Measurement, Thermometry, Bimetallic Thermometers, thermocouple, Pressure and Velocity Measurement, deadweight tester, Flow measurement, Orifice meter, Venturi meter, roto-meter. **Chapter 4** provide us complete information regarding various strain Measurement technique, Motion, Force and Torque Measurement technique, Velocity measurement technique, moving coil transducers; angular velocity measurement, stroboscopic measurement, Force measurement, measurement of torque on rotating shafts, Power estimation from rotational speed and torque, load cells, Potentiometers, Linear variable differential transformers, rotary variable differential transformer, resistance strain gauges etc. **Chapter 5** Open loop and closed loop control, Transfer function, impulse response function, block diagram of closed loop system, block diagram reduction, Transient and steady state response analyses: First order systems, unit step and unit impulse response of first order systems, second order systems, unit step and unit impulse response of second order systems, transient response specifications, modeling of mechanical systems, modeling of electrical systems, signal flow graphs, modeling of fluid systems, liquid level systems, **Chapter 6** deals with the possible recommend lab work structure regarding to mechanical measurement & control. This text book is flexible and can be used in a variety of course structures at both the undergraduate and graduate levels.

Arvind Kaushal

Rekha Arvind Kaushal

Ajay Vardhan

Acknowledgement

Any kind of manuscript or book preparation is very complicated and difficult task which could not be achieved by a single person. At very initial it is a constant supports & motivation of God for giving me the strength and thought for this task. We are heartly expressing our sincerest appreciation to my parents and others family members, teachers, colleagues, as well as discussion and question form under graduate's students throughout the teaching session except it many others individuals have also been contributed to the preparation of this book. I am also extending my sincere thanks to Professor **Dr.R.S. Rajput** who is Director of **UIT** (University Institute of Technology) Rajiv Gandhi Proudyogiki Vishwavidyalaya Bhopal **(MP)** for their constant supports and encouragement. Last but not the least I am thanks to all teachers who have taught me thereby I have been able to achieve this state of study. Finally, this work is wholly dedicated to **GOD.**

Arvind Kaushal

Rekha Arvind Kaushal

Ajay Vardhan

Section–I

Chapter	Contents	Page No

Section–II

Lab Work

Section–III

Formula for Numerical Problem

Section–I

CHAPTER 1

BASIC CONCEPTS OF MEASUREMENT

1.0. Introduction

The basic concept of Measurement is to achieve a reading with respect to a particular reference for safe and effective operation of a particular process that is generally adopted by engineers in mechanical industries or any other industries, scientists, design engineers, research scholars and by students. In general, We make measurements every day everywhere. Consider an example like measurement of outdoor temperature by thermometer for any day, measurement of length, the indication of fuel volume of the fuel tank of vehicle, required proportion of ingredients for cooking a recipe. In the above cases we always perform the selection of instruments for these routine measurements on the basis of direct data except the accuracy of applied method. But when accuracy and performance is on prime consideration then attention would be based on the selection of measuring equipment and techniques with the required interpretation because the basic objective of measurements is to recognize the deviation of some variable with respect to true value, the results of which are addressed as an output of the measuring device. Thus we can say that *"An measuring instrument is a device that transforms a physical variable of interest (the measurand) into a form that is suitable for recording (the measurement). It is common to employ a standard system of units by which the measurement from one instrument can be compared with the measurement of another "*

In mechanical system, the measurement may be applied to fulfill the purpose like length (in meters, inches, etc), weight or load, temperature, displacement, flow rate, pressure measurement, velocity measurement etc. The objective of any measurement is based on the output of the measurement device or system.

There are important issues to be addressed to ensure that the output of the measurement device is a reliable indication of the true value of the measured variable so therefore we must address the following important questions regarding the reliability of measuring system these are:

- How can be a measurement or test plan is performed by **Engineer / Operator** so that we can achieve a unique single output with great reliability?
- How can a measurement system be used so that the engineer can easily interpret the measured data according to the process?

In general, Engineers design physical systems in the form of machines or mechanism to perform some specified task. The behavior of the parts of the machine during the operation of the machine needs to be examined or analyzed or designed such that it is functioning reliably or not, in order to get a safe and effective operation. Such an activity needs data regarding the machine parts in terms of material properties, For example temperature, displacement, or flow rate and creating a voltage that is proportional to it. This voltage can then be measured with a voltmeter, a digital multi-meter, an oscilloscope or a computer equipped with analog to digital conversion boards. There are many options available when assembling a measurement system and it is possible to design different systems that will achieve the same objectives.

The design of a good MEASUREMENT system is an iterative process, partly because initially you will probably not have all the information that you require for the design, especially if the quantity that you are measuring has not been measured on this particular machine before.

i. **Measuring instrument:** The device that is used for measurement of certain physical quantity is called as measuring instrument. The measuring instruments are used frequently in our day-to-day life for the measurement of various quantities like length, weight, temperature, pressure, current, voltage, torque, velocity etc. The instruments indicate the value of these quantities, based on which we get some understanding and also take appropriate actions and decisions. Usually there are two type of measuring instrument are Applied, These are Analogue instruments and Digital measuring instruments. The brief description is given below.

ii. **Analogue Instruments:** The analogue instruments indicate the magnitude of the quantity in the form of the pointer movement. One has to learn reading such instruments since there are certain markings on the scale. They usually indicate the values in the whole numbers, though one can get the readings up to one or two decimal places also. The readings taken in decimals places may not always be entirely correct, since some human error is always involved in reading.

iii. **Digital measuring instruments:** The digital measuring instruments indicate the values of the quantity in digital format that has numbers for easy read out in the numerical form. They can give the readings in one or more decimal places. Since there is no human error involved in reading these instruments, they are more accurate than the analogue measuring instruments.

1.1. Basic Concept of Measurement

Measurement is the process of comparing unknown magnitude of certain parameter with the known predefined standard of that parameter. For instance, if we have to measure the temperature of the body, we measure it with the thermometer that has predefined scale indicating different values of the temperature. If we want to measure the length of the wall , we measure it with the measuring tape that has predefined markings on it. The measurement enables us obtaining magnitude of certain parameters whose value is not known by comparing them with the standards whose value is predefined. Measurement is the process of systematically assigning numbers to objects and their properties to facilitate the use of mathematics in studying and describing objects and their relationships. For instance, measuring a person's weight in pounds or kilograms or his height in feet and inches or in meters. Note that the particular system of measurement used is not as important as the fact that we apply a consistent set of rules like we can easily convert a weight expressed in kilograms to the equivalent weight in pounds, as long as the system has a consistent relationship with the property being measured, we can use the results in calculations. Measurement is not limited to physical qualities such as height and weight. Tests to measure abstract constructs such as intelligence or scholastic aptitude are commonly used in education and psychology, and the field of psychometrics is largely concerned with the development and refinement of methods to study these types of constructs. Establishing that a particular measurement is accurate and meaningful is more difficult when it can't be observed directly. Although you can test the accuracy of one scale by comparing results with those obtained from another scale known to be accurate.

1.1.1. Methods of Measurement

There are two methods of measurement:

i. Direct comparison with the standard, and

ii. Indirect comparison with the standard.

(i) Direct Comparison with the Standard: In the direct comparison method of measurement, we compare the quantity directly with the primary or secondary standard. For instance, if we have to measure the length of the bar, we will measure it with the help of the measuring tape or scale that acts as the secondary standard. Here we are comparing the quantity to be measured directly with the standard. Even if you make the comparison directly with the secondary standard, it is not necessary for you to know the primary standard. The primary standards are the original standards made from certain standard values or formulas.

The secondary standards are made from the primary standards, but most of the times we use secondary standards for comparison since it is not always feasible to use the primary standards from accuracy, reliability and cost point of view. There is no difference in the measured value of the quantity whether one is using the direct method by comparing with primary or secondary standard. The direct comparison method of measurement is not always accurate. In above example of measuring the length, there is limited accuracy with which our eye can read the readings, which can be about 0.01 inch. Here the error does not occur because of the error in the standards, but because of the human limitations in noting the readings. Similarly, when we measure the mass of any body by comparing with some standard, it's very difficult to say that both the bodies are of exactly the same mass, for some difference between the two, no matter how small, is bound to occur. Thus, in direct method of measurement there is always some difference, however small, between the actual value of the quantity and the measured value of the quantity.

(ii) Indirect Method of Measurement : In the indirect method of measurements some transducing devise, called transducer, is used, which is coupled to a chain of the connecting apparatus that forms the part of the measuring system. In this system the quantity which is to be measured (input) is converted into some other measurable quantity (output) by the transducer. The transducer used is such that the input and the output are proportional to each other. The readings obtained from the transducer are calibrated to as per the relations between the input and the output thus the reading obtained from the transducer is the actual value of the quantity to be measured. Such type of conversion is often necessary to make the desired information intelligible. The indirect method of measurements comprises of the system that senses, converts, and finally presents an analogues output in the form of a displacement or chart. This analogues output can be in various forms and often it is necessary to amplify it to read it accurately and make the accurate reading of the quantity to be measured. The majority of the transducers convert mechanical input into analogues electrical output for processing, though there are transducers that convert mechanical input into analogues mechanical output that is measured easily.

(i) Requirements of Measurement: For the measurements results to be accurate , two point must be followed

- The standard which is used for comparison must be well defined accurately and it should be universally accepted. For instance, the weight cannot be just light or heavy. It should be light or heavy in comparison to some standard weight and should be measured accurately against it.

- Another condition for measurements is that the procedure applied for the measurements should be provable and there should be provable instruments for measurements. This means the methods for making the measurements and the instruments used for them should be reliable enough to make the correct measurements.

(ii) Working of measurement system: The Generalized measurement system is a system that is comprised of the typical elements of a measurement system. It helps to understand how a measurement system works. The diagram given below shows the Basic Measuring Process.

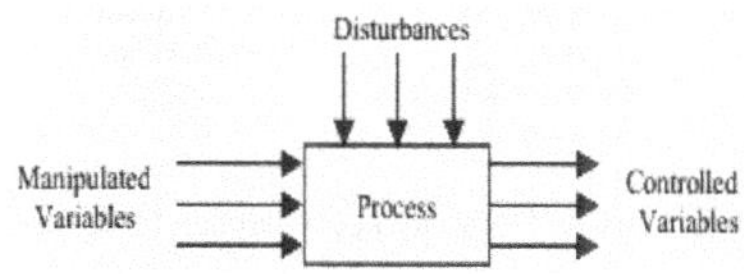

Diagram 1.1: (a) Process Control Variables

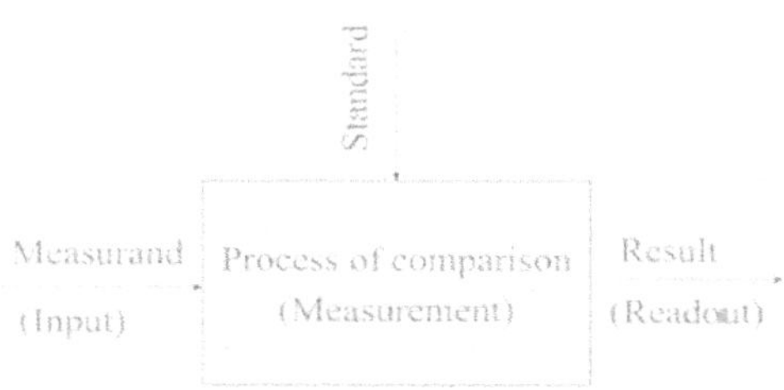

Diagram 1.1: (b) Process of Measurement

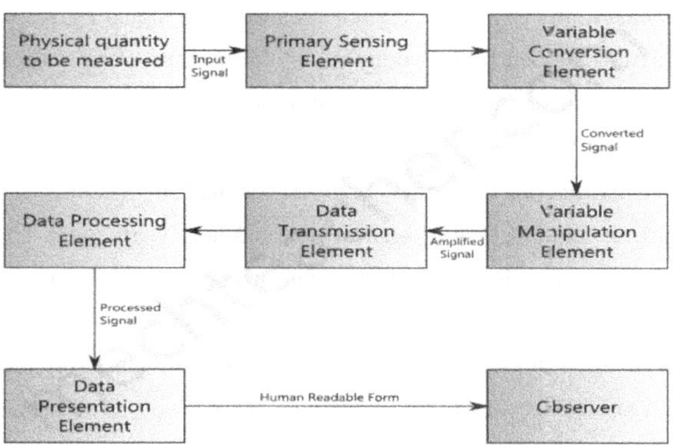

Diagram 1.1: (c) Basic Measuring Process

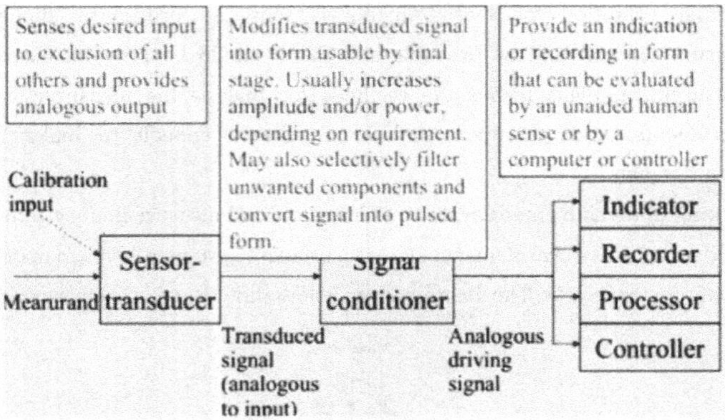

Diagram 1.2: Generalized Measuring System

A brief description of all those elements of measuring system are given below

(i) **Primary Sensing Element:** The primary sensing element receives signal of the physical quantity to be measured as input. It converts the signal to a suitable form like electrical, mechanical or other form), so that it becomes easier for other elements of the measurement system, to either convert or manipulate it.

(ii) **Variable Conversion Element:** Variable conversion element converts the output of the primary sensing element to a more measurable suitable form.

(iii) **Variable Manipulation Element:** Variable manipulation element, manipulates and amplifies the output of the variable conversion element.

(iv) **Data Processing Element:** Data processing element is an important element used in many measurement systems. It processes the data signal received from the variable manipulation element and produces suitable output. Data processing element may also be used to compare the measured value with a standard value to produce required output.

(v) **Data Transmission System:** Data Transmission System is simply used for transmitting data from one element to another. Work as a communication path between different elements of the measurement system. The data transmission elements may be cables, wireless, transducers or may be others.

(vi) **Data Presentation Element:** It is used to present the measured physical quantity in a human readable form to the observer. It receives processed signal from data processing element and presents the data in a human readable form.

1.2. Functions of the Measuring Instruments

Most of the measuring instruments indicate the value of parameter in the form of the indicator movement, which gives us the magnitude of that quantity. Following are very important function of measuring instrument.

Indicating the value of the physical quantity: The instruments are calibrated against the standard values of the physical quantities. The movement of the pointer directly indicates the magnitude of the quantity, which can be whole numbers or also fractions. for example in many digital instruments numerical as well as decimals reading may be observed.

- **Measuring instruments used as the controllers:** Sometimes the measuring instruments worked as measuring instruments as well as controlling device, for instance compressor work as, pressure measuring & controlling device and the thermostat starts or stops the compressor of the refrigeration system depending on the temperature achieved in the evaporator.
- **Recording the data:** Some measuring instruments can also be used to record and also store the data. In present time the application of computer helps to storing, recording & retrieving data can easily handle. For example chrono-cycle graph used for measuring the hand motion of the operator, strain–stress curve achieved during tensile testing on universal testing machine.
- **Transmitting the data:** The measuring instruments can also be used to transfer the data to some distance places. The instruments kept in unsafe locations like high temperature can be connected by wires and their output can be taken at some distant places which are safe for the human beings. The signal obtained from these instruments can also be used for operating some controls.
- **Do calculations:** Some measurement is also carried out a number of calculations like Addition, Subtraction, Multiplication, Division, etc. Some can also be used to find solutions to highly complex equations.

1.3. Advantages & Disadvantages of Measuring Instruments

1.3.1. Advantages of the Digital Instruments

- They are very easy to read.
- Since there are very few moving parts in the electronic instruments, they are usually more accurate than the analogue instruments. Even the human error involved in

reading these instruments is very less, which adds to the accuracy of digital instruments.

- The electronic items tend to be cheaper than the mechanical items.
- The data from the instruments can be recorded for future reference.
- The output of the digital devices can be obtained in the computer

1.3.2. Disadvantages of the Digital Instruments

- Sometimes they tends to indicate undesirable values due to faulty electronic circuit or damaged display.
- In case of high humidity and corrosive atmosphere the internal parts may get damaged and indicate the faulty values.
- Sometimes these instruments show some readings even though there is no applied measurable parameter.

1.3.3. Advantages of the Digital Instruments over the Analogue Instruments

- They are very easy to readout the output.
- Since there are very few moving parts in the electronic instruments, they are usually more accurate than the analogue instruments. Thus the effect of human error may be considered negligible or very less, which adds to the accuracy of digital instruments.
- The electronic items tend to be cheaper than the mechanical items.
- The data from the instruments can be recorded for future reference.
- The output of the digital devices can be obtained in the computer

1.4. Errors in Measuring Instruments

In order to understand the concept of errors in measurement, we should know about the two terms that helps to define the various possible error which can be obtained throughout the measurement.

(i) True Value: True value may be defined as the average value of an infinite number of measured values when average deviation due to various contributing factor will approach to zero.

(ii) Measured Value: It may be defined as the approximated value of true value. It can be find out by taking means of several measured readings during an experiment, by applying suitable approximations on physical conditions. Now we are in a position to define static error. **Static error is defined as the difference of the measured value and the true value of the quantity.**

There are three types of errors usually achieved throughout the measuring process whenever measurement is performed these are Assembly Errors, Environmental Errors, and Random Errors is comes on the front line. All these have been discussed below.

(i) Assembly Errors: The assembly errors are the errors in the measuring instrument due to improper manufacturing of the instruments. Various components of the instrument are made separately and then they are assembled together. There may be errors in the individual components or in the assembly of the components. Here are some of the possible assembly errors:

- **Displaced scale:** This is the incorrect fitting of the measuring scale. For instance the zero of pointer may not coincide with actual zero on the scale. Sometimes the scale gets cracked, thus showing the faulty readings.
- **Non-uniform scale:** If the scale division of the measuring instrument is inexact & non – uniform may also a cause of error in measurement.
- **The pointer is bent:** This happens in many cases. The pointer may get bent in either horizontal direction or the vertical direction, in either case, is shows errorful reading.
- **Manufacturing errors in the components:** The instruments are made up of a number of small components, which may be manufactured in different places. Sometimes there are manufacturing errors in some of these components like gear, lever, links , hinges etc.

(ii) Environmental Errors: The measuring instruments are assembled and calibrated in certain environmental conditions and are designed to be used for specific set of conditions, like temperature, pressure, humidity. Then measurement is imposed certain error, which is considered as an environmental errors. Hence most of the instruments are designed to be used within certain limits of temperature, pressure, humidity, altitude etc but the errors due to change in environmental conditions could not be evaluate. These errors can be sometimes positive, negative and combination of the both. Here are some precautions must be taken to reduce the environmental errors in the instruments:

- Use the instruments within the specified limits of temperature, pressure and humidity for which the instrument has been designed. These limits are mentioned in the instruments instructions manual.
- If you have to use the instrument beyond the specified limits of environmental conditions, there must be apply suitable corrections to the recorded data.
- Calibrate the instrument time to time, if necessary.

- Error compensations should be automatically in the measuring instruments.

Except assembly and environmental errors some other error may be possible which could not be evaluated called as random errors.

- **Frictional errors:** There are number of moving mechanical parts in the analogue measuring instruments which imposed frictional error in measurement.
- **Mechanical vibrations:** When the instrument is used in vibrating place the parts of the instrument start vibrating also giving faulty readings.

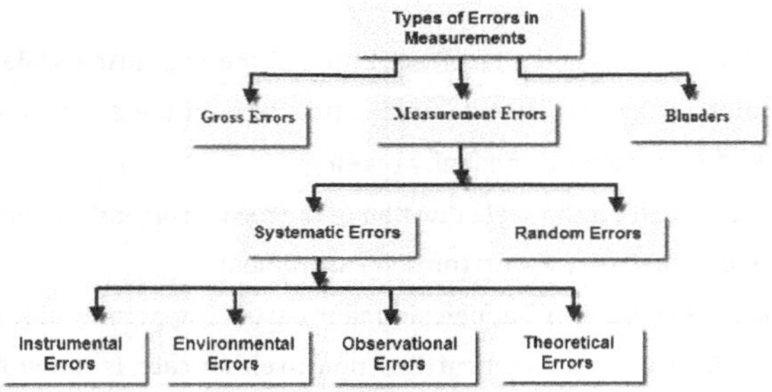

Diagram 1.3: Various Types of Errors in Measuring Instruments

(iii) Random Errors : The random Errors are due to the following reasons

- **Frictional errors:** Since measuring instruments are assembly of various moving parts or components thereby its movements creates friction between them and wear and tear is achieved. This frictional resistance leads to errors throughout the measurement. In order to prevents such kind of error the analogue measuring instruments must be inspected & calibrated periodically or if necessary it may be replaced by new one.
- **Mechanical vibrations:** Whenever the measuring instrument is operated in place where vibration exists then components of the measuring instrument also starts vibration hence we read out error full quantity.
- **Backlash in the movement:** This is the error due to time lag between the application of the parameter and the instrument actually showing reading.
- **Hysteresis of the elastic members:** Due to long time application the of measuring instrument or It becomes obsolete then the elastic members tend to lose some elasticity so that we have also errors in the indicated value of the instrument.
- **Finite scale divisions:** The scale marking can be made only up to certain limits and always try to make proper division if not it gives faulty reading.

(iv) Systematic Error Sources or Bias: The systematic error or Bias can be imposed by measuring device throughout the measurement process are due to the following reasons.

- One of the Most Common & Popular cause for Systematic Error or Bias are mis-calibration that occurs due to the change in input–output response of a sensor, For example, An improperly calibrated thermometer which consistently reads 99ºC in boiling water instead of 100ºC.
- If input modification & interference of input takes place then sensor mis-calibration occurs.
- Aging of the measuring device components & Damage of the sensor is also a reason of instrument mis-calibration.
- Systematic errors can also be introduced if the measurement process Itself changes the intended measurand.
- If the signal is modified in some way, the indicated measurement will be different from the sensed value.
- Systematic errors or bias can also be achieved due to by the observers when he or she readout the indicated dial because the apparent reading will be shifted from the correct value.

Example of Systematic Error Sources Bias

- In physical signal paths such as mechanical systems that transmit force or displacement, friction can modify the propagation of the signal.
- If temperature is a modifying input, using the sensor at a temperature other than the calibrated temperature will result in a systematic error.
- In electrical circuits, Resistance or attenuation can also modify the Signal, Resulting in a Systematic error.
- A meter, Micro-meter, Vernier calliper, or other instrument which was not properly zeroed or for which the zero correction factor was not considered.
- Theoretical errors due to a Simplified Mathematical Model for the system which consistently gives a calculated value different from the calculated value predicted from a more accurate mathematical model.

In order to eliminate or minimise these systematic errors, sensors should be check & recalibrated against a predefined standardtime to time if necessary, on the other hand, if systematic error source has been detected then corrective action should be performed by applying compensation methods in which correction factor can compensate or reduce the error .

Precaution while operating the instruments:

- **Calibration (systematic):** In case of precise measurement, the calibration of an instrument should be checked before starting. If it does not possible then, the accuracy of the instrument should be checked by comparing with another instrument that is at least as precise.
- **Zero offset (systematic):** When making a measurement with a Micrometer caliper, Electronic balance or Electrical meter, always check the zero reading first or Re-zero it .If does not device will results in a constant error that is more significant for smaller measured values than for larger ones.
- **Parallax (systematic or random):** This error can occur whenever there is some distance between the measuring scale and the indicator used to obtain a measurement. If the observer's eye is not squarely aligned with the pointer and scale, the reading may be too high or low.
- **Instrument drift (systematic):** Most electronic instruments have readings that drift over time. The amount of drift is generally not a concern, but occasionally this source of error can have significant.
- **Lag time and hysteresis (systematic):** Some measuring devices require time to reach equilibrium, and taking a measurement before the instrument is stable will result in a measurement that is too high or low. For example

A common example is taking temperature readings with a thermometer that has not been reached thermal equilibrium with its environment.

A similar effect is hysteresis where the instrument readings lag behind and appear to have a "memory" effect, as data are taken sequentially moving up or down through a range of values.

(v) Gross Errors : Such kind of error is involve due to carelessness while taking reading or recording by human operator , For example if operator read out by fault 29 as 21. Then these types of errors falls under this category. Hence A proper care must be taken throughout the reading, recording and during calculation of error.

(vi) Instrumental Errors: These errors may be due to wrong construction, calibration of the measuring instruments. These types of error may be arising due to friction or may be due to hysteresis. These types of errors also include the loading effect and misuse of the instruments. Misuse of the instruments results in the failure to the adjust the zero of instruments. In order to minimize the gross errors in measurement various correction factors must be applied and in extreme condition instrument must be re-calibrated carefully.

(vii) Observational Errors: As the name suggests these types of errors are due wrong observations. The wrong observations may be due to parallax. In order to minimize the parallax error highly accurate meters are required, provided with mirrored scales.

(viii) Errors in Calculated Quantities: Always in scientific experimentation we often use the measured values of particular quantities to calculate a new quantity. The error in the new quantity depends upon the errors involved in the previously measured quantities.

- **Addition & Subtraction:** When two or more quantities are added or subtracted to calculate a new quantity, we add the maximum probable errors in each quantity to obtain the maximum probable error in the new quantity. **For Example:** When heating water we may measure the starting temperature to be (35.0 ± 0.5)°C and the final temperature to be (85 ± 0.5)°C. The change in temperature is therefore (85.0 – 35.0)°C ± (0.5+0.5)°C or (50.0 ± 1.0)°C. Thus we add the MPE's in the measurements to obtain the MPE in the result.
- **Multiplication & Division:** When two or more quantities are multiplied or divided to calculate a new quantity, we add the percentage errors in each quantity to obtain the percentage error in the new quantity. **For Example:** Let us assume we are to determine the volume of a spherical ball bearing. After performing a series of measurements of the radius using a micrometer screw gauge, the mean value of the radius is found to be 9.53 mm ± 0.05 mm. Thus, the percentage error in the radius is 0.5%.

[% error = (0.05/9.53)x100]

The formula for the volume of a sphere is :

$$V = 4/3\ p\ r^3 \quad \text{...} \ 1$$

Using this formula, the value for the volume of the ball bearing is found to be 3625.50 mm^3. Note that the only measured quantity used in this calculation is the radius but it appears raised to the power of 3.

The formula is really:

$$V = 4/3\ p\ r \times r \times r \quad \text{...} \ 2$$

Then, % Error in volume = % Error in r + % Error in r + % Error in r

Therefore, % Error in volume = 0.5% + 0.5% + 0.5% = 1.5%

Hence the volume of the ball bearing will be therefore 3625.50 mm^3 ± 1.5% or 3625.50 mm^3 ± 54.38 mm^3.

(x) Limiting Errors or Guarantee Errors: Limiting error is an error which is offered by a manufacturer for any measuring instruments indicating that the maximum possible error range. For example suppose any manufacturing company promise that the error in the ammeter is selling not greater the limit he sets. This limit of error is known as limiting errors or guarantee error.

(xi) Relative Error or Fractional Error: It is defined as the ratio of the error and the specified magnitude of the quantity. Mathematically,

$$\text{Relative Error} = \frac{dA}{A}$$

where, dA is the error and A is the magnitude.

(xii) Error due to Other Factors

- **Temperature Variation:** Variation in temperature not only changes the values of the parameters but also brings changes in the reading of the instrument. For a consistent error, the temperature must be constant. Effect of the service time on Instruments which creates the change in calibration of the instrument with time. This may be called ageing of the instrument.
- **Effect of External Electrostatic and Magnetic Fields:** These electrostatic and magnetic fields influence the readings of instruments. These effects can be minimized by proper shielding.
- **Mechanical Error:** Friction between stationary and rotating parts and residual torsion in suspension wire cause errors in instruments.

1.5. Important Terms of Measurement Process

(i) Accuracy: It is defined as the degree of closeness with which the reading approaches the true value of the quantity to be measured.

The accuracy can be expressed in following ways:

- **Point Accuracy:** Such type of accuracy is obtained at a one particular point of scale.
- **Accuracy at % of scale span:** When an instrument as uniform scale , its accuracy may be expressed in terms of scale range.
- **Accuracy at % of true value:** It is defined as the true value of the quantity being measured in %.

- **Precision:** It is defined as a measure of the degree of agreement to which an instrument gives constant results when repeated measurements is performed for same true value with the same instrument. Diagram shows a relationship between precision and accuracy.

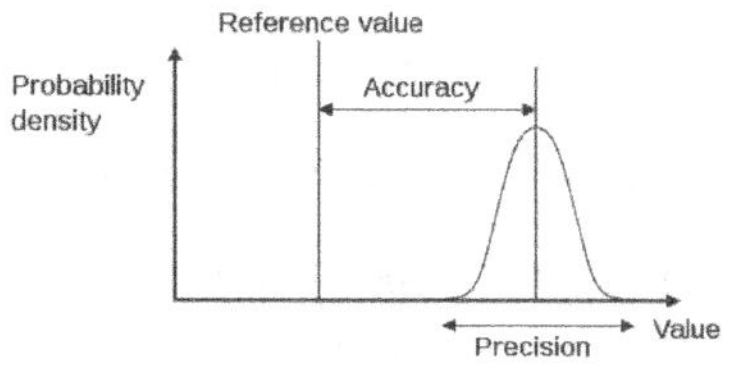

Relationships between Precision and Accuracy

Given curve shows a reference value for any particular measurement, the area under the distribution curve represents the precision value of the measuring instruments. The center line of the distribution curve and the area between the referenced values indicating the accuracy of the measuring devices.

- **Number of significant figures:** The precision of the measurement is obtained from the number of significant figures, in which the reading is expressed. The significant figures convey the actual information about the magnitude & the measurement , precision of the quantity. The precision can be mathematically expressed as:

$$P = 1 - \left| \frac{X_n - \overline{X_n}}{\overline{X_n}} \right| \quad \text{..} 3$$

Where ,

P = Precision

X_n = Value of nth measurement

$\overline{X_n}$ = Average value the set of measurement values

(ii) Sensitivity: It is defined as the ratio of the changes in the output of an instrument to a change in the value of the quantity to be measured.

- If the calibration curve is linear, the sensitivity of the instrument is the slope of the calibration curve.
- If the calibration curve is not linear , then the sensitivity varies with the input. Inverse sensitivity or deflection factor is defined as the reciprocal of sensitivity.

Inverse sensitivity or deflection factor = **1 / sensitivity =** $\Delta q_i \,/\, \Delta q_o$

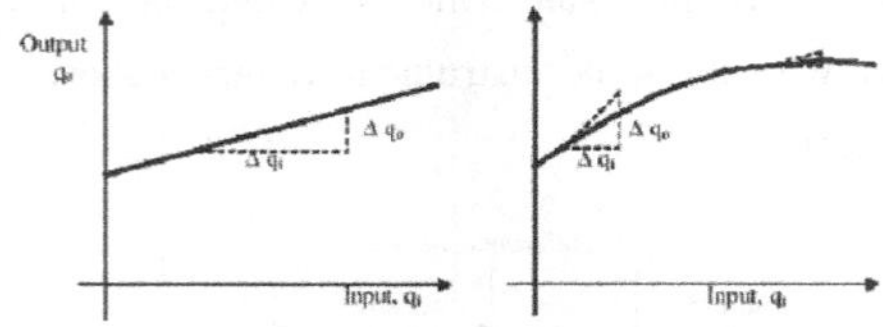

Diagram 1.4: Sensitivity Curve

$$\text{Sensitivity} = \frac{\text{Infinitesimal change in output}}{\text{Infinitesimal change in input}}$$

$$= \frac{\Delta qo}{\Delta qi}$$

(v) Linearity: Linearity is defined as the ability to produce the input characteristics symmetrically and linearly. Which can be expressed by straight line equation.

$$Y = mx + c \quad \text{..} \; 4$$

where ,

y is output ,

x = input ,

m = slope ,

c = intercept

$$\% \text{ Non- Linearity} = \frac{\text{maximum deviation of output from idealized straight line}}{\text{full scale reading}}$$

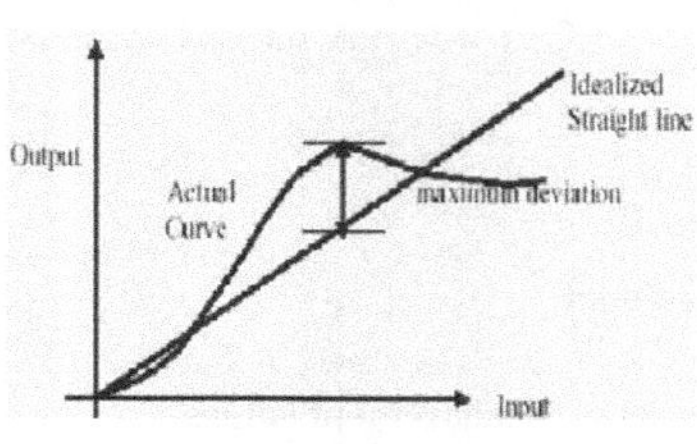

Diagram 1.5: Linearity Curve

- **Reproducibility:** It is defined as the degree of closeness with which a given value is being measured in repeated manner under given set of condition.
- **Repeatability**: It is defined as the variation of scale reading which is random in nature which can be carried out by proper checking and maintenance of the measuring instruments.

- **Drift:** Drift may be classified into three categories:
- **Zero drift:** If the whole calibration gradually shifts due to slippage, permanent set, or due to undue warming up of electronic tube circuits, zero drift sets in.

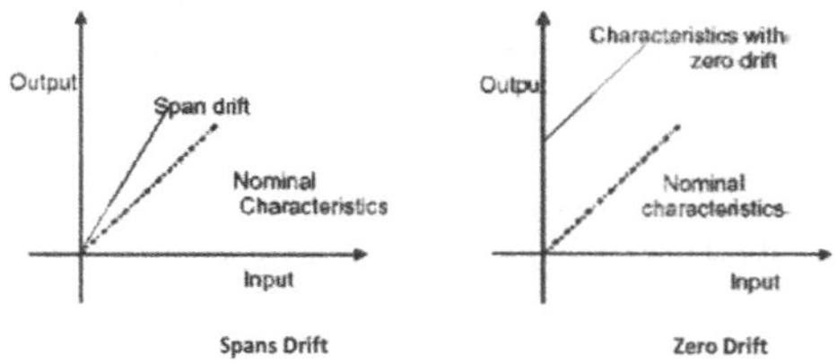

Diagram 1.6: Span & Zero Drift

- **Span drift or sensitivity drift:** If there is proportional change in the indication all along the upward scale, the drifts is called span drift or sensitivity drift.
- **Zonal drift:** In case the drift occurs only a portion of span of an instrument, it is called zonal drift.
- **Resolution:** If the input is slowly increased from some arbitrary input value, it will again be found that output does not change at all until a certain increment is exceeded. This increment is called resolution.
- **Threshold:** If the instrument input is increased very gradually from zero there will be some minimum value below which no output change can be detected. This minimum value defines the threshold of the instrument.
- **Stability:** It is the ability of an instrument to retain its performance throughout is specified operating life.
- **Tolerance:** The maximum allowable error in the measurement is specified in terms of some value which is called tolerance.
- **Range or span:** The minimum & maximum values of a quantity for which an instrument is designed to measure is called its range or span.
- **Calibration Interval:** This is the time between consecutive calibrations for the measuring instrument.
- **Confidence Level:** The probability that a measured value will be inside of the tolerance Limits. The confidence level for standards is usually set at 95%.
- **Linearity:** A measure of the degree to which a response maintains a linear relationship with given input stresses.

- **Measurand**: A quantity subjected to measurement or the characteristic that is measured.
- **Measurement Standard**: A material measure, instrument, system or reference intended to define or reproduce known values of a quantity. Standards are used to check the accuracy and precision of measuring instruments and lower level standards
- **Precision:** The extent to which an instrument gives constant results when making repeat measurements on the same unit.
- **Primary Standards:** These are the highest level standards and are usually maintained by an internationally recognized standards organization. All other standards must have tractability to the primary standards.
- **Uncertainty**: This is the error associated with a standard together with the confidence level.
- **Working Standards**: It is a standard usually applied to examine the measuring device which is calibrated with transfer standards or calibrated with primary standards.
- **Hysteresis:** Hysteresis is a difference in the output curve of a sensor when the direction of the input has been reversed. Common causes of hysteresis are mechanical strain and friction.
- **Noise:** A signal generated by internal circuit or due to external interference that is combined to the output signal.
- **Distortion:** It is difference between actual output to the predefined result called distortion.
- **Monotonicity:** A monotonic curve is one in which the dependent variable always increases or decreases as the independent variable increases

1.6. Modification and Interference of Inputs Signal

Consider the following diagram to understand the concept of Modification and Interference of Inputs signal,

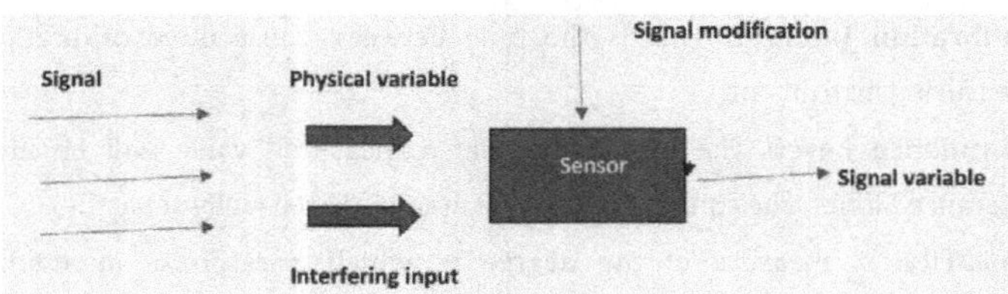

Diagram 1.7: Modification and Interference

In this figure

X is the intended measurand (physical variable)

Y is an interfering input

Z is a modifying input

We know that the measured signal output is a combination of X and Y but interfering signal affecting the intended measurand X, Hence error will achieved in measurement. Similarly if Modifying inputs is done than the behaviour of the sensor or measurement system, thereby modifying the input/output relationship and calibration of the device.

1.7. Normal Distribution Curve

A normal distribution curve or **"Gaussian curve"**, sometimes called the bell shape curve, is a distribution curve that occurs naturally in many situations and it is defined as, a graph of the frequencies of different values of a variable in a statistical distribution. The highest point on the curve, or the top of the bell, represents the most probable event in a series of data, while all other possible occurrences are equally distributed around the most probable event, creating a downward-sloping line on each side of the peak. Means the bell curve is symmetrical about central axis. Half of the data will fall to the left of the mean , half will fall to the right. It's widely used in business, statistics and in government bodies.

- Heights of people.
- Measurement errors.
- Roll of two dice
- Size of things produced by machines
- Scores in Particular Examination.
- Salaries of employs.
- In banking system

The empirical rule: Following steps must be followed whenever data analyses is performed,

- The total area under the curve is equal to 1 (100%)
- The mean, median, and mode of a normal distribution are equal.
- Normal distributions are denser in the centre and less dense in the tails.
- Normal distributions are defined by two parameters, the mean (μ) and the standard deviation (σ).

- One standard deviation- 68% of the data falls within one standard deviation of the mean.
- Two standard deviations- 95% of the data falls within two standard deviations of the mean
- Three standard deviations - 99.7% of the data falls within three standard deviations of the mean.

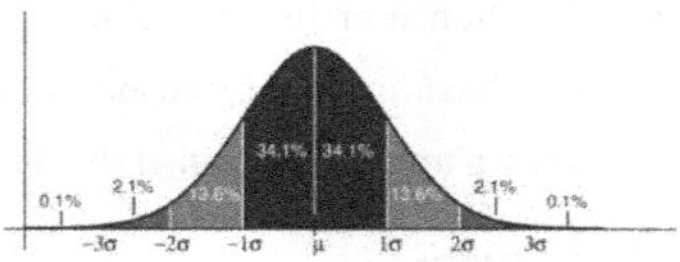

Diagram 1.8: (a) Normal Distribution Curve

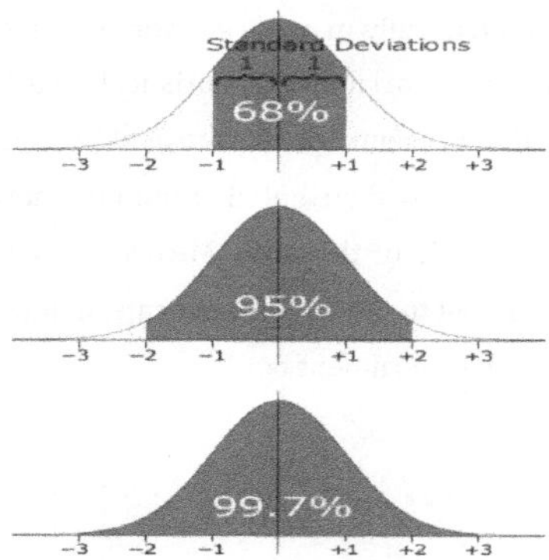

Diagram 1.8: (b) Normal Distribution Curve

The density of the normal distribution is shown below. The parameters **μ** and **σ** are the mean and standard deviation, respectively, and define the normal distribution. The symbol ***e*** is the base of the natural logarithm and **π** is the constant.

$$\frac{1}{\sqrt{2\pi\sigma^2}} e^{\frac{-(x-\mu)^2}{2\sigma^2}} \qquad \text{.. } 5$$

The standard deviation controls the spread of the distribution. A smaller standard deviation indicates that the data is tightly clustered around the mean; the normal distribution will be taller. A larger standard deviation indicates that the data is spread out around the mean, the normal distribution will be flatter and wider.

1.7.1. Distribution Curves

In distribution curve readings or measured values of a quantity indicating along the x-axis and the frequencies of these measured values plot along the y-axis. The Normal Curve is a smooth, continuous curve and is symmetrical about a central **"x"** value. The peak in frequency occurs at this central **x** value. The basic idea here is that if we could make an infinite number of readings of a quantity and graph the frequencies of readings versus the readings themselves, random errors would produce as many readings above the **"actual" or "true"** value of the quantity as below the **"true"** value and the graph that we would end up with is the Normal Curve. The value that occurs at the Centre of the Normal Curve , called the mean of the normal distribution, can then be taken as a very good estimate of the **"true"** value of a measured quantity. So, we can start to answer the question we asked above. The effect of random errors on a measurement of a quantity can be largely nullified by taking a large number of readings and finding their mean. The formula for the mean is, of course, as shown below:

$$\overline{X} = \frac{\Sigma x}{n}$$

$\overline{X}$ = mean, Σx = sum of readings, n = total no. of readings

Examine the Set of micrometer readings we had for the diameter of the copper wire. Let us calculate their mean, the deviation of each reading from the mean and the squares of the deviations from the mean.

Reading x (mm)	Deviation From Mean	Squares of Deviations From Mean
0.73	+0.01	0.0001
0.71	-0.01	0.0001
0.75	+0.03	0.0009
0.71	-0.01	0.0001
0.70	-0.02	0.0004
0.72	0.00	0.0000
0.74	+0.02	0.0004
0.73	+0.01	0.0001
0.71	-0.01	0.0001

$$\overline{X} = \frac{\Sigma x}{n} = \frac{7.23}{10} = 0.723$$

The mean is calculated as **0.723 mm.**

For example: Let's say we use a micrometer screw gauge to measure the diameter of a piece of copper wire. The micrometer allows us to read down to 0.01mm. We may obtain a set of readings in mm such as: 0.73, 0.71, 0.75, 0.71, 0.70, 0.72, 0.74, 0.73, 0.71 and 0.73.If we had taken say 50 readings of the diameter of the wire instead of just 10, we could use our knowledge of Statistics to draw a frequency histogram of our measurements, showing the number of times, each particular value occurs. This would be very helpful to anyone reading our results since at a glance they could then see the nature of the distribution of our readings. If the number of readings we take is very high, so that a fine sub division of the scale of readings can be made, the histogram approaches a continuous curve and this is called a distribution curve. If the errors are truly random, the particular distribution curve we will get is the bell-shaped Normal Distribution curve as shown below.

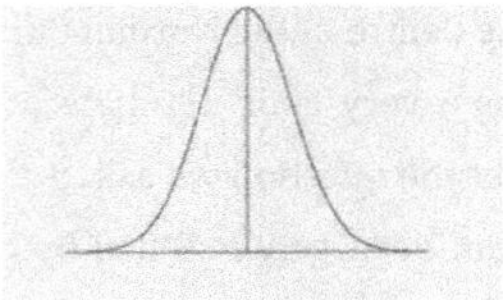

Diagram 1.9: Distribution Curve

1.8. Variables

Variables can be defined as it is those parameters or factors which affects the output of a measurement process or **In other word Variables are entities that influence the test plan or measurement. except all those variable some others variable is also involve which affects the output, all the known variables based on the cause-and effect concept.**

Types of Variables

a. **Independent Variable:** If a change in one variable will not affect the value of some other variable, the two are considered independent of each other. A variable that can be changed independently of other variables is known as an independent variable.

b. **Dependent Variable:** A variable that is affected by changes in one or more other variables is known as a dependent variable. The cause-and-effect relationship between the independent variables and the dependent variable is found by controlling the values of the independent variables while measuring the dependent variable.

c. **Controlled Variable:** A controlled variable or constant variable is a variable that does not change during an experiment. Example: In the experiment measuring the effect of temperature on solubility, controlled variable could include the source of water used in the experiment, the size and type of containers used to mix chemicals, and the amount of mixing time allowed for each solution.

d. **Extraneous Variables:** Variables that could not be controlled during the measurement process but affects the value of the variable measured are called extraneous variables. Extraneous variables can introduce differences in repeated measurements of the same measured variable taken under seemingly identical operating conditions. Extraneous variables are **"extra"** variables that may or may not be influenced the outcome of an experiment but aren't taken into account during measurement. Ideally, these variables won't impact the final conclusion drawn by the experiment, but they may introduce error into the final output scientific results.

e. **Discrete and Continuous Variables:** A continuous variable has an infinite number of possible values between any two points on the measurement scale. For example, such as stress under a changing load or temperature in a room , A discrete variable on the other hand can only take on a limited number of values and all the categorical variables are discrete , the value of the role of dice or a test run by a single operator.

f. **Categorical Variable:** A Categorical Variable is one which contains nominal or ordinal level data. This data can also be referred as qualitative data or non-metric data. Statistical methods that can be applied on categorical variable are called non-parametric statistics. They are limited compared to what can be done with numeric variable. Many times the templates used to collect and store data could use numeric values to denote categories.

g. **Numeric Variable:** A Numerical Variable is one which contains interval or ratio level data. This data can also be referred as quantitative or metric data. Statistics that can be applied on this type of data are called parametric statistics.

1.9. Measurement Scale

The type of unit on which a variable is measured is called a scale. Generally there are four types of measurement scales:

- Nominal scale
- Ordinal scale
- Interval scale
- Ratio scale

(i) Nominal Scales: A nominal scale is a lowest level scale of measurement used to assign events or objects into discrete categories for analyzing qualitative variables . Variables on a nominal scale are often called categorical variables. This form of scale does not require the use of numeric values or categories ranked by class, but simply unique identifiers to label each distinct category and from a statistics point of view nominal scale is one of the easiest to understand measurement scale. For example, Colors on the car do not have any assigned order.

Characteristics of Nominal Scale

- In nominal scale a variable is falls into two or more categories like **Agree / Disagree, Yes or No.**

- Nominal scale is always qualitative in nature but if necessary the numbers are assigned for indicating the relationship to categorize or identify objects.

- In nominal scale, numbers don't define the characteristics related to the object, which means each number is assigned to one object.

(ii) Ordinal Scale: Ordinal data is quantitative data in which ordinal variables are used to represent observations that can be categorized and based on rank ordered of the data without considering the degree of variation between them. Hence, an ordinal scale is used to make comparative study between parameter or variable to understand whether the variables are greater or lesser than one another using sorting. The central tendency of the ordinal scale is Median.

Ordinal Scale Characteristics

- It is sorting technique which is based on identification and the magnitude, the ordinal scale shows the relative rank of variables.
- Measurement of non-numeric attributes such as frequency, satisfaction, etc.
- In addition to the information provided by nominal scale, ordinal scale identifies the rank

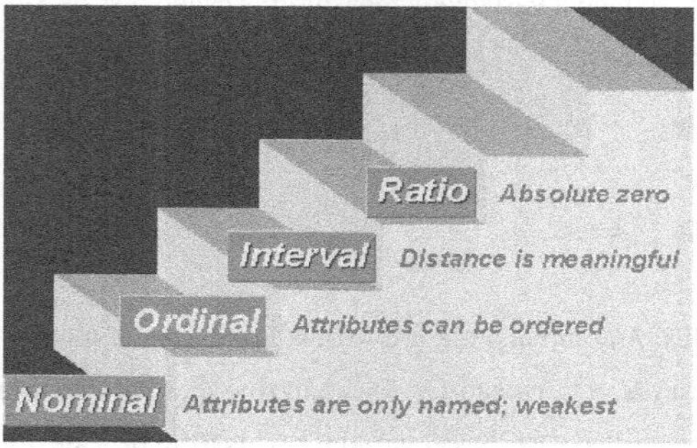

Diagram 1.10: Relationships amongst Measurement Scale

(iii) Interval scale: The interval scale is defined as a quantitative measurement scale where the difference between two variables are meaningful and order does matter where the presence of zero is arbitrary.

Properties of Interval Scales

- Identity
- Magnitude
- Equal distance

The equal distance between scale points helps in knowing how many units greater than , or less than, one case is from another. The meaning of the distance between 25 and 35 is the same as the distance between 65 and 75.

For example: temperature (in Fahrenheit or Celsius) is interval data. The difference between 10 degrees and 20 degrees is 10 degrees. The difference between 80 degrees and 90 degrees is 10 degrees. The scale at any given point is constant, while a measurement of 0 degrees does not reflect a true **"lack of temperature"**

(iv) Ratio Scale: Ratio scale is a type of variable measurement scale which is also quantitative one. Ratio scale have a zero as a indication of origin representing a unique feature of ratio scale. For example A person with age of 20 years can be considered twice as old as another with 10 years. 50 cm is half as tall as 100 cm. Similarly, Weight and Income can be subject to mathematical operations like addition, subtraction, multiplication and division.

Properties of Ratio Scale

- Identity
- Magnitude
- Equal distance
- Absolute/true zero

Characteristics of Ratio Scale

- Ratio scale has an absolute zero characteristic
- Ratio scale doesn't have a negative characteristic
- Ratio scale based on statistical analysis and mathematical operation of variables can be possible like addition, subtraction, multiplication and division. All the statistical analysis including mean, mode, the median can be calculated using ratio scale.

1.10. Noise

Noise may be defined as any unwanted signal that interferes with the communication, measurement or processing of an information-bearing signal. Noise is present in various degrees in almost all environments which can affects the all type of measurement process, **For example** A coustic noise, thermal noise, electromagnetic radio-frequency noise. Noise can cause transmission errors and may even disrupt a communication process, The success of a noise processing method depends on its ability to characterize and model the noise process, and to use the noise characteristics advantageously to differentiate the signal from the noise. Depending on its source, a noise can be classified into a number of categories , indicating the broad physical nature of the noise , as follows:

(i) **Acoustic noise:** Acoustic noise is imposed by moving, vibrating, or colliding sources and most common type of noise present in environments. Following are some come reasons by which it is achieved like moving machinery, Air-conditioners, Computer fans, Traffic, wind, rain, etc.

(ii) **Electromagnetic noise:** The electromagnetic noise is generated by all kind of electric devices, such as radio and television transmitters and receivers, generate electromagnetic noise.

(iii) **Electrostatic noise:** The electrostatic noise generated by the presence of a voltage due to current flow. Fluorescent lighting is one of the more common sources of electrostatic noise.

(iv) **Processing noise:** It is due to the results from the digital/analog processing of signals, e.g. quantization noise in digital coding of speech or image signals, or lost data packets in digital data communication systems.

Depending on its frequency or time characteristics, a noise process can be classified into one of several categories as follows:

i. **Narrowband noise:** A noise process with a narrow bandwidth such as a 50/60 Hz from the electricity supply.

ii. **White noise:** Purely random noise that has a flat power spectrum. White noise theoretically contains all frequencies in equal intensity.

iii. **Band-limited white noise:** A noise with a flat spectrum and a limited bandwidth that usually covers the limited spectrum of the device or the signal of interest.

iv. **Colored noise:** A non-white noise or any wideband noise whose spectrum has a non-flat shape , examples are pink noise , brown noise and auto-regressive noise.

v. **Impulsive noise:** It consists of short-duration pulses of random amplitude and random duration.

vi. **Transient noise pulses:** It consists of relatively long duration noise pulses.

vii. **Thermal Noise:** Thermal noise also referred to as Johnson noise (after its discoverer J. B. Johnson), is generated by the random movements of thermally energized particles. The concept of thermal noise has its roots in thermodynamics and is associated with the temperature-dependent random movements of free particles such as gas molecules in a container or electrons in a conductor. Although these random particle movements average to zero, the fluctuations about the average constitute the thermal noise. For example

- The random Movements and collisions of gas molecules in a confined space produce random fluctuations about the average pressure. As the temperature increases, the kinetic energy of the molecules and the thermal noise increase.
- An electrical conductor contains a very large number of free electrons, together with ions that vibrate randomly about their equilibrium positions and resist the movement of the electrons. The free movement of electrons constitutes random spontaneous currents, or thermal noise, that average to zero since in the absent of a voltage electrons move in all different directions.

viii. **Shot Noise:** It is generated due to the analysis of random variations in the emission of electrons from the cathode of a vacuum tube. Discrete electron particles in a current flow arrive at random times, and therefore there will be fluctuations about the average particle flow. The fluctuations in the rate of particle flow constitutes the shot noise. Other instances of shot noise are the flow of photons in a laser beam , the flow and recombination of electrons and holes in semiconductors , and the flow of photoelectrons emitted in photodiodes.

ix. **Electromagnetic Noise:** All electrical devices that generates , consumes or transmits power is a potential source of electromagnetic noise and interference for other systems , In general, the higher the voltage or the current level, and the closer the proximity of electrical circuits/devices, the greater will be the induced noise. The common sources of electromagnetic noise are transformers, radio and television transmitters, mobile phones, microwave transmitters, Ac power lines, Motors and motor starters, Generators, Relays, oscillators, Fluorescent lamps, and Electrical storms. Electromagnetic noise are divided as,

Interference: It produces undesirable deterministic trends on measured values, such as a sinusoidal wave superimposed on the signal path.

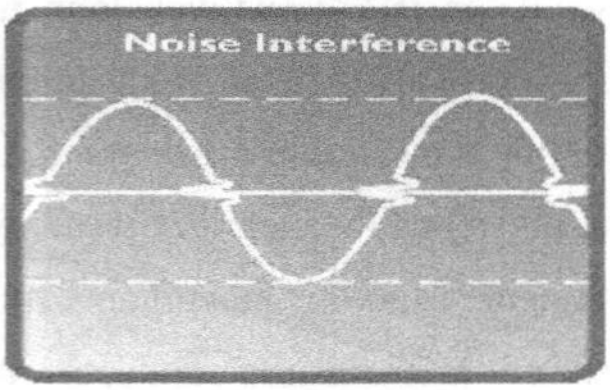

Diagram 1.11: Noise Interference

- **Parameters:** It represent the functional relationships between variables. For example, a moment of inertia or a Reynolds number has its value determined from the values of a grouping of variables.
- **Control Parameter:** It has an effect on the behavior of the measured variable and should be controlled during test period called a control parameter.
- **Repetitions:** It is the repeated measurement of dependent variable during a single test run or on a single batch are called repetitions. Repetition helps to quantify the variation in a measured variable as it occurs during any one test or batch while the operating conditions are held under nominal control. (diameter of ball bearing batch - multiple measurements on same batch/day)
- **Replication:** It is an independent duplication of a series of test measurements under similar operating conditions. Replication allows for quantifying the variation in a measured variable as it occurs between different tests, each having the same nominal values of operating conditions. (ball bearing batch on separate day under similar conditions)

1.11. Calibration

Calibration is a comparison between a known measurement (the standard) and the measuring device. Typically, the accuracy of the standard should be ten times the accuracy of the measuring device being tested. However, accuracy ratio of **3:1** is acceptable by most standards organizations. Calibration of your measuring instruments has following objectives.

- It checks the accuracy of the instrument and it determines the traceability of the measurement
- Calibration also helps in repair of the device if it is out of calibration ,
- To explain how calibration is performed.

(i) Necessity of calibration: The accuracy and functioning of all the measuring devices reduced over time or ageing of the device. This is typically caused due to wear and tear between the moving parts. However, changes in accuracy can also be caused by electric or mechanical shock/vibration, Manufacturing Environment or may be due to other factors. Depending upon the type of the instrument and the environment in which it is being used. The basic of calibration is to improve the accuracy of the measuring device. Accurate measuring devices improve product quality. A measuring device should be calibrated for the following purpose:

- According to recommendation of the manufacturer
- After any mechanical or electrical shock
- Periodically (annually, quarterly, monthly) inspected to accuracy
- To ensure readings from an instrument are consistent with other measurements
- To determine the accuracy of the instrument readings
- To check the reliability and quality of the instrument

(ii) Calibration procedure: Following steps must be taken for the proper calibration of measuring devices.

- Identification of the instrument, gauge or test equipment, or group of such items to which the procedure is applicable
- Identification of all measurement standards/Reference materials and associated equipment used to perform the calibration
- The procedures to be adopted for handling, transporting, storing and preparing items for calibration
- The environmental conditions that must be used, the limits applicable, the procedure for any corrections that may have to be made as a result of the environmental conditions and, where relevant, the minimum period of stabilization before calibration
- The method or procedure for calibration should be adopt in the form of written instructions and diagrams where necessary
- Details of the measurement or calibration data to be recorded
- The limits of acceptance for the calibration data for the item or type of item being calibrated.
- To estimate the uncertainty of measurement with the help of calibration process
- The procedures to be adopted for selecting calibration intervals when the equipment /reference material is being used by the Laboratory to perform calibration or tests

- The procedures for checking equipment and reference materials between calibrations

1.11.1. Calibration of Sensor

whenever any kind of mechanical measuring device is operated there may be possibility of either direct output or indirect output ,In case of indirect output we putts the sensor whose function is to convert physical variables to signal variables , these sensors are sometimes also called transducers that convert input energy of one form into output energy of another form.

- **Passive sensor:** Passive sensors do not add energy as part of the measurement process but may remove energy from the operation. for example thermocouple, which converts a physical temperature into a voltage signal, pressure gauge (diaphragm, aneroid, Borden tube pressure gage)
- **Active sensor:** Active sensors add energy to the measurement environment as part of the measurement process. for example–Radar

(i) Calibration curve: The relationship between the physical measurement variable input and the signal variable (output) for a specific sensor is known as the calibration of the sensor. A sensor is calibrated by providing a known physical input to the system and recording the output. Then the data are plotted on a calibration curve. For example, consider the following curve which is drawn between input & output.

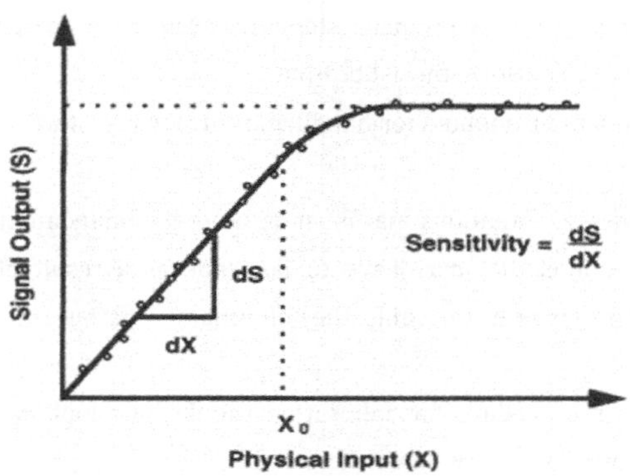

Diagram 1.12: Calibration Curve

- In the given curve , physical input is proportional to the signal variable (output) up to **(X_0)** , which is a limiting value of measuring device hence we have got linear response
- Beyond the point **(X_0)**, calibration curve indicating less sensitivity between input & output called saturation

- The sensitivity is slope of the calibration curve

$$\text{Sensitivity} = \frac{dS}{dX}$$

- Dynamic range of the device = **(Largest Physical Inputs - Smallest Physical Inputs)**

Type of calibration: The overall performance of an instrument is based on its static and dynamic characteristics. It indicates how well the instrument measures the desired input and rejects the undesired inputs. If the sensor is made to measure constant or slowly varying quantities, its performance can be evaluated with only the static characteristics. The dynamic description is necessary if we need to measure rapidly varying quantities. This involves the study of the instrument input-output relations and the use of differential equations.

(i) Static Calibration: Static calibration refers to the input-output relations obtained when only one input of the instrument is varied at a time, all other inputs being kept constant. To each constant input, which is measured with another instrument, we associate the sensor output. The plot of these points is the sensor static characteristic for that specific input.

Steps for Calibration

- Identify all the possible inputs of the instrument.
- Decide which of the inputs will be significant in your application.
- Determine the apparatus and methods to control all significant inputs over the desired range.
- By varying one input and holding the other inputs constant, develop the sensor input-output relations.

Example: Calibration of the pressure gage sensor, the objective in this example is to determine the relationship between the desired input (pressure) and the output (scale reading).

- The first step of the calibration process requires identifying the desired, interfering and modifying inputs of the pressure gage.
- In the second step you must determine how, or in what conditions, you are going to use the sensor.

(ii) Static Sensitivity: The sensitivity of an instrument refers to its ability to detect changes in the measured quantity. It can be defined as the slope of the calibration curve if the input/output relationship is linear. The output quantity must be taken as the actual physical output.

For example: An angular displacement for the pressure gage is 5 degrees/ k Pa, if the spacing between two marks is 5 degrees. Thus the static sensitivity of the pressure gage is 5 x 1.08 (slope of the calibration curve) = 5.4°/ kPa.

(iii) Threshold: The threshold is the minimum value of the input below which no output is obtained . It defines the smallest measurable input, starting from rest.

(iv) Resolutions: The resolutions is the minimum increment of input quantity (starting from non-zero) below which no output is readout (zoom on the figure). It defines the smallest measurable input value .It may be indicated by least count of the output readout indicator.

(v) Static characteristic: The static characteristics and parameters of measuring instruments describe the performance of the instruments related to the steady-state input/output variables only. **Or** the measuring quantities which are slowly varying with time or mostly constant, i.e., do not vary with time, is called 'static characteristics' Following are the static characteristics of measuring instruments

(i) Accuracy	(ii) Precision	(iii) Sensitivity
(ii) Linearity	(v) Reproducibility	(vi) Repeatability
(iii) Resolution	(viii) Threshold	(ix) Drift
(iv) Stability	(xi) Tolerance	(xii) Range or span

All those terms has been also defining in previous article.

- **Dynamic characteristics:** Dynamic characteristics tell us about how well a sensor responds to changes in its input. For dynamic signals, the sensor or the measurement system must be able to respond fast enough to keep up with the input signals.

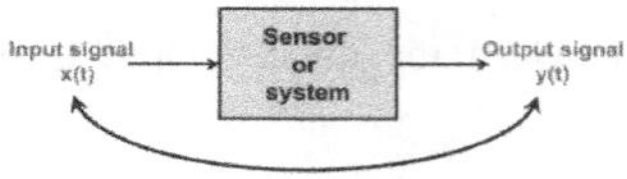

Diagram 1.13: Dynamic Characteristics

Or In other word the measuring quantities which are varying with time i.e., is called **'dynamic characteristics'**. Following are some important dynamic characteristics of measuring system.

- **Speed of Response:** It is defined as the rapidity with which a measurement system responds to changes in the measured quantity.
- **Measuring Lag:** It is the retardation or delay in the response of a measurement system to Changes in the measured quantity. The measuring lags are of two types:

- **Retardation Type:** In this case the response of the measurement system begins immediately after the change in measured quantity has occurred.
- **Time Delay Lag:** In this case the response of the measurement system begins after a dead time after the application of the input.
- **Fidelity:** It is defined as the degree to which a measurement system indicates changes in the measurand quantity without dynamic error.

(iii) Dynamic error: It is the difference between the true value of the quantity changing with Time & the value indicated by the measurement system if no static error is assumed, It is also called measurement error. Dynamic characteristics are determined by analyzing the response of the sensor with variable input waveform.

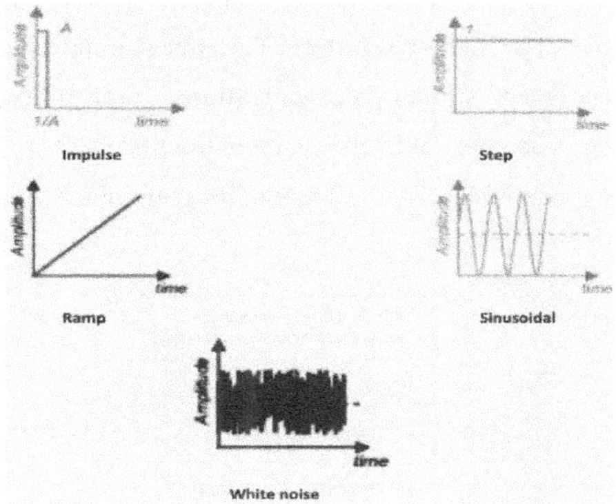

Diagram 1.14: Dynamic Error

Here we are define, some others Dynamic Characteristics of measuring system.

- **Rise time:** This is the length of time It takes the output to reach **90%** of full response when a step is applied to the input.
- **Time constant:** It is the time for the output to change by **63.2%** of its Maximum possible change.
- **Settling time:** This is the time it takes from the application of the input step until the Output has settled within a specific band of the final value.
- **Transfer Function**: It is a simple, concise and complete way of describing the sensor or system performance

  H (s) = Y(s) / X(s) 6

  where **Y(s)** and **X(s)** are the Laplace Transforms of the input and output respectively.

(iv) Frequency Response: It describe how the ratio of output and input changes with the input frequency.

(v) Dynamic error δ (ω) = { M(ω) - 1 } a measure of the inability of a system or sensor to Adequately reconstruct the amplitude of the input for a particular frequency Band width the frequency band over which **M (ω) ≥ 0.707**

(vi) Cutoff frequency: The frequency at which the system response has fallen to **0.707 (-3 dB)** of the stable low frequency.

(vii) Traceability: Traceability is a term used to describe the unbroken chain of comparisons that qualify measurement equipment to national or international standards. A quantity is measured with measurement equipment. Measurement equipment is calibrated with standards. A standard can be calibrated with a primary standard **or** it may be calibrated with another standard, known as a transfer standard. Transfer standards are calibrated by NIST or other internationally recognized standards organizations. Traceability certificates should state which standards have been used to calibrate the instruments. Primary standards are standards of the highest accuracy available. The following diagram illustrates the hierarchy of the traceability chain.

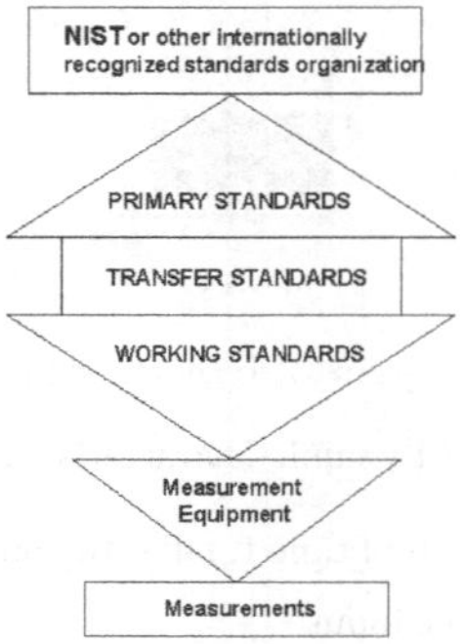

Diagram 1.15: Traceability Chain

1.12. Dynamic Calibrations or Frequency Response

A sensor is a filter that affects the frequency content of the data that passes through it. It is up to the experimentalist to make sure that these filtering effects are understood and do not affect the results that he is trying to obtain. To measure the frequency response of a sensor, one could input an impulse, measure the response, transform it and obtain the frequency response function. An impulse is a good input since it contains all frequencies. To understand it , consider the gate function

$$\delta(t) \approx \mathbf{1/\tau \prod (t/\tau) \supset sinc(f\tau) = sinc(\prod f \tau) / (\prod f \tau)} \quad \text{.....................} \; 7$$

As $\boldsymbol{\tau} > \mathbf{0}$, the gate function goes to an impulse. The first zero crossing of the **sinc** function occurs at $\mathbf{f = 1/\tau}$, where the **Sin** goes to **zero**. This is also observed to be the inverse of the length of the gate function, **τ.** So as τ goes to zero, the zero crossing moves to higher and higher frequency and the **sinc** function broadens. See graph below represent the width gate functions and associated **sinc** function. In the limit of a delta function, the **sinc** function goes to constant function of frequency. The transform of the impulse, is a non zero constant so contains equal energy at all frequencies

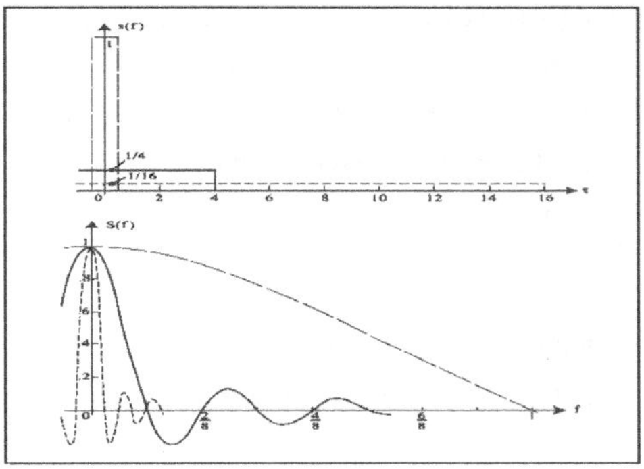

Diagram 1.16: Frequency Response

The gate function with varying widths (16, 4, 1) on top, and the associated transform (**sinc** function) on the bottom, showing that the **sinc** gets broader as the gate gets narrower. If we are considering a temperature sensor, It is difficult to input an impulse, because it is an infinitely short, infinitely high pulse. However, to get around this problem, consider the case of the step function.

X (t) = 1 for t > 0 .... **8**

X (t) = 0 for t<0.................................. **9**

This is easy to make, just plunge the sensor quickly from warm air into cold water or between a warm and cold-water bath in a time much faster than the response time of the sensor. In this case we have a scaled step function where we can write

$\mathbf{T(t) = T_{air} \; for \; t < 0}$ **10**

$\mathbf{T(t) = T_{water} \; for \; t > 0}$ **11**

Now the derivative of the step function is an impulse which has an amplitude , an equal to the change in temperature. We assume that the time for the sensor to move through the interface is short compared with the response time of the sensor, so relative to the sensor's response time, the change in temperature does look like a true step. Thus, we have a known input or forcing function. If the sample interval, δ(t) , is small compared with the response time of the sensor, then we can resolve the sensor's response which looks a bit more like a slow ramp, rather than a step. Hence, with a known input, and a measure of the output as a function to time, y(t), we can calculate the response function. From the definition of the convolution product we have

$$Y(t) = [T(t) \times 1(t)] \quad \text{.......12}$$

$$Y(t) = [T(t) \times L(\omega)] \quad \text{.......13}$$

and transforming into frequency space, the convolution theorem allows us to replace the convolution product in time space with a multiplication in frequency space. Therefore, we can divide the transforms to get the frequency response function,

$$Y(t) = \frac{L(\omega)}{T(t)} \quad \text{.......14}$$

Note that we want the impulse narrow so that it contains energy at all frequencies. If our step were indeed a true impulse, then its transform would be just 1 and the response would be the transform of the output, y(t). However, our step function, and hence our first difference, has an amplitude.

(i) Time constant : In physics and engineering, the time constant, usually denoted by the Greek letter τ (tau), is the parameter characterizing the response to a step input of a first-order , linear time-invariant (LTI) system. The time constant is the main characteristic unit of a first-order LTI system. In the time domain , the usual choice to explore the time response is through the step response to a step input , or the impulse response to a Dirac delta function input. In the frequency domain, the time constant also determines the bandwidth of a first-order time-invariant system, that is, the frequency at which the output signal power drops to half the value it has at low frequencies. The time constant is also used to characterize the frequency response of various signal processing systems - magnetic tapes, radio transmitters and receivers, record cutting and replay equipment, and digital filters - which can be modeled or approximated by first-order LTI Systems. For example

- Time constant used in control systems for integral and derivative action controllers, which are often pneumatic, rather than electrical.
- Time constants provide basic foundation for a lumped system analysis**(lumped capacity analysis method)** for thermal systems, used when objects cool or warm uniformly under the influence of convective cooling or warming.

- In case of radioactive decay the time constant is called the decay constant (λ), and it represents both the mean lifetime of a decaying system before it decays, or the time it takes for all but **36.8%** of the atoms to decay. For this reason, the time constant is longer than the half-life, which is the time for only **50%** of the atoms to decay.

Or in other word we can say that, the time constant represents the elapsed time required for the system response to decay to zero if the system had continued to decay at the initial rate, because of the progressive change in the rate of decay the response will have actually decreased in value. In an increasing system, the time constant is the time for the system's step response to reach of its final (asymptotic) value (say from a step increase). Here the first order **LT I** systems are characterized by the following differential equation,

$$\tau \frac{dV}{dt} + V = f(t) \quad \text{..............................15}$$

Where τ represents the exponential decay constant and V is a function of time t

$$V = V(t) \quad \text{..16}$$

The right-hand side is the forcing function **f(t)** describing an external driving function of time, which can be regarded as the system input, to which ***V(t)*** is the response, or system output. Classical examples for ***f(t)*** are : The Heaviside step function, often denoted by ***u(t)***

$$u(t) = \begin{cases} 0, & t < 0 \\ 1, & t \geq 0 \end{cases} \quad \text{.......................... 17}$$

the impulse function, often denoted by $\delta(t)$, and also the sinusoidal input function:

$$f(t) = A\sin(2\pi ft) \quad \text{..............................18}$$

$$f(t) = Ae^{j\omega t}, \quad \text{.................................... 19}$$

Where $A$is the amplitude of the forcing function, f is the frequency in Hertz and **ω = 2π f**is the frequency in radians per second.

For Example: Calculation of Time Constant for Thermal System

Time constants are a feature of the lumped system analysis (lumped capacity analysis method) for thermal systems, used when objects cool or warm uniformly under the influence of convective cooling or warming. In this case, the heat transfer from the body to the ambient at a given time is proportional to the temperature difference between the body and the ambient.

$$F = hA_s\,(T(t) - T_a)\,, \quad \text{... 20}$$

Where h is the heat transfer coefficient, and A_s is the surface area, T(t) = body temperature at time t, and T_a is the constant ambient temperature. The positive sign indicates the convention that *F* is positive when heat is *leaving* the body because its temperature is higher than the ambient temperature (F is an outward flux). If heat is lost to the ambient, this heat transfer leads to a drop in temperature of the body given by

$$\rho c_p V \frac{dT}{dt} = -F, \qquad \text{.......... 21}$$

Where ρ = density, c_p = specific heat and V is the body volume. The negative sign indicates the temperature drops when the heat transfer is outward from the body (that is, when *F* > 0). Equating these two expressions for the heat transfer

$$\rho c_p V \frac{dT}{dt} = -hA_s \left(T(t) - T_a\right). \qquad \text{.......... 22}$$

Evidently, this is a first-order **LT I** system that can be cast in the form:

$$\frac{dT}{dt} + \frac{1}{\tau} T = \frac{1}{\tau} T_a, \qquad \text{.......... 23}$$

With

$$\tau = \frac{\rho c_p V}{hA_s}. \qquad \text{.......... 24}$$

In other words, the time constant says that larger masses ρ V and larger heat capacities c_p lead to slower changes in temperature, while larger surface areas A_s and better heat transfer h lead to faster temperature changes. Comparison with the introductory differential equation suggests the possible generalization to time-varying ambient temperatures T_a. However, retaining the simple constant ambient example, by substituting the variable $\Delta T \equiv (T - T_a)$, one finds.

$$\frac{d\Delta T}{dt} + \frac{1}{\tau} \Delta T = 0. \qquad \text{.......... 25}$$

Systems for which cooling satisfies the above exponential equation are said to satisfy Newton's law of cooling. The solution to this equation suggests that, in such systems, the difference between the temperature of the system and its surroundings ΔT as a function of time t, is given by:

$$\Delta T(t) = \Delta T_0 e^{-t/\tau}, \qquad \text{.......... 26}$$

Where ΔT_0 is the initial temperature difference, at time t= 0. In words, the body assumes the same temperature as the ambient at an exponentially slow rate determined by the time constant.

(ii) Settling time: The settling time of an amplifier or other output device is the time elapsed from the application of an ideal instantaneous step input to the time at which the amplifier output has entered and remained within a specified error band. Settling time includes a propagation delay, plus the time required for the output to slew to the vicinity of the final value, recover from the overload condition associated with slew, and finally settle to within the specified error. Systems with energy storage cannot respond instantaneously and will exhibit transient responses when they are subjected to inputs or disturbances.

According to Tay, Mareels and Moore settling time is defined as **"The time required for the response curve to reach and stay within a range of certain percentage (usually 5% or 2%) of the final value.** Settling time depends on the system response and frequency. The settling time for a second order, under damped system responding to a step response can be approximated if the damping ratio ζ **<< 1** by

$$T_s = -\frac{\ln(\text{tolerance fraction})}{\text{damping ratio} \times \text{natural freq}}$$

A general form is

$$T_s = -\frac{\ln(\text{tolerance fraction} \times \sqrt{1-\zeta^2})}{\text{damping ratio} \times \text{natural freq}}$$

Thus, settling time to within 2% = 0.02 is

$$T_s = -\frac{\ln(0.02)}{\zeta\omega_n} \approx \frac{3.9}{\zeta\omega_n} \qquad \text{.......... 27}$$

1.13. Measurement System Model

The description of measurement system model is given below.

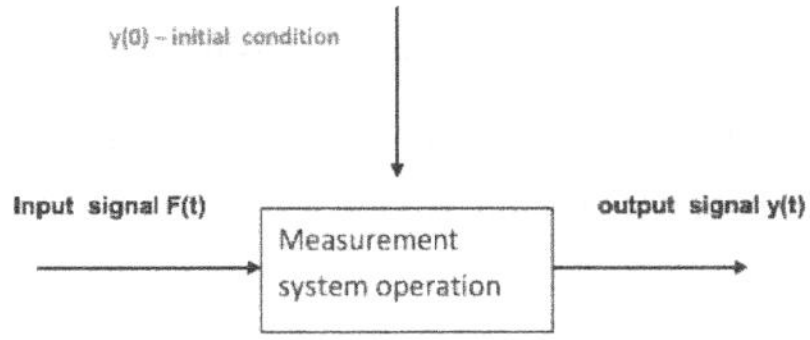

Diagram 1.17: Measurement System Model

Measurement system operation on an input signal , F(t) , provides the output signal, y(t)

If the measurement system operation performed on the input signal, F(t), in figure is an nth-order linear differential equation then the output signal, y(t), can be represented with the equation:

$$a_n \frac{d^n y}{dt^n} + a_{n-1} \frac{d^{n-1} y}{dt^{n-1}} \ldots + a_1 \frac{dy}{dt} + a_0 y = F(t) \qquad \ldots\ldots\ldots\ldots\ldots 28$$

where the coefficients, a_0, a_1, a_2, ..., a_n represent the physical system parameters whose properties and values will depend on the measurement system itself. The forcing function, F(t), can also be generalized into an m_{th}-order equation of the form:

$$F(t) = b_m \frac{d^m x}{dt^m} + b_{m-1} \frac{d^{m-1} x}{dt^{m-1}} \ldots + b_1 \frac{dx}{dt} + b_0 x \qquad m \le n \qquad \ldots\ldots\ldots\ldots\ldots 29$$

Where b_0 , $\mathbf{b_1,\ldots, b_m}$ also represent physical system parameters. The nature of these equations should reflect the governing equations of the pertinent fundamental physical laws of nature that are relevant to the measurement system

Zero-Order System :- If all the derivatives in Equation - 28 are zero then the most basic model of a measurement system is obtained , the zero-order differential equation:

a_0 y = F(t) **30**

From this equation it is easy to see that any input, **F(t),** is instantly reflected in the output y with only a factor, $\mathbf{a_0}$, modification. If the input is a dynamically varying signal $\mathbf{b_0 x}$ then $\mathbf{y = b_0 / a_0 x}$ or **y = Kx.** The factor **K** is often times referred to as the static sensitivity found during a static calibration.

1.13.1. Zero- Order Systems

Imagine a thermometer that measures the temperature in a room of an office building. For practical purposes, the thermometer will indicate the current temperature at the location where it has been installed. The fact that the output of the instrument follows the input **"exactly"** is the defining characteristic of $zero_{th}$ order systems.

Mathematically, if we let f(t) be the input to the system as a function of time and y(t) be the output, then the relationship between them is

y(t) = K f(t) **31**

where

f(t) is actual room temperature

y(t) is indicated room temperature

K is a constant that multiplies the input to generate the output

If **y(t)** is the temperature as displayed in a readout device and the thermometer is calibrated correctly, then **K** would ideally be equal to one. For example, an electrical signal , then ***K*** would be a constant with units of **Volts / Fahrenheit**. The electrical signal could be used to actuate a valve that directs either cold or hot air from the air-conditioning system to the room. **K** is usually called the static sensitivity. Note that the output of $zero_{th}$ order systems is not affected by the speed at which **f(t)** changes. Hence Equation- 28 will always be valid for the results of static calibration are sufficient to characterize the response of the system.

1.13.2. First Order System

In order to understand the concept of first order system, consider the case of an oral thermometer which is used to measuring the body temperature increased due to fever. Initially the thermometer is at room temperature but when the thermometer is held in the patient's mouth it experiences & adopt a sudden increment in temperature. Generally, we have to wait for a while before reading the temperature.

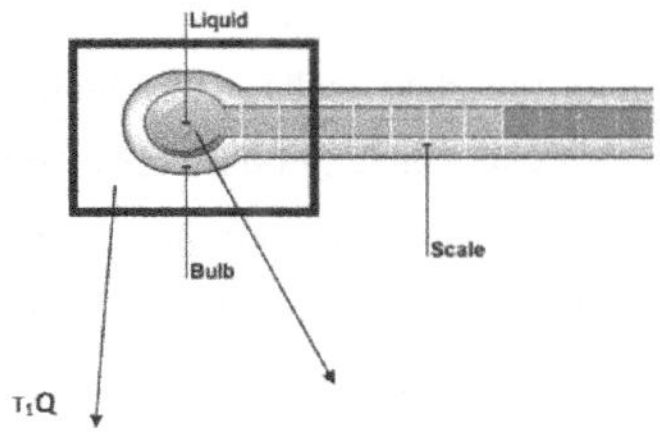

Glass Bulb Thermometer

Diagram 1.18: First Order System

Let's us consider thermometer as shown in **Diagram- 1.18** was originally kept at room temperature, which will be denoted by temperature $\mathbf{T_o}$. It is then put in the mouth of a patient represented by the area inside the dashed line which is at temperature $\mathbf{T_1}$. In order for the thermometer to work, the mercury in the bulb must be heated to T_1. The thermal expansion of the mercury will cause the column of mercury in the stem of the thermometer to increase in length. Measuring this length with the scale marked in the glass gives a temperature reading **T(t).** It takes a while, however, for the temperature of the mercury to reach the value T_1, so we must wait that long before the thermometer indicates the correct temperature. In order to write an equation to model the response of the thermometer can be explained with the help of thermodynamics and heat transfer concept.

- Heat is a form of energy and it is represented by **Q** which flows from a hot body to a cold body.
- The energy in the mercury in the bulb of the thermometer denoted by **E** and indicating increased heat level which travels from the mouth of the patient to the bulb. Due to conservation of energy, the heat flow rate from the bulb with respect to time can be represents by following mathematical equation ,

$$\frac{dE}{dt} = \frac{dQ}{dt} \quad \text{..} 32$$

Where " **t** " represents time

- As E increases, the temperature of the mercury, **T,** rises in proportion. How fast the temperature increases depends upon the specific heat,' c_v of the mercury and how much mercury the bulb holds. The increase in energy is related to the increase in temperature by

$$\frac{dE}{dt} = mc_v \frac{dT}{dt}. \quad \text{..................................} 33$$

- Heat must travel through the glass walls of the bulb on its way from the patient's mouth to the mercury. How fast heat can flow through the walls of the bulb depends on a property of glass called the convection heat transfer coefficient, **h,** the surface area of the bulb, A, and the current temperature difference between the mercury and the mouth of the patient. In equation form,

$$\frac{dQ}{dt} = hA[T_1 - T]. \quad \text{..................................} 34$$

From equation 32 , 33 , & 34 , we have got following equation,

$$mc_v \frac{dT}{dt} = hA[T_1 - T] \quad \text{..............................} 35$$

Which can be re-written as

$$\frac{mc_v}{hA} \frac{dT}{dt} + T = T_1 \quad \text{..................................} 36$$

This is a differential equation that governs what the temperature of the mercury is at any time. Since the length of the column of mercury is proportional the temperature, the overall system can be represented with the help of first order system which is given as,

$$\tau\frac{dy}{dt} + y = Kf(t). \qquad \text{.................................37}$$

For the example above, we have

$$y = T$$

$$f(t) = T_1$$

$$\tau = \frac{mc_v}{hA} \qquad \text{....................................... 38}$$

And $K = 1$

The above equation governing the behavior of a first order system is a first order differential equation, so called because the highest derivative of the output variable in the equation is the first with respect to time.

(iii) Second Order Systems

Suppose, you go to grocery store for purchasing sugar which can be weighted with the help of weight scales. The weight scales having a pointer, which indicates weight on a big dial. If you drop a bag with sugar on the platform of the scale then there may be chances of the pointer which will oscillate a bit before settling and indicating the correct weight. The reason for the oscillation of the pointer is that the scale has mass, hence

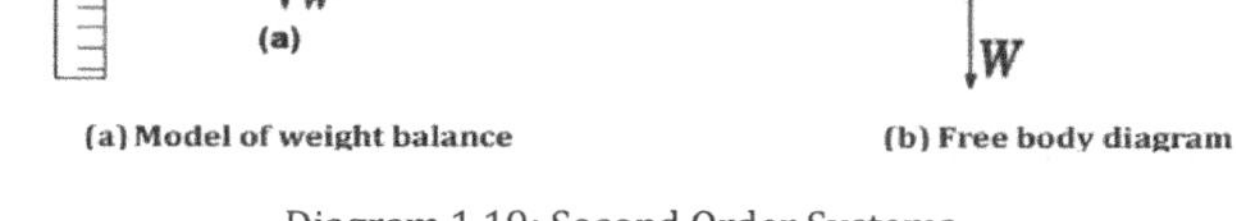

(a) Model of weight balance (b) Free body diagram

Diagram 1.19: Second Order Systems

Inertia and its behavior is dictated by Newton's laws of motion. You may now see where the second order label comes from if you recall that acceleration is the second derivative of displacement. Let's analyze the balance and see what we get. The balance can be represented by the mass-spring-dashpot system shown in Diagram 1.19(a). To simplify things a bit, the dial has been replaced by the straight scale at the left. We can write the equation of motion for the balance using basic dynamics as follows:

- The mass **m** represents the object weighted so that the weight **W** is equal to **mg** where **g** is the acceleration of gravity.
- The spring, attached near the left edge of the mass, is a mechanical element which develops force in proportion to how much it has been stretched Since the top of the spring is fixed to the ceiling, the force is proportional , to the displacement ***y*** of the mass. If the zero value of y corresponds to the position of the spring when it is unloaded, then the force Fs required to stretch the spring a distance y is given by

$$F_s = k\,y \qquad \text{.. 39}$$

Where **k** is called the spring constant.

The balance needs a way to damping the oscillations of the pointer after a weight is dropped. The damping is provided by the dashpot, which is attached to the mass near the right. Dashpots are like shock absorbers in cars. You can imagine one as a piston inside a closed cylinder. The cylinder is filled with a viscous fluid. As the piston moves, a small gap between the piston and the cylinder allows the fluid to flow from one side of the piston to the other. A force is needed to move the piston due to the viscosity of the fluid. The net result is that dashpots produce a force, F_d which is proportional to the speed of the piston relative to the cylinder. This force-speed relation can be written as

$$F_d = c\frac{dy}{dt}$$

....40

Where **c** is called the damping coefficient.

The free body diagram of the mass is shown in Diagram 1.19 (b). Applying Newton's second law we obtain the equation

$$m\frac{d^2y}{dt^2} + c\frac{dy}{dt} + ky = W,$$

..............................41

Which can be re-written as

$$\frac{m}{k}\frac{d^2y}{dt^2} + \frac{c}{k}\frac{dy}{dt} + y = \frac{1}{k}W.$$

..............................42

This is the differential equation that governs the motion of the scale. Since the weight is indicated by the displacement of the scale, the equation also governs the indicated weight. In general, the equation for a second order system is given by

$$\frac{1}{\omega_n^2}\frac{d^2y}{dt^2} + \frac{2\zeta}{\omega_n}\frac{dy}{dt} + y = Kf(t) \quad \text{.......................... 43}$$

In the example of the scale above we have

$$\omega_n = \sqrt{\frac{k}{m}},$$

$$\zeta = \frac{c}{2\sqrt{km}}$$

$$K = \frac{1}{k}. \quad \text{.............................. 44}$$

1.14. Solution to the Differential Equations

(i) First Order Systems

Step Input Response

A step input is used to represent situations when the input changes from one, constant, value to another constant value in a very short period of time. The two examples above the thermometer in a patients' mouth and the potatoes dropped on the weight scale can be thought of as step inputs. Mathematically a step input in **f(t)** from **Fo** to F_1applied at **t = 0** is given by:

$$f(t) = \begin{cases} F_o & : \quad t < 0 \\ F_1 & : \quad t \geq 0 \end{cases} \quad \text{.............................. 45}$$

The application of the step input causes the output to change from a initial value y_o to a final value y_1. Then , the solution of the equation will be as follows ,

$$y(t) = KF_1 - (KF_1 - y_o)e^{-t/\tau}. \quad \text{........................ 46}$$

Note that at **t = 0**, **y(t)** = y_o, and that as $t \rightarrow$**y(t)** , KF_1 = y_1. The transition between the initial and the final values of **y (t)** is exponential as shown in **Diagram – 1.20**. The parameter τ is called the time constant of the system. It indicates how fast the system responds to a change in the input. Note that after a time of one time constant, **y(t) is 63.2%** of the way to its final value. Generally, **y(t)** is considered to have achieved its final value after five time constants. As an example, suppose that an oral thermometer takes at least 3 minutes to indicate a patient's temperature. Thus the time constant of the thermometer belongs to order of **0.6** minutes.

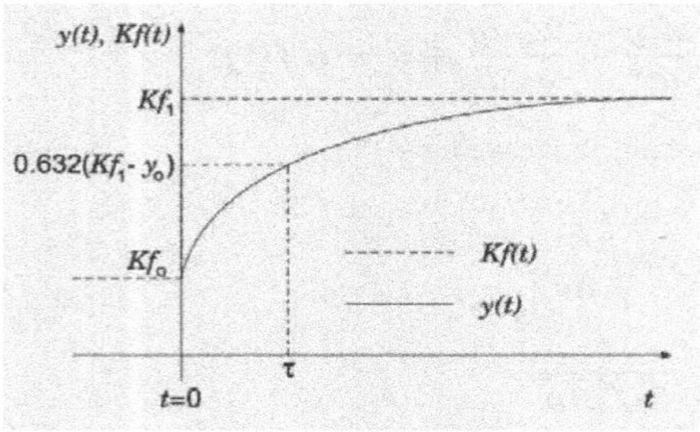

Diagram 1.20: Response of a First Order System to a Step Input Response

(ii) Harmonic Input Response

Harmonic input response indicate that the input is a sine wave with constant frequency and amplitude. In other words the input is of the form**(s)**

$$f(t) = A\sin\frac{2\pi t}{T} = A\sin 2\pi f t = A\sin\omega t. \quad \text{.................. } \mathbf{47}$$

All three definitions are equivalent. In the first, T is the period in seconds. In the second, f is the frequency in cycles per second or Hertz. In the third, **ω** is the angular frequency in radians per second. For example, vibration of a motor, the movement of pistons in an internal combustion engine, the turning of wheels, the air conditioning system at your house turning on and off . The steady state solution to will be

$$y = MKA\sin(\omega t - \phi) \quad \text{.................................... } \mathbf{48}$$

$$M = \frac{1}{\sqrt{1 + (\tau\omega)^2}} \quad \text{....................................}\mathbf{49}$$

Where M is called the magnitude ratio and

$$\text{Ø} = \arctan(\tau\omega) \quad \text{..}\mathbf{50}$$

Ø is called the phase angle.

Conclusion

- The magnitude ratio indicate that how the amplitude of the output of the system depends on the frequency of the input.
- The phase angle indicate that how far the system output is falling behind the input. further we have

$$(\omega t - \phi) = \omega(t - \frac{\phi}{\omega}) = \omega(t - \Delta t). \quad \text{.................................... } \mathbf{51}$$

Therefore, the time lag, Δt is given by

$$\Delta t = \frac{\phi}{\omega}.$$

The response characteristics measured in a static calibration are valid. As the frequency increases, the magnitude ratio decreases and the phase angle increases rapidly, indicating that dynamic effects become important.

(ii) Second Order Systems

Step input response

Now, let us consider the response of a second order system to a step input having a solution with equation 47. since we have **three** cases regarding to **damping ratio** with assumption that the system is at rest with **output $y = 0$** when the step input is applied , these are given below

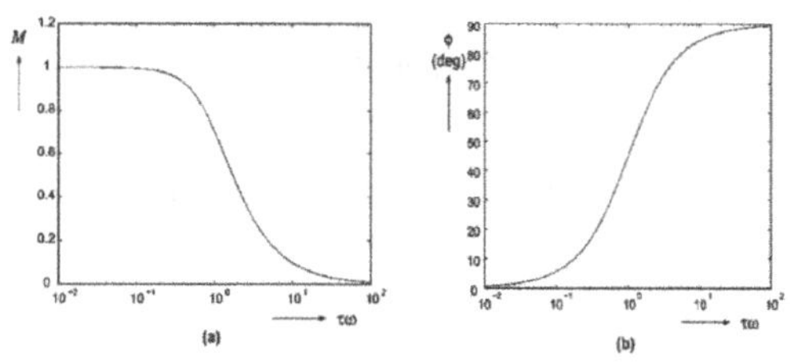

Diagram 1.21: Step Input Response

Where

(a) Magnitude ratio for a first order system.

(b) Phase angle.

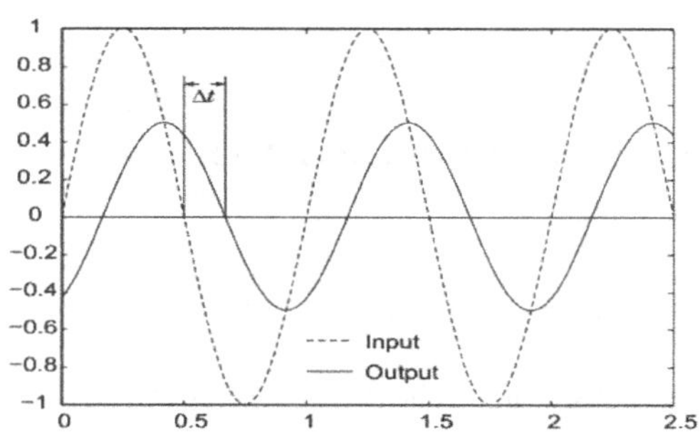

Diagram 1.22: Step Input Response

Case -1when system is under damped means ζ < 1

Thus the solution will be ,

$$y(t) = KF_1 \left[1 - e^{-\zeta\omega_n t}\left(\frac{\zeta}{\sqrt{1-\zeta^2}}\sin\omega_d t + \cos\omega_d t\right)\right] \qquad \text{.............. 52}$$

And frequency of oscillation ω_d is given as ,

$$\omega_d = \omega_n\sqrt{1-\zeta^2}. \qquad \text{..................................53}$$

Where

ω_d = Oscillation frequency

ω_n= Natural frequency at **ζ = 0**

$e^{-\zeta\omega_n t}$ = coefficient represents the damped oscillation system

Whenever system is subjected to step input then output will represents in the form of sine and cosine function that has oscillating nature.

Here the function of **ζ (damping factors)** is to control and overcome the damping effects which is create due to oscillation

If

$t \to \infty,\ y \to KF_1$ **and ζ = 0**

then system will oscillate forever

Case - 2 when system is critically damped means ζ = 1

Thus the solution will be given as ,

$$y(t) = KF_1\left[1 - (1+\omega_n t)e^{-\omega_n t}\right]. \qquad \text{.. 54}$$

Case - 3 when the system is over damped means ζ > 1

Thus the solution will be given as ,

$$y(t) = KF_1\left[1 - e^{-\zeta\omega_n t}\left(\frac{\zeta}{\sqrt{\zeta^2-1}}\sinh\omega_n\sqrt{\zeta^2-1t} + \cosh\omega_n\sqrt{\zeta^2-1t}\right)\right] \qquad \text{......................... 55}$$

Whenever over damped system is subjected to step input then output of the over damped system do not oscillate and system achieve the final value **y = K F₁**

Example 1: Suppose we want to measure a temperature which fluctuates with a Frequency of 0.1 Hz with a minimum of 98% amplitude reduction. basic first-order temperature sensor like a thermocouple FIND: Magnitude ratio of at least 0.98

$$M(\omega) \geq 0.98, \text{ or dB} = 20\log 0.98 = -0.175$$

$$M(\omega) = \frac{B}{KA} = \frac{1}{\left[1+(\omega\tau)^2\right]^{1/2}}$$

rearranging gives

$$\omega\tau = \left[1/M(\omega)^2 - 1\right]^{1/2}$$

$$\text{so for } M(\omega) \geq 98\%, \ \omega\tau \leq 0.2$$

$$\text{or, } \tau \leq 0.2/\omega = 0.2/2\pi f = 0.2/\ 2\times 3.142\times 0.1$$

$$\tau \leq 0.31 \text{ sec}$$

Example 2: A thermocouple has a time constant of 20 ms. Determine its 90% rise time.

$$90\% = 1 - \Gamma(t) = 1 - e^{-t/\tau}$$

$$.9 - 1 = -.1 = -e^{-t/\tau}$$

$$ln(0.1) = -t/\tau$$

$$t_{90} = 2.3\tau = 46\ ms$$

Example 3: Suppose a bulb thermometer originally indicating 20ºC is suddenly exposed to a fluid temperature of 37 ºC. Develop a simple model to simulate the thermometer output response. Assume for Normal first-order response. **FIND : T(t)** if Given that T_0= 20º$_C$, T_∞= 37º$_C$, F(t) = [T_∞- T_0]U(t)

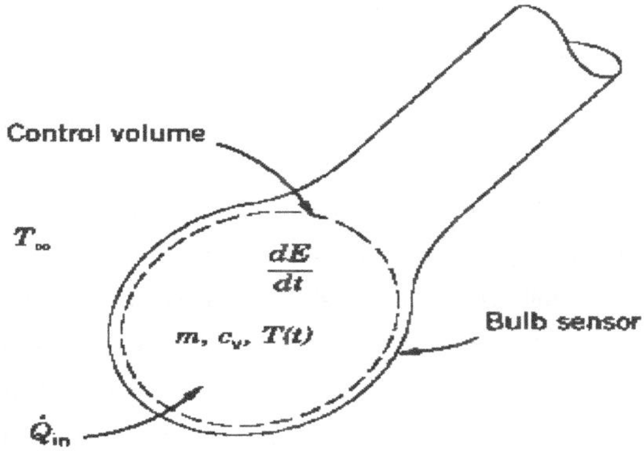

Diagram 1.23: Bulb Thermometer

Solution: The rate at which energy is exchanged between the sensor and the environment through convection, must be balanced by the storage of energy within the thermometer, **dE/dt.**

$$\frac{dE}{dt} = \dot{Q}$$

For a constant mass temperature sensor,

$$\frac{dE(t)}{dt} = mc_v \frac{dT(t)}{dt} = hA\Delta T = hA_s[T_\infty - T(t)]$$

This can be written in the form

$$mc_v \frac{dT(t)}{dt} + hA_s[T(t) - T(0)] = hA_s[T_\infty - T(0)]U(t)$$

Dividing by hAs

$$\frac{mc_v}{hA_s}\frac{dT(t)}{dt} + T(t) = T_\infty$$

Therefore:

$$\tau = \frac{mc_v}{hA_s}, \quad K = \frac{hA_s}{hA_s} = 1$$

The thermometer response is therefore:

$$T(t) = T_\infty + [T(0) - T_\infty]e^{-t/\tau}$$
$$= 37 - 17e^{-t/\tau}\ [^oC]$$

.... Ans

1.15. Phase Linearity

The phase linearity is defined as the maximum deviation from a straight line of the plot of the phase versus frequency. The measurement is performed in steps. First, the multi tone signal is generated, passed through the **DUT,** and acquired on the input channel. The phase of each tone in the multi tone signal is measured and plotted versus frequency to yield a graph similar to the one in the following front panel. Phase linearity is one way to measure the amount of group delay distortion that a device under test **(DUT)** introduces. The phase linearity measurement uses a multi tone signal to measure the relative delay of each component through the **DUT.** The multi tone test signal must be constructed with zero relative phases between the individual frequency components.

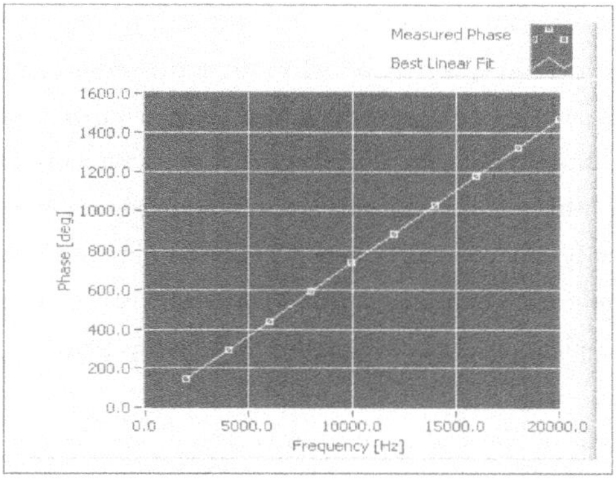

Diagram 1.24: Phase and Frequency Curve

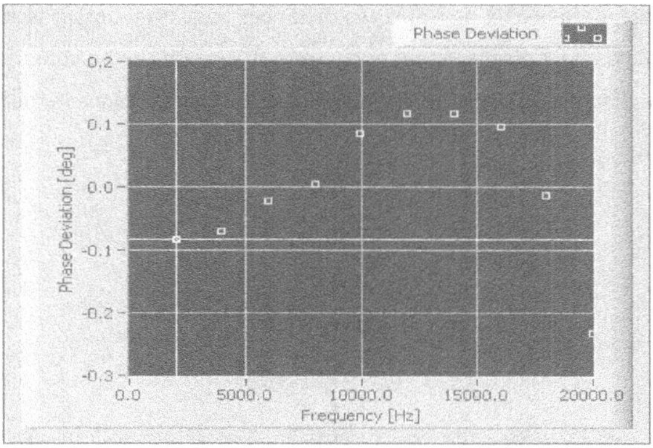

Diagram 1.25: Phase Deviation and Frequency Curve

Figure shows the maximum absolute value of the phase deviation is the value returned as the phase linearity. In this case the maximum phase deviation value is 0.215 degrees. Phase linearity measurements always are dependent on the bandwidth of the measurement. To completely report the result, specify that the phase linearity is 0.215 degrees with a 20 kHz bandwidth, 1 V_{pk} test amplitude, and unity gain set for the DUT.

1.16. Data Presenting Technique

Data presenting technique is nothing but it is data plotting format representing the relationship between available variables which communicate information regarding the context. there are two method of data presentation,

- Software data presenting technique which is applied for high accuracy,
- Manual data presenting technique which is drawn on graph paper manually. with respect to the context , here we will only described manual data presenting technique

(i) Rectangular Coordinates format : The rectangular coordinate system is also known as the Cartesian coordinate system , The rectangular coordinate system is based on a grid, and every point on the plane can be identified by unique **x** and **y** coordinates , both the ordinate and the abscissa have uniformly sized divisions on the linear scale. This is the most common format used for constructing plots and establishing the relationship between the independent and dependent variable.

- **Axes:** Locations on the grid are measured relative to a fixed point, called the origin, and are measured according to the distance along a pair of axes. The **x** and **y** axes are just like the number line, with positive e distances to the right and negative to the left in the case of the **x** axis, and positive distances measured upwards and negative down for the **y** axis. Any displacement away from the origin can be constructed by moving a specified distance in the **x** direction and then another distance in the **y** direction.

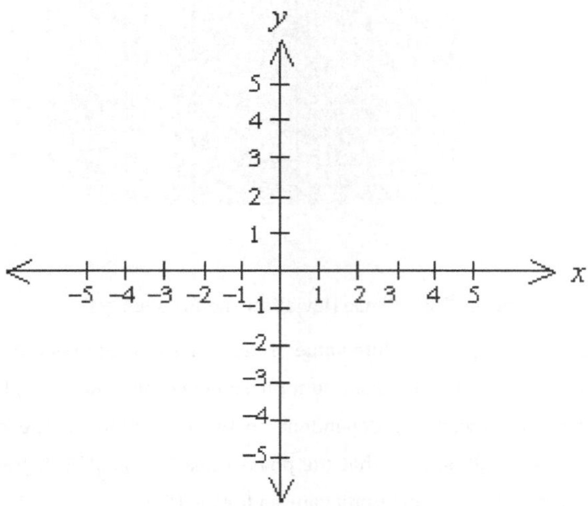

Diagram 1.26: Coordinates, Graphing Points

We specify the location of a point by first giving its **x** coordinate (the left or right displacement from the origin), and then the y coordinate (the up or down displacement from the origin). Thus, every point on the plane can be identified by a pair of numbers **(x, y)**, called its coordinates.

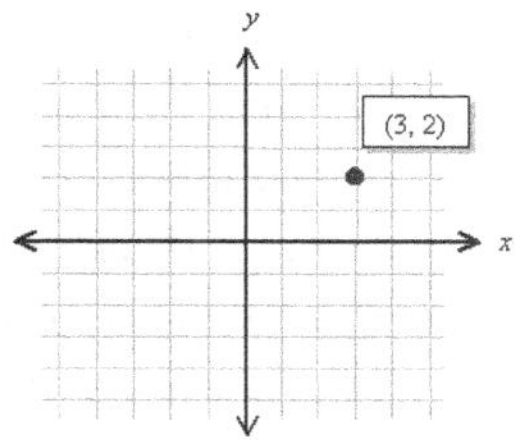

(a)

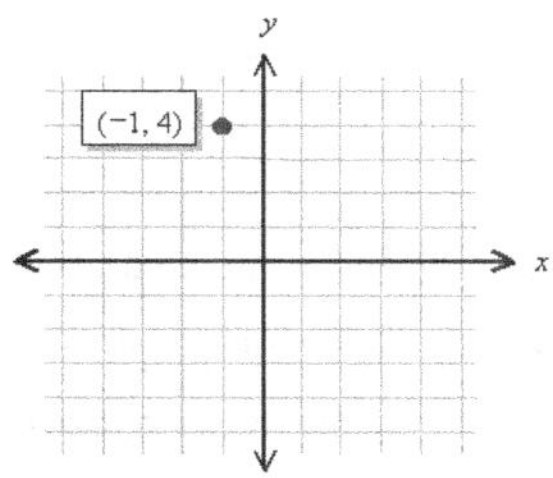

(b)

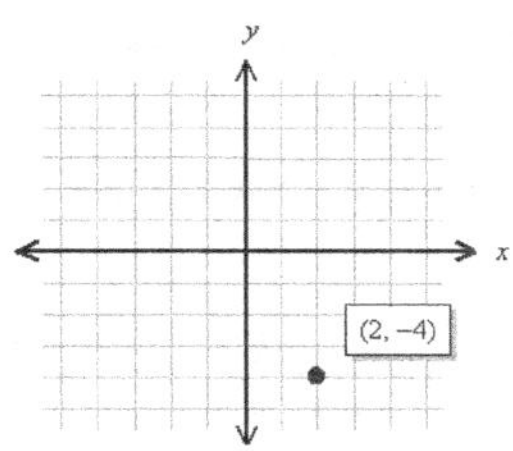

(c)

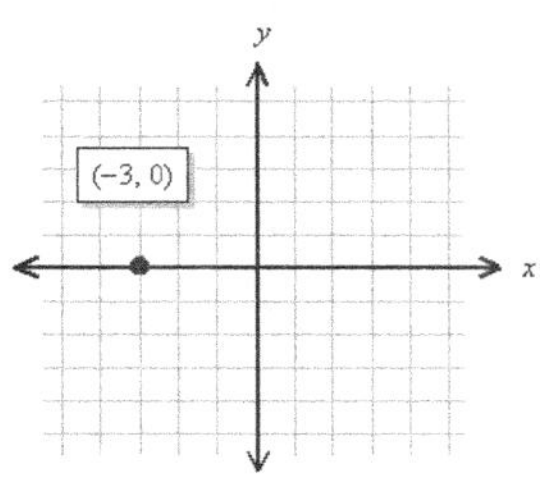

(d)

Diagram 1.26: Representation of Graphing Points

- **Quadrants:** Sometimes we just want to know what general part of the graph we are talking about. The axes naturally divide the plane up into quarters. We call these quadrants, and number them from one to four. Notice that the numbering begins in the upper right quadrant and continues around in the counter-clockwise direction. Notice also that each quadrant can be identified by the unique combination of positive and negative signs for the coordinates of a point in that quadrant.

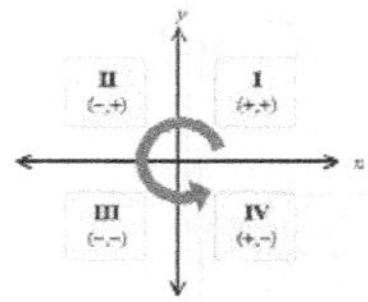

Diagram 1.27: Quadrants Representation

(ii) Semi-Logarithmic Axes: In a semilogarithmic graph, one axis has a logarithmic scale and the other axis has a linear scale. In the following set of axes, the vertical scale is logarithmic (equal scale between powers of 10) and the horizontal scale is linear (even spaces between numbers). There are no negative numbers on the **y-axis**, since we can only find the logarithm of positive numbers. Logarithmic scales are advantageous when one of the variables spans more than one order of magnitude. In particular, the semi log format always follow a relationship of the form $\mathbf{y = ae^x}$ or $\mathbf{y = a10^x}$ sometimes A natural logarithmic may also be operated .

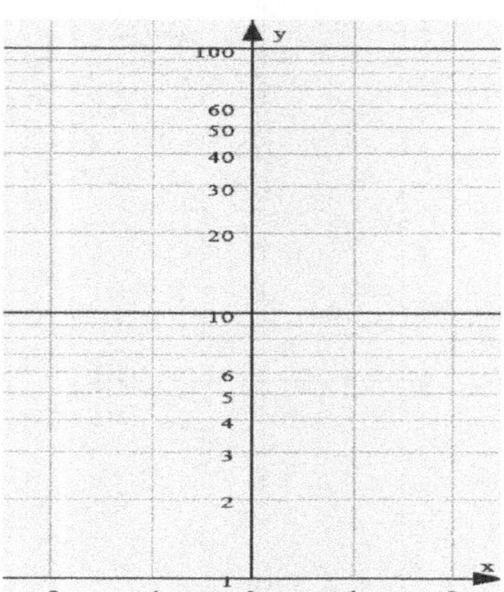

Diagram 1.28: Semi-logarithmic Graph

Example 4: Graph of **y = x** Let's see what the simple graph **y = x** looks like on different axis types.

1. **Y = x on Linear Axes On ordinary linear axes,** the graph of **y = x** is a straight line, passing through ,(−2,−2) , (−1,−1) , (0,0) , (1,1) , (2,2) etc.

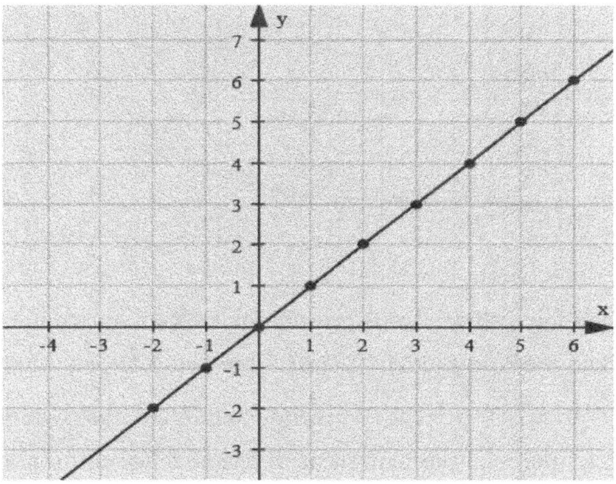

Diagram 1.29: Simple Semi-logarithmic Graph

2. **Y = x on Semi-Logarithmic Axes (vertical axis logarithmic, horizontal axis linear)**

On semi-logarithmic axes, the graph of **y = x** is a curve, not a straight line. It still passes through, (1,1), (2,2), (3,3), etc, but you'll notice there are no negative values for **y** (and so in this case, no negative values for **x** either) since we can't find the log of a negative number.

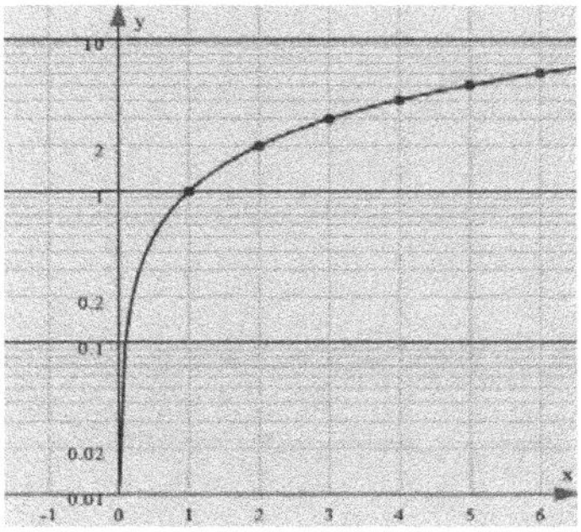

Diagram 1.30: Semi-logarithmic Graph (vertical axis logarithmic, horizontal axis linear)

Marked the points (1,1), (2,2), (3,3), (4,4) (5,5), (6,6)

Y = x on Semi-logarithmic Axes (vertical axis linear, horizontal axis logarithmic)

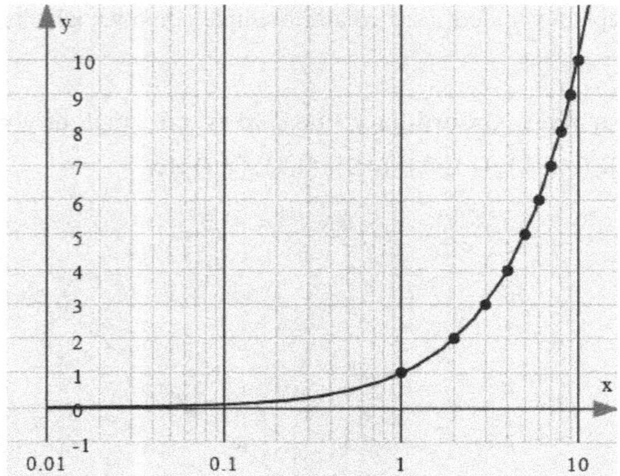

Diagram 1.31: Semi-logarithmic Graph (vertical axis linear, horizontal axis logarithmic) Marked the Points (1,1), (2,2), (3,3) up to (10,10) on the Curve

Example 5: For Variable Exponent, We conducted some observations in an experiment involving growth of a microbial population at different temperatures and obtained data as follows

T (^{0}c)	-2	-1	0	1	2	3	4
P	0.02	0.14	1	7	49	343	2401

We plot the data on linear T-P axes as follows

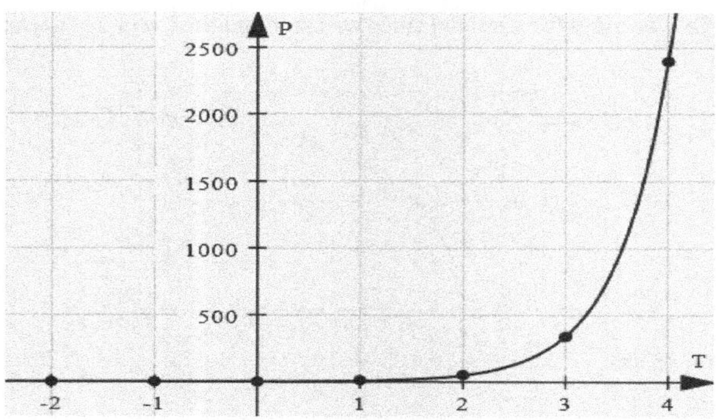

Our plot shows reasonable detail for values of x greater than 1, but doesn't tell us much for smaller values of x or y. The points are too close to the x-axis for us to see what is going on, Let's now plot the same data on semi-logarithmic axes

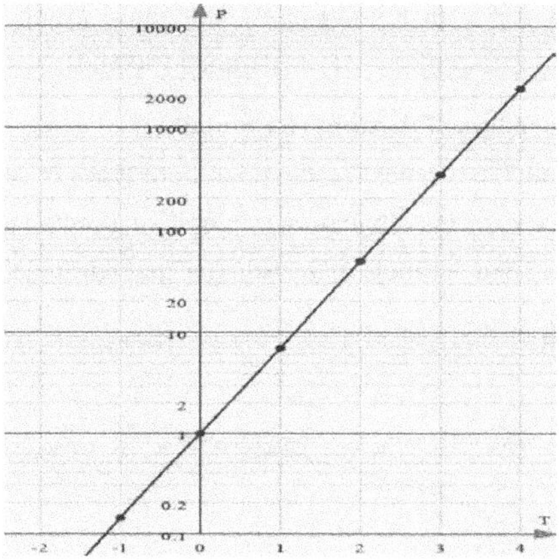

Diagram 1.32: Semi-logarithmic Graph for Variable Exponent

Now we can see a lot more information for smaller values of x and y. This is the beauty of semi-logarithmic axis plots - you can see more detail in graphs where there is a very wide range of values, but some of the data is close together.

(iii) Full-Log Coordinate Format or log-log format: The full-log or log-log format has logarithmic scales for both axes and representing the plot between log y and log x on rectangular axes. a linear curve is always achieved in log-log format. Such a format is preferred when both variables contain data values that span more than one order of magnitude. With data that follow a trend of the format of $\mathbf{y = a\,x^n}$**,**

In log-log graphs, both axes have a logarithmic scale. Here's the graph of y = x on log-log axes

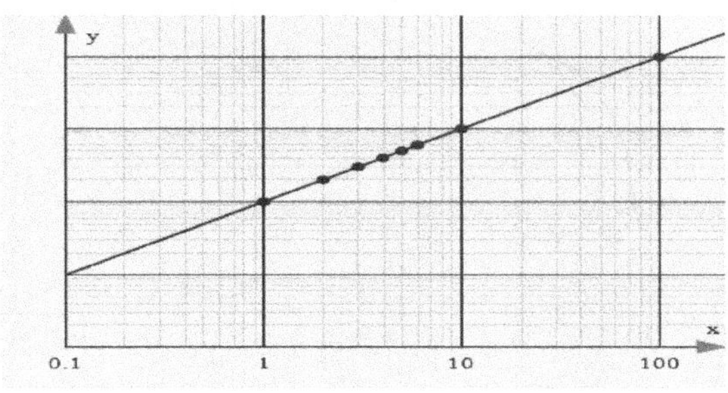

Diagram 1.33: Log-log Graph or Full Log Graph

Points along the line **y = x** on **log-log axes**. The above graph has the following points highlighted for clarity : 1,1), (2,2), (3,3),..., (10,10), and (100,100).

Example 6: Variable Raised to a Fractional Exponent

Graph **$y = x^{1/2}$** using all 4 axis types (rectangular, both types of semi-log, and log-log). This function is equivalent to **$y = \sqrt{x}$** Rectangular axes plot .Using rectangular axes, we can see that the graph of **$y = x^{1/2}$** is half of a parabola on its side (i.e. that parabola's axis is vertical) :

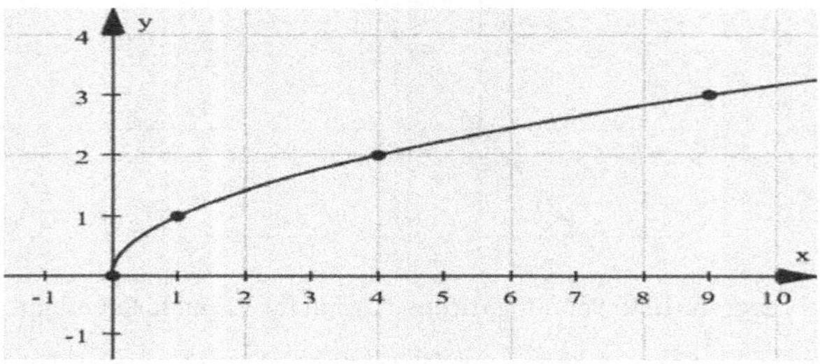

Graph of $y = \sqrt{x}$ on linear axes. We've seen this curve before , in The Parabola section.

Note 1: The detail near (0,0) is not so good using a rectangular grid.

Note 2: The curve passes through (0,0), (1,1), (4,2) and (9,3). In each case, the y-value is the square root of the x-value, which is to be expected. Let's now see the curve using semi-logarithmic plots. Logarithmic vertical axis, linear horizontal axis

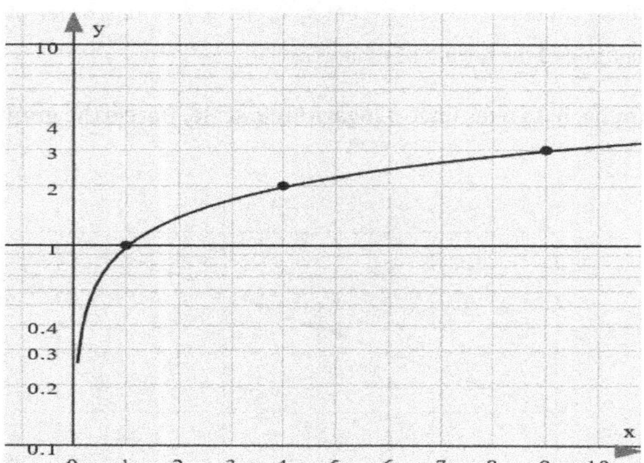

Graph of $y = \sqrt{x}$ on semi logarithmic (log – lin) axes

Now we have a lot better detail for small y. The lowest value of y that the graph indicates is y = 0.1. We can go lower than this, but cannot show **y = 0,** since the logarithm of 0 is not defined. We can see that the curve still passes through (1,1), (4,2) and (9,3). **Linear vertical axis, logarithmic horizontal axis.**

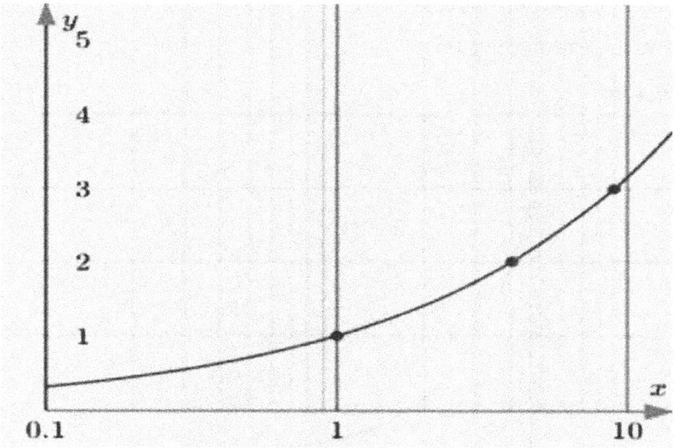

Points along the curve y = √x using line – log axes Logarithmic vertical axis, logarithmic horizontal axis (log-log) plot

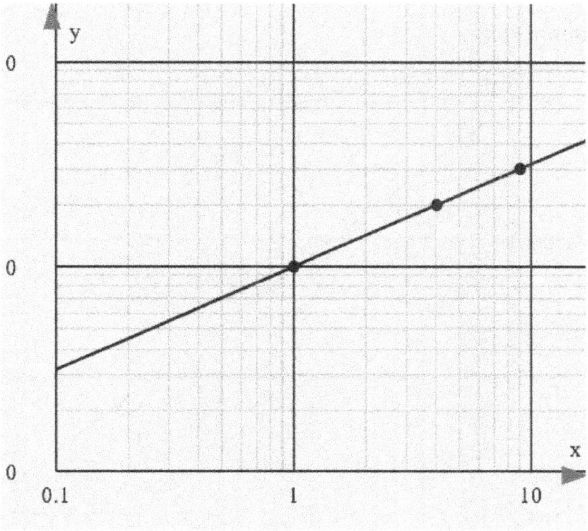

Points along the Curve y = √x on log – log axes

We observe that the graph of **y = x** $^{1/2}$ is a straight line when graphed on log-log axes.

Application 1: Air pressure

1. By pumping, the air pressure in a tank is reduced by 18% each second. So the percentage of air pressure remaining at time t is given by p = 100(0.82) t.Plot p against t for 0 < t < 30s on

   (a) A rectangular co-ordinate system

   (b) A semi logarithmic system

(a) Rectangular Plot

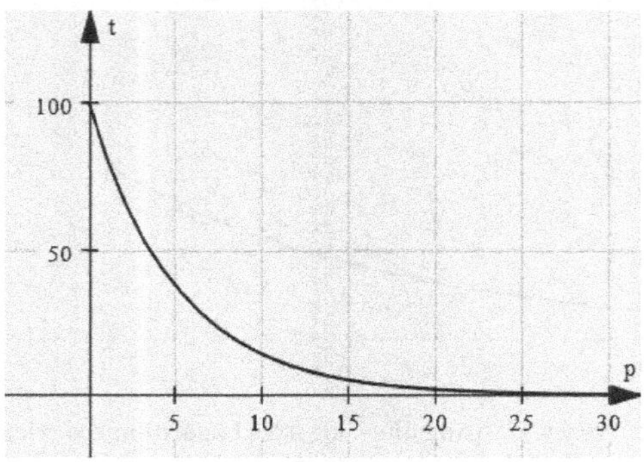

Graph of y = 100 (0.82) Ton Linear Axes

(b) Semi logarithmic Plot

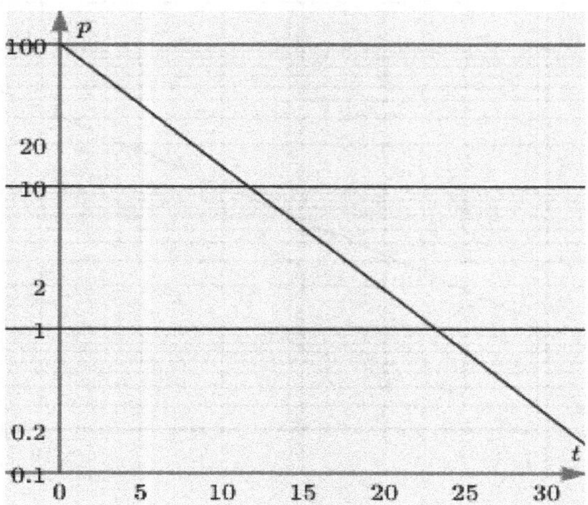

Graph of y = 100 (0.82) Ton-semi Logarithmic Axes

Question for Discussion

1. What are the various possible sources of errors in measurement? Define static error and explain any two static errors with example.
2. Differentiate the followings briefly.
   (i) Precision and Accuracy
   (ii) Line and End Standards
   (iii) Systematic Error and Random Error
   (iv) Instrument and Measurement
3. Explain the various modes of measurements with two examples.
4. Explain the following characteristics of measurement system:
   (i) Dead zone
   (ii) Drift
   (iii) Sensitivity
   (iv) Threshold
   (v) Fidelity
   (vi) Linearity
   (vii) Overshoot
5. Classify measurement methods. Discuss Primary, Secondary and tertiary methods of measurement with suitable examples.
6. State necessity and objectives of mechanical measurement. Also explain in brief precision and accuracy.
7. Define the following terms
   (i) Precision
   (ii) Accuracy
   (iii) Errors
   (iv) Dead zone
   (v) Drift.
   (vi) Indirect Method of measurement
8. Define the following terms
   (i) Resolution
   (ii) Threshold
   (iii) Range
   (iv) Span

9. Write short notes on:
   (i) Instrumental error
   (ii) Environmental error
   (iii) Calibration
   (iv) Full interchange ability
   (v) systematic error
   (vi) random error.
10. Describe the different types of errors in measurement and their causes.
11. Give the structure of generalized measuring system. Explain any four types with neat sketch.
12. Define the following terms
    (i) Primary sensing elements
    (ii) Repeatability.
    (iii) Calibration
    (iv) Range of measurement
    (v) Resolution
13. Explain the following terms with respect to the measurement system:
    i. Threshold and Resolution
    ii. Sensitivity and Drift
    iii. Hysteresis
14. Describe the basic concepts of mechanical measurement and errors in measurements.
15. Described about the Direct method and Indirect method of measurement
16. What is the need of calibration? Explain the classification of various measuring methods.
17. What are elements of a measuring system? How they affect accuracy and precision? How error due to these elements are eliminated
18. Explain in detail various types of errors that may arise in engineering measurements and the ways to control it.
19. Draw the block diagram of generalized measuring system and explain the different stages with suitable example.
20. Define accuracy, explain how the accuracy of an instruments can be specified.
21. Explain the purpose of calibrating the instruments and discuss the primary and secondary calibration.
22. Classify standards of measurement on the basis of their function and application.

23. Each of the following equations can be represented as a straight line on an x-y plot by choosing the appropriate axis scales. Plot them both in rectangular coordinate format and then in an appropriate format to yield a straight line. Explain how the plot operation yields the straight line. Variable y has units of volts. Variable x has units of meters (use a range of 0:01 < x < 10:0). Note: This is easily done using a spreadsheet program where you can compare the use of different axis scales.

a. $y = x^2$
b. $y = 1.1x$
c. $y = 2x^{0.5}$
d. $y = 10x^4$
e. $y = 10e^{-2x}$

Plot $y = 10e^{-5x}$ volts on in semilog format (use three cycles). Determine the slope of the equation at $x = 0$; $x = 2$; and $x = 20$.

OR

Plot the following data on appropriate axes, estimate the static sensitivity K at each X .

Y [V]	*X* [V]
2.9	0.5
3.5	1.0
4.7	2.0
9.0	5.0

24. For the calibration data given, plot the calibration curve using suitable axes. Estimate the static sensitivity of the system at each X. Then plot K against X. Comment on the behaviour of the static sensitivity with static input magnitude for this system.

Y [cm]	*X* [kPa]
4.76	0.05
4.52	0.1
3.03	0.5
1.84	1.0

25. Consider the function

$$y = \frac{1}{x}$$

a. Plot the function on rectangular coordinates.
b. Plot the function on semilog coordinates.
c. Plot the function on log-log coordinates.
d. Discuss the implications of each plot as it would relate to analyzing experimental data.

26. Explain about zero order, first order, and second order system in measurement.
27. What is the significance of time constant, settling time and phase linearity.
28. Explain in detail data presentation techniques, described any one of them.

References

[1] R. S. Figliola and D.E. Beasley, Theory and Design for Mechanical Measurements, 2nd ed., New York: Wiley, 1995.

[2] P.H. Sydenham, Handbook of Measurement Science, Chichester, U.K.: John Wiley & Sons, 1983.

[3] E.O. Doebelin, Measurement Systems, Application and Design, 4th ed., New York: McGraw-Hill, 1990.

[4] I.C. Gupta, Engineering Metrology, Dhanpat Rai Publications, Delhi.

[5] Mechanical Measurements, R.K. Jain Khanna Publishers, 1994.

[6] Mechanical Measurements, Beckwith Marangoni and Lienhard, Pearson Education, 6th Ed., 2006.

[7] P.W. Bridgeman, Dimensional Analysis, 2nd ed., Yale University Press, New Haven, CT, 1931.

[8] W.J. Duncan, Physical Similarity and Dimensional Analysis, Arnold, London, 1953.

[9] B.S. Massey, Units, Dimensions and Physical Similarity, Van Nostrand-Reinhold, New York, 1971.

[10] C.M. Close and D.K. Frederick, Modeling and Analysis of Dynamic Systems, 2nd ed., Wiley, Boston, 1994.

[11] E.O. Doebelin, System Modeling and Response, Theoretical and Experimental Approaches, Wiley, New York, 1980.

[12] K. Ogata, System Dynamics, 4th ed., Prentice-Hall, Englewood Cliffs, NJ, 2003.

[13] Mechanical Measurements by Thomas G. Beckwith and N. Lewis Buck.

CHAPTER 2

STATISTICS & ERROR UNCERTAINTY

2.0. Linear Regression

In general, we know that throughout the measurement process, the controlling of measured variable is performed by sampling which may be a function of one or more than one independent variables under a given set of condition, further this process is again repeated to establishing a relationship between the measured variable and an independent variable. In regression analysis we will try to established a functional relationship between the dependent variable and the independent variable. Linear regression requires that model is linear in regression parameters. Thus regression analysis is a statistical technique that attempts to explore and modeled the relationship between one or more response variables sometimes it is also called dependent variables , explained variables , predicted variables, or regress variable which is denoted by (*y*) and the predictors called independent variables, explanatory variables, control variables, or regressors generally denoted by x_1 , x_2 , ... x_p) .

2.0.1. Type of Regression Analyses

There are three types of regression, these are given below,

Case-I Simple Linear Regression

The simple linear regression is applied for modeling the linear relationship between two variables. One of them is the dependent variable y and another is the independent variable. For example, the simple linear regression can model the relationship between muscle strength (y) and lean body mass (x). The simple regression model is often written as the following form

$$y = \beta_0 + \beta_1 x + \varepsilon, \quad \text{..............................} \ 1$$

OR $\quad y = E(y) + \varepsilon$

independent variable, x , also known as, regressors or predictor variables and ,y, is called response variable.

where

β_1 = Slope,

β_0 = Intercept or slop of regression line,

ε = Random Error

$E(y) = \beta_0 + \beta_1 x$ = Regression Coefficients

Straight line formula : In general, a simple linear regression equation is represented by the equation of straight line that can be given by Y = MX + C or y =a +bx , the identification of inherent elements is given below ,

$$Y = MX + C \quad \text{..............................} \quad 2$$

Where ,

Y = Predicted Value / Dependent Variable / Response Variable

M = Regression coefficient

C = Regression Intercept

X = Predictor / Independent Variable / Explanatory variable

The betas form of such a equation can be given as ,

$$y = \beta_0 + \beta_1 x \quad \text{..............................} \quad 3$$

whenever we will plotting the data the dependent variables are put on x axis and independent data on the y respectively,

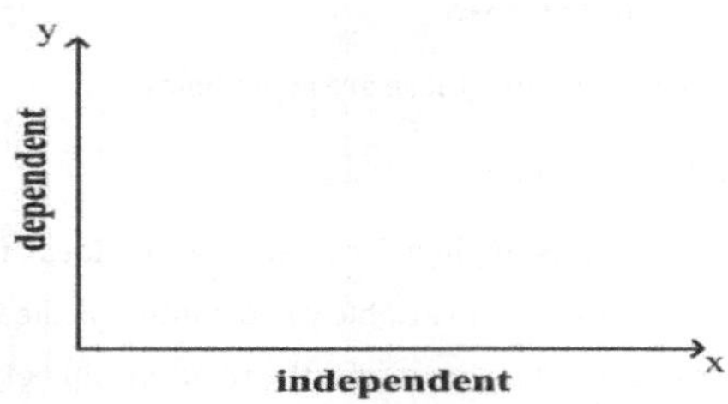

Diagram 2.1: Correlations between Dependent and Independent Variables

For instance, the line **y =1+0.5x** has an intercept of **1** and a gradient of **0.5.** Its graph is as follows:

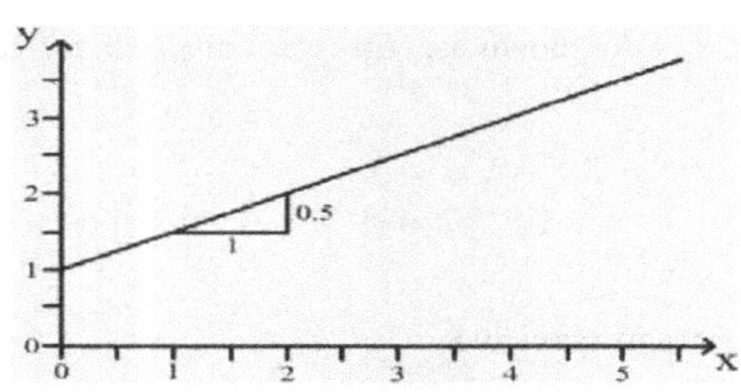

Diagram 2.2: Straight-line Representation of Variables

For example, suppose that height was the only determinant of body weight. If we will plot height the independent or 'predictor' variable along X-axis as a function of body weight , the dependent or 'outcome' variable representing along Y–axis, we have following curve between the variables, as shown below.

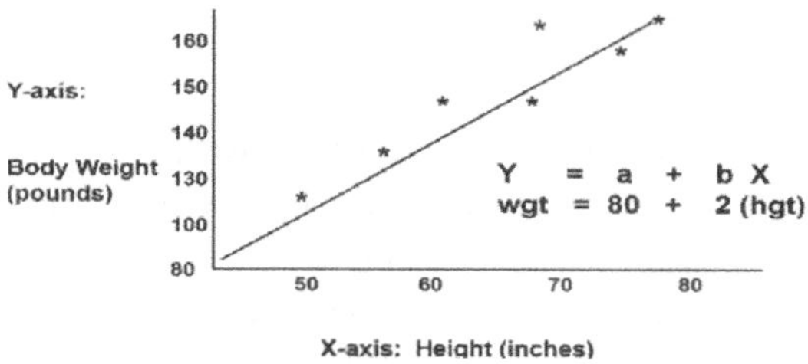

Diagram 2.3: Straight-line Representation of Weight and Height

In this simple linear regression, we determines the impact of one independent variable on the outcome

2.0.2. Model Assumptions

Consider the simple linear regression model which is being used for analyses and assume that it has zero mean and variance σ_2 ,The equation is given below,

$$Y_i\ \beta=\beta_0+\beta_1\ X_i\varepsilon I \quad \ldots\ldots\ldots\ldots\ldots\ldots\ldots\ldots\ldots\ldots \ 4$$

The deterministic component for such a model consists of regression coefficients as β_0 , β_1 and Residual Error (ε_i). These deterministic component indicating that it is a straight line which gives relationship between dependent & independent variable and the difference between both are considered as a residual error. In the Diagram 2.4 , It is concluded that the points has randomly scattered around a straight line having a consistence variation.

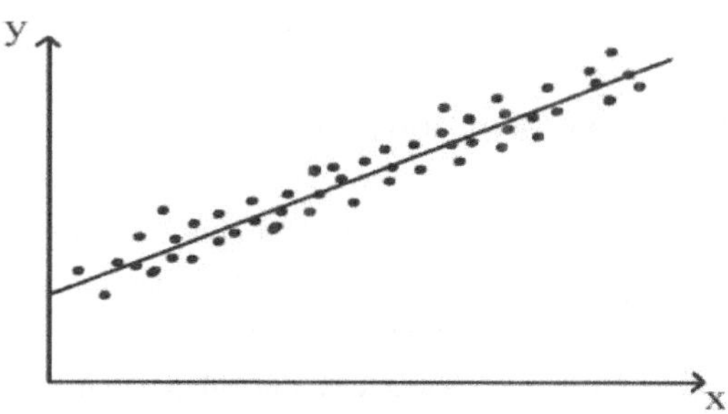

Diagram 2.4: Straight-line Representation of Variables

For "**Best fit** " or Feasible & Favorable result we will minimize the residual error by implementing the least squares linear regression . The curve fitting gives a clear estimation of the error variance which is a key concept of significance test regarding the regression coefficients and for producing confidence/prediction intervals

Effects on Model assumptions: Following parameters are generally affected the procedure of model analyses.

Outliers: It is a very common picture about data scattering that either all the data may or may not be set to mean value or fit the tendency of the data set or we can say that Data that lie outside the probability of normal variation incorrectly offset the sample mean value estimate , inflate the random uncertainty estimates , and influence a least-squares correlation. Statistical techniques can be used to detect such data points, which are known as outliers. So therefore, initially we have to estimate the value of outlier for best fit &results ,Check it for extreme values should form part of the initial data screening process and should be performed on both the response and explanatory variables. In case of Univariate outliers it is detected by box plots on the other hand Multivariate outliers by residual scatter plots.

Estimation of outliers: In general the following two common methods are generally applied for outliers is estimations ,

(i) Chauvenet's criterion, which identifies outliers having less than a **1/2N** probability of occurrence.

$$z_0 = |x_i - \bar{x}/s_x|,$$

where $\mathbf{x_i}$ is a suspected outlier in a data set of N values.

$$(1 - 2 \times P(z_0)) < 1/2N$$

(ii) In case of large data points, we can apply the three-sigma test to detect those data points which is lie outside the range of 99.73% probability of occurrence, $\bar{x} \pm t_{v,99.7}s_x$, as potential outliers.

- **Multi-collinearity and singularity:** Multi-collinearity exists when there are high correlations among the explanatory variables. Singularity exists when there is perfect correlation between explanatory variables. The presence of either affect the interpretation of the explanatory variables effect on the response variable. Also it can lead to numerical problems in finding the regression solution.
- **Normality, linearity, homo-scedasticity and independence of residuals :** The first three of these assumptions are checked using residual diagnostic plots after having fit

a multiple regression model. The independence of residuals is usually assumed to be true if we have indeed collected a random sample from the relevant population .

2.0.3. *Graphical Analysis of Simple Regression Model*

The following key points must be taken into consideration while simple regression model is drawn for any particular problem,

- Scatter plot: It is a plot applied for visualizing the linear relationship between the predictor and response variable.
- Box plot: To spot any outlier observations in the variable. Having outliers in your predictor can drastically affect the predictions as they can easily affect the direction/ slope of the line of best fit.
- Density plot: It is the distribution of the predictor variable, for best result it should be a close to normal distribution, without being skewed to the left or right is preferred.

2.0.4. *Residual Checking*

The basic concept of Residual Analysis is to check the favourability & feasibility to creating the regression model for any particular problem. A positive residual indicates a data point higher than their expected mean value, and a negative residual indicates a point lower than expected value. "**A residual is the deviation of an outcome from the predicated mean value for all subjects with the same value for the explanatory variable**" A plot of all residuals on the y-axis vs. the predicted values on the x-axis, called a residual vs. fit plot , The residual vs. fit plot can be used to check the non-linearity and unequal variance. In order to understand the concept of residual analyses, consider the following graph,

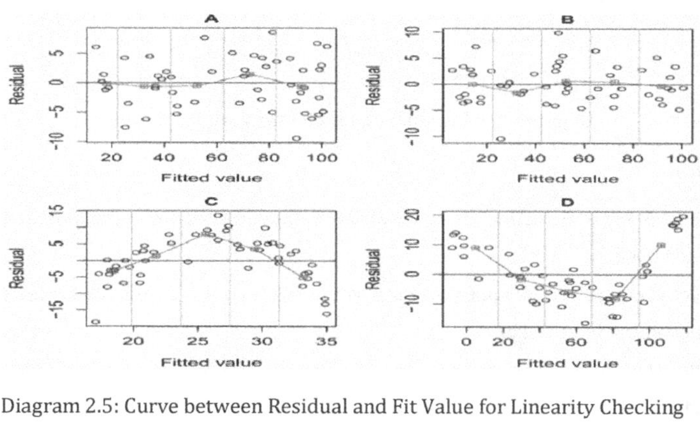

Diagram 2.5: Curve between Residual and Fit Value for Linearity Checking

Conclusion

- For each value of x, the mean of Y falls on a straight line, then the residuals have a mean of zero.
- If the data follow a simple curve, we will tend to have either a pattern of high then low then high residuals or the reverse.
- If the curve is close to a horizontal line at 0 on the y-axis, then we have a linear curve.
- Plots A and B show no obvious pattern away from a horizontal line other than the small amount of expected "noise".
- Plots C and D show clear deviations from normality, because the lines connecting the mean residuals of the vertical bands show a clear from (C) and smile (D) pattern, rather than a flat line.
- Untransformed linear regression is inappropriate for the data that produced plots C and D.

Consider another curve between residual vs. fit which shows good linearity.

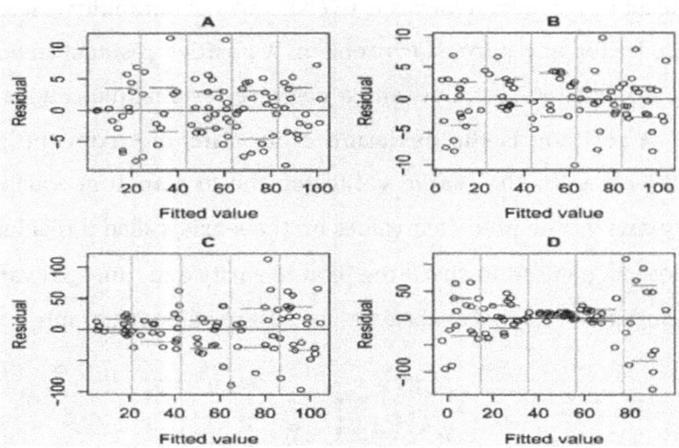

Diagram 2.6: Residual vs. Fit Plot for Variance Testing

Conclusion

- Plots A and B indicating the equal variance, the red lines are a similar distance apart in each band.
- plot C indicate to unequal variance, as the red lines increase in distance apart as you move from left to right.
- Plot D show a pattern with unequal variance in which the smallest variance is in the middle of the range of predicted values, with larger variance at both ends.

2.0.5. *Regression Analysis by Least-Squares Method*

Least square method is generally applied for most of the engineering problem in regression analyses, the least square method helps in minimizing the sum of the square of the deviation between actual data and polynomial.

Consider the case of linear regression analyses for single variable for equation y = f (x) , the polynomial fit can be given as,

$$y_c = a_0 + a_1 x + a_2 x^2 + \cdots + a_m x^m \qquad \text{.......... 5}$$

Suppose there are N values of x and y, which is referred as , x_i and y_i where i=1,2, ... N. then deviation between dependent variable y_i and polynomial y_i-y_{ci} where y_{ci} is value of polynomial at the data point (x_i and y_i), hence deviation may be given as,

$$D = \sum_{i=1}^{N} (y_i - y_{ci})^2 \qquad \text{.......... 6}$$

From equation 5 &6, we get

$$D = \sum_{i=1}^{N} [y_i - (a_0 + a_1 x + \cdots + a_m x^m)]^2 \qquad \text{.......... 7}$$

Standard deviation may be given as ,

$$s_{yx} = \sqrt{\frac{\sum_{i=1}^{N} (y_i - y_{ci})^2}{\nu}} \qquad \text{.......... 8}$$

where,

v=degree of freedom which is equal to N–(m+1) If variability is considered then uncertainty may be given as,

$$t_{\nu,P} s_{yx} \left[\frac{1}{N} + \frac{(x - \bar{x})^2}{\sum_{i=1}^{N} (x_i - \bar{x})^2} \right]^{1/2} \quad (P\%) \qquad \text{.......... 9}$$

Where ,

$$\bar{x} = \sum_{i=1}^{N} x_i / N \qquad \text{.......... 10}$$

The confidence interval is given by the following equation,

$$y_c(x) \pm t_{v,P} s_{yx} \left[\frac{1}{N} + \frac{(x - \overline{x})^2}{\sum_{i=1}^{N} (x_i - \overline{x})^2} \right]^{1/2} \quad (P\%) \qquad \text{.................................. 11}$$

In case of Linear Polynomials we will consider the following set of equation. then Correlation Coefficient (r) can be given as ,

$$r = r_{xy} = \frac{N \sum_{i=1}^{N} x_i y_i - \sum_{i=1}^{N} x_i \sum_{i=1}^{N} y_i}{\sqrt{N \sum_{i=1}^{N} x_i^2 - \left(\sum_{i=1}^{N} x_i \right)^2} \sqrt{N \sum_{i=1}^{N} y_i^2 - \left(\sum_{i=1}^{N} y_i \right)^2}} \qquad \text{.................................. 12}$$

The slope can be calculate by the following equation,

$$s_{a_1} = s_{yx} \sqrt{\frac{I}{\sum_{i=1}^{N} (x_i - \overline{x})^2}} \qquad \text{.................................. 13}$$

The value of zero intercept can be calculate by the following equation,

$$s_{a_0} = s_{yx} \sqrt{\frac{\sum_{i=1}^{N} x_i^2}{N \sum_{i=1}^{N} (x_i - \overline{x})^2}} \qquad \text{.................................. 14}$$

Example 1: The following data are suspected to follow a linear relationship. Find an appropriate equation of the first-order form.

x (cm)	y (V)
1.0	1.2
2.0	1.9
3.0	3.2
4.0	4.1
5.0	5.3

Where ,

x = Independent variable

y = Dependent measured variable

N = 5 = No of reading

Solution: Here we have linear equation which can be given as ,

$$y_c = a_0 + a_1 x \qquad \text{.. (i)}$$

Then Standard deviation will be

$$D = \sum_{i=1}^{N} (y_i - y_{ci})^2 \qquad \text{.. (ii)}$$

For minimizing terms , we will indicate the derivative as a zero ,

$$\frac{\partial D}{\partial a_0} = 0 = -2\left\{\sum_{i=1}^{N} [y_i - (a_0 + a_1 x)]\right\}$$

$$\frac{\partial D}{\partial a_1} = 0 = -2\left\{\sum_{i=1}^{N} [y_i - (a_0 + a_1 x)]\right\} \qquad \text{..................... (iii)}$$

Hence the value of coefficient will be given as ,

$$a_0 = \frac{\sum x_i \sum x_i y_i - \sum x_i^2 \sum y_i}{(\sum x_i)^2 - N \sum x_i^2}$$

$$a_1 = \frac{\sum x_i \sum y_i - N \sum x_i y_i}{(\sum x_i)^2 - N \sum x_i^2} \qquad \text{............................... (iv)}$$

We get , the value of coefficient

$a_o = 0.02$

$a_1 = 1.04$

then final equation becomes ,

$y_c =$ **0.02 + 1.04** x

In case of Linear Polynomials we will consider the following set of equation .

2.1. Multiple Regression Method

Multiple regression is nothing but it is an extended form of simple linear regression analyses in which problem is modeled with one dependent variable and several independent variables or predictor variables instead of establishing a correlations between single independent variable & single dependent variable as in case of linear regression model. In multiple regression analyses the independent variables may be continuous or categorical and response variable is measured on continuous scale. The general form of the multiple linear regression can be given as , where $\mathbf{X_1}$ and $\mathbf{X_2}$ are independent variables.

$$Y = \beta_0 + \beta_1 X_1 + \beta_2 X_2 + \varepsilon \quad \ldots\ldots\ldots\ldots\ldots\ldots\ldots 15$$

and

β_0 ,β_1 and β_2 are known as , regression coefficients and ε is residual (error) term .

2.1.0. Residual Error (ε)

It is defined as the difference between independent variables **&** observed values of individuals with assumption that all the variables are independently and identically distributed normally having a zero mean value and variance σ_2 . Consider the case of two continuous independent variable then the deterministic component will be in the form of plane , For example we have two specific values x_1 and x_2 in the $X_1 - X_2$ plane then the resulting **Y** value will be the expected value (**Orthogonal Projection**) from this explanatory variable combination , which can be represented as ,

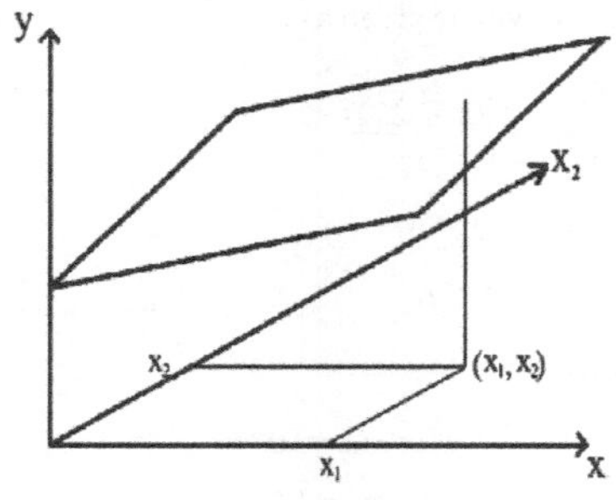

Diagram 2.7: Multiple Regression for Continuous Variable

Observed values for this combination of explanatory variables are drawn from a normal distribution with variance σ_2 centered on the expected value point:

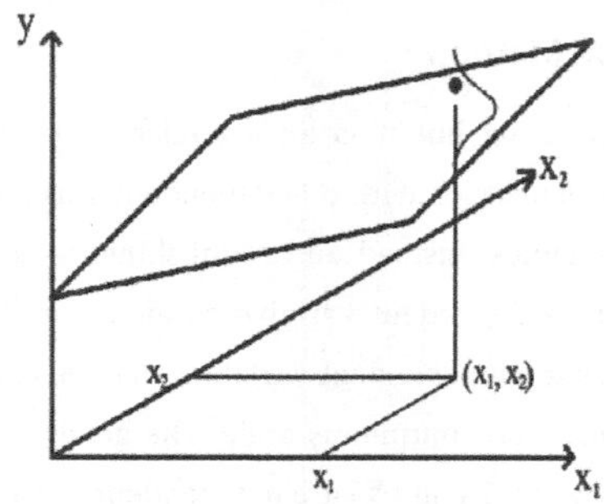

Diagram 2.8: Multiple Regression for Explanatory Variable

Our data should thus appear to be a collection of points that are randomly scattered with constant variability around the plane.

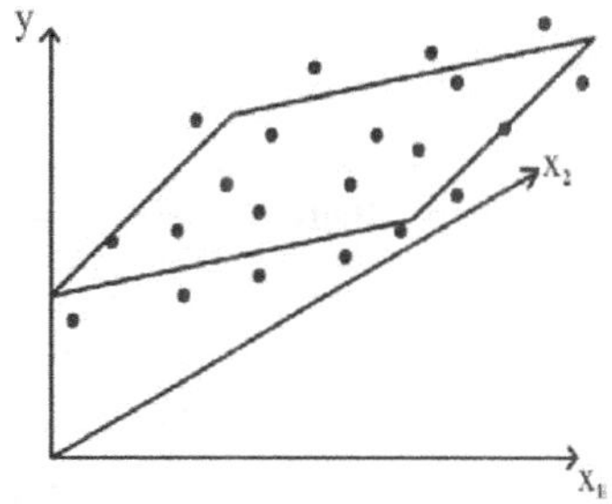

Diagram 2.9: Multiple Regression Model Fitting

The multiple regression model fitting process takes such data and estimates the regression coefficients **(β_0 ,β_1 and β_2)** that yield the plane that has best fit amongst all planes.

Case III Nonlinear Regression

In nonlinear regression, the relationship between dependent variable and independent variables is not linear in regression parameters. Example of nonlinear regression model may be written as

$$y = \frac{\alpha}{1 + e^{\beta t}} + \varepsilon, \quad \text{.......... 16}$$

Where y is the growth of a particular organism as a function of time t ,β and α are model parameters , and ε is the random error. Nonlinear regression model is more complicated than linear regression model in terms of estimation of model parameters ,model selection, model diagnosis ,variable selection, outlier detection, or influential observation identification.

2.2. Steps of Regression Analysis

Following steps must be taken by the students in regression analyses

- Determine independent and dependent variables .
- Look that the assumptions for dependent variables are satisfied : Residuals analysis .

(a) Linearity (assumption 1)

(b) Normality (assumption 3)-draw histogram for residuals (dependent variable) or normal P-P plot

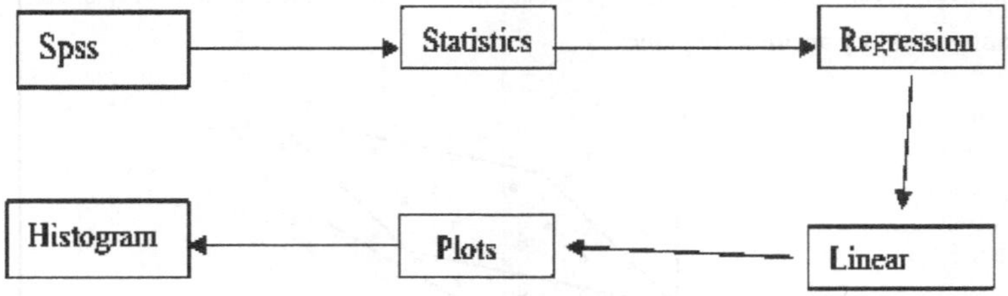

'Normal P-P plot of regression standardized'

Diagram 2.10: P-P Plot Regression

(c) Equal variance (homo-scedasticity : assumption 4) - Draw scatter plot for residuals

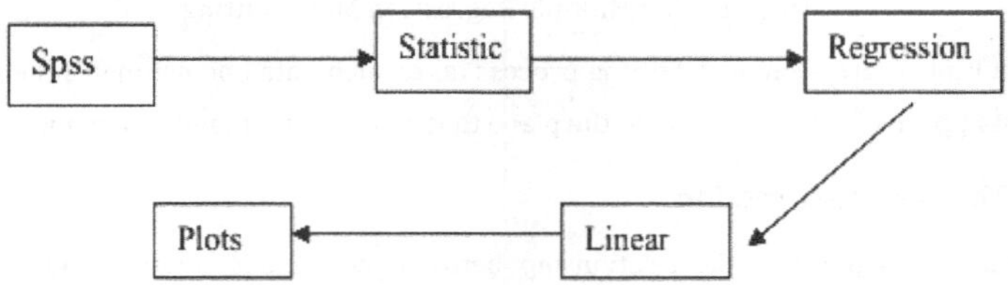

Diagram 2.11: Equal Variance Scatter Plot for Residuals

(d) Independence (Assumption 5,6: no autocorrelation between the disturbances, zero covariance between error term and X) - Each individual should be independent .

- Look at the correlation between two variables by drawing scatter graph

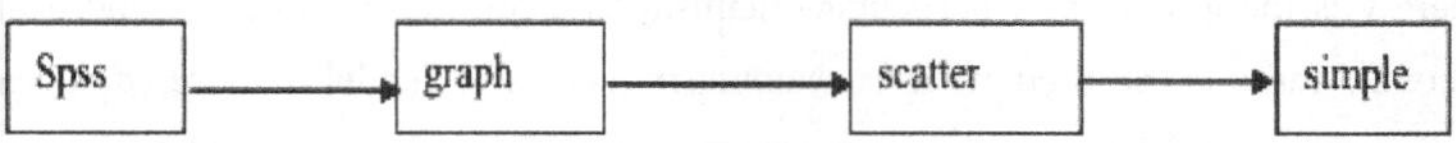

- Is there any correlation ?
- Is there any linear relation?
- Are there outlier ?If yes, clarify the reason and modify it .
- Are there separated groups. If yes, it means those data came from different population.
- Established a model by using statistical packages (SPSS)

Model Testing

1. Perform the **F-Test** ,find the **F- value** and **P–value** from **ANOVA table** , If **p-value** is smaller than alpha , the model is significant . if it does not significance then , " There is no linear relationship between **X** variable and **Y** variable."

2. Test the goodness of fit of the model. In this condition we will established a ' **Model Summary** , determine the value of **R-square** which indicate the percentage of the total variation in Y , **If the model is significant but R-square is small , It means that observed values are widely spread around the regression line. Test that the slope is significantly different from zero**
   - Obtained t-value in the 'Coefficients' table and find **P-value.**
   - T-square should be equal to **F-value.**

- If there is the significance of the model, **Show the model and interpret it**
  (i) Show the SRF
  (ii) In **"Model Summary "** Interpret R-square .
  (iii) In **"ANOVA"** table Æ Show the table, interpret **F-value** and the null hypothesis.
  (iv) In "**Coefficients"** table Æ Show the table and interpret beta values.
  (v) Show the residuals statistics and residuals' scatter plot
- **Mean estimation** (prediction) and **individual prediction**: We can predict the mean, individuals and their confidence intervals.

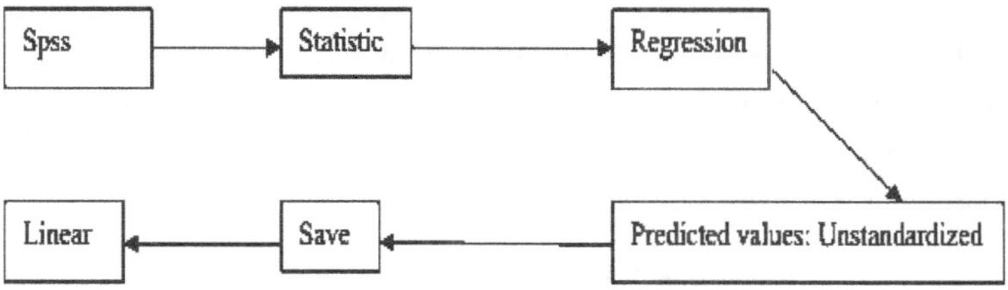

Diagram 2.12: Mean Estimation and Individual Prediction

2.2.1. Goals of Regression Analysis

Regression analyses is a statistical methods not only applied in engineering filed but also it is equally applicable in medicine, biology, agriculture, economics, sociology and geology, etc.

Purpose of Regression Method

1. Establish a causal relationship between response variable y and regressors $x_1, x_2 \dots\dots x_n$.
2. Predict y based on a set of values of $x_1, x_2 \dots\dots , x_n$.
3. To identify which variables are more important than others to explain the response variable y

Procedures of the Regression Analysis

(i) Detect the problem, is the problem belongs to a certain specific field & regression analysis can be best fitted to the problem.

(ii) Define a regression model which may be written as,

Response variable = Repressors + Random error

$$y = f(x_1, x_2 \dots\dots, x_p) + \varepsilon \quad \dots\dots\dots\dots\dots\dots\dots\dots\dots\dots\dots 17$$

(iii) Collect the data from the experiments that will be used in the statistical analysis.

(iv) Create at a data sets in a particular format that are easy to be read in to chosen software.

(v) Evaluate the model whether is it appropriate to modeled problem if not model diagnosis should be performed to establishing the sound statistical model.

Example 2 :- A marketer is interested in the relation between the width of the shelf space for her brand of coffee (x) and weekly sales (y) of the product in a suburban supermarket (assume the height is always at eye level) . Marketers are well aware of the concept of 'compulsive purchases', and know that the more shelf space their product takes up, the higher the frequency of such purchases. She believes that in the range of 3 to 9 feet, the **mean weekly sales** will be linearly related to the width of the shelf space. Further, among weeks with the same shelf space, she believes that sales will be normally distributed with unknown standard deviation σ (that is, σ measures how variable weekly sales are at a given amount of shelf space). Thus, she would like to fit a model relating weekly sales y to the amount of shelf space x her product receives that week. That is , she is fitting the model:

$$y = \beta_0 + \beta_1 x + \varepsilon, \quad \dots\dots\dots\dots\dots\dots\dots\dots\dots\dots\dots\dots\dots\dots\dots\dots 18$$

$$\text{so that } y|x \sim N(\beta_0 + \beta_1 x, \sigma).$$

One limitation of linear regression is that we must restrict our interpretation of the model to the range of values of the predictor variables that we observe in our data. We cannot assume this linear relation continues outside the range of our sample data. We often refer to **($\beta_0 + \beta_1$ x)** as the systematic component of **y** and **ε** as the random component.

- **Least Squares Estimation of β_0 and β_1:** We now have the problem of using sample data to compute estimates of the parameters **β_0**and **β_1**. First, we take a sample of **n** subjects , observing values **y** of the response variable and **x** of the predictor variable. We would like to choose as estimates for β_0 and β_1, the values b_o and b_1 that **'best fit'** the sample data. Suppose the market re-conducted the experiment over a twelve week period (4 weeks with 3' of shelf space , 4 weeks with6', and 4 weeks with 9'), and observed the sample data in Table 2.2

Table 2.2: Coffee Sales Data for n = 12 Weeks

Shelf Space x	Weekly Sales y	Shelf Space x	Weekly Sales y
6	526	6	434
3	421	3	443
6	581	9	590
9	630	6	570
3	412	3	346
9	560	9	672

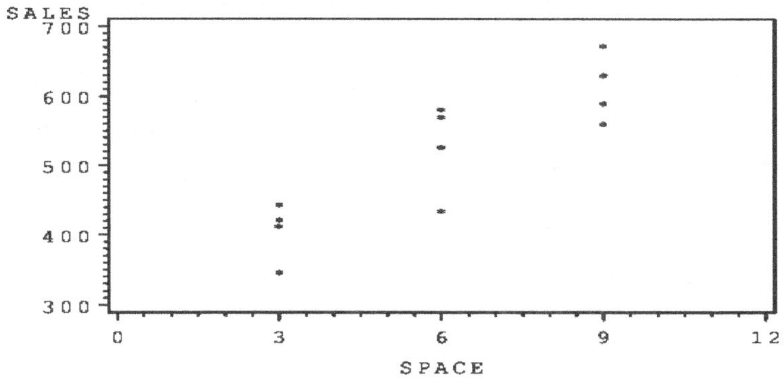

Diagram 2.13: Plot of Coffee Sales vs Amount of Shelf Space

Now, look at Figure 2.13 Note that while there is some variation among the weekly sales at 3', 6', and 9', respectively, there is a trend for the mean sales to increase as shelf space increases. If we define the **fitted equation** to be an equation:

$$\hat{y} = b_0 + b_1 x, \quad \text{.. 19}$$

we can choose the estimates b_0 and b_1 to be the values that minimize the distances of the data points to the fitted line. Now, for each observed response y_i, with a corresponding predictor variable x_i, we obtain a **fitted value** $\hat{y}_i = b_0 + b_1 x_i$. **So,** we would like to minimize the sum of the squared distances of each observed response to its fitted value. That is, we want to minimize the **error sum of squares,** SSE, where:

$$SSE = \sum_{i=1}^{n}(y_i - \hat{y}_i)^2 = \sum_{i=1}^{n}(y_i - (b_0 + b_1 x_i))^2. \quad \text{............................... 20}$$

A little bit of calculus can be used to obtain the estimates:

$$b_1 = \frac{\sum_{i=1}^{n}(x_i - \overline{x})(y_i - \overline{y})}{\sum_{i=1}^{n}(x_i - \overline{x})^2} = \frac{SS_{xy}}{SS_{xx}}, \qquad \text{.. 21}$$

And

$$b_0 = \overline{y} - \hat{\beta}_1\overline{x} = \frac{\sum_{i=1}^{n} y_i}{n} - b_1\frac{\sum_{i=1}^{n} x_i}{n}. \qquad \text{....................................... 22}$$

An alternative formula , but exactly the same mathematically, is to compute the sample covariance of *x* and *y*, as well as the sample variance of *x*, then taking the ratio.

$$cov(x,y) = \frac{\sum_{i=1}^{n}(x_i - \overline{x})(y_i - \overline{y})}{n-1} = \frac{SS_{xy}}{n-1} \qquad s_x^2 = \frac{\sum_{i=1}^{n}(x_i - \overline{x})^2}{n-1} = \frac{SS_{xx}}{n-1} \qquad b_1 = \frac{cov(x,y)}{s_x^2}$$

Some shortcut equations , known as the corrected sums of squares and cross products, that while not very intuitive are very useful in computing these and other estimates are ,

- $SS_{xx} = \sum_{i=1}^{n}(x_i - \overline{x})^2 = \sum_{i=1}^{n} x_i^2 - \frac{(\sum_{i=1}^{n} x_i)^2}{n}$
- $SS_{xy} = \sum_{i=1}^{n}(x_i - \overline{x})(y_i - \overline{y}) = \sum_{i=1}^{n} x_i y_i - \frac{(\sum_{i=1}^{n} x_i)(\sum_{i=1}^{n} y_i)}{n}$
- $SS_{yy} = \sum_{i=1}^{n}(y_i - \overline{y})^2 = \sum_{i=1}^{n} y_i^2 - \frac{(\sum_{i=1}^{n} y_i)^2}{n}$ **.................... 23**

2.3. Uncertainty Analysis

Uncertainty analysis is nothing but it is a quality based analyses of the result or achieved data from any particular measurement or test plan . In actual practice uncertainty is the negative aspects of measurement process which could not be interpreted but their estimation is possible by limiting the errors value. Uncertainty is arisen where we have measured values only but don' t know about the true value of the measurement hence we could not achieved conformation toward estimation a range of probable error **i.e.** how much it is ?

Uncertainty can be defined as It is an interval about the measured value within which we believed that the true value must fall with a stated probability. Thus , we can say that the uncertainty is an assigned numerical value that quantifies the probable range of these errors called the uncertainty

Uncertainty analysis is the process of identifying , quantifying , and combining the errors. Uncertainty is a outcome of a measurement process which has been quantifies the quality of that result. Uncertainty analysis provides a powerful design tool for evaluating different measurement systems and methods , designing a test plan , and reporting uncertainty.

Error and Uncertainty: From the context point of view, Uncertainty analyses is considered in to two category ,

- **Engineering Classification:** It is widely used approach applied for estimating the Uncertainty imposed by random and systematic error to calculate a standard deviation or not for a particular error or uncertainty source or groups errors and their uncertainties by their effect on the experiment or test, The relationship between the true value of the population and the measured data set, containing both systematic and random errors , can be given by the following diagram,

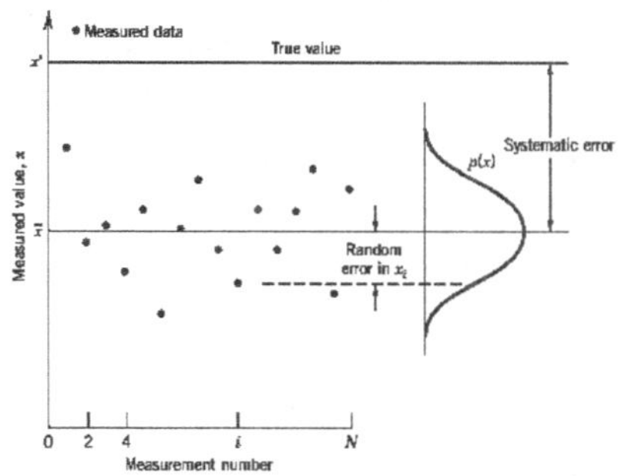

Diagram 2.14: Random and Systematic Errors

From the diagram It is concluded that , The systematic errors shift the sample mean away from the true mean by a fixed amount, and within a sample of many measurements, the random errors bring about a distribution of measured values about the sample mean. Hence accurate measurement contains small amounts of systematic and random errors.

ISO classifications: In this groups errors and their uncertainties is depending upon the data availability to calculate the sample standard deviation for a particular error and its uncertainty.

2.3.1. Types of Uncertainty

Measurement uncertainties may be classified as either random or systematic depending on how the measurement was obtained (an instrument could cause a random uncertainty in one situation and a systematic uncertainty in another).

i. **Random uncertainties are** statistical fluctuations (in either direction) in the measured data. These uncertainties may have their origin in the measuring device, or in the fundamental physics underlying the experiment. The random uncertainties may be masked by the precision or accuracy of the measurement device. Random uncertainties can be evaluated through statistical analysis and can be reduced by averaging over a large number of observations

ii. **Systematic uncertainties** are reproducible inaccuracies that are consistently in the "**same direction,**" and could be caused by an artifact in the measuring instrument, or a flaw in the experimental design (because of these possibilities, it is not uncommon to see the term "systematic error"). These uncertainties may be difficult to detect and cannot be analyzed statistically. If a systematic uncertainty or error is identified when calibrating against a standard,applying a correction or correction factor to compensate for the effect can reduce the bias. Unlike random uncertainties, systematic uncertainties cannot be detected or reduced by increasing the number of observations. When making careful measurements, the goal is to reduce as many sources of uncertainty as possible and to keep track of those that cannot be eliminated. It is useful to know the types of uncertainties that may occur, so that we may recognize them when they arise.

Contributions to Uncertainty of Measurand (1)

- **Sampling Uncertainty:** The undisturbed value is one of a large population of possible values for the undisturbed quantity of interest. Sampling Uncertainty is due to random errors.
- **System Sensor Uncertainty:** System disturbance changes the state of the system The sensor experiences interactions with the environment, e.g., radiation and conduction effects on thermocouple measurement. System Sensor Uncertainty is due to a fixed error.

Contributions to Uncertainty of Measurand (2)

- **Instrument Uncertainties:** Estimates of instrumentation uncertainties are readily made, though not all such estimates reflect the true uncertainty. **Moffat** claims that with modern instrumentation, the instrument uncertainties are often the least significant. Instrument Uncertainties are due to random errors for a properly calibrated instrument.
- **Centering Uncertainties:** Each centering correction has an associated uncertainty. These must be included in a final estimate of the uncertainty in the value of the measured quantity. Centering Uncertainties are due to fixed errors. We accept these fixed errors after we have done our best to minimize them.

Estimating Uncertainties (1)

System Disturbance Errors : The presence of a sensor changes the system

- Can be estimated, but not with any certainty
- Use estimates of disturbance errors to design the experiment, and to inform the installation of sensors in the experimental apparatus. For example, locate a probe to minimize flow blockage, or leakage.
- Goal is to design, build and instrument the system to minimize system disturbance errors.

System /Sensor Interaction Errors: System phenomena alter value sensed by the sensor.

- **Example:** Radiation from the walls of a duct alter the temperature of an exposed thermocouple junction used to measure the temperature of a flowing gas ,
- Should be estimated using engineering models of heat transfer and fluid flow ,
- Uncertainty in the estimate of this error needs to be included in final uncertainty estimate for the quantity of interest .

Estimating Uncertainties (2)

Instrumentation Errors

- Use manufacturer's data or traceable calibration.
- Best solution is to perform simple benchmark experiments to *measure the instrument* errors *in situ*

Uncertainties in Centering Corrections

- Report corrected values
- Uncertainty in the correction should be smaller than the correction itself.

Estimating Uncertainties (3)

Conceptual Errors

- Can be very large
- Not subjected to laws of physics
- Minimize by using good practice

Overall Errors

- Combine with methods presented below
- Use root-mean-squared procedure

Estimation Of Uncertainty Due to Random Error: we know that the concept of error limitation is perform with the help Uncertainty Analysis which is defined by confidence level , in any kind of engineering problem we considered two type of errors these are random and systematic error , known as random uncertainty and systematic uncertainties , In following section we will described here both in mathematically form .

For Random Uncertainty

The standard deviation for random uncertainty will be expressed mathematically as ,

$$S_R = \left[\sum_{\alpha=A}^{B}\sum_{i=1}^{N}(\phi_i D_{XT})^2\right]^{1/2} = \left[\sum_{\alpha=A}^{B}\sum_{i=1}^{N}[(\phi_i D_{XT}/\sqrt{M_T})^2]\right]^{1/2} \quad \text{......................} 24$$

D_{XT} = The sample standard deviation of the **i**[th] random error source of Type **T**

$\check{D}_{XT}$= The random uncertainty (standard deviation average) of the **i**[th] parameter random error source of Type

S_R = The random uncertainty of the measurement or test result

N = The total number of random uncertainties, Types **A** and **B,** combined

M_T = The number of data points averaged for the **i**[th] error source, Type **A** or **B**

ϕ_i = The sensitivity of the test or measurement result to the **i**[th] random uncertainty

For systematic Uncertainty

Systematic Uncertainty: The correlation for systematic uncertainty is given by following equation which is demonstrated previously by various authors ,

$$B_R = \left\{\sum_{T=A}^{B}\sum_{i=1}^{N_T}\left[\left(\theta_i B_{i,T}\right)^2 + \sum_{j=1}^{N_T}\theta_i\theta_j B_{(i,T),(j,T)}\left(1-\delta_{i,j}\right)\right]\right\}^{1/2}$$

and 25

B(i,T),(j,T) representing the covariance of **Bi** on **Bj** can be given as ,

$$= \sum_{l=1}^{M} B_{i,T}(l)\, B_{j,T}(l)$$

.... 26

Where ,

B_{iT}= The **i**[th] parameter elemental systematic uncertainty of Type **T**

B_R = The systematic uncertainty of the measurement or test result

N = The total number of systematic uncertainties

θ_i = The sensitivity of the test or measurement result to the ith systematic uncertainty

θ$_j$= The sensitivity of the test or measurement result to the j thsystematic uncertainty

l = An index or counter for common uncertainty sources

δ $_{i,j}$ = The Kro Necker delta. $\delta_{i,j} = 1$ if i = j, and $\delta_{i,j} = 0$

Hence the **total uncertainty** is combination of **Random uncertainty & Systematic uncertainty**

Total uncertainty = Random Uncertainty + Systematic Uncertainty = $S_R + B_R$

where ,

$$S_R = \left[\sum_{\kappa=A}^{B}\sum_{i=1}^{N}(\phi_i S_{\bar{X}_i})^2\right]^{1/2} = \left[\sum_{\kappa=A}^{B}\sum_{i=1}^{N}(\phi_i S_{X_i}/\sqrt{N_i})^2\right]^{1/2}$$

$$B_R = \left\{\sum_{T=A}^{B}\sum_{i=1}^{N_T}\left[(\theta_i B_{i,T})^2 + \sum_{j=1}^{N_T}\theta_i\theta_j B_{(i,T),(j,T)}(1-\delta_{i,j})\right]\right\}^{1/2} \quad \text{.. 27}$$

Total uncertainty will be given as ,

$$= t_{95}\left[(B_R/2)^2 + (S_R)^2\right]^{1/2} \quad \text{.. 28}$$

2.4. Classification of Errors

The error related to Uncertainty Analysis is of two types , but before going through the error we have to consider some points . These points are define as ,

- **Accuracy : Accuracy** is the closeness of agreement between a measured value and a true or accepted value. Measurement error is the amount of inaccuracy.
- **Precision : Precision** is a measure of how well a result can be determined with respect to true value. It is the degree of consistency and agreement among independent measurements of the same quantity, also the reliability or reproducibility of the result. the accuracy and precision are related as,

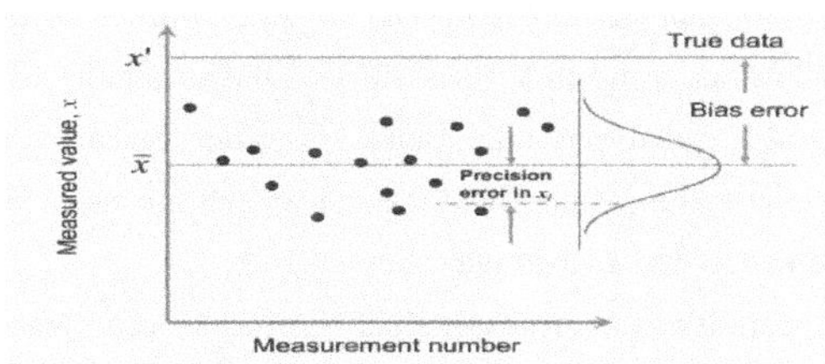

Diagram 2.15: Relationship Diagram between Bias, Precision & True Data

- **Bias (Systematic) Errors:** Bias error is a systematic inaccuracy caused by a mechanism that we can control. We might be able to adjust the way measurement are taken in an attempt to reduce bias errors. We can try to correct bias errors by including adjustments in our data analysis after the measurements are taken.
- **Precision (Random) Errors:** Random errors have different magnitude during subsequent measurements of the same quantity. For well-behaved systems , random errors in a measurement cause a cluster of observed values. We will assume that random errors are normally distributed. This is reasonable for large sample sizes and truly random errors

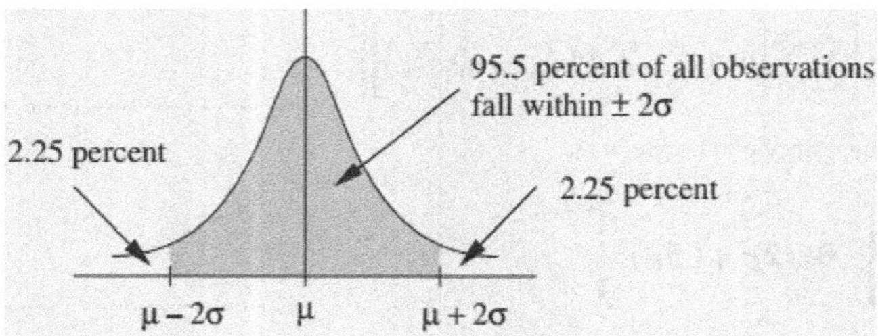

Diagram 2.16: Normal Distribution Curve for Precision Errors

A large sample (say, 30 or more observations) of the system at the same operating point tends toward a normal (Gaussian) distribution of values about some mean value if the population is Gaussian.

- **Zero offset (systematic)** : When making a measurement with a micrometer caliper, electronic balance , or electrical meter , always check the zero reading first. Re-zero the instrument if possible , or at least measure and record the zero offset so that readings can be corrected later. It is also a good idea to check the zero reading throughout the experiment. Failure to zero a device will result in a constant offset that is more significant for smaller measured values than for larger ones.
- **Physical variations (random)** : It is always wise to obtain multiple measurements over the widest range possible. Doing so often reveals variations that might otherwise go undetected. These variations may call for closer examination , or they may be combined to find an average value.
- **Parallax (systematic or random)**: This error can occur whenever there is some distance between the measuring scale and the indicator used to obtain a measurement. If the observer's eye is not squarely aligned with the pointer and

scale, the reading may be too high or low (some analog meters have mirrors to help with this alignment).

- **Instrument drift (systematic)**: Most electronic instruments have readings that drift over time. The amount of drift is generally not a concern , but occasionally this source of uncertainty can be significant.
- **Lag time and hysteresis (systematic)**: Some measuring devices require time to reach equilibrium , and taking a measurement before the instrument is stable will result in a measurement that is too high or low. A common example is taking temperature readings with a thermometer that has not reached thermal equilibrium with its environment. A similar effect is hysteresis, wherein the instrument readings lag behind and appear to have a **"memory"** effect, as data are taken sequentially moving up or down through a range of values. Hysteresis is most commonly associated with materials that become magnetized when a changing magnetic field is applied. **Last but not least,** some uncertainties are the result of carelessness, poor technique, or bias on the part of the experimenter. The experimenter may use a measuring device incorrectly, or may use poor technique in taking a measurement, or may introduce a bias into measurements by expecting (and inadvertently forcing) the results to agree with the expected outcome. Gross uncertainties of this nature can be referred to as **mistakes** or **blunders**, and should be avoided and corrected if discovered. As a rule, these uncertainties are excluded from any uncertainty analysis discussion because it is generally assumed that the experimental result was obtained by following correct and well-intentioned procedures - there is no point to performing an experiment and then reporting that it was known to be done incorrectly. The term **human error** should be avoided in uncertainty analysis discussions because it is too general to be useful.

2.5. Variable but Deterministic Error

Some errors that appear to be random can be caused by faulty measurement techniques (e.g. aliasing) or the errors may be variable but deterministic.

- Errors change even though the system is at the same nominal operating point
- Errors may not be recognized as deterministic: variations between tests , or test conditions , may seem random.
- Cause of these errors are initially hidden from the experimenter

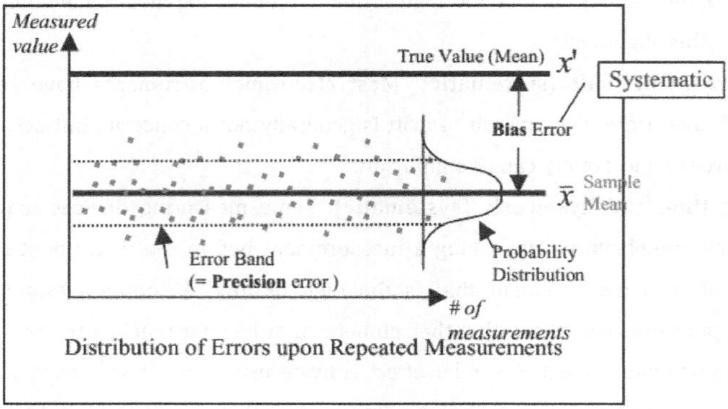

Diagram 2.17: Error Distribution Due to Repeated Measurement

Fixed Error: All repeatable errors are fixed errors. A fixed error is the same for each nominal operating point of the system

Sources of Fixed Error

- sensor calibration
- non-random disturbance to system

2.6. Combining Errors or Root Sum Squares (RSS) Method

Any experiment is composed of multiple instruments and sensors, all of which introduce errors and thus uncertainties is obtained . A strategy to combine errors or uncertainties is the RSS method. It combines the errors by adding the squares of each error and taking the square root of the sum. The RSS method is based on the assumption that the errors are not correlated (errors combine randomly).

$$u_x = \pm\sqrt{u_1^2 + u_2^2 + \ldots + u_K^2} \qquad (P\%\ probability) \qquad \ldots\ldots 29$$

In the above equation it is important that **all error terms** are based on the **same units and with respect to the same reference**. A common reference is to convert all errors to a percentage of the output signal. For example, A **26lbm** calibrated weight (accurate to within0.1%) is placed on a scale. **20 repeated** measurements are performed and the data is shown on the right A statistical analysis of the data gives a mean value **x** of **25.72** and a sample standard deviation **Sx =0.34**.

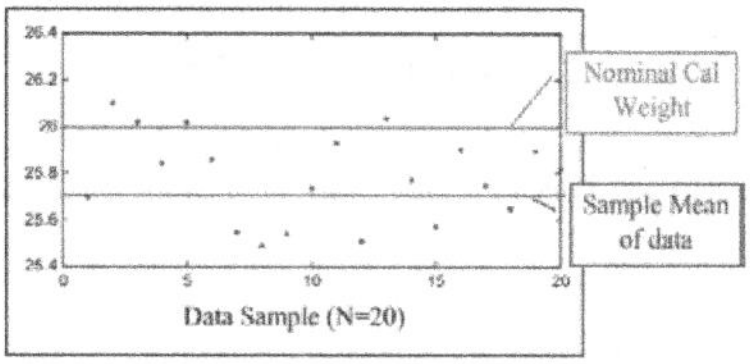

Diagram 2.18: Estimate the Uncertainty of the Mass Measurement

The mass of the calibration weight itself has some uncertainty. It is given as 0.1%. It can be assumed that this percentage has a confidence level of 95%, in other words 95% of the calibration weights will be with in a 0.1% range of the nominal value (26lbm).Therefore

$$m_{cal} = m_{cal-nominal} \pm u_{cal} = 26 \pm 0.026 \ \ lb_m \ \ (\text{Prob.} = 95\%) \quad \text{.......... 30}$$

And

$$u_{cal} = 0.026 \ \ lb_m \quad \text{.......... 31}$$

From the data we recognize that there is an offset between the measured mean value and the nominal calculated weight. This is a bias or systematic error in the measurement.

$$u_{bias} = 26 - 25.72 = 0.28 \ \ lb_m \quad \text{.......... 32}$$

Finally, the data indicates that there we have limited precision when the measurement is repeated due to the scatter of the data around the mean. This is the precision or random error. We can also assign an uncertainty to it based on the student-t distribution for 20 data points and a 95% probability.

$$u_{\text{Precision}} = t_{v,P} \, S_x = t_{19,95\%} \, S_x = 2.093 * 0.34 = 0.71 \ \ lb_m \quad \text{.......... 33}$$

Combining all the uncertainties based on the RSS method we get the total uncertainty in the mass measurement .

$$u_m = \sqrt{u_{cal}^2 + u_{bias}^2 + u_{\text{Precision}}^2} = \sqrt{.026^2 + .28^2 + .71^2} = 0.77 \ lb_m \quad \text{.......... 34}$$

The analysis shows that for the given measurement system we can only estimate the mass with an uncertainty of 0.77 lbm. The result also shows the major contributors to the uncertainty, the bias and the precision error. The error in the calculated weight is much smaller (as it should be for a good calibration reference). Using the calibration process , one can identify and thus remove the systematic bias error un bias from the data by subtracting the average bias error. In that case the uncertainty in the mass measurement can be reduced to

$$u_m = \sqrt{u_{cal}^2 + 0 + u_{Precision}^2} = \sqrt{.026^2 + 0 + .71^2} = 0.71\ lb_m \quad \text{.................................... 35}$$

Note that the total uncertainty has not been significantly reduced due to the relative large precision error.

2.7. Calibration (Systematic)

Whenever possible, the calibration of an instrument should be checked before taking data. If a calibration standard is not available, the accuracy of the instrument should be checked by comparing with another instrument that is at least as precise, or by consulting the technical data provided by the manufacturer. Calibration errors are usually linear, so that larger values result in greater absolute errors.

Calibration Errors: Calibration in itself does not eliminate system errors but it can help to quantify the uncertainty in the particular pieces of equipment used. Calibration errors include those elemental errors that enter the measuring system during its calibration. Calibration errors tend to enter through three sources:

(i) The standard or reference value used in the calibration,

(ii) The instrument or system under calibration, and

(iii) The calibration process.

For example, the laboratory standard used for calibration contains some inherent uncertainty, and this is passed along with the input value on which the calibration is based. Measuring system errors, such as linearity, repeatability, hysteresis, and so forth, contribute uncertainty. Depending on how the calibration is done, there can be a difference between the value supplied by the standard and the value actually sensed by the measuring system , in this section we have following error source group , Standard or reference value errors , Instrument or system errors , Calibration process errors and Calibration curve fit

2.8. Design-Stage Uncertainty Analysis

In the design stage, we always try to perform the analyses with the help of mathematical formula in design stage before starting the test initiation. there by we can get the clear image of uncertainty regarding the measurement process & what will be the limiting value ? In the design stage if achieved estimated value have minimum uncertainty then it is favourable on the other hand if it is too large then it is opposites to the above notion. the basic concept of the design – stage is to helps in evaluations and selection of instruments and applied measurement techniques. in the design stage , it is very impossible to detect the error by which uncertainty is being imposed but we will only consider the sources of error and their assigned uncertainty in measurement. We know that the measuring instruments does not a single unit but it composed of so many elements like , transducer , sensor etc. thus we also accumulate the respective contributions to system uncertainty.

Zero order uncertainty: In this section if all errors are '**Zero**' . the zero-order uncertainty of the instrument ,u_0 (which is random uncertainty involve by data scattering) due to instrument resolution, assumes that the variation expected in the measured values will be only that amount due to instrument resolution which is being controlled. In this case zero order uncertainty is equal to the half of the resolution .

$$u_0 = \frac{1}{2}\text{resolution} = 1\ \text{LSD}$$

where **LSD** is called least significant digit of the measuring instruments. If we will consider the normal distribution then equation becomes ,

$$u_0 = \pm\frac{1}{2}\ \text{resolution}\ (95\%)$$

Except above factors we also consider the instrument error that is due to manufacturing which is also considered in uncertainties .

2.8.1. *Design-Stage Uncertainty*

The design-stage uncertainty is nothing but it is combination of the interval of both the uncertainties ie . zero-order uncertainty & instrumental uncertainty and providing a minimum value criteria for design stage . throughout the Design-Stage Uncertainty we will consider the following error in the RSS method , as shown in figure. which is denoted by u_d. mathematically we can expressed as ,

$$u_d = \sqrt{u_0^2 + u_c^2} \quad (P\%) \quad \textbf{RSS method for combining error} \quad \text{.................................. 36}$$

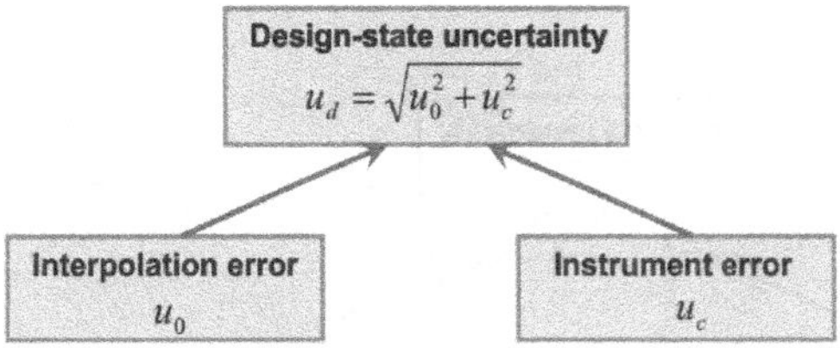

Diagram 2.19: Design-stage Uncertainty Procedure in Combining Uncertainties

(i) Zero-Order Uncertainty (Interpolation Error) :- Even when all errors are zero, the value of the measurand must be affected by the ability to resolve the information provided by the instrument. This is called zero-order uncertainty. At zero-order, we assume that the variation expected in the measurand will be less than that caused by the instrument resolution. And that all other aspects of the measurement are perfectly controlled .

$$U_0 = \pm 1/2 \text{ resolution } (95\%)$$

Instrument Uncertainty, u_c This information is available from the manufacturer's catalog.

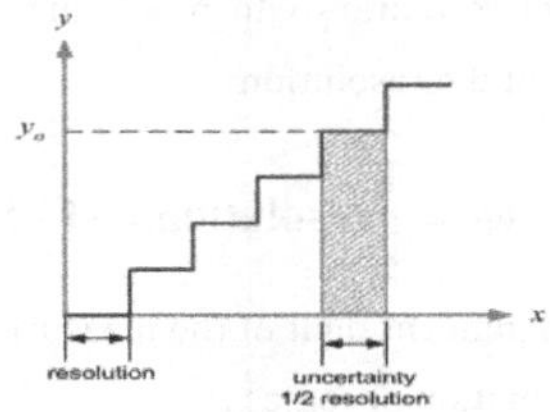

Diagram 2.20: Zero-Order Uncertainty

Example 3: Consider the force measuring instrument described by the catalog data that follows. Provide an estimate of the uncertainty attributable to this instrument and the instrument design state uncertainty.

Force Measuring Instrument

Resolution : 0.25N

Range : 0-100N

Linearity : within 0.20 N

Over range Repeatability : within 0.30 N over range

Known: Instrument specifications, Assume: Values representation of instrument 95% probability

Solution

Design-state uncertainty

$$u_d = \sqrt{u_0^2 + u_c^2}$$

$$u_d = \pm\sqrt{0.125^2 + 0.36^2} = \pm 0.38 \text{ N}$$

u_0

½ Resolution = 0.125 N

u_c

$$\sqrt{e_l^2 + e_r^2} = \pm\sqrt{0.2^2 + 0.3^2} = \pm 0.36 \text{ N}$$

Example 4: A voltmeter is used to measure the electrical output signal from a pressure transducer. The nominal pressure is expected to be about 3 psi (3 lb/in.[2] = 0.2 bar). Estimate the design-stage uncertainty in this combination . The following information is available:

Voltmeter	
Resolution:	10 μV
Accuracy:	within 0.001% of reading
Transducer	
Range:	±5 psi (~±0.35 bar)
Sensitivity:	1 V/psi
Input power:	10 VDC ± 1%
Output:	±5 V
Linearity error:	within 2.5 mV/psi over range
Sensitivity error:	within 2 mV/psi over range
Resolution:	negligible

Calculate ,u_c for each device and u_d for the measurement system .

The uncertainty in the voltmeter at the design stage (u_d) is given by the following relationship ,

$$(u_d)_E = \sqrt{(u_o)_E^2 + (u_c)_E^2}$$

Since we have ,

$$(u_0)_E = 5\ \mu\text{V}$$ then

For a Nominal Pressure of 3 psi , we expect to measure an output of 3 V. Then,

$$(u_c)_E = (3\text{ V} \times 0.00001) = 30\ \mu\text{V}$$

So , design stage uncertanity will be ,

$$(u_d)_E = 30.4\ \mu\text{V}$$

The uncertainty in the pressure transducer output at the design stage can be given as ,

$$\begin{aligned}(u_c)_p &= \sqrt{u_1^2 + u_2^2} \\ &= \sqrt{(2.5\text{ mV/psi} \times 3\text{ psi})^2 + (2\text{ mV/psi} \times 3\text{ psi})^2} \\ &= 9.61\text{ mV}\end{aligned}$$

Finally, u_d for the combined system is found by using the RSS method for the design-stage uncertainties of the two devices. The design-stage uncertainty is estimated as ,

$$
\begin{aligned}
u_d &= \sqrt{(u_d)_E^2 + (u_d)_p^2} \\
&= \sqrt{(0.030\ \text{mV})^2 + (9.61\ \text{mV})^2} \\
&= \pm 9.61\ \text{mV} \quad (95\%)
\end{aligned}
$$

But since the sensitivity is 1 V/psi, the uncertainty in pressure can be stated as,

$$u_d = \pm 0.0096\ \text{psi} \quad (95\%)$$

2.9. Estimating Experimental Uncertainty for a Single Measurement

Any measurement will have some uncertainty associated with it, no matter the precision of the measuring tool. How is this uncertainty determined and reported ? The uncertainty of a single measurement is limited by the precision and accuracy of the measuring instrument, along with any other factors that might affect the ability of the experimenter to make the measurement. For example, if you are trying to use a meter stick to measure the diameter of a tennis ball, the uncertainty might be ±5 mm, but if you use a Vernier caliper, the uncertainty could be reduced to maybe ± 2 mm. The limiting factor with the meter stick is parallax, while the second case is limited by ambiguity in the definition of the tennis ball 's diameter (it's fuzzy!). In both of these cases , the uncertainty is greater than the smallest divisions marked on the measuring tool (likely 1 mm and 0.05 mm respectively). Unfortunately, there is no general rule for determining the uncertainty in all measurements. The experimenter is the one who can best evaluate and quantify the uncertainty of a measurement based on all the possible factors that affect the result. Therefore, the person making the measurement has the obligation to make the best judgment possible and report the uncertainty in a way that clearly explains what the uncertainty represents:

Measurement = **(Measured Value ± Standard Uncertainty) (unit of measurement)** where "**standard uncertainty**" indicates approximately a **68%** confidence interval .

2.9.1. Estimating Uncertainty in Repeated Measurements

Suppose you time the period of oscillation of a pendulum using a digital instrument (that you assume is measuring accurately) and find that T = 0.44 seconds. This single measurement of the period suggests a precision of ±0.005 s , but this instrument precision may not give a complete sense of the uncertainty , and you should avoid reporting the uncertainty in this fashion if possible. If you repeat the measurement several times and examine the variation among the measured values, you can get a better idea of the uncertainty in the period. For example , here are the results of **5** measurements, in seconds : **0.46, 0.44, 0.45, 0.44, and 0.41.** For this situation, the best estimate of the period is the **average**, or **mean**:

$$\text{Average (mean)} = \frac{x_1 + x_2 + \ldots + x_N}{N} \quad \ldots\ldots 37$$

2.10. Outliers

Outliers are data points which lie outside the general linear pattern of which the midline is the regression line or In statistics, an outlier is a data point that differs greatly from other values in a data set , In pre-employment testing , the most common data that is observed are test scores , usually plotted against a measure of employee performance. Examining the data will help determine how correlated test scores are with test performance.

A rule of thumb is that outliers are points whose standardized residual is greater than 3.3 (corresponding to the .001 alpha level) .The removal of outliers from the data set under analysis can at times dramatically affect the performance of a regression model. Outliers should be removed if there is reason to believe that other variables not in the model explain why the outlier cases are unusual – that is , these cases need a separate model. Alternatively, outliers may suggest that additional explanatory variables need to be brought into the model (that is, the model needs re-specification). Another alternative is to use robust regression .

The leverage statistic, h, also called the *hat-value,* is available to identify cases which influence the regression model more than others. Leverage assesses how far away a value of the independent variable value is from the mean value: the farther away the observation the more leverage it has.

Belsley, Kuh, and Welsch, define the leverage (**h_i**) of the i_{th} observation as

$$h_i = \frac{1}{n} + \frac{(x_i - \bar{x})^2}{(n-1)S_x^2} \quad \ldots\ldots 38$$

- 0 <**h** < 1 The leverage statistic varies from **0 to 1**
- A rule of thumb is that cases with leverage under **0.2** are not a problem, but if a case has

  leverage over **0.5**, the case has undue leverage and should be examined for the possibility of measurement error or the need to model such cases separately. STATA command predict h, hat.

(i) Cook's distance, D, is another measure of the influence of a case. Cook's distance measures the effect of deleting a given observation.

- Observations with larger **D** values than the rest of the data are those which have unusual leverage.
- **D > 4/n** the criterion to indicate a possible problem. STATA command predict D, cooksd

(ii) df betas, is another statistic for assessing the influence of a case ,

- If dfbetas > 0, the case increases the slope , if < 0, the case decreases the slope.
- The case may be considered an influential outlier if **|dfbetas| > 2 / % n. STATA** commands dfbeta creates dfbeta's for all variables. or predict DFx1, dfbeta(x1) for individual variables

(iii) df Fit :- Df Fit measures how much the estimate changes as a result of a particular observation being dropped from analysis.

dfits is defined as

$$\text{dfits}_i = \text{Rstudent}\sqrt{\frac{h_i}{1-h_i}} \qquad \text{... 39}$$

Where **R** student is the studentized residual.

- The case may be considered an influential outlier if DFITS > 2 / % k / n **STATA** command predict DFITS, dfits .

(iv) Studentized residuals: Studentized residuals and deleted studentized residuals are also used to detect outliers with high leverage. A "studentized residual" is the observed residual divided by the standard deviation. The "studentized deleted residual" also called the "jac-knife residual" is the observed residual divided by the standard deviation computed with the given observation left out of the analysis. Analysis of outliers usually focuses on deleted residuals some time it is also called externally studentized residual or, misleadingly, standardized residual.

(v) Partial regression plots: Partial regression plots, also called partial regression leverage plots or added variable plots, are yet another way of detecting influential sets of cases. Partial regression plots are a series of **bi-variate** regression plots of the dependent variable with each of the independent variables in turn. The plots show cases by number or label instead of dots. One looks for cases which are outliers on all or many of the plots.

(vi) Why study outliers: Outliers can provide useful information about the process. An outlier can be created by a shift in the location (mean) or in the scale (variability) of the process. Though an observation in a particular sample might be a candidate as an outlier, the process might have shifted. Sometimes, the spurious result is a gross recording error or a measurement error. Measurement systems should be shown to be capable for the process they measure. Outliers also come from incorrect specifications that are based on the wrong distributional assumptions at the time the specifications are generated.

(vii) How to handle outliers: Once an observation is identified by means of graphical or visual inspection as a potential outlier, root cause analysis should begin to determine whether an assignable cause can be found for the spurious result. If no root cause can be determined, and a retest can be justified, the potential outlier should be recorded for future evaluation as more data become available. Often, values that seem to be outliers are the right tail of a skewed distribution. When reporting results, it is prudent to report conclusions with and without the suspected outlier in the analysis. Removing data points on the basis of statistical analysis without an assignable cause is not sufficient to throw data away. Robust or nonparametric statistical methods are alternative methods for analysis. Robust statistical methods such as weighted least-squares regression minimize the effect of an outlier observation .There are various approaches to outlier detection depending on the application and number of observations in the data set. Visual inspection alone cannot always identify an outlier and can lead to mis-labeling an observation as an outlier. Using a specific function of the observations leads to a superior outlier labeling rule. Because data are used in estimation with classical measures such as the mean being highly sensitive to outliers, statistical methods were developed to accommodate outliers and to reduce their impact on the analysis. Some of the more commonly used identification methods are discussed in this article.

2.11. Box-plot

A box plot is a graphical representation of dispersion of the data. The graphic represents the lower quartile (Q_1) and upper quartile (Q_3) along with the median .The median is the 50th percentile of the data. A lower quartile is the 25th percentile, and the upper quartile is the 75th percentile. The upper and lower fences usually are set a fixed distance from the inter-quartile range (Q3 - Q1). Diagram, shows the upper and lower fences to be set at 1.5 times the inter-quartile range. Any observation outside these fences is considered a potential outlier. Even when data are not normally distributed, a box plot can be used because it depends on the median and not the mean of the data.

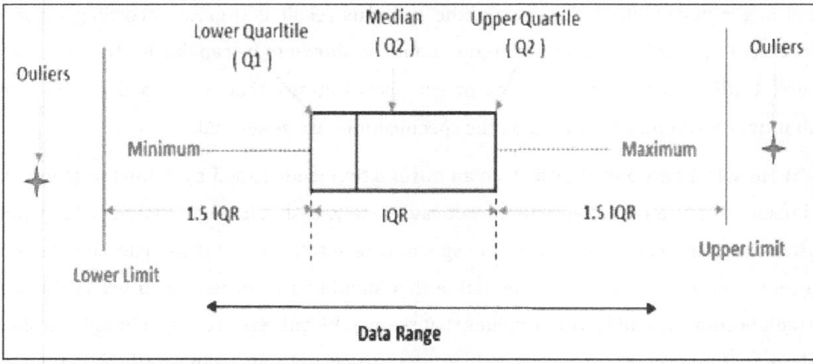

Diagram 2.21: Box-plot Representation

2.12. Trimmed Means

A trimmed is calculated by discarding a certain percentage of the lowest and the highest scores and then computing the mean of the remaining scores. The trimmed mean has the advantage of being relatively resistant to outliers. When outliers are present in the data , trimmed means are robust estimators of the population mean that are relatively insensitive to the outlying values. After viewing the box plot, a potential outlier might be identified. If the upper and lower 5% of the data are removed, then it creates a 10% trimmed mean. An example of trimmed means is the recent change in the Olympic scoring system for ice skating , in which the highest and lowest scores are eliminated, and the mean of the remaining scores is used to assess skaters' scores. If a trimmed mean is presented, the untrimmed mean should be presented for comparison.

2.13. Extreme Studentized Deviate

The extreme studentized deviate (ESD) test is quite good at identifying a single outlier in a normal sample. The maximum deviation from the mean:

$$R_i = \frac{max|x_i - \bar{x}|}{s_n}$$

where:

x_i = individual observation from the data set
$\bar{x}$ = sample mean
s_n = sample standard deviation of the n observations

is calculated and compared with a tabled value (see Table I). If the maximum deviation is greater than the tabled value, then the observation is removed, and the procedure is repeated. If no observation exceeds the tabled value, then we cannot claim there is a statistical outlier. A downfall of this method is that it requires the assumption of a normal data distribution.

Usually, this assumption holds true as the sample size gets larger . Table- I is an example of 10 observations (raw data), the critical value for **N = 10** at an **α** level of **0.05 is 2.29.** Therefore, the data value 16.3 is an outlier because it corresponds to a studentized deviation of 2.49, which exceeds the 2.29 critical value.

Table 1: Standard Outliers Value

Number of observations	Number of potential outliers					
	0	1	2	3	4	5
9	2.2119	2.1239	2.0177	1.8855	1.7140	1.4809
10	2.2865	2.2119	2.1239	2.0177	1.8855	1.7140
11	2.3510	2.2865	2.2119	2.1239	2.0177	1.8855
12	2.4076	2.3510	2.2865	2.2119	2.1239	2.0177
13	2.4579	2.4076	2.3510	2.2865	2.2119	2.1239
14	2.5030	2.4579	2.4076	2.3510	2.2865	2.2119
15	2.5438	2.5030	2.4579	2.4076	2.3510	2.2865
16	2.5811	2.5438	2.5030	2.4579	2.4076	2.3510
17	2.6152	2.5811	2.5438	2.5030	2.4579	2.4076
18	2.6468	2.6152	2.5811	2.5438	2.5030	2.4579
19	2.6760	2.6468	2.6152	2.5811	2.5438	2.5030
20	2.7032	2.6760	2.6468	2.6152	2.5811	2.5438

2.14. Dixon-type Tests

Dixon-type tests are based on the ratio of the ranges. These tests are flexible enough to allow for specific observations to be tested. They also perform well with small sample sizes. Because they are based on ordered statistics, there is no need to assume normality of the data. Depending on the number of suspected outliers, different ratios are used to identify potential outliers. The first class of ratios, r_{10}, is used when the suspected outlier is the largest or smallest observation. The second set of ratios, r_{11}, is used when the potential observation is the second smallest or second largest. Situations like these arise because of masking. Masking occurs when several observations are close together, but the group of observations is still outlying from the rest of the data. Masking is a common phenomenon especially for bimodal data (*i.e.,* data from two distributions). There are additional sets of ratios depending on how many marked points are excluded. The following equations are the r_{10} and r_{11} ratios :

Number of observations	If you suspect the highest value	If you suspect the lowest value
3 to 7	$\tau = \dfrac{x_n - x_{n-1}}{x_n - x_1}$	$\tau = \dfrac{x_2 - x_1}{x_n - x_1}$
8 to 10	$\tau = \dfrac{x_n - x_{n-1}}{x_n - x_2}$	$\tau = \dfrac{x_2 - x_1}{x_{n-1} - x_1}$
11 to 13	$\tau = \dfrac{x_n - x_{n-2}}{x_n - x_2}$	$\tau = \dfrac{x_3 - x_1}{x_{n-1} - x_1}$
14 to 25	$\tau = \dfrac{x_n - x_{n-2}}{x_n - x_3}$	$\tau = \dfrac{x_3 - x_1}{x_{n-2} - x_1}$

Number of observations	r_{10}		r_{11}	
	α = 0.01	α = 0.05	α = 0.01	α = 0.05
3	0.988	0.941		
4	0.889	0.765	0.991	0.955
5	0.780	0.642	0.916	0.807
6	0.698	0.560	0.805	0.689
7	0.637	0.507	0.740	0.610
8	0.590	0.468	0.683	0.554
9	0.555	0.437	0.635	0.512
10	0.527	0.412	0.597	0.477

2.15. Outliers in Regression

Regression analysis or least-squares estimation is a statistical technique to estimate a linear relationship between two variables. This technique is highly sensitive to outliers and influential observations. Outliers in regression can overstate the coefficient of determination (R_2), give us no error values for the slope and intercept, and in many cases lead to false conclusions about the model. Outliers in regression usually are detected with graphical methods such as residual plots including deleted residuals.

2.16. Data Acquisition

Data acquisition is the process of taking measurements 0f physical parameters such as temperature, strain ,force & motion , pressure, flow, humidity, geographic position, vibration, salinity or a thousand and one other parameters. The measurements may be for later analysis, for the control of a process or for simply displaying to interested personnel. Then Test and measurement to input and output data to and from a computer.

Data Acquisition and control system consists of following components:

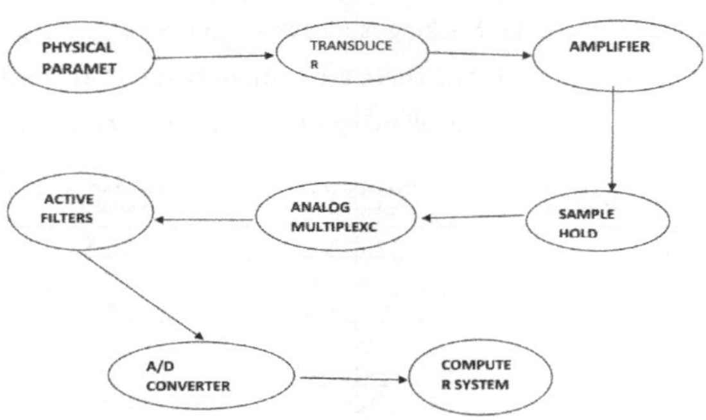

Diagram 2.22: Data Acquisition Components

The brief description of the components are given below

(i) Physical Parameters : Physicals parameters are those whose measurements is being performed , like temperature , strain , force & motion , pressure , flow, humidity , geographic position, vibration , salinity or a thousand and one other parameters.

(ii) Transducers : A transducer is a device that converts a signal from one physical form to a corresponding signal having a different physical form. Therefore, it is an energy converter. This means that the input signal always has energy or power, i.e. signals consist of two component quantities whose product has energy or power dimension. But in measurement systems, one of the two components of the measured signal is usually so small that it is negligible, and thus only the remaining component is measured. Since there are six different kinds of signals—mechanical, thermal, magnetic, electric, chemical and radiation (corpuscular and electromagnetic, including light)—any device converting signals of one kind to signals of a different kind is a transducer. The resulting signals can be of any useful physical form. Devices offering an electric output are called sensors. Most measurement systems use electric signals, and hence rely on sensors. Diagram shows a diagrammatic arrangement of transducers,

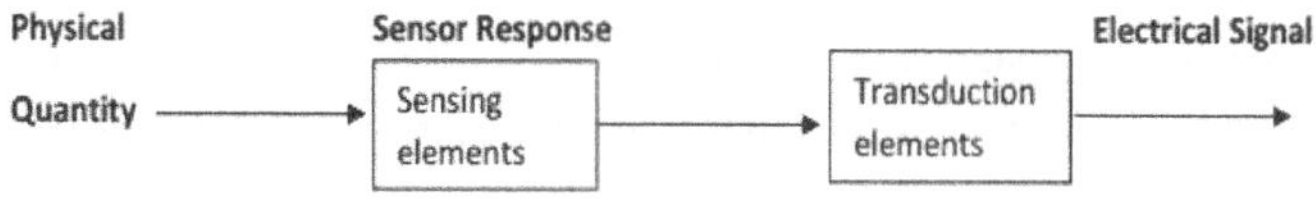

Diagram – 2.23 Function of Transducers

A transducer will have basically two main components. They are

- **Sensing Element:** The physical quantity or its rate of change is sensed and responded to by this part of the transistor.
- **Transduction Element:** The output of the sensing element is passed on to the transduction element. This element is responsible for converting the non-electrical signal into its proportional electrical signal .There may be cases when the transduction element performs the action of both transduction and sensing. The best example of such a transducer is a thermocouple. A thermocouple is used to generate a voltage corresponding to the heat that is generated at the junction of two dissimilar metals. Transducers sense physical phenomena and provide electrical signals that the DAQ system can measure. For example, thermocouples, RTDs, thermistors, and IC sensors convert temperature into an analog signal that an ADC can measure .Other examples include strain gauges, flow transducers, and pressure transducers, which measure

force, rate of flow, and pressure, respectively. In each case, the electrical signals produced are proportional to the physical parameters they are monitoring.

(iii) Signal Conditioning & Amplification: The electrical signals generated by the transducers must be optimized for the input range of the DAQ board. Signal conditioning accessories can amplify low-level signals, and then isolate and filter them for more accurate measurements. In addition, some transducers require voltage or current excitation to generate a voltage output. Many sensors deliver signals that must be transformed before they can be digitized. **For example,** a micro electrode pipette may be used to measure current, while the digitizer measures potential (voltage). The patch clamp amplifier provides a current-to-voltage amplification, usually measured in **mV** of output per $\mathbf{p_A}$ of input. This transformation of the sensor signal is called signal conditioning. Signal conditioning may be more complex. An input signal from a non-linear sensor may be converted to a voltage that is linear in the quantity being measured , compensation may be made for second-order effects such as temperature, or an indirect effect such as a frequency shift may be converted to a voltage.

- **Preamplifier:** Many instrumentation systems are built with a preamplifier located as close to the sensor as possible. A separate amplifier converts the preamplifier output to a high-level signal. Placing the preamplifier close to the sensor reduces noise, by allowing the signal to be amplified before being sent over a cable. Since physical space near the sensor is limited, the preamplifier is as small as possible, with the bulk of the electronics being located in the amplifier. For example, in a patch clamp setup, the sensor is a solution-filled pipette, the preamplifier is the head stage, and the amplifier is the patch-clamp amplifier itself .
- **Amplification:** The most common type of conditioning is amplification. Low-level thermocouple signals, for example , should be amplified to increase the resolution and reduce noise. For the highest possible accuracy, the signal should be amplified so that the maximum voltage range of the conditioned signal equals the maximum input range of the analog-to digital converter **(ADC)**
- **Filtering :** The purpose of a filter is to remove unwanted signals from the signal that you are trying to measure. A noise filter is used on DC-class signals such as temperature to attenuate higher frequency signals that can reduce the accuracy of your measurement.

(iv) Multiplexing : A common technique for measuring several signals with a single measuring device is called multiplexing , Signal conditioning devices for analog signals often provide multiplexing for use with slowly changing signals such as temperature.

(v) Sample hold : A digitizer consists of an A/D (analog to digital) converter that samples an analog input signal and converts it to a sequence of digital signal values.

(vi) What is A/D converter: An Analog-Digital Converter **(ADC)** is a widely used electronic component that converts an analog electric signal **(usually a voltage)** into a digital representation. The ADCs are at the front-end of any digital circuit that needs to process signals coming from the exterior world. Its schematic symbol is:

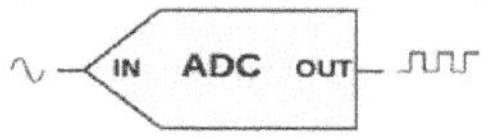

2.16.1. How Does the ADC Convert a Signal

All ADCs work under the same principle: they need to convert a signal to a certain number of bits. The sequence of bits represents the number and each bit has the double of the weight of the next, starting from the Most Significant Bit (MSB) up to the Least Significant Bit (LSB). In a nutshell, we want to find the sequence of bits that represents the analog value, The MSB has weight, the next [Math Processing Error], etc., and the LSB has weight [Math Processing Error]. Therefore, more bits leads to more precision in the digital representation. Here we simplify the range to be between 0 and [Math Processing Error], although the range may be between any two values.

Following ADCs are generally used in present time

- Flash
- Pipelined
- Successive-Approximations Register (SAR)
- Integrating or Dual-slope
- Sigma Delta

(i) Analog Signal Graphs: Because a signal varies over time, it's helpful to plot it on a graph where time is plotted on the horizontal, ***x***-axis, and voltage on the vertical, ***y***-axis. Looking at a graph of a signal is usually the easiest way to identify if it's analog or digital , a time-versus-voltage graph of an analog signal should be smooth and continuous.

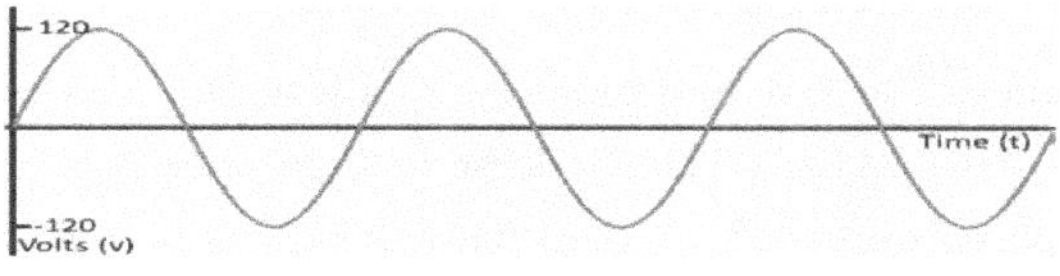

Diagram 2.24: Analog Signal Graphs

While these signals may be limited to a **range** of maximum and minimum values, there are still an infinite number of possible values within that range. For example, the analog voltage coming out of your wall socket might be clamped between -120V and +120V, but, as you increase the resolution more and more, you discover an infinite number of values that the signal actually have.

(ii) Digital Signals : Digital signals must have a finite set of possible values. The number of values in the set can be anywhere between two and a-very-large-number-that's-not-infinity. Most commonly digital signals will be one of two values – like either 0V or 5V. Timing graphs of these signals look like square waves.

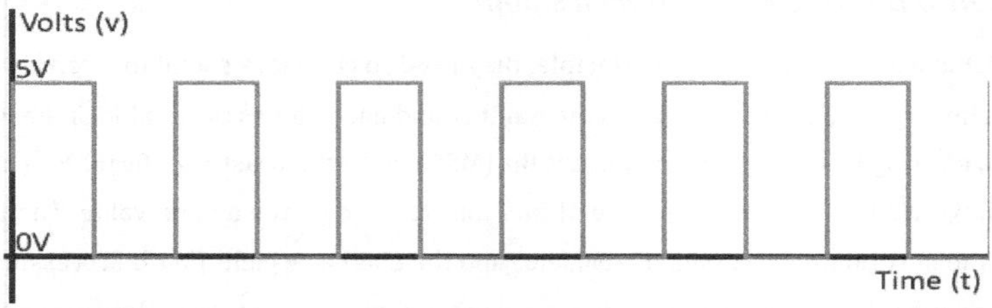

Diagram 2.25: Digital Signal Graphs

Or a digital signal might be a discrete representation of an analog waveform. Viewed from afar, the wave function below may seem smooth and analog, but when you look closely there are tiny discrete **steps** as the signal tries to approximate values:

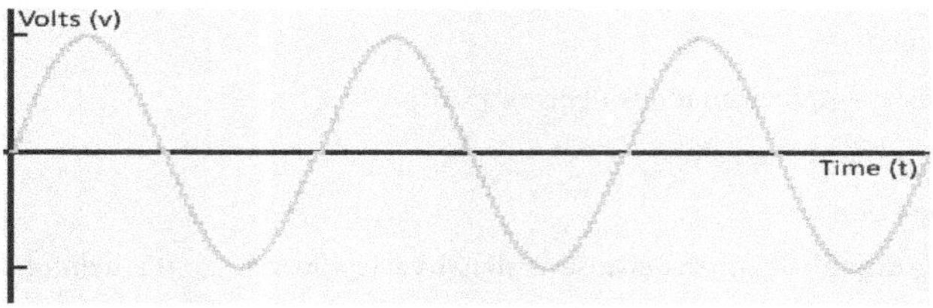

Diagram 2.26: Discrete Signal Graphs

That's the big difference between analog and digital waves. Analog waves are smooth and continuous, digital waves are stepping, square, and discrete.

(iii) Digital Codes: One final consideration in data acquisition and conversion systems is the digital coding of the data at the output of the A/D converter. Data is usually encoded in either binary or binary-coded-decimal (BCD) form. Binary encoded data formats are most commonly employed for digital computer-oriented applications where the processing is normally performed in binary notation. BCD data encoding is usually required in applications where the data is fed to decimal end devices such as digital readouts and printers. The majority of applications require binary encoding.

The most commonly used binary codes in A/D converters are:

- Uni-polar Straight Binary (USB)-used for uni-polar analog signal ranges, i.e., 0 to ±5 V, 0 to ±10 V,etc.
- Bipolar Offset Binary (BOB)-used for bipolar analog signal ranges, i.e., ±5 V, ±10 V, etc
- Bipolar Two's Complement (BTC)-used for bipolar analog signal ranges in many digital computer applications.

2.16.2. A/D Converter Points: There are Important Terms related to A/D Converter

- **Accuracy :** All analog values are presumed to exist at the input to the A/D converter. The A/D converter quantizes or encodes specific values of the analog input into equivalent digital codes as an output. These digital codes have an inherent uncertainty or quantization error of **± ½ LSB.** That is, the quantized digital code represents an analog voltage that can be anywhere within ± ½ LSB from the mid-point between adjacent digital codes. An A/D converter can never be more accurate than the inherent **±1/2LSB** quantizing error. Analog errors such as gain, offset, and linearity errors also affect A/D converter accuracy. Usually, gain and offset errors can be trimmed to zero, but linearity error is not adjustable because it is caused than **± ½ LSB** linearity error. Another major error consideration is differential linearity error. The size of steps between adjacent transition points in an ideal A/D converter is one LSB. Differential linearity error is the difference between adjacent transition points in an actual A/D converter and an ideal one LSB step. This error must be less than one LSB in order to guarantee that there are no missing codes. An A/D converter with **± ½ LSB** linearity error does not necessarily imply that there are no missing codes. by the fixed-value ladder resistor network and network switch matching. Most quality A/D converters have less .
- **Selecting the Resolution :** The number of bits in the A/D converter determines the resolution of the system. System resolution is determined by the channel(s) having the widest dynamic range and/or the channel that measures pressure has a dynamic range of 4000 psi that must be measured to the nearest pound. This will require an A/D converter with a minimum resolution of 4000 digital codes. A 12-bit A/D

converter will provide a resolution of 212 or 4096 codes—adequate for this requirement. The actual resolution of this channel will be 4000/4096 or 0.976 psi. The A/D converter can resolve this measurement to within **± 0.488 psi (± ½ LSB).**

- **Resolution:** The number of bits in an A/D converter determines the resolution of the data acquisition system. A/D converter resolution is defined as:

$$\text{Resolution} = \text{One LSB} = \frac{V_{FSR}}{2^n}, \text{ for binary A/D converters} = \frac{V_{FSR}}{10^D}, \text{ for decimal A/D converters}$$

Where ,

LSB = Least significant bit

VFSR = Full scale input voltage range

N = Number of bits ,

D = Number of decimal digits

- The number of bits defines the number of digital codes and is 2^n discrete digital codes for A/D converters

  (i) Computer : Once data has been digitized , It must be transferred to a computer. Usually a digitizer is built as a computer plug in board , so transfers take place over the computer bus. Digitizers used for high-speed measurement can feed data to the computer at a high and constant rate. For example, a digitizer running on one channel at 100k samples/second will typically produce 200k bytes/second of data continuously. This is a large stream of data. The continuous nature of much data acquisition requires some kind of buffering. For example, if the computer stops for 30 ms to write data to disk or to update a display, 6000 bytes of data will accumulate. The data must be stored somewhere, or it will be lost.

2.17. Data-Acquisition Systems

The basic concept of a data-acquisition is to perform the quantification and storage the data throughout the measurement process. the data-acquisition system can be classified in two category,

(i) Human assisted data-acquisition systems : Human assisted data–acquisition system is based on the fact that data collection is perform with the help of pointer on to the dial position then recorded these data in a log book . Figure given below shows that how a data-acquisition system (DAS) might fit into the general measurement scheme between the actual measurement and the subsequent data reduction

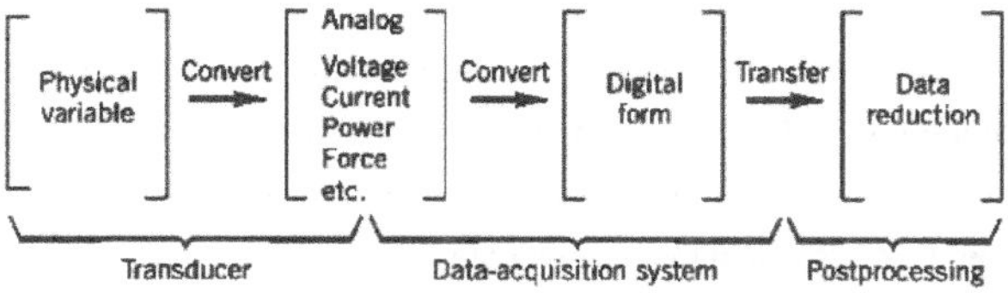

Diagram 2.27: Data-Acquisition Systems

(ii) Automated data-acquisition systems: Automated data-acquisition systems is operated by microprocessor which has performed automatic data quantification and storage . The schematic diagram for microprocessor-based data-acquisition systems or automated data-acquisition systems is given below,

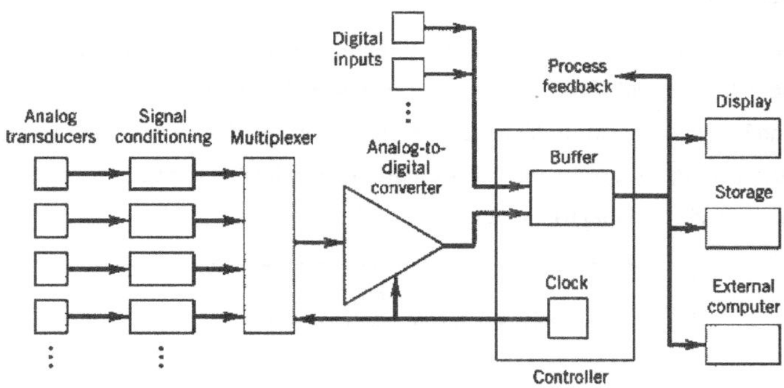

Diagram 2.28: Automated Data-Acquisition Systems

In the above system, we have multiple signal flow input operated by the single microprocessor which has been automatically perform data quantification, their interpretation, storage and control over the process . A cost-effective microprocessor system may be derived from a programmable logic controller (PLC).

(iii) Computer-based data-acquisition systems are hybrid systems combining a data-acquisition package with both the microprocessor and human interface capability of a personal computer (PC).

Components of data-acquisition system : The data-acquisition system required Signal Conditioning which is performed by two components i.e. Filters and Amplification .

- **Filter :** Analog filters are used to control the frequency content of the signal being sampled. Anti-alias analog filters remove signal information above the Nyquist frequency prior to sampling. Digital filters are used to signal analysis, digital filter system initiate with the Fourier transform of the sampled signal, multiplying the signal amplitude of frequency domain to attain the desired frequency response , A simple & popular digital filter is a smoothing filter, which is used for removing noise , sometimes it is also called moving average .

- **Amplifiers:** the function of amplifier is to provide amplification to input signals . thus in all kind of data-acquisition systems the input signal are maximize for further processing , Thus some transducer signals need amplification or attenuation prior to conversion.

2.18. Error Considerations: Following Errors are Associated with Data Acquisition

(i) Aliasing Error : From the Ny-quist sampling theorem, a minimum of two samples per cycle of the data bandwidth is required in an ideal sampled data system to reproduce sampled data with no loss of information. Thus, the first consideration for determining system sampling rate is aliasing error, i.e., errors due to information being lost by not taking a sufficient number of samples per cycle of signal frequency

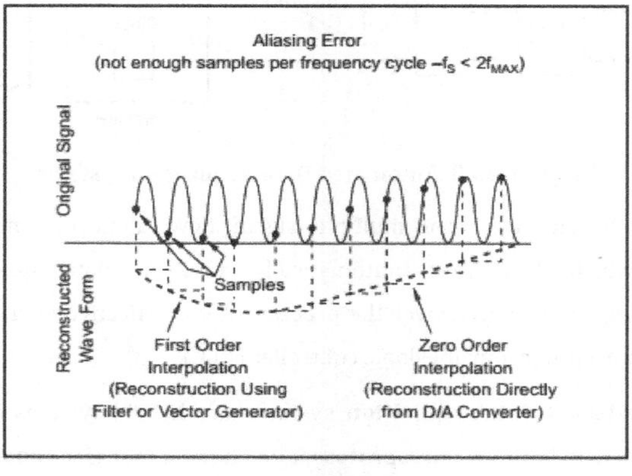

Diagram 2.29: Representation of Aliasing Error

(ii) Aperture Error : Aperture error is defined as the amplitude and time errors of the sampled data points due to the uncertainty of the dynamic data changes during sampling. In data acquisition and conversion systems, aperture error can be reduced or made insignificant either by the use of a sample/hold or with a very fast A/D converter. For sinusoidal data, maximum aperture error occurs at the zero crossing where the greatest dV/dt occurs, and is expressed mathematically as:-

$$\text{Aperture Error} = d\frac{(A \sin 2\pi ft)}{dt} \times t_A \times 100\% = 2\pi\, ft_A \times 100\%\max \qquad \text{................40}$$

Where ,

f = Maximum data frequency.

t_A = Aperture time of system (This can be the conversion time of the A/D converter with no sample/hold, or the aperture time of a sample/hold if one is in front of an A/D converter.)

(iii) System Throughput Accuracy : The most common method used to describe data acquisition and conversion system accuracy is to compute the **root-sum squared(RSS)** errors of the system components. The RSS error is a statistical value which is equivalent to the standard deviation **(1σ),** and represents the square root of the sum of the squares of the peak errors of each system component, including ADC quantization error:

$$\varepsilon_{RSS} = \sqrt{\varepsilon_{MUX}^2 + \varepsilon_{AMP}^2 + \varepsilon_{SH}^2 + \varepsilon_{ADC}^2} \qquad \text{..........................41}$$

Where ,

- ε_{MUX} = Analog multiplexer error
- ε_{AMP} = Input amplifier error
- ε_{SH} = Sample / hold error
- ε_{ADC}= A/D converter error

2.19. Data Reduction

Data reduction is the process of converting raw data to final results. Data reduction methods are designed to ensure that data are accurately and systematically reduced into a usable form. The data generated at site and/or in the laboratory will be used to support the professional judgment-based decisions and the risk evaluations. The laboratories will provide their standard report package format. These data will be detailed in tabular form (e.g., a summary spreadsheet format), identifying all "hits" (detections greater than detection limit) .

(i) Supplementary data : Supplementary data produced for internal records and not reported as part of the analytical data may include laboratory worksheets, laboratory notebooks, sample tracking system forms, instrument logs, standards records, maintenance records, calibration records, and associated quality control records. These data will be available for inspection during audits and when needed to determine the validity of data.

In the areas of data reduction and scientific analysis, consider following two steps

- Generate and maintain all ground software and procedures needed for operating the instrument, and for performing monitoring and trend analysis;
- Provide all software and procedures required for error correction, calibration and generally for the scientific processing of the data from the instrument.

(ii) Functional description : Figure shows the data reduction , it consists of five main headings: monitoring of instrument operations, instrument calibration, quality monitoring, science data processing, and user support.

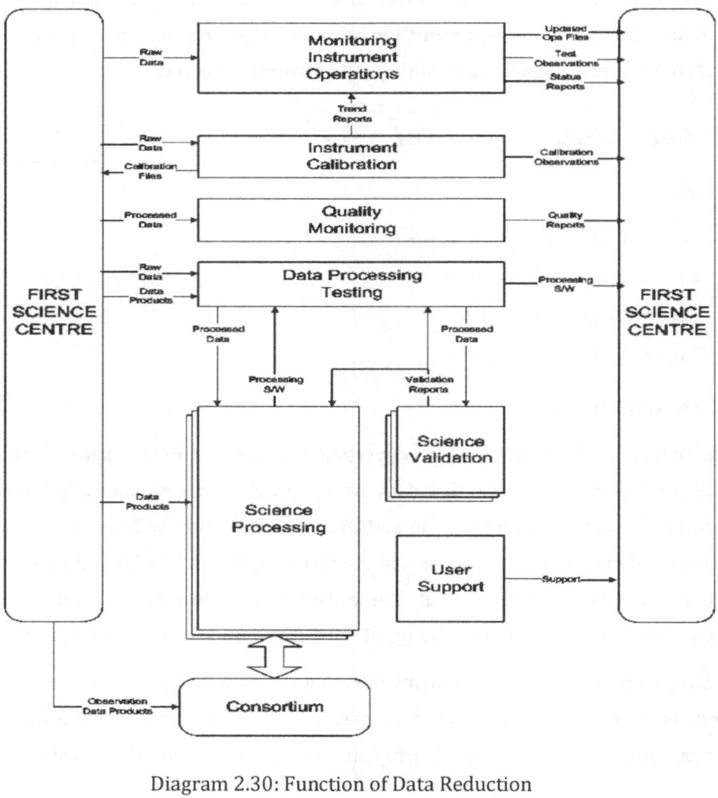

Diagram 2.30: Function of Data Reduction

2.20. Principle of Data Reduction

We study three principles of data reduction.

- The **Sufficiency Principle** promotes a method of data reduction that does not discard information about θ while achieving some summarization of the data.
- The **Likelihood Principle** describes a function of the parameter determined by the observed sample , that contains all the information about θ that is available from the sample.
- The **Equivariance Principle** prescribes yet another method of data reduction that still preserves some important features of the model.

(i) Sufficiency Principle : If T (**X**) is a sufficient statistic for θ, then any inference about θ should depend on the sample **X** only through the value T (**X**). That is, if **x** and **y** are two sample points such that **T (x) = T (y)**, then the inference about θ should be the same whether **X** = **x** or **Y** = **y** is observed.

Definition: A statistic T (**X**) is a sufficient statistic for θ if the conditional distribution of the sample **X** given the value of T (**X**) does not depend on θ. To use this definition to verify that a statistic T (**X**) is a sufficient statistic for θ, we must verify that P (**X** = **x**|T (**X**) = T (**x**)) does not depend on θ. Since {**X** = **x**} is a subset of {T (**X**) = T (**x**)},

$$\begin{aligned} P_\theta(\boldsymbol{X} = \boldsymbol{x} | T(\boldsymbol{X}) = T(\boldsymbol{x})) &= \frac{P_\theta(\boldsymbol{X} = \boldsymbol{x} \text{ and } T(\boldsymbol{X}) = T(\boldsymbol{x}))}{P_\theta(T(\boldsymbol{X}) = T(\boldsymbol{x}))} \\ &= \frac{P_\theta(\boldsymbol{X} = \boldsymbol{x})}{P_\theta(T(\boldsymbol{X}) = T(\boldsymbol{x}))} = \frac{p(\boldsymbol{x}|\theta)}{q(T(\boldsymbol{x}))|\theta)}, \end{aligned} \qquad \text{.................... 42}$$

where **p(x / θ)** is the joint pmf / pdf of the sample **X** and q(t / θ) is the pmf/pdf of T (**X**). Thus, **T** (**X**) is a sufficient statistic for **θ** if and only if, for every **x**, the above ratio is constant as a function of **θ**

(ii) likelihood principle: If **x** and **y** are two sample points such that **L(θ / x)** is proportional to **L(θ / y)**, that is, there exists a constant **C (x, y)** such that

L(θ / x) = C(x, y)L(θ / y)

for all **θ**, then the conclusions drawn from x and y should be identical.

Note that **C (x, y)** may be different for different **(x, y)** pairs but **C (x, y)** does not depend on θ.

Definition: Let **f(x / θ)** denote the joint pdf or pmf of the sample **X** = (X1, . . . , Xn). Then, given that **X** = **x** is observed, the function of θ defined by

$$L(\theta / x) = f(x / \theta) \qquad \text{.................................... 43}$$

is called the likelihood function.

This definition almost seems to be defining the likelihood function to be the same as the pdf or pmf. The only distinction between these two functions is which variable is considered fixed and which is varying. When we consider the pdf or pmf $f(x / \theta)$, we are considering θ as fixed and x as the variable; when we consider the likelihood function $L(\theta / x)$, we are considering x to be the observed sample point and θ to be varying over all possible parameter values. If we compare the likelihood function at two parameter points and find that $L(\theta_1 / x) > L(\theta_2 / x)$, Then the sample we actually observed is more likely to have occurred if $\theta = \theta_1$ than if $\theta = \theta_2$, which can be interpreted as saying that θ_1 is a more plausible value for the true value of θ than is θ_2. We carefully use the word "plausible" rather than "probable" because we often think of θ as a fixed value. Furthermore, although $f(x / \theta)$, as a function of x, is a pdf, there is no guarantee that $L(\theta / x)$, as a function of θ, is a pdf

(iii) Equivariance Principle: If $Y = g(X)$ is a change of measurement scale such that the model for Y has the same formal structure as the model for X, then an inference procedure should be both measurement equi-variant and formally equivariant.

(iv)The Equi-variance Principle: The first type of equi-variance might be called measurement equivariance. It prescribes that the inference made should not depend on the measurement scale that is used. The second type of equi-variance, actually an invariance, might be called formal invariance. It states that if two inference problems have the same formal structure in terms of the mathematical model used, then the same inference procedure should be used in both problems.

2.21. Data Reduction Procedure

In order to verify the accuracy of data reduction, the following procedures will be implemented:

- Technical staff will document and review their own work and will be accountable for the accuracy of that work.
- Major calculations will be subject to an independent technical review by a Technical Lead or other suitably experienced party (internal to Foster Wheeler) to ensure that both the methods and the calculations are correct(i.e., check the formula and the math) and consistent with the approved work plan and applicable policies in the Corporate Reference Library.

- The Project Manager will be responsible for ensuring that data reduction is conducted in a manner that produces high quality data via review and approval of concepts, methods, assumptions, and calculations

2.22. Data Verification

All results, conclusions, and recommendations will be based upon verified(validated) data. The purpose of data verification is to ensure that all data used for subsequent evaluations and calculations are scientifically valid, of known and documented quality, and legally defensible. The following items will be reviewed to verify the data as applicable:

- Sample holding times
- Documentation that the analytical results are within the control limits
- Documentation that data and calculations were checked by the supervisor who was not involved in the performance of sampling, analysis, or data reduction
- Documentation that a final review of the data was made by the laboratory manager for correctness and validity of the data
- Calibration of methods and instruments
- Routine instrument checks (noise levels, drift, linearity, etc.)
- Documentation on traceability of instrument standards, samples, and data
- Documentation on analytical methodology and QC methodology
- Results of performance audits with appropriate audit materials
- The control for interference contaminants in analytical methods (use of reference blanks and check standards for method accuracy and precision)
- Documentation of routine maintenance activity to ensure analytical reliability
- Documentation of sample preservation and transport; and
- Documentation of inventory control of chemicals and items used for testing (e.g., shelf life).

2.23. Standard Deviation and Normal Distribution

(i) Standard deviation is a widely used measurement of variability or diversity used in statistics and probability theory. It shows how much variation or "dispersion" there is from the "average" (mean, or expected value). A low standard deviation indicates that the data points tend to be very close to the mean, whereas high standard deviation indicates that the data is spread out over a large range of values. From a set of data with n values, where x_1 represents the first term and x_n represent the n_{th} term, if x_m represents the mean, then the standard deviation can be found as follows:

$$SD = \frac{\sqrt{(x1-xm)2+(x2-xm)2+....+(xn-xm)2}}{N} \quad .... 44$$

A normal distribution is a very important statistical data distribution pattern occurring in many natural phenomena. Random variation conforms to a particular probability distribution known as the normal distribution, which is the most commonly observed probability distribution. Fifty percent of the distribution lies to the left of the mean and fifty percent lies to the right of the mean

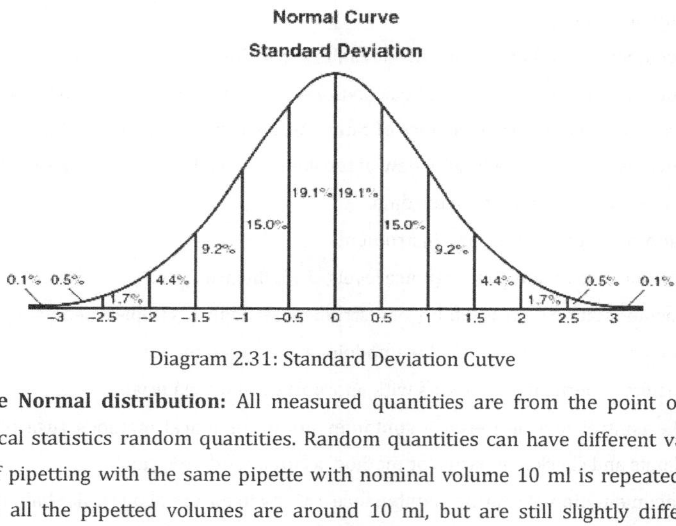

Diagram 2.31: Standard Deviation Cutve

(ii) The Normal distribution: All measured quantities are from the point of view of mathematical statistics random quantities. Random quantities can have different values. For example, If pipetting with the same pipette with nominal volume 10 ml is repeated multiple times then all the pipetted volumes are around 10 ml, but are still slightly different. If a sufficiently large number of repeated measurements are carried out and if the pipette volumes are plotted according to how frequently they are encountered then it becomes evident that although random, the values are still governed by some underlying relationship between the volume and frequency: the maximum probability of a volume is somewhere in the range of 10.005 and 10.007 ml and the probability gradually decreases towards smaller and larger volumes. This relationship is called **distribution function.**

There are numerous distribution functions known to mathematicians and many of them are encountered in the nature, i.e. they describe certain processes in the nature. In measurement science the most important distribution function is the **normal distribution(also known as the Gaussian distribution).** Its importance stems from the so-called **Central limit theorem.**

In a simplified way it can be worded for measurements as follows: if a measurement result is simultaneously influenced by many uncertainty sources then if the number of the uncertainty sources approaches infinity then the distribution function of the measurement result approaches the normal distribution, irrespective of what are the distribution functions of the factors/parameters describing the uncertainty sources. In reality the distribution function of the result becomes indistinguishable from the normal distribution already if there are 3-5 (depending on situation) significantly contributing uncertainty sources. hence so many cases measured quantities have either normal distribution which has a mathematical base of measurement science and measurement uncertainty estimation is predicted with the help of normal distribution.

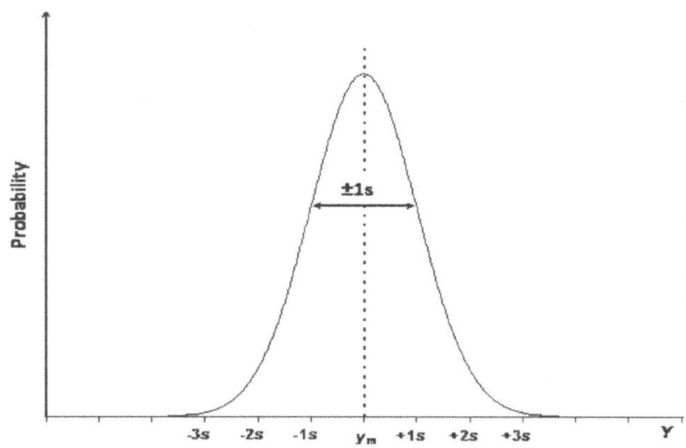

Diagram 2.32: Normal Distribution Curve

The normal distribution curve of quantity *Y* with mean value y_m and standard deviation *s*.

The normal distribution curve has the bell-shaped appearance, and is expressed by following equation ---

$$f(y) = \frac{1}{s\sqrt{2\pi}} e^{-\frac{(y-y_m)^2}{2s^2}} \qquad \text{........................ 45}$$

In this equation $f(y)$ is the probability that the measurand *Y* has value *y*. y_m is the **mean** value of the **population** and *s* is the **standard deviation** of the population. y_m characterizes the position of the normal distribution on the *Y* **- axis** , characterizes the width (spread) of the distribution function, which is determined by the scatter of the data points. The mean and standard deviation are the two parameters that fully determine the shape of the normal distribution curve of a particular random quantity. The constants 2 and p are normalization factors, which are present in order to make the overall area under the curve equal to 1. The word **"population"** here means that we would need to do an infinite number of measurements in order to obtain the true y_m and s values. In reality we always operate with a limited number of measurements, so that the mean value and standard deviation that we have from our experiments are in fact **estimates** of the true mean and true standard deviation. larger is the number of repeated measurements the more reliable are the estimates. The number of parallel measurements is therefore very important and we will return to it in different other parts of this course.

The normal distribution and the standard deviation are the basis for definition of **standard uncertainty**. Standard uncertainty, denoted by *u*, is the uncertainty expressed at standard deviation level, i.e., uncertainty with roughly 68.3% coverage probability (i.e. the probability of the true value falling within the uncertainty range is roughly 68.3%). The probability of 68.3% is often too low for practical applications. Therefore uncertainty of measurement results is in most cases not reported as standard uncertainty but as **expanded uncertainty**. Expanded uncertainty, denoted by *U*, is obtained by multiplying standard uncertainty with a **coverage factor**, denoted by *k*, which is a positive number, larger than 1. If the coverage factor is e.g. 2 (which is the most commonly used value for coverage factor) then in the case of normally distributed measurement result the coverage probability is roughly **95.5%.** These probabilities can be regarded as fractions of areas of the respective segments from the total area under the curve as given in **Diagram 2.33**,

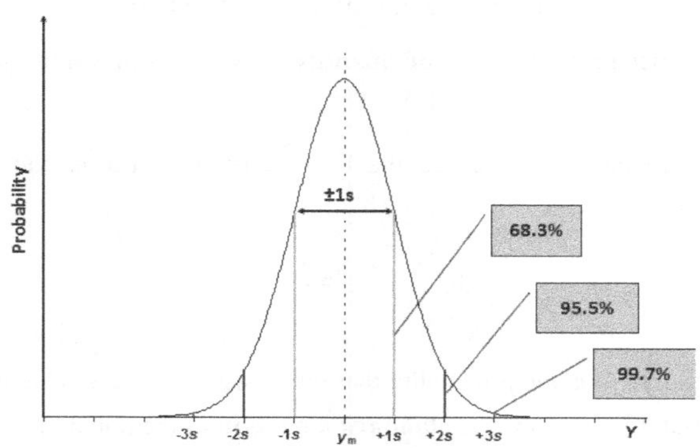

Diagram 2.33: Normal Distribution Curve with 2s and 3s Segments Indicator

Properties of a Normal Distribution

- The normal curve is symmetrical about the mean y_m
- The mean is at the middle and divides the area into halves
- The total area under the curve is equal to 1
- It is completely determined by its mean and standard deviation **σ (or variance σ^2)**

Example 5: Say **y_m= 2,** and **σ = 1/ 3**in a normal distribution. The graph of the normal distribution is as follows

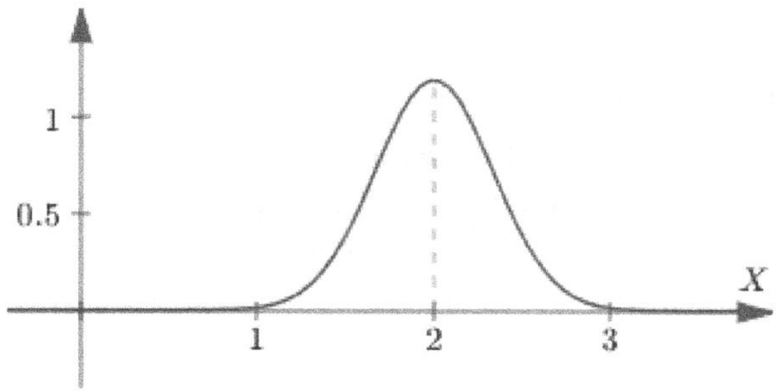

Diagram 2.34: Normal Curve for y_m= 2, and σ = 1/ 3

The following graph represents the same information, but it has been **standardized** so that **y_m = 0 and σ = 1**

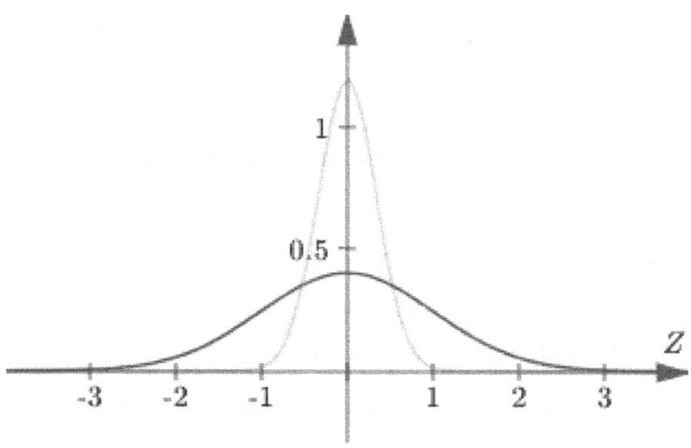

Diagram 2.35: For Normal Curve y_m= 0, and σ = 1/ 3

The two graphs have different ***μ*** and ***σ*,** but have the same area. The new distribution of the normal random variable Z with mean and variance 1 (or standard deviation) is called a **standard normal distribution.**

Formula for the Standardized Normal Distribution

If we have mean y_m and standard deviation **σ,** then **standard normal distribution** is given by following relationship

$$Z = (x - y_m) / \sigma \quad \text{...} \quad 46$$

Since all the values of *X* falling between x_1 and x_2 have corresponding *Z* values between z_1 and z_2, it means: The area under the *X* curve between $X = x_1$ and $X = x_2$ equals the area under the *Z* curve between $Z = z_1$ and $Z = z_2$.

Hence, we have the following equivalent probabilities : $P(x_1 < X < x_2) = P(z_1 < Z < z_2)$

2.24. Chi-squared Distributions

Considered the $Z1$, Z_2, ..., $Z_v$are independent random variables and each has the standard normal distribution then the sum $(z_1)^2 + (z_1)^2 + ... (z_v)^2$ has a chi-squared (or chi-square) distribution with **v** degrees of freedom. **A** chi squared distribution, like a log-normal distribution, is asymmetric and does not include negative values. For large **v,** the chi-squared distribution is approximately normal. Given diagram shows the densities for chi-square distributions with 1, 2, 3 and 10 degrees of freedom. Following **Diagram shows the** chi-squared distributions for 1, 2, 3 and 10 degrees of freedom.

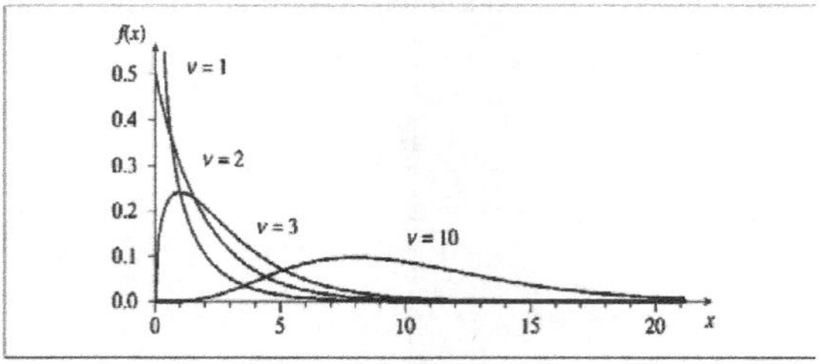

Diagram 2.36: Chi-squared Distributions for 1, 2, 3 and 10 Degrees of Freedom

Chi-squared distributions are applied frequently in hypothesis testing of the variances of normally distributed data and least-squares analysis.

A sum of independent chi-squared random variables is also chi-squared. Specifically, if *X* and *Y* are independent chi-squared random variables with v_1 and v_2 degrees of freedom, respectively, then ***X+Y*** has a chi-squared distribution with v_1+v_2 degrees of freedom. The mean of a chi-squared distribution equals the number of degrees of freedom ***v*,** and the variance equals ***2v*.**

2.25. Error Propagation

Error Propagation is defined as the effects on a function by a variable's uncertainty which is a observational estimation of derived quantities that are functions of raw data in order to provide an accurate measurement of uncertainty.

Partial Derivatives of the function :- In error propagation we will always draw a" **partial derivatives**" of a function which has a certain variables that can be understand by the following example , Suppose we have a function f , where $f = f(x, y, z)$.

- The partial derivative of f with respect to x is a ordinary derivative where y and z behave as constants. Then the notation of this derivative can be given as,

$$\frac{\partial f}{\partial x} \quad \text{...} \textbf{(i)}$$

- The partial derivative of f with respect to y is a ordinary derivative where x and z behave as constants. Then the notation of this derivative can be given as,

$$\frac{\partial f}{\partial y} \quad \text{...} \textbf{(ii)}$$

- The partial derivative of f with respect to z is a ordinary derivative where x and y behave as constants. Then the notation of this derivative can be given as,

$$\frac{\partial f}{\partial z} \quad \text{...} \textbf{(iii)}$$

For a function $f = f(x, y, z)$, the absolute error in f, δf is defined as : then ,

(i) Absolute Errors

$$\delta f = \sqrt{\left\{\left[\left(\frac{\partial f}{\partial x}\right)\delta x\right]^2 + \left[\left(\frac{\partial f}{\partial y}\right)\delta y\right]^2 + \left[\left(\frac{\partial f}{\partial z}\right)\delta z\right]^2\right\}} \quad \text{.....................} \textbf{(iv)}$$

(ii) Relative Errors ,

$$\frac{\delta f}{f} = \frac{1}{f}\sqrt{\left\{\left[\left(\frac{\partial f}{\partial x}\right)\delta x\right]^2 + \left[\left(\frac{\partial f}{\partial y}\right)\delta y\right]^2 + \left[\left(\frac{\partial f}{\partial z}\right)\delta z\right]^2\right\}} \quad \text{.....................} \textbf{(v)}$$

(iii) Percentage Error ,

$$f = \frac{\delta f}{f} \cdot 100 \quad \text{...} \textbf{(vi)}$$

(iv) Percent Discrepancy

$$X = \left| \frac{X_{accepted} - X_{experimental}}{X_{accepted}} \right| \cdot 100 \qquad \text{.......................................(vii)}$$

For example : Suppose **X ± ΔX** is a measurement and we want to calculate the derived quantity **Z = *f* (X) ,** Then we will try to analyse the error bar **± ΔX** " propagates " through the function **f(X)** to give an error bar **±ΔZ** for the final number **Z** , hence the error propagation follows the given rule ,

$$Z = f(X) \quad \Longrightarrow \quad \Delta Z = \left| \frac{df}{dX} \right| \Delta X$$

Where

***df*/dX** = Limiting value of the ratio **Δf / ΔX ,**

Δf = Change in the value of the function that is due to small change **ΔX** in the value of **X ,**

ΔX = Measurement Error .

Basic Rules of Propagation of Errors: Suppose we have two measured quantities **x** and **y** having uncertainties or errors , Δx and Δy then measured quantity will be given as **X ± ΔX** and **Y ± ΔY ,** the final measurement is such a manner that **z = *f* (*x* , *y*) .** thus all the possible guiding principle will be described as ,

(a) Addition and Subtraction : z = x + y or z = x – y

In this case we will always try to arranged the uncertainties in such a manner that **z** will possible close to true value as P.

(i) Average deviations **Δz = |Δx| + |Δy|** in both cases

(ii)

$$Z = X + Y \quad \Longrightarrow \quad \Delta Z = \sqrt{(\Delta X)^2 + (\Delta Y)^2}$$

$$Z = X - Y \quad \Longrightarrow \quad \Delta Z = \sqrt{(\Delta X)^2 + (\Delta Y)^2} \qquad \text{.....................47}$$

Multiplication and Division: **z = x y** or **z = x / y**

Z = X x Y

log z = log x + log y then ,

$$Z = X \times Y \quad \Longrightarrow \quad \Delta Z = |Z| \times \sqrt{\left(\frac{\Delta X}{X}\right)^2 + \left(\frac{\Delta Y}{Y}\right)^2} \qquad \text{.................................48}$$

For division

$$Z = X \div Y \quad \Longrightarrow \quad \Delta Z = |Z| \times \sqrt{\left(\frac{\Delta X}{X}\right)^2 + \left(\frac{\Delta Y}{Y}\right)^2} \qquad \text{.......................49}$$

(iii) Products of powers :

$$z = x^m y^n \quad = \quad \Delta z = \sqrt{\left(\frac{m\Delta x}{x}\right)^2 + \left(\frac{n\Delta y}{y}\right)^2 + \cdots} \qquad \ldots\ldots 50$$

2.25.1. How Error Propagated

Here one question is arises ie. , how an error is propagates throughout the measurements system. In order to understand the phenomenon of error propagation lets be suppose that we have a relationship ***y = f(x)***between a dependent(output) variable *y* and a measured variable *x*. How will the uncertainty u_x associated with *x* be reflected in the uncertainty $\mathbf{u_y}$ of ***y***? the schematic diagram of error propagation is as shown in **Diagram – 2.37 ,** In equation form we can write the relationship between a nominal input ***x*** and its uncertainty $\boldsymbol{u_x}$ and output ***y*** and the uncertainty $\boldsymbol{u_y}$ due to the propagation of $\mathbf{u_x}$ as

$$\bar{y} = f(\bar{x})$$
$$\bar{y} \pm u_y = f(\bar{x} \pm u_x) \qquad \ldots\ldots \text{(i)}$$

Our objective is to determine $\boldsymbol{u_y}$, the uncertainty associated with the dependent variable ***y.*** Expansion of the second equation into a Taylor series yields

$$\bar{y} \pm u_y \approx f(\bar{x}) \pm \left.\frac{\partial y}{\partial x}\right|_{x=\bar{x}} u_x \qquad \ldots\ldots \text{(ii)}$$

Substitution of ***y* = *f*** **(*x*)** yields the uncertainty $\boldsymbol{u_y}$ as

$$u_y = \left.\frac{\partial y}{\partial x}\right|_{x=\bar{x}} u_x \qquad \ldots\ldots \text{(iii)}$$

The uncertainty u_y of the dependent variable *y* is therefore the uncertainty u_x associated with the measured variable *x* multiplied by the sensitivity (slope) of *y* with respect *x* at the operating point *x*.

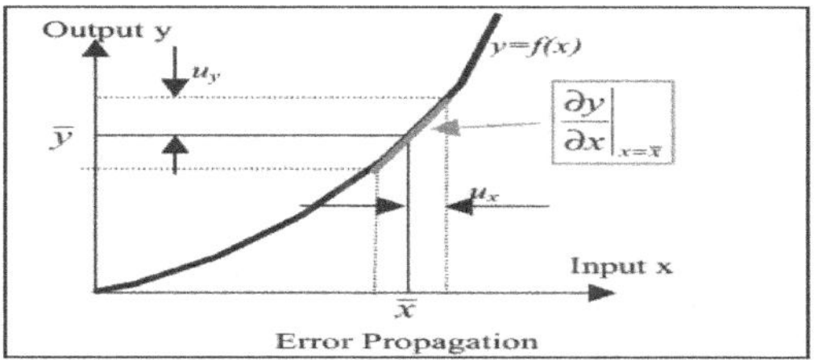

Diagram 2.37: Representing an Error Propagation

If the dependent variable y depends on several inputs x_i use the RSS method to combine the uncertainties

$$u_y = \left[\left(\left. \frac{\partial y}{\partial x_1} \right|_{x_1=\bar{x}_1} \right)^2 u_{x_1}^2 + \left(\left. \frac{\partial y}{\partial x_2} \right|_{x_2=\bar{x}_2} \right)^2 u_{x_2}^2 + \ldots \right]^{\frac{1}{2}}$$

.... (iv)

To estimate the relative uncertainty (percentage) associated with the operating point y normalize (divide)the above equation with respect to y

$$\frac{u_y}{\bar{y}} = \left[\left(\left. \frac{\partial y}{\partial x_1} \right|_{x_1=\bar{x}_1} \right)^2 \frac{u_{x_1}^2}{\bar{y}^2} + \left(\left. \frac{\partial y}{\partial x_2} \right|_{x_2=\bar{x}_2} \right)^2 \frac{u_{x_2}^2}{\bar{y}^2} + \ldots \right]^{\frac{1}{2}}$$

.... (v)

The relative uncertainty, which is propagated from each independent variable x_i to y is defined as

$$e_i = \frac{1}{\bar{y}} \left. \frac{\partial y}{\partial x_i} \right|_{x_i=\bar{x}_i} u_{x_i}$$

.... (vi)

therefore the total relative uncertainty $e_y$is

$$e_y = \sqrt{e_1^2 + e_2^2 + \ldots + e_i^2}$$

...................................(vii)

Uncertainty for dependent variables y, which are proportional to powers of x_i For functions, which are powers of the independent variables x_i, the relative uncertainty can be quickly calculated. Assume that y can be written as

$$y = f(x_1, \ldots, x_i, \ldots) = g(x_1, \ldots) x_i^n$$

.... (viii)

The relative uncertainty was defined earlier as

$$e_i = \frac{1}{\bar{y}} \left. \frac{\partial y}{\partial x_i} \right|_{x_i=\bar{x}_i} u_{x_i}$$

.... (ix)

Differentiate y with respect to the variable x_i.

$$\frac{\partial y}{\partial x_i} = \frac{\partial (g() x_i^n)}{\partial x_i} = g() n x_i^{n-1}$$

..................................(x)

and substitute the result into the equation for e_i

$$e_i = \frac{1}{\bar{y}}\frac{\partial y}{\partial x_i}u_{x_i} = \frac{1}{g()x_i^n}g()nx_i^{n-1}u_{x_i} = \frac{nx_i^{n-1}}{x_i^n}u_{x_i} = n\frac{u_{x_i}}{x_i} \quad \text{.....................(xi)}$$

In summary we find that if the variable x_i is a power of n, its relative uncertainty e_i is simply

$$e_i = n\frac{u_{x_i}}{x_i} \quad \text{.. (xii)}$$

Example 6: Determine the value of **z = x + y - w** and its uncertainty. Considering the following parameters **w** = (4.52 ± 0.02) cm ,**x** = (2.0 ± 0.2) cm , **y** = (3.0 ± 0.6) cm.

Solution

z = x + y - w = 2.0 + 3.0 - 4.5 = 0.5 cm

uncertainty can be determine as,

Δz =Δx + Δy + Δw = 0.2 + 0.6 + 0.02 = 0.8 cm

Solution with Standard Deviations

$$\Delta z = \sqrt{(\Delta x)^2 + (\Delta y)^2 + \ldots}$$

$$\Delta z = 0.633$$

Example 7: The radius of a circle is **x = (3.0 ± 0.2) cm**. Find the circumference and its uncertainty.

Solution

The circumference of the circle is given as ,

R = 2 πΔx= 2 x 3.14 x 3.2 = 18.850 cm

Calculation of uncertainty will be given as ,

ΔR = 2 πΔx = 2 x 3.14 x 0.2 = 1.257 cm = Equivalent to **1.3**

R = **(18.8 ± 1.3)** cm

Example 8 :- Calculate the value of **z = x - 2y** and its uncertainty , if has following data ,

x = (2.0 ± 0.2) cm , y = (3.0 ± 0.6) cm.

Solution

z = x - 2y = 2.0 – 2 (3.0) = - 4.0 cm

Uncertainty

$$\Delta z = \Delta x + 2,\ \Delta y = 0.2 + 1.2 = 1.4 \text{ cm}$$

Example 9 : Deduce an expression for Data acquisition and Reduction errors in case of Multiple temperature sensors .

Solution

Consider that we have Multiple temperature sensors or thermocouple say **M** and resultant of such a system is measured and stored through M-channel data acquisition system also considered that all the sensors are operated at normal temperature with same referenced junction and **20** reading for each **M** thermocouples are observed say it is **N** . then ,

N_i = 20 Readings Taken For Each Thermocouple,

Where, i = 1 , 2 , 3 , ... N

M_j = Total No of Thermocouple .... say **M**

Where, **j = 1 , 2 , 3 , ...N**

Suppose that all the thermocouple is of T-type , then

The mean value of all the reading (N) may be given by following relation,

$$T_j = \frac{1}{N}\sum_{i=1}^{N} T_{ij}$$

And the pooled mean for **'M' thermocouples** is given by the following relationship ,

$$\langle \overline{T} \rangle = \frac{1}{M}\sum_{j-1}^{M} \overline{T}_j$$

Conclusion

- The Estimated Systematic Uncertainty during data acquisition cab be given as ,

Systematic Uncertainty = (Pooled Mean Temperature – Normal Known Temperature)

- The Estimated Stanard Random Uncertainty during data acquisition & reduction cab be given as

$$\textbf{Random Uncertainty} = \{ \overline{T}_j - \langle \overline{T} \rangle \}$$

$$= \left[T_j = \frac{1}{N}\sum_{i=1}^{N} T_{ij} \; - \; \langle \overline{T} \rangle = \frac{1}{M}\sum_{i-1}^{M} T_j \right]$$

The standard random uncertainty for the data acquisition and reduction instrumentation system can be deduced as ,

$$s_{\overline{T}} = \frac{\langle s_T \rangle}{\sqrt{M}}$$

Where ,

$$\langle s_T \rangle = \sqrt{\frac{\sum_{j=1}^{M} \sum_{i=1}^{N} \left(T_{ij} - \overline{T}_j\right)^2}{M(N-1)}}$$

And Degree of Freedom for such a system can be given as ,

Degree of Freedom = M (N - 1)

Example 10: Find the value of $z = w\,y^2 / \sqrt{A}$ **and relative errors**

For the following values , w = (4.52 ± 0.02) cm , A = (2.0 ± 0.2) , y = (3.0 ± 0.6) cm.

Solution

$z = w\,y^2 / \sqrt{A}$ = [4.5 x (3.02)^2] / (2)$^{0.5}$ = 28.63 cm^2

Relative Errors due to ' z '

Δ z = [(0.2 / 4.5) + 2 (0.6 / 0.3) + 0.5 (0.2 / 2.0)] x 28.63 = **0.49**

Relative Errors due to ' A '

(ΔA / A) x 0.5 = (0.2 / 2.0) x 0.5 = **0.05 cm**

(It is multiplied by **0.05** , because It **A** has a power of **0.05**)

Relative Errors due to ' Y '

(Δy / y) x 2 = (0.6 / 3.0) x 2 = **0.4 cm**

(It is multiplied by **2** , because It **y** has a power of **2**)

Thus , **Δ z** becomes

Δ z = 0.49 x28.63 cm^2 = 14 cm^2

Then value of **Z** will be

Z = (29 ± 14) cm ^{2}

Example 11: To find the volume of a certain cube, you measure its side as **2 .00 ± 0 . 02 cm.** Convert this uncertainty to a percent and then find the volume with its uncertainty.

Solution : The volume ***V*** is given in terms of the side ***s*** by

$$V = s^3$$

$$\Delta V = 3\ S^2\ \Delta S = 0.24$$

and the volume is **(8.0 ± 0. 2) cm³.**

Example 12: Find the value of **q**& also associated uncertainty for following equation ,

$$q = \frac{x+2}{x+y\cos 4\theta}.$$

If the value function has three independent variable which is given as ,

$$x = 10 \pm 2, \quad y = 7 \pm 1, \quad \theta = 40 \pm 3^\circ$$

Solution : At initial condition we will calculate the partial derivative of **q**

$$\frac{\partial q}{\partial x} = \frac{x + y\cos 4\theta - (x+2)}{(x+y\cos 4\theta)^2}$$

$$= \frac{y\cos 4\theta - 2}{(x+y\cos 4\theta)^2} = -0.732$$

$$\frac{\partial q}{\partial y} = -\frac{(x+2)\cos 4\theta}{(x+y\cos 4\theta)^2} = 0.963$$

$$\frac{\partial q}{\partial \theta} = \frac{4(x+2)y\sin 4\theta}{(x+y\cos 4\theta)^2} = 9.813.$$

Hence , **q** will be = 3.507

$$\begin{aligned} \delta q^2 &= (0.732 \times 2)^2 \\ &\quad +(0.963 \times 1)^2 \\ &\quad +(9.813 \times 3 \times \pi/180)^2 \\ &= 3.3. \end{aligned}$$

Then the value of q becomes = **(3.5 ± 2)**

Calculation of Uncertainty

$$q = \frac{x+2}{x+y\cos 4\theta}.$$

The above equation can be written as ,

$$\frac{1}{q} = 1 + \frac{y\cos 4\theta - 2}{x+2}.$$

Following steps will be applied for solving the uncertainty of the equation,

(i) The Relative Uncertainty in y cos 4θ is

$$\Delta_1 = \sqrt{\left(\frac{\delta y}{y}\right)^2 + \left(\frac{4\delta\theta \sin 4\theta}{\cos 4\theta}\right)^2} = 0.16.$$

(ii) The Absolute Uncertainty in y cos 4θ is

$$\Delta_2 = |y \cos 4\theta| \times \Delta_1 = 1.1$$

(iii) The Relative Uncertainty in 1 / (q - 1) is

$$\begin{aligned}\Delta_3 &= \sqrt{\left(\frac{\Delta_2}{y\cos 4\theta - 2}\right)^2 + \left(\frac{\delta x}{x+2}\right)^2}\\ &= 0.21.\end{aligned}$$

(iv) The Absolute Uncertainty in 1 / (q - 1) is

$$\Delta_4 = |1/q - 1| \times \Delta_3 = 0.15.$$

(v) The relative uncertainty in (1/q)is (q x Δ4) which is also the relative uncertainty in q. Therefore the absolute uncertainty in q is

$$\delta q = q^2 \times \Delta_4 = 2$$

Example 13: Calculate the % Error for the following

$x = 3.0cm \pm 0.1cm$, $y = 5.2cm \pm 0.1cm$, and $z = 2.4cm \pm 0.1cm$ experimental data's, If the partial derivatives equation is given as ,

$$\delta f = \sqrt{(10\, xyz^3\, \delta f)^2 + (5\, x^2z^3\, \delta y)^2 + (15\, x^2yz^2\, \delta z)^2}$$

Solution: The following formula will be applied for the error calculation, Formula for % Errors

$$f = \frac{\delta f}{f} \cdot 100$$

Simplified the Equation

$\delta f = \sqrt{(10\, xyz^3\, \delta f)^2 + (5\, x^2z^3\, \delta y)^2 + (15\, x^2yz^2\, \delta z)^2}$ to achieve desired form ,We get ,

$$\frac{\delta f}{f} = \sqrt{\left\{\left(\frac{10xyz^3}{5x^2yz^3}\delta x\right)^2 + \left(\frac{5x^2z^3}{5x^2yz^3}\delta y\right)^2 + \left(\frac{15x^2yz^2}{5x^2yz^3}\delta z\right)^2\right\}}$$

$$\frac{\delta f}{f} = \sqrt{\left\{\left(\frac{2\delta x}{x}\right)^2 + \left(\frac{\delta y}{y}\right)^2 + \left(\frac{3\delta z}{z}\right)^2\right\}}$$

Then % error will be calculated as,

$$\frac{\delta f}{f}=\sqrt{\left\{\left(\frac{(2)(0.1)}{3.0}\right)^2+\left(\frac{0.1}{5.2}\right)^2+\left(\frac{(3)(0.1)}{2.4}\right)^2\right\}}=0.143\approx 15\%$$

Example 14: Calculate the % error for the following experimental data, ***a* = (8.2 ± 0.1) *cm*, *b* = (6.5 ± 0.1) *cm* , *c* = (5.1± 0.1) *cm* , if**

$$V=\frac{3a^2+5b^2}{c}$$

Solution : simplify the above equation , in desired form

$$V=\frac{3a^2+5b^2}{c}$$

$$\delta V=\sqrt{\left\{\left[\left(\frac{\partial V}{\partial a}\right)\delta a\right]^2+\left[\left(\frac{\partial V}{\partial b}\right)\delta b\right]^2+\left[\left(\frac{\partial V}{\partial c}\right)\delta c\right]^2\right\}}$$

Where ,

$$\frac{\partial V}{\partial a}=\frac{6a}{c},\quad \frac{\partial V}{\partial b}=\frac{10b}{c},\quad \frac{\partial V}{\partial c}=-\frac{3a^2+5b^2}{c^2}$$, then

$$\delta V=\sqrt{\left\{\left[\left(\frac{6a}{c}\right)\delta a\right]^2+\left[\left(\frac{10b}{c}\right)\delta b\right]^2+\left[\left(\frac{3a^2+5b^2}{c^2}\right)\delta c\right]^2\right\}}$$

$$\frac{\delta V}{V}=\sqrt{\left\{\left[\left(\frac{6a}{3a^2+5b^2}\right)\delta a\right]^2+\left[\left(\frac{10b}{3a^2+5b^2}\right)\delta b\right]^2+\left[\frac{\delta c}{c}\right]^2\right\}}$$

$$\frac{\delta V}{V}=\sqrt{\left\{\left[\left(\frac{(6)(8.2)}{(3)(8.2)^2+(5)(6.5)^2}\right)0.1\right]^2+\left[\left(\frac{(10(6.5))}{(3)(8.2)^2+(5)(6.5)^2}\right)0.1\right]^2+\left[\frac{0.1}{5.1}\right]^2\right\}}=5.5\%$$

Hence % Error & Volume will be as,

Percent error = 6 %

***V* = 81 *cm* ± 6 %**

Example 15: The standard deviation of measurements of a linear dimension of a mechanical part is 0.20 mm. What sample size is required if the standard error of the mean must be no more than (a) 0.06 mm, (b) 0.04 mm ?

Solution: The population can be calculated as, because the linear measurement is performed under repetitions and hence population size will be infinite

$$\sigma_{\bar{x}}=\frac{\sigma}{\sqrt{n}}$$

Case – I When $\sigma_{\bar{x}}$ = **0.06** mm and σ= **0.20** mm **, then**

$$\sqrt{n} = (0.20 / 0.06) = 3.33$$

$$n = 11.08$$

hence , when $\sigma_{\bar{x}} \leq$ **0.06** , the minimum value of sample size will be belongs to **12** .

Case – II When $\sigma_{\bar{x}}$ = **0.04** mm and σ= **0.20** mm **, then**

$$\sqrt{n} = (0.20 / 0.04) = 5$$

$$n = 25$$

Hence , when $\sigma_{\bar{x}} \leq$ **0.04** , the minimum value of sample size will be belongs to **25**

Then for $\sigma_{\bar{x}} \leq 0.04$ mm, the minimum sample size is 13.

Example 16: The standard deviation of percentage iron in analyses is 0.12 , the sample size is 6, and the normal distribution applies with no systematic error. The sample mean is determined from measurements to be 32.56 percent iron. Find the 95% confidence interval for the true mean iron content of the population from which the sample came.

Solution : For 95% confidence level the interval is still from

$z = -1.96$ to $z = +1.96$.

$x = 32.56$.

Then the interval estimates for the population mean with 95% confidence is

from $32.56 - 1.96\left(\frac{0.12}{\sqrt{6}}\right)$ to $32.56 + 1.96\left(\frac{0.12}{\sqrt{6}}\right)$, or from **32.46** to **32.66** percent iron and concluded that the sample mean falls between See 32.56 ± **0.10,** or within **±0.10** . the standard deviation at 95% confidence interval is given below ,

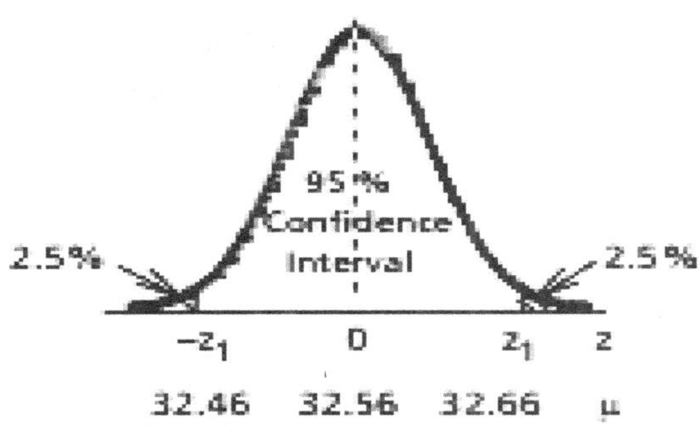

Example 17: A large population is normally distributed with a standard deviation of 0.12. A random sample will be taken from this population and the sample mean will be calculated. We require at least 98% confidence that the true population mean will be within ±0.05 of the sample mean, assuming there is no systematic error. What sample size is required ? assume that at 98% confidence level the value of $z_1 = 2.33$, $\mathbf{1\text{-}\phi(z_1)} = 0.01$, $\phi(-z_1) = 0.01$

Solution : In the problem , it is given that ,

$$\bar{x}_1 - \mu = 0.05\,, \quad \mu - \bar{x}_2 = 0.05 \text{ with}\,, \quad \sigma = 0.12$$

We know that,

$$z_1 = \frac{\bar{x}_1 - \mu}{\sigma_{\bar{x}}} \quad \textbf{And} \quad \sigma_{\bar{x}} = \frac{\sigma}{\sqrt{n}} \quad \textbf{then ,}$$

$$z_1 = \frac{\bar{x}_1 - \mu}{(\sigma/\sqrt{n})}$$ Subtituting the data we have got ,

***Z*₁ = 2.33**

$$2.33 = \frac{0.05}{\left(\frac{0.12}{\sqrt{n}}\right)}$$

$$\frac{0.12}{\sqrt{n}} = \frac{0.05}{2.33}$$

$$\sqrt{n} = \frac{(0.12)(2.33)}{0.05} = 5.59$$

and $$n = (5.59)^2 = 31.3.$$

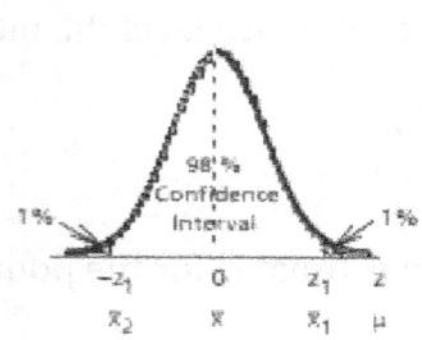

The Confidence Level

Example 18: The electrical resistances of components are measured as they are produced. A sample of six items gives a sample mean of 2.62 ohms and a sample standard deviation of 0.121 ohms. At what observed level of significance is this sample mean significantly different from a population mean of 2.80 ohms? Is there less than 2% probability of getting a sample mean this far away from 2.80 ohms or farther purely by chance when the population mean is 2.80 ohms?

Solution

Ho : μ = 2.80

H_a: μ ≠ 2.80 (two-sided test)

The test statistic is $t = \frac{(\bar{x}-\mu)}{\left(\frac{s}{\sqrt{n}}\right)} = \frac{(\bar{x}-2.80)}{\left(\frac{0.121}{\sqrt{6}}\right)}$.

Large values of $|t|$ indicate that H_0 is unlikely to be correct.

With $\bar{x} = 2.62$, $t_{observed} = \frac{(2.62-2.80)}{\left(\frac{0.121}{\sqrt{6}}\right)} = \frac{-0.18}{0.0494} = -3.64$

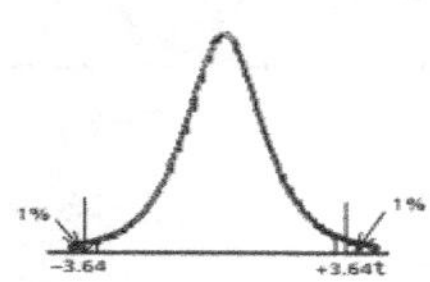

Level of Significance

Thus, there *is* less than a 2% probability of getting a sample mean this far from 2.80 ohms or farther purely by chance when the population mean is 2.80 ohms.

Example 19: We want to fit the following set of data to an equation of the form, ln $\boldsymbol{y = a + b\,x}$, by the method of least squares.

x	0	1	2	3	4	5	6	7	8	9	10	11
y	1.178	1.142	1.273	1.354	1.478	1.737	1.842	1.778	2.160	2.418	2.339	2.931

Solution : The first step is to calculate ln y for each value of y. Then x^2, $(\ln y)^2$, and the product (x) $(\ln y)$ are calculated.

	x	ln y	x^2	$(\ln y)^2$	$(x)(\ln y)$
	0	0.164	0	0.027	0.000
	1	0.133	1	0.018	0.133
	2	0.241	4	0.058	0.482
	3	0.303	9	0.092	0.910
	4	0.391	16	0.153	1.562
	5	0.552	25	0.305	2.760
	6	0.611	36	0.373	3.666
	7	0.575	49	0.331	4.027
	8	0.770	64	0.593	6.161
	9	0.883	81	0.780	7.946
	10	0.850	100	0.722	8.496
	11	1.075	121	1.157	11.830
Totals	66	6.548	506	4.607	47.973

$$\text{Then } \bar{x} = \frac{66}{12} = 5.5 \text{ and } \overline{\ln y} = \frac{6.548}{12} = 0.5457.$$

$$S_{xx} = \sum_{i=1}^{12} x^2 - \frac{\left(\sum_{i=1}^{12} x\right)^2}{12} = 506 - \frac{(66)^2}{12} = 506 - 363 = 143$$

$$S_{\ln y,\ln y} = \sum_{i=1}^{12} (\ln y)^2 - \frac{\left(\sum_{i=1}^{12} \ln y\right)^2}{12} = 4.607 - \frac{(6.548)^2}{12} = 4.607 - 3.573 = 1.034$$

$$S_{x,\ln y} = \sum_{i=1}^{12} (x)(\ln y) - \frac{\left(\sum_{i=1}^{12} x\right)\left(\sum_{i=1}^{12} \ln y\right)}{12} = 47.973 - \frac{(66)(6.548)}{12} = 47.973 - 36.014 = 11.958$$

$$\text{Then } b = \frac{S_{x,\ln y}}{S_{xx}} = \frac{11.958}{143} = 0.08362$$

$$\text{and } a = \overline{\ln y} - (b)(\bar{x}) = 0.5457 - (0.08362)(5.5) = 0.5457 - 0.4599 = 0.0858$$

Then the fitted equation is

$$\ln y = 0.0858 + 0.08362\,x$$

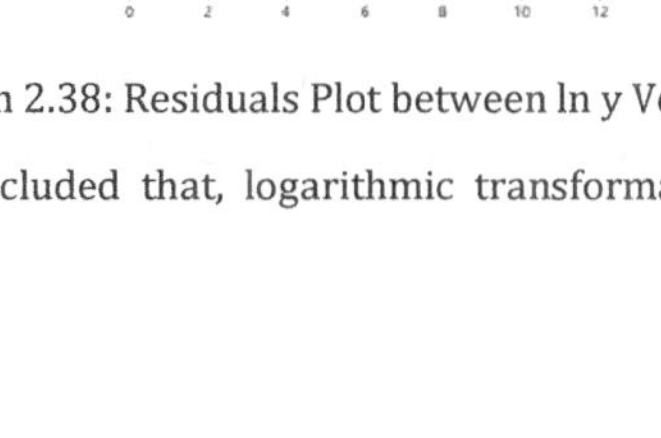

Diagram 2.38: Residuals Plot between ln y Versus x

From the curve it is concluded that, logarithmic transformation has constant variance between both the plots.

2.26. Advanced-Stage Uncertainty Analysis

when design-stage uncertainty analysis is conducted we always concentrating the following two key elements

- Uncertainty imposed by error due to measurement system resolution and calibration errors.
- In designing stage of measurement system It analyze that how the resultant will be affected if applied technique or equipment were changed ?

In advance design stage uncertainty, we will design the measuring instruments with considering the possible procedure error throughout the measurement and test control errors that affect the measurement hence an advanced-stage analysis , also known as single measurement uncertainty analysis because It is applied for such kind of a test plan where large data availability &repeating measurements is impossible and the applied method assesses different aspects of the main test by quantifying potential errors through various well focused verification tests , for identifying and estimating the uncertainty associated with errors.

Object of Advanced design stage uncertainty: The following objectives are applied in advanced design stage uncertainty analyses.

- The advanced design stage of a test is to estimate the expected uncertainty, beyond the initial design stage estimate.
- The results of a test program that involved measurements over a range of one or more parameters but with no or relatively few repeated measurements of the pertinent variables at each test condition.

For example, calibration errors that specify only a range or operating condition errors that are related to establishing a controlled set point are well modelled by a rectangular distribution. The standard uncertainty for a rectangular distribution defined by the interval **(b-a)** is given by

$$u_x = (b-a)/\sqrt{12} \qquad \text{.. (i)}$$

(i) Zero-Order Uncertainty : A zero-order uncertainty is based on the assumption that all the variables and parameters that affect the outcome of the measurement, including time , are to be fixed except for the physical act of observation itself. Under such circumstances, any data scatter introduced upon repeated observations of the output value is the result of instrument resolution alone.

(ii) First -Order Uncertainty: In the first order uncertainty analyses , measurement is based of analyses of extraneous variable with considering the time factor . suppose we want to analyse the test measurement, at initial, starting the test under given set of operating condition with time factor, if a variation achieved that is due to some extraneous influence affecting process control or simply inherent in the behavior of the variable being measured. The first-order uncertainty is to estimate the true value of a measured value can be estimated as ,

$$u_1 = t_{\nu,P} S_{\bar{x}} \quad \text{.................................... (ii)}$$

Where,

u1 = uncertainty due to Error &Resolution

but the first-order uncertainty analyses do not give the feasible & favourable test results.

(iii)Second -Order Uncertainty: In the second Order uncertainty analyses, we will identified another factor which is affecting the measured value & introduced into the analysis , thus giving a higher but more realistic estimate of the uncertainty under a given set of operating conditions .

(iv) Higher-Order Uncertainty: In Higher-order uncertainty estimation we will try to control the test operating conditions and the variability of all measured variables.

(v) Nth-Order Uncertainty: At the Nth-order estimate, instrument calibration characteristics are entered into the scheme through the instrument uncertainty u_c. A practical estimate of the Nth-order uncertainty u_N is given by

$$u_N = \left[u_c^2 + \left(\sum_{1}^{N-1} u_i^2\right)\right]^{1/2} \quad (P\%) \quad \text{.................................... (iii)}$$

Uncertainty estimates at the Nth order allow for the direct comparison between results of similar tests obtained either using different instruments or at different test facilities. The procedure for a single-measurement analysis is outlined as ,

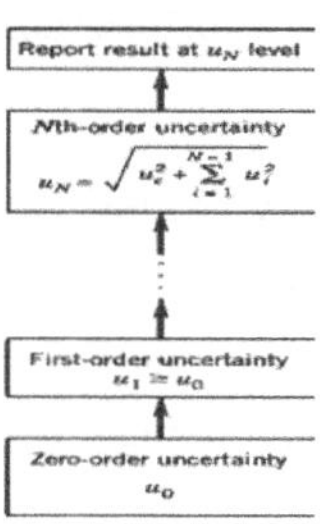

Diagram 2.39: Advanced-stage and Single-measurement Uncertainty

The Nth-order uncertainty estimate provides the uncertainty value sought in advanced design stage or in single-measurement analyses. It is an appropriate value to be used to report the results from single-measurement tests.

2.27. Higher Order Uncertainty

Consider controllability of test operating conditions ,

For First order "u_1" involves influence of time

- What is variability if all factors are held constant and you collect data over period of time?
- Collect N ≥ 30 samples, $u_1 = \pm t_{v,93} s_{x_i}$
- If $u_1 = u_0$ then time has no influence

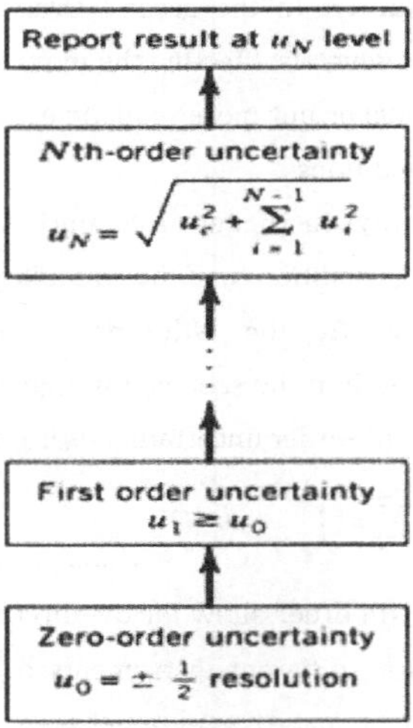

Diagram 2.40: Uncertainty Level

For Higher Order

- Each successive order identifies a factor that influences the outcome, thus giving a more realistic estimate of uncertainty
- At each stage, collect N ≥ 30 samples ,$u_i = \pm t_{v,93} s_{x_i}$
- For example, at the second level it might be appropriate to assess the limits of the ability to duplicate the exact operating conditions and the consequences on the test outcome

For N_{th} order

- Instrument calibration uncertainty u_L is entered in

$$u_N = \sqrt{(u_c)^2 + \sum_{i=1}^{N-1} u_i^2} \qquad (95\%) \qquad \text{.. (iv)}$$

- Allows for direct comparison between results obtained by either different instruments or different test facilities.
- Approximate value to use in reporting results.
- Note the difference between design stage and advanced stage is $\mathbf{u_1, u_2, \ldots, u_{n-1}}$ terms that consider the influence of time and operating conditions

Question for Discussion

1. Discuss how systematic uncertainty can be estimated for a measured value. How is random uncertainty estimated? What is the difference between error and uncertainty?
2. Explain the critical difference(s) between a design-stage uncertainty analysis and an advanced-stage uncertainty analysis
3. Determine the required uncertainty in the measurement of temperature to yield an uncertainty in efficiency of 1%. Assume errors are uncorrelated. Use T_h = 1000 K and T_c = 300 K.
4. Suppose the scale was somehow offset by one least count of resolution. How would this affect its uncertainty ? Explain in terms of random and systematic error.
5. How would the resolution of the display scale of an instrument affect its uncertainty?
6. A particular flow meter allows the volumetric flow rate Q to be calculated from a measured pressure drop Δp Theory predicts $Q \propto \sqrt{\Delta p}$, and a calibration provides the following data:

Q (m^3/min)	Δp(Pa)
10	1,000
20	4,271
30	8,900
40	16,023

   What uncertainty in the measurement of Δp is required to yield an uncertainty of 0.25% in Q over the range of flow rates from 10 to 40 m3/min?
7. Estimate the required level of uncertainty in the measurement of resistance for a platinum RTD ifit is to serve as a local standard for the calibration of a temperature measurement system for an uncertainty of 0.005C. Assume R(0C) = 100 V.
8. What is the need of calibration ? Explain
9. Estimate the design-stage uncertainty in determining the voltage drop across an electric heating element. The device has a nominal resistance of 30 V and power rating of 500 W. Available is an ohmmeter (Accuracy : within 0.5%; resolution: 1 V) and ammeter (accuracy: within 0.1%; resolution: 100 mA). Recall E = IR
10. Estimate the uncertainty at 95% confidence in the strength of a metal alloy. Six separate specimens are tested with a mean of 448.1 MPa and standard deviation of 1.23 MPa in the results. A systematic standard uncertainty of 1.48 MPa with n_b = 19 is estimated by the operator.
11. Explain about Higher order uncertainty analysis.

12. A pressure transducer outputs a voltage to a readout device that converts the signal back to pressure. The device specifications are:

Resolution:	0.1 psi
Sensitivity error:	0.1 psi
Linearity error:	within 0.1% of reading
Drift:	less than 0.1 psi/6 months (32–90°F)

The transducer has a claimed accuracy of within 0.5% of reading. For a nominal pressure of 100 psi at 70F, estimate the design-stage uncertainty in a measured pressure

13. Described the terms Error propagation.
14. Explain about
    (i) Data acquisition
    (ii) Data reduction
    (iii) Calibration
    (iv) Bias & Precision errors

References

[1] International Standards Organization, Guide to the Expression of Uncertainty Measurement, 1993.

[2] International Standards Organization, Guide to the Expression of Uncertainty in Measurement, 1993, 10 and 11.

[3] R.H. Dieck, Measurement Uncertainty, Methods and Applications, ISA, Research Triangle Park, NC, 1992.

[4] E.O. Doebelin, Measurement Systems, Application and Design, 4th ed., New York: McGraw-Hill, 1990.

[5] M.G. Kendal and A. Stuart, Advanced Theory of Statistics, Vol. 2, Griffin, London, 1961.

[6] J. Bendat and A. Piersol, Random Data Analysis, Wiley, New York, 1971.

[7] C. Lipson and N.J. Sheth, Statistical Design and Analysis of Engineering Experiments, McGraw-Hill, New York, 1973.

[8] I.C. Gupta, Engineering Metrology, Dhanpat Rai Publications, Delhi.

[9] Mechanical Measurements, R.K. Jain Khanna Publishers, 1994.

[10] Beckwith Marangoni and Lienhard, Mechanical Measurements, Pearson Education, 6th Ed., 2006.

[11] I. Miller and J.E. Freund, Probability and Statistics for Engineers, 3rd ed., Prentice Hall, Englewood Cliffs, NJ, 1985.

[12] K. Bury, Statistical Distributions in Engineering, Cambridge Press, Cambridge, UK, 1999.

[13] S.B. Vardeman, Statistics for Engineering Problem Solving, PWS Publishing Company, Boston, 1994.

[14] International Organization for Standardization (ISO), Guide to the Expression of Uncertainty in Measurement, Geneva, Switzerland, 1993

[15] H. Coleman and W.G. Steele, Experimentation and Uncertainty Analysis for Engineers, 2nd ed., Wiley-Interscience, New York, 1999.

[16] R. Dieck, Measurement Uncertainty:Methods and Applications, 3rd ed., Instrumentation, Systems and Automation Society (ISA), Research Triangle Park, NC, 2002.

[17] A. Vandoren, Data-Acquisition Systems, Reston (a Prentice Hall company), Reston, VA, 1982 Beauchamp, K.G., and C.K. Yuen, Data Acquisition for Signal Analysis, Allen & Unwin, London, 1980.

CHAPTER 3

TEMPERATURE MEASUREMENT

3.0. Temperature Measurement

Temperature is defined as a degree of the hotness or coldness of an object or substance or body with reference to some standard value. The temperature of two systems is the same when the systems are in thermal equilibrium. The atoms and molecules in a substance do not always travel at the same speed. This means that there is a range of energy (the energy of motion) among the molecules. So therefore, we can also have defined as the temperature is a measure of the average heat or thermal energy or kinetic energy of the particles in a substance due to that temperature variation takes places.

Some Others Possible Definition of Temperature

1. It is defined as degree of hotness or coldness of body that is measured on a definite scale.
2. It is defined as the degree of heat that is natural to the body of a living being. Usually about 37.0°C (98.6°F) in humans.
3. Temperature is also a measure of the kinetic energy indicated by molecules or atoms in a substance or body. The greater this energy, the faster the particles are moving, and the higher the reading is represented by an instrument.

> In general Molecules have a rotational as well as translational motion and the molecules of solids have collection of modes of both the motion, if degree of motion is higher than temperature will rise. The kinetic energy will be transferred by collision between molecules with high kinetic energy to low kinetic energy. thus, it may be defined as the average translational kinetic energy of the molecules, the so called kinetic temperature

4. Temperature is a measure of the internal energy of the system. It is roughly described by the kinetic theory, at least for gases and fluids.
5. A measure of the ability of a substance, or more generally of any physical system, to transfer heat energy to another physical system.
6. Temperature can be defined as in macroscopic terms, using concepts of thermodynamics, as an intrinsic property of matter that quantifies the ability of one body to transfer thermal energy (heat) to another body.
7. Temperature is expressed as the inverse of the rate of change of entropy with internal energy, with volume V and number of particles **N** held constant.

3.1. Temperature Scales

There are several scales used to measure temperature. Three of the most common are Kelvin, Celsius, and Fahrenheit. Temperature scales may be relative or absolute. A relative scale is based on the kinetic behavior relative to a certain material. Celsius and Fahrenheit scales are relative scales based on the freezing point (or triple point) of water and its boiling point . The Kelvin scale is an absolute scale, which has no degrees.

1. **Kelvin Temperature Scale Definition :** The Kelvin temperature scale is an absolute temperature scale based on the definition that the volume of a gas at constant (low) pressure is directly proportional to temperature and that 100 degrees separates the freezing and boiling points of water. The Kelvin scale is used often in physics, and is adjusted so that 0 degrees Kelvin is absolute zero, in theory, the coldest possible temperature, in which all kinetic motion ceases. Kelvin temperatures are written with a capital letter **'K'** Note that **0**K is 'absolute zero' and there are no negative Kelvin temperatures.
2. **Celsius scale :** A centigrade thermometer has 100 divisions or intervals between the normal freezing point and the normal boiling point of water. Today, the centigrade scale is known as the Celsius scale.
3. **Fahrenheit temperature scale :** There are 180 divisions or intervals between these two temperatures when using the Fahrenheit scale. the normal freezing point of water is designated as **32 degrees** and the normal boiling point of water is designated as **212 degrees** in the Fahrenheit scale. Fahrenheit is abbreviated as **°F.** Temperatures expressed by the Fahrenheit scale can be converted to the Celsius scale equivalent using the equation below:

   °C = (°F - 32°) / 1.8 1
4. **Kelvin temperature scale:** The Kelvin temperature scale is similar to the Celsius temperature scale in the sense that there are 100 equal degree increments between the normal freezing point and the normal boiling point of water. The temperature of 0 Kelvin is equivalent to a temperature of **-273.15 °C** .Temperatures expressed by the Celsius temperatures can be converted to Kelvin temperatures scale equivalent using the equation below:

   K = °C + 273.15 2

The zero point on the Kelvin scale is known as absolute zero. It is the lowest temperature that can be achieved

3.2. Thermometry

Thermometry is nothing but it is a process or method by which we can measure the temperature difference of any body or object with known standard .The word thermometer is derived from two smaller word fragments: thermo from the Greek for heat and meter from Greek, meaning to measure. A thermometer has two important elements, the temperature sensor (bulb) in which some physical change occurs with temperature, and scale .A change in temperature causes some kind of mechanical motion , due to that materials expand with a rise in temperature and depressed when it falls such kind of device is called thermometers . in general thermometer is of three types depends upon the temperature sensitive material , the temperature sensitive material may be liquids, solids, or may be gases .

(i) The Liquid in Glass Thermometer: It consists of sealed glass tube which contains liquid, when temperature rises the liquid expanded on the other hand it falls when temperature decreases. Liquid in Glass thermometers have been used in science, medicine, metrology and in an industry. Commonly the liquids may be Mercury or Alcohol.

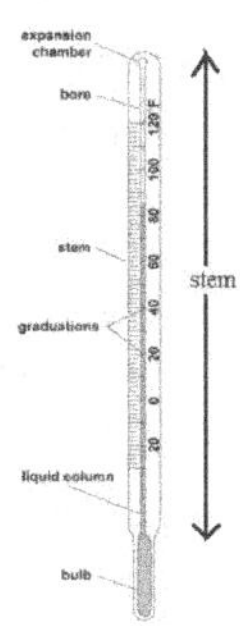

Diagram 3.1: Liquid in Glass Thermometer

Construction of Thermometer

- **The bulb:** Acting as a reservoir holding the liquid whose volume changes with Temperature. The Bulb also acts as a sensor or gauge which is inserted in the body whose temperature is to be measured.
- **The Stem:** The Stem containing the scale that is measuring the temperature and a capillary through which the liquid can accordingly expand and contract.
- **Scale:** the scale is graduated in degrees, fractions or multiples of degrees.

- **Contraction Chamber:** an enlarged capillary bore which serves to reduce a long length of capillary, or prevent contraction of the entire liquid column into the bulb.
- **Expansion Chamber:** an enlargement of the capillary bore at the top of the thermometer to prevent build up of excessive pressures in gas-filled thermometers.

Advantages: Following are Some Important Advantages of Liquid Thermometer

1. They are comparatively cheaper than other temperature measurement device.
2. It can be handled and use easily.
3. Observation of reading is also very simple.

(ii) The mercury in glass thermometer: A mercury-in-glass thermometer, also known as a mercury thermometer, was invented by German physicist Daniel Gabriel Fahrenheit in 1724 and is a thermometer consisting of mercury in a glass tube. It consists of a bulb containing mercury attached to a glass tube of narrow diameter, the volume of mercury in the tube is much less than the volume of the bulb. The volume of mercury changes slightly with temperature; the small change in volume drives the narrow mercury column a relatively long way up the tube. Calibrated marks on the tube allow the temperature to be read by the length of the mercury within the tube, which varies according to the heat given to it. The space above the mercury may be filled with nitrogen or it may be at less than atmospheric pressure, a partial vacuum.

The mercury-in-glass thermometer can be replaced by Alcohol Thermometer which has similar functions as that of mercury-in-glass thermometer as well as the alcohol thermometer are less toxic .The **ethanol** version is also applied due to its low cost and low hazard properties. the others liquid may also be used like **pure ethanol, toluene, kerosene or Iso-amyl acetate.**

- Operating range for mercury thermometer - it can measure temperature is about **-38 °C to 356 °C.**
- Operating range for alcohol thermometer - it can measure temperature from**-200°C to 80°C**
- Operating range for gas thermometer - it can measure temperature beyond **600 °C.**

Advantages

- Mercury is a naturally opaque liquid of Silver colour and can be utilized in pure form.
- It has cohesive properties and does not wet the glass.
- It is very temperature sensitive liquid.

Disadvantages

- It is toxic in nature

(iii) Gas thermometer: The working principle of gas thermometer is depends upon the variation in volume or pressure of the gas due to the effects of temperature. The constant volume gas thermometer consists of a bulb connected by a capillary tube to a manometer. The bulb is filled with a gas and the type of gas is depends upon application of device as well as accuracy required such that the volume of the gas in the bulb remains constant. The volume is related to temperature by k, known as Charles law. When the temperature of gas increases its expansion takes place. Thus, the deviation in scale reading will be proportional to the rise in temperature.

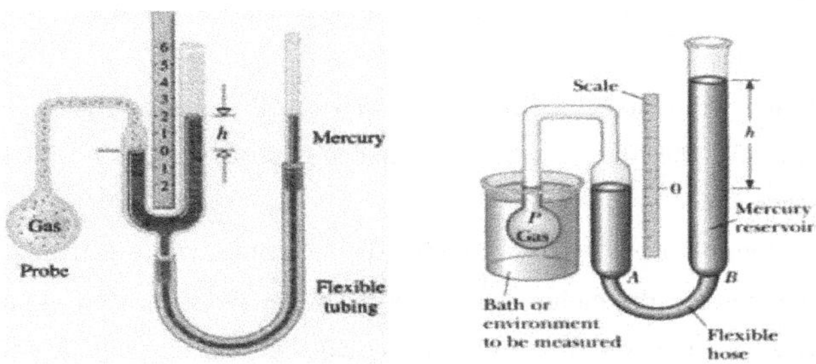

(a) (b)

Diagram 3.2: Constant Volume Gas Thermometer

For constant volume gas thermometer – According to Charles law **P** is proportional to **T** for constant volume.

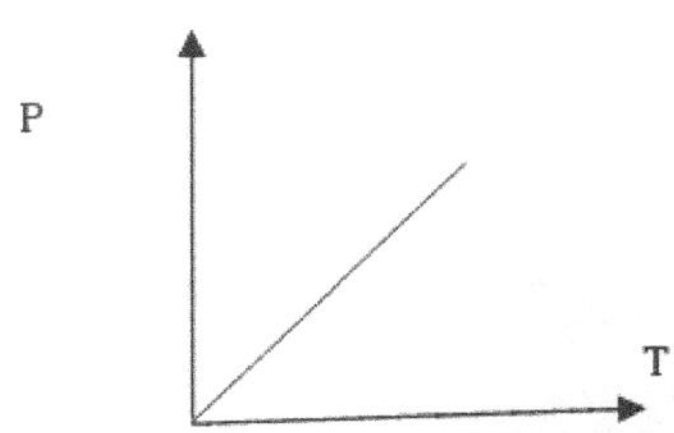

P/T = constant

There are two type of gas thermometer generally used in industrial application

(i) Constant volume gas thermometer

(ii) Constant pressure gas thermometer

Gas Used in Gas Thermometer

(i) Hydrogen gas – temperature range - 200 °C to 500 °C

(ii) Nitrogen gas – temperature range - 500 °C to 1500 °C

(iii) Helium gas – temperature range - 270 °C to 1500 °C

3.3. Bimetallic Thermometer

In an industry, there is always a need to measure and monitor temperature of a particular spot, field devices or locality. There are so many equipments are available for measuring the temperature among them bimetallic thermometer is one. Bimetallic thermometer are based on two principle

1. A metal tends to undergo a volumetric dimensional change (expansion/contraction), according to the change in temperature.
2. Different metals have different co-efficient of thermal expansion for a given temperature. The rate of volumetric change depends on this co-efficient of thermal expansion.

In bimetallic thermometer the two different kind of metal strips are arranged in cantilever beam form, the bimetallic thermometer consists of a bimetallic strip. A bimetallic strip is made of two thin strips of metals which have different coefficients of expansion. The two metal strips are joined together by brazing, welding or riveting so that the relative motion between them is arrested. The cantilevers form of thermometer means its one end is fixed and other ends is free. The strips are kept in such a manner that they overlap to each other. An increase in temperature will result in the deflection of the free end of the strip as shown in diagram because both the metal has different coefficient of thermal expansion. This deflection is linear and can be related to temperature changes. Such type of thermometer recommended between temperature range of 400 ^{0}C to 471^{0}C.

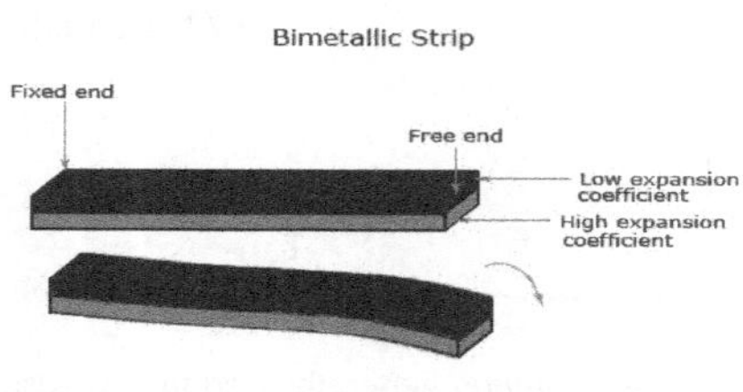

Diagram 3.3: Bimetallic Thermometer

Blue strips – Metal (1), low coefficient of thermal expansion (k_1)

Red strips - Metal (2), high coefficient of thermal expansion (k_2)

Position-1 when the strips (thermometer) are kept at room temperature (T_0) it has in linear form

Position-2 when the strips (thermometer) are kept at temperature zone (**T**) it will be bent in curve

Hence the deflection in strips is directly proportional to the change in temperature. The radius of the curvature of the bimetallic strip is calculated by following relation –

$$R = t\{3(1+m)^2 + (1+mn)[m^2+1/m\,n]\} / 6(\alpha_2 - \alpha_1)(T_2-T_1)(1+m)^2 \quad \text{.................... 3a}$$

Where ,

R = Radius of the curvature at the temperature T_2.

T = Total thickness of the bimetallic strip = $(t_1 + t_2)$

m = t_1 / t_2

α_1 = Coefficient of expansion of lower expansion metal.

α_2 = Coefficient of expansion of higher expansion metal.

T_1 = Initial temperature.

T_2 = Temperature

n = No of strips

Bimetallic Strips thermometers are generally applied in control devices, air conditioners, refineries, oil burners, tyre vulcanizing process etc .

Following are Some Advantages of Bimetallic Thermometers

- They are simple, robust and inexpensive and easy to use.
- Their accuracy is between ±2% to 5% of the scale.
- They can with stand 50% over range in temperatures.
- Bimetallic thermometers is accurate than glass thermometer.

3.4. Type of Bimetallic Thermometers

In general, the cantilever type of bimetallic thermometers are applied in practical application except it following are also used in temperature measurement

(i) Helix type. (ii) Spiral type.

(ii) Cantilever type. (iv) Flat type.

(i) Helix Bimetallic Thermometer: Following are principle parts of Helix Bimetallic Thermometer, a brief description is given below.

- A bimetallic helix coil which is fixed at one end to the body of the instrument and free at its other end.
- The free end of the bimetallic helix coil is attached to the shaft.
- The free end of the bimetallic helix coil is connected to a pointer which sweeps over a temperature calibrated circular dial , graduated in degrees of temperature.

The schematic diagram of helix type bimetallic thermometer is given below

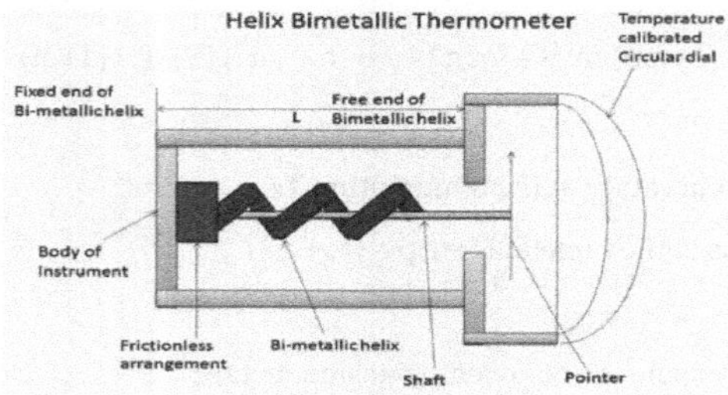

Diagram 3.4: Front View of Helix Bimetallic Thermometer

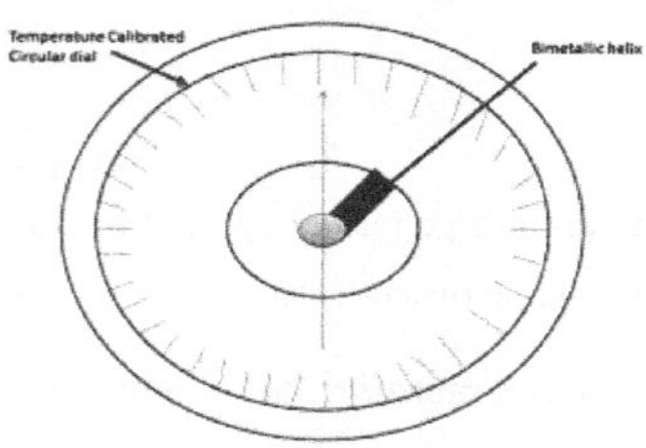

Diagram 3.5: Side View of Helix Bimetallic Thermometer

When we want to measure the temperature of particular system or area, we will kept or installed a Helix type Bimetallic Thermometer. When the thermometer experienced the effect of temperature then expansion in bimetallic helix takes places and twisted the shaft since the free end is connected to the pointer, it moves over the temperature calibrated dial indicating the change in temperature. The free end of the bimetallic helix deflects proportional to the change in temperature.

(ii) Spiral Strip Bimetallic Thermometer : In this thermometer bimetallic strips is converted into spiral shaped , when it experienced a temperature spiral spring coil twists , due to this twisting movement of spring , a pointer attached to the dial moves over the temperature calibrated dial indicates the temperature, the deflection will be a proportional to the temperature change .

3.5. Resistance thermometry

The basic principle of resistance thermometer is that if the temperature of conductor is vary then resistance of that conductor will also be varied, the measurement of resistance represent the proportional temperature. the conductor of such kind of thermometer is generally pure metals, alloys or semiconductors which depends upon the application of thermometer . A variety of sensing materials are generally used in such kind of thermometer like Copper, Nickel and Platinum except these material *Carbon thermometers* are used to measure low temperatures in superconducting magnetic systems because of their high sensitivity and small dependence of R on the presence of a magnetic field. The another very important resistance thermometer that's widely used in industrial application called Semiconductor thermometers *or germanium resistance thermometers used at a temperature below 20 K.*

(i) Resistive Temperature Detectors (RTD): There are many types of resistance thermometer, often called RTD's or PRT's (**Platinum Resistance Thermometer**) and they all measure temperature through a change in electrical resistance of a length of wire. Resistance Temperature Detector or RTD are precision temperature thermometer made from pure conducting metals such as platinum, copper or nickel wound into a coil form whose electrical resistance changes when temperature varies. it has a positive temperature coefficients (PTC). RTD's are passive resistive devices, have a base resistance of about 100Ω at 0°C, increasing to about 140Ω at 100°C with an operating temperature range of between -200 to +600°C. In order to remove the error from the RTD it is connected into a Whetstone Bridge network which has additional connecting wires for lead-compensation and/or connection to a constant current source, the schematic diagram of RTD is given below.

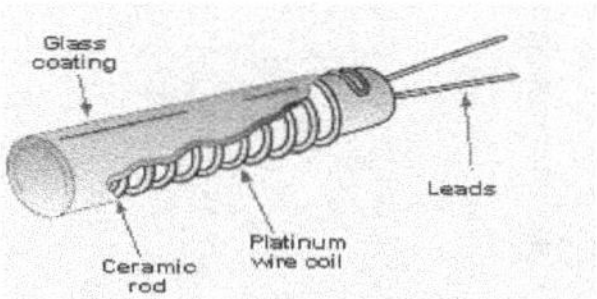

Diagram 3.6: Resistive Temperature Detectors (RTD) - Platinum Resistance Sensor

The sensing element of a platinum resistance thermometer consists of wire or strip wound on a rigid rod made up of ceramics or mica which enclosed in an envelope of glass tube at the sealed end of thermometer two terminals are created that's connecting the measuring devices. Platinum resistance thermometers are used in the temperature range from –263 to 1000°C.

Platinum has been chosen as the standard material for resistance thermometers as it gives an extremely linear resistance change in relation to temperature and is very stable. It is very to easy to operate and control and gives an accurate temperature measurement. Other metals are also used to manufacture resistance thermometers such as Copper resistance thermometers applied in industrial application which operates between temperature ranges of -150 to 600°C. RTDs are manufactured from metals whose resistance increases with temperature increments. Each metal has a specific and unique resistivity. This resistance is directly proportional to a metal wire's length, and inversely proportional to the cross-sectional area.

$R \propto L$ and $R \propto (1/A)$ thus we have

$$R = k L / A \quad \text{..3}$$

Where,

R = resistance (ohm, Ω)

k = constant of proportionality or resistivity of the material (ohm, Ω)

L = length of conductor (m)

A = cross sectional area of conductor (m^2)

Resistivity and temperature can be expressed as

$$\mathbf{K_t = k_o [1 + \alpha (t - t_o)]} \quad \text{..4}$$

Where ,

$\mathbf{k_t}$ = resistivity at temperature **t** (ohm, Ω)

$\mathbf{k_o}$= resistivity at standard temperature $\mathbf{t_o}$ (ohm, Ω)

α = temperature coefficient of resistance (1/ °C, Ω / Ω °C)

t = temperature (°C)

$\mathbf{t_o}$ = standard temperature (°C)

From equation 1 & 2 , we have

$$\mathbf{R / R_o = \alpha t + 1} \quad \text{..5}$$

This equation used for temperature measurement.

Temperature range of some important Resistance Thermometer

- Platinum resistance thermometers is operated between temperature range of **- 263 °C to 1000°C**

- copper resistance thermometers is operated between temperature range **of -150 °C to 600°C**
- Rhodium-iron alloy **(Rh + 0.5% Fe)**, resistance thermometers is operated between temperature range of **0.5 to 300 K** .
- *Semiconductor thermometers* or germanium resistance thermometers used at a temperature below **20 K.**
- *Carbon thermometers* are used to measure low temperatures.
- Posistor resistance thermometers are made from ferroelectric ceramics based on titanates, zirconates that's operated between temperature range of **20 to 100°C,** widely used in automatic control and protection systems.

3.6. Thermistor

A thermistor is a temperature sensitive resistor whose prime function is to indicating the change in electrical resistance when the body is subjected to change in temperature. A thermistor's electrical resistance vary may be due to conduction, radiation or self heating by power dissipation within the device. When a thermistor is used in a circuit where the power dissipated within the device is not sufficient to cause "self heating", environment is responsible for temperature change on the other hand self heating is achieved then the thermistor's body temperature depending upon the environmental thermal conductivity as well as its temperature.

Type of Thermistor: There are two type of Thermistor usually applied in measuring devices

- Negative Temperature Coefficient thermistors **(NTC)**
- Positive Temperature Coefficient thermistors **(PTC)**

Thermistors are ceramic semiconductors which is made from metal oxides. The types of oxides and additives denote their characteristic behavior. In case of Negative Temperature Coefficient thermistors cobalt, nickel, iron, copper or manganese are common oxides. For Positive Temperature Coefficient thermistors barium, strontium or lead Titanates are used.

(I) Negative Temperature Coefficient thermistors (NTC) : Thermistors are made from certain metal oxides whose resistance decreases with increasing temperature. Because the resistance characteristic falls off with increasing temperature they are called negative temperature coefficient (NTC) sensors. The operating temperature range of such kind of thermistors is about -55°C to 200°C .The temperature sensitivity of an NTC sensor is expressed as "percentage change per degree C". the typical values of temperature sensitivities range vary from -3% to -6% per °C.

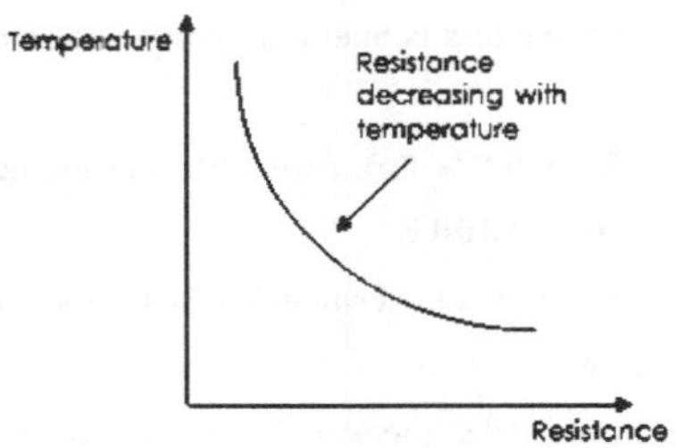

Diagram 3.7a: Negative Temperature Coefficient Thermistors (NTC)

(ii) Positive Temperature Coefficient thermistors: This type has the property where the resistance increases with increasing temperature, i.e. the value of k is positive. the operating temperature range of such kind of thermistors is about 60°C to 120°C.

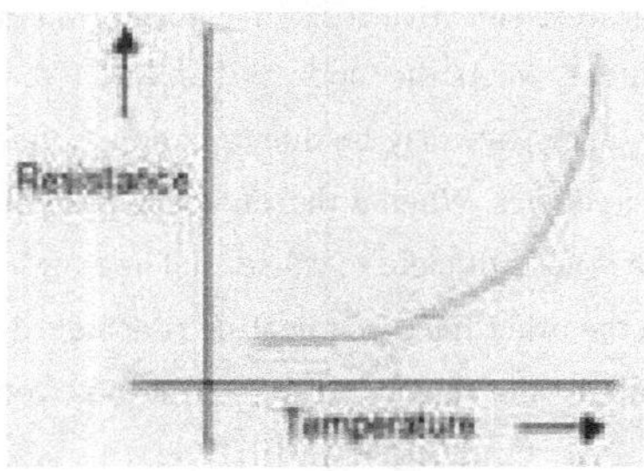

Diagram 3.7b: Positive Temperature Coefficient Thermistors

Thermistor is represented by the following symbols

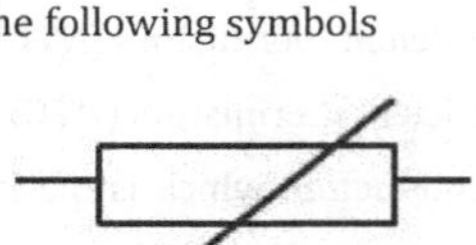

Common Symbol for Thermistor

The relationship between thermistor resistance and temperature is expressed as

$$\frac{R(T)}{R(T_o)} = e^{-\beta\left[\frac{1}{T}-\frac{1}{T_o}\right]}$$

Where ,

R(T) = Resistance at some temperature (in oK)

R(To) = Resistance at an initial measurement (reference) temperature, T (in oK);

β = Dissipation or thermistor material constant (usually expressed in mW /°C or the amount of power required to induce a temperature rise of 1°C .The other very simplified equation for thermistor is given below

$$\Delta R = k \times \Delta T$$

Where ,

ΔR = change in resistance.

ΔT = change in temperature.

k = First-order temperature coefficient of resistance

(iii) Construction of NTC thermistors : Materials typically involved in the fabrication of NTC resistors are platinum, nickel, cobalt, iron and oxides of silicon, used as pure elements or as ceramics and polymers. NTC thermistors can be classified into three groups.

- **Bead thermistors :** These NTC thermistors are made from platinum alloy lead wires directly sintered into the ceramic body. It has fast response times, better stability and allow operation at higher temperatures than Disk and Chip NTC sensors, however they are more fragile. The typical sizes range from 0.075 – 5 mm in diameter.

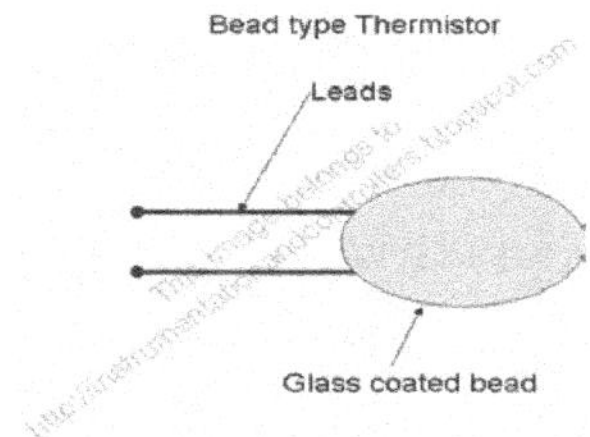

Diagram 3.8: Bead Thermistors

Probe Thermistor: Beads may be sealed in the tips of solid glass rods to form probes. Glass probe have a diameter of about 2.5mm. the probes are used for measuring temperature of liquids.

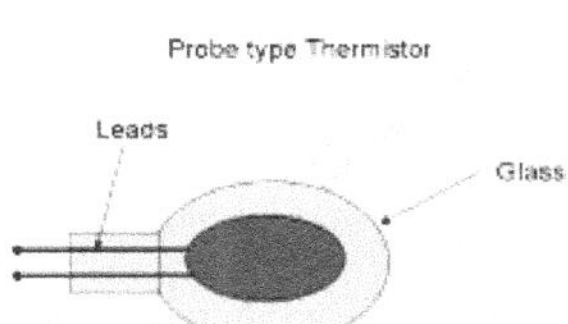

Diagram 3.9: Probe Thermistor

Disk type thermistors: Disk type thermistors are made by pressing a blend of oxide powders into a round die, which are then sintered at high temperatures. Chips are usually fabricated by a tape-casting process where a slurry of material is spread out as a thick film, dried and cut into shape. The typical sizes range from 0.25-25mm in diameter.

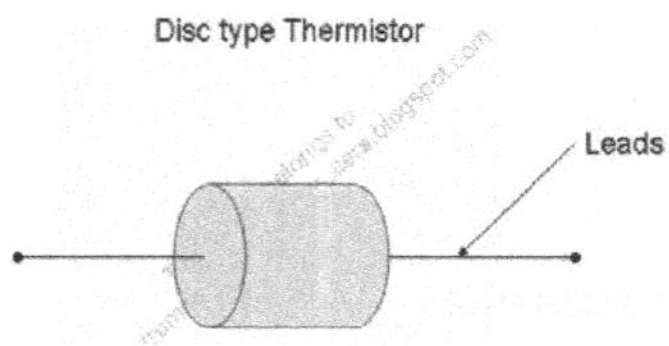

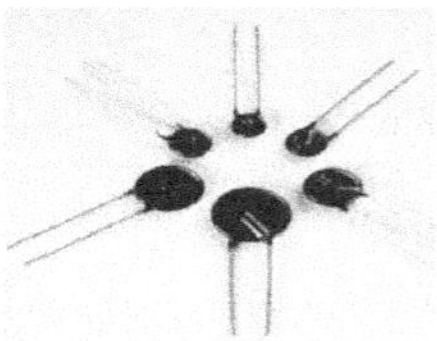

Diagram 3.10: Chip type Thermistor

- **Washer type thermistor :** Washer type is usually long cylindrical units. Leads are attached to the ends of the rods. The advantage of this type is, it produce high resistance under moderate power.

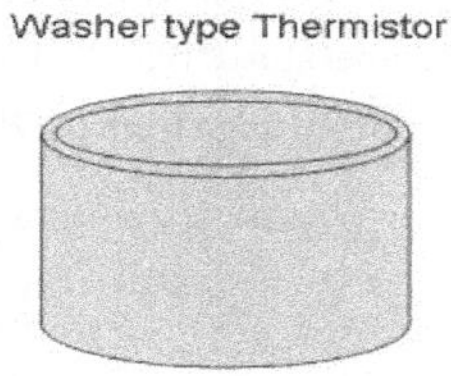

Diagram 3.11: Washer Type Thermistor

- **Glass thermistors :** They are made by hermetically sealing bead type NTC resistors into a glass container. The operating temperatures range is above 150°C, or for printed circuit board mounting, where ruggedness is a must. The typical sizes range from 0.4-10mm in diameter

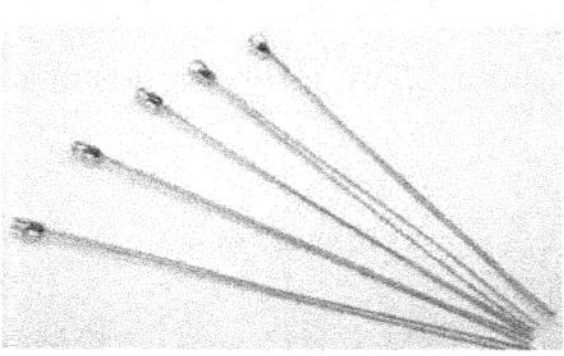

Diagram 3.12: Glass Thermistors

Thermistor applications: Thermistor are found in many applications. They provide very cheap, yet effective elements in circuits and as such they are very attractive to use.

Applications of Thermistors

1. **Digital thermostats:** These thermistors are also commonly used in modern digital thermostats.
2. **Battery pack monitors:** NTC thermistors are also used to monitor the temperature of battery packs while charging.
3. **In-rush protection devices:** NTC thermistors can be used as in-rush-current limiting devices in power supply circuits. They present a higher resistance initially which prevents large currents from flowing at turn-on, and then heat up and become much lower resistance to allow higher current flow during normal operation.
4. **Current limiting devices :** PTC thermistors can be used as current limiting devices in electronic circuits, where they can be used as an alternative to a fuse. Current flowing through the device under normal conditions causes a small amount of heating which does not give rise to any undue effects. However if the current is large, then it gives rise to more heat which the device may not be able to lose to the surroundings and the resistance goes up.

   They're also widely used in a variety of electronic applications, most often as temperature sensors. Additional uses of thermistors include current limiters, current protectors, and heating elements.

3.7. Thermocouple

Thermocouples are thermoelectric sensors that basically consists of two junctions of dissimilar metals, such as copper and constantan that are welded together. One junction is kept at a constant temperature called the reference (Cold) junction, while the other the measuring (Hot) junction. When the two junctions are at different temperatures, a voltage is developed across the junction which is used to measure the temperature sensor as shown below.

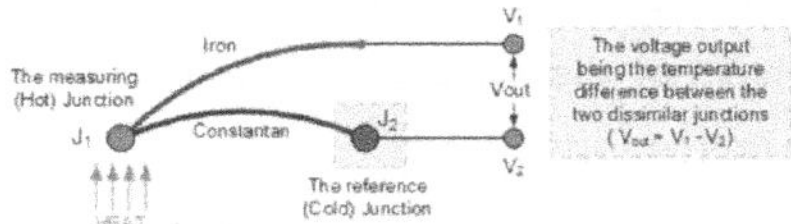

Diagram 3.13: Thermocouple

If both the junctions are at the same temperature the potential difference across the two junctions will be zero in other words, no voltage output as $V_1 = V_2$. However, when the junctions are connected within a circuit and are both at different temperatures a voltage output will be detected relative to the difference in temperature between the two junctions, $V_1 - V_2$. This difference in voltage will increase with temperature until the junction's peak voltage level is reached and this is determined by the characteristics of the two dissimilar metals used. The inherent procedure of thermocouple is given below

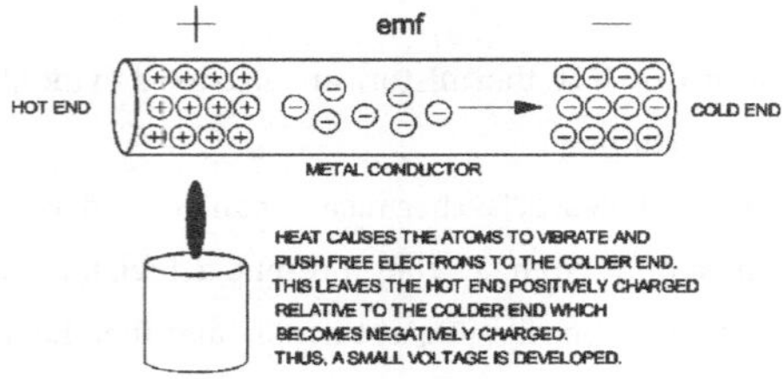

Diagram 3.14: Thermocouple

By application of heat energy to one end of a conductor, the atoms that make up the metal will vibrate more. The kinetic energy produced by the rapidly vibrating atoms will spread along the wire and conduct heat from the hotter end to the colder end. These rapidly vibrating atoms at the heated end of the metal also drive free orbiting electrons to the colder end of the wire, leaving the hotter end more positively charged than the colder end. The magnitude of the resultant charge separation, or voltage, varies with different conductor materials. Further Because different materials shows different degrees of charge separation for the same temperature gradient, by tying these two conductors together at one end, we can measure a net voltage difference at the open end of this conductor loop that is directly proportional to the difference in temperature between the hot end , and the cold end .The three most common thermocouple materials used above for general temperature measurement are Iron-Constantan (Type J), Copper-Constantan (Type T), and Nickel-Chromium (Type K). Thermocouples have operating temperature range from below -200oC to well over 2000oC.

Three Fundamental Principles of Thermocouple

- A thermoelectric current cannot be sustained in a circuit composed of a single homogeneous material by the application of heat alone, and regardless of how the material may vary in cross-section.
- The algebraic sum of the thermoelectric emf's in a circuit composed of any number of dissimilar materials is zero if all the junctions are maintained at a uniform temperature
- If two dissimilar homogeneous materials produce a thermo electric voltage V_1 when its junctions are at T_1 and T_2 , and then produce thermo electric voltage V_2 when the junctions are at T_2 and T_3 , then the voltage produced when the junctions are at T_1 and T_3 will be $V_1 + V_2$, as long as $T_1 < T_2 < T_3$

3.7.1. Thermocouple Materials

Thermocouples are manufactured from various combinations of the base metals copper and iron, the base-metal alloys of Alumel (Ni/Mn/Al/Si), Chromel (Ni/Cr), Constantan (Cu/Ni), Nicrosil (Ni/Cr/Si) and Nisil (Ni/Si/Mn), the noble metals platinum and tungsten, and the noble-metal alloys of platinum/rhodium and tungsten / rhenium. Only certain combinations of these are used as thermocouples . The below table is shown the some of the material types and their characteristics. The curves show the linearity of various combination.

Type	Positive lead	Negative lead	Temperature range	Temperature coeff.variation μv/°C	Most linear range and sensitivity in the range
R	Platinum-Rhodium (87% Pt. 13% Rh)	Platinum	0-1500°C	5.25-14.1	1100-1500°C 13.6-14.1 μv/°C
S	Platinum-Rhodium (90% Pt. 10% Rh)	Platinum	0-1500°C	5.4-12.2	1100-1500 °C 13.6-14.1 μv/°C
K	Chromel (90%Ni, 10% Cr)	Alumel ($Ni_{94}Al_2$ Mn_3Si)	-200-1300°C	15.2-42.6	0-1000 °C 38-42.9 μv/°C
E	Chromel	Constantan (57%Cu, 43%Ni)	-200-1000°C	25.1-80.8	300-800 °C 77.9-80.8 μv/°C
T	Copper	Constantan	-200-350°C	15.8-61.8	nonlinear
J	Iron	Constantan	-150-750°C	21.8-64.6	100-500 °C 54.4-55.9

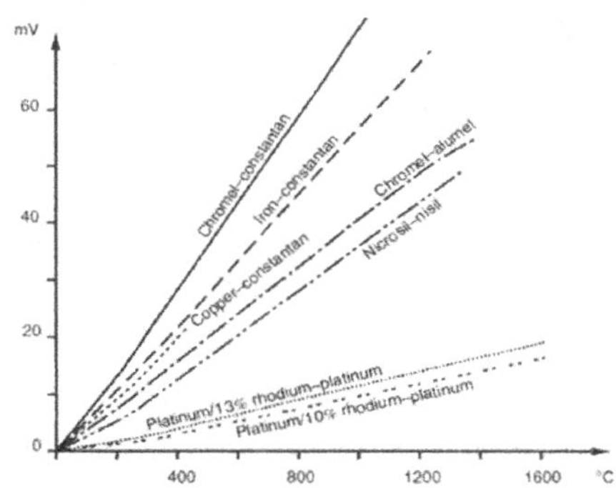

EMF – temperature characteristics for some standard thermocouple materials

Diagram 3.15: Thermocouple Temperature Curve

Advantages

- It has practical temperature ranges, from cryogenics to jet-engine exhaust, a thermocouple is capable of measuring temperature in the range -200°C to + 2500°C.
- Thermocouples is also suitable to shock, vibration and hazardous environments. means it is unaffected from shock, vibration and hazardous environments.
- They can respond to rapidly changing temperatures within a short period of time.
- No self-heating, fully depends upon the temperature difference.

Disadvantages

- signal conditioning is required to avoid errors and achieving desired level of accuracy hence designing cost is increased.
- inaccuracies may be imposed in thermocouples due to their metallurgical properties of metals used as well as corrosion of metals.
- noise problem may be comes on the front line due to the electrical and magnetic fields generation.

3.7.2. International Recognized Standards for Thermocouple

Following colour codes and temperature are used to allow the user to choose the correct thermocouple sensor for a particular application.

Code Type	Conductors (+/-)	Temperature range	BritishBS 1843:1952	Standard
E	Nickel Chromium/ Constantan	-200 to 900oC		EN60584
J	Iron / Constantan	0 to 750oC		EN60584
K	Nickel Chromium / Nickel Aluminium	-200 to 1250oC		EN60584
N	Nicrosil / Nisil	0 to 1250oC	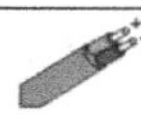	EN60584
T	Copper / Constantan	-200 to 350oC		EN60584
U	Copper / Copper Nickel	0 to 1450oC	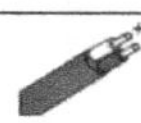	EN60584

- The insulation provided for the negative lead is white for all thermocouples.
- The insulation provided for the positive lead has a colored as above.

3.8. Pressure Measurement

Pressure is defined as force per unit area that a fluid exerts on its surroundings.

$$P = F/A \quad \text{..........} \quad 7$$

where ,

P is Pressure

F is force per unit area

A is area upon which force is being applied

The SI units of pressure is **pascal or N/m²**. but others measuring units may also be possible such as **pounds/square inch (psi) , Atmospheres (atm) , Bars, Inches of mercury (in. Hg), Millimeters of mercury (mm Hg) , and torr .**

Units of pressure

Pressure units	Other unit of pressure
1 Pascal (1N / m²)	= 10 dyne / cm²
1 mm Hg	= 133.32pascals = 13.595mm Water
Standard Atmosphere	= 1.013 * 105N/m2
1 millibar	= 1000 dyn/cm2
1 micron	= 10-6mHg
1 torr	= 1 mmHg = 1000 micron

A pressure measurement can be considered on the following two basis

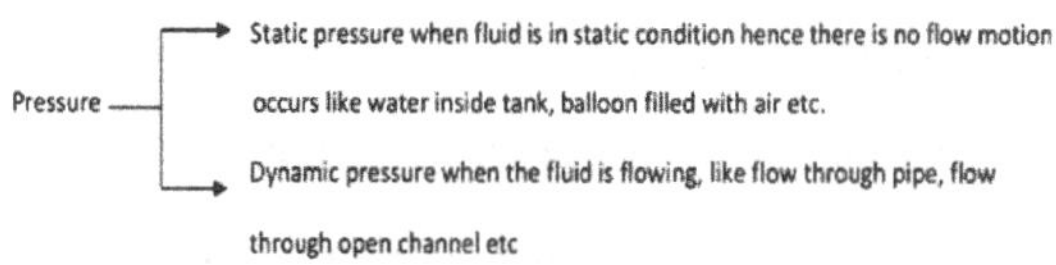

Pressure measurements can be perform on the concept of following types

Pressure	Description
Absolute pressure	Force / unit area exerted on a surface by a fluid , It is a difference between the pressure at a given point in a fluid and pressure at absolute zero or a perfect vacuum
Gage pressure	Gage pressure is the difference between the absolute pressure and the atmospheric pressure The atmospheric pressure depending upon the on atmospheric temperature, altitude (height) and weather condition of that particular place
Differential pressure	It is the difference between the two unknown measured pressures. For example pressure drop measurement in a fluid system.

Figure under here indicating the relationship amongs absolute pressure, gage pressure and vacuum pressure.

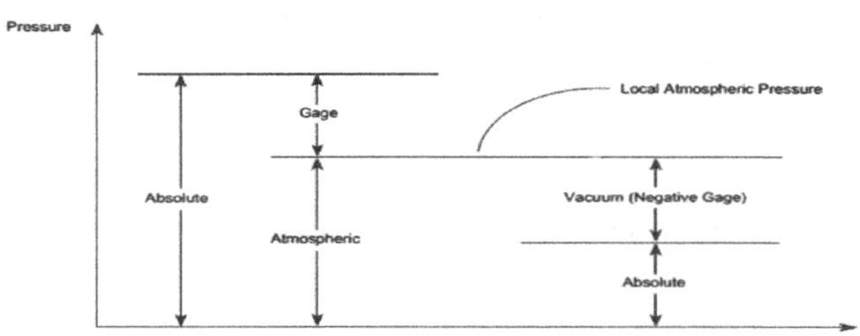

Diagram 3.16a: Relationship Between Various Pressure

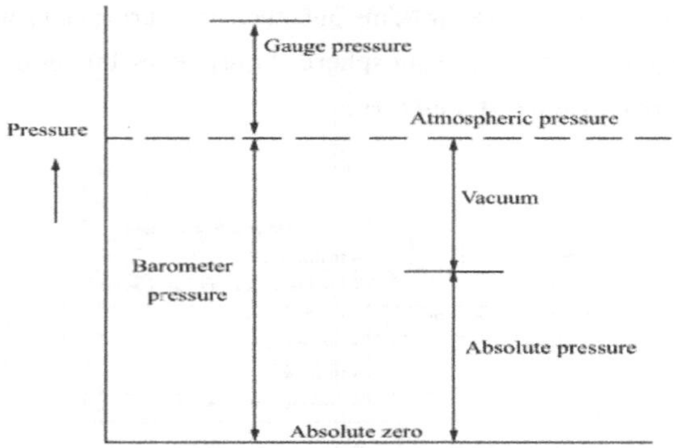

Diagram 3.16b: Pressure Relationship Diagram

Relationship between various pressure

Absolute Pressure = Gauge Pressure + Atmospheric Pressure

3.8.1. The Important Elements of Pressure is Summarized Below

(i) Absolute Pressure : The absolute is measured with reference to 0 (zero) which is a static pressure in vacuum . Actually Absolute Pressure is nothing but it is a measurement of pressure imposed by atmosphere in definite state. Sometimes it is also defined as

Absolute pressure is a algebraic summation of barometer pressure & manometer pressure which will indicating the gauge pressure or measurement of pressure with reference to absolute zero pressure

The abbreviation **'abs'** refers to an absolute measurement.

(ii) Gauge Pressure : *Gage pressure* is the measurement of the difference between the absolute pressure and the local atmospheric pressure. Local atmospheric pressure can vary depending on ambient temperature, altitude and local weather conditions. A gauge pressure by convention is always positive. A 'negative' gage pressure is defined as vacuum. For example tire pressure inside the vehicle tire and blood pressure of human body .Vacuum is the measurement of the amount by which the local atmospheric pressure exceeds the absolute pressure. A perfect vacuum is zero absolute pressure.

(iii) Differential pressure: Differential pressure is simply the measurement of one unknown pressure with reference to another unknown pressure. The pressure measured is the difference between the two unknown pressures. This type of pressure measurement is commonly used to measure the pressure drop in a fluid system.

3.9. Barometric Pressure Measurement

A barometer is an instrument for measuring the pressure of the air, due to the weight of the column of air above it. Such type of air pressure is called barometric pressure. As the earth's atmosphere gets thinner with increasing height, it follows that as we ascend from sea level and low-lying parts of the earth's surface, the weight of the air will decrease. That is, the pressure will fall. It is one of the basic sensors included in weather stations. Generally there are two main types are used in measurement & meteorology

- Mercury barometer and
- Aneroid barometer

But here we will only study the measurement of pressure and construction of Mercury barometer. Air pressure can be measured in milli-bars or inches of mercury.

(i) Mercury barometers: Mercury barometer is generally applied for measuring atmospheric pressure in which mercury barometer balances the weight of mercury in a glass tube against the atmospheric pressure above a reservoir of mercury at its base to provide a measurement of atmospheric pressure.

Mercury barometers consists of following elements

- It has a reservoir which contains mercury at desired level.
- A long glass tube which is filled with mercury, close at one ends and open at other end.
- Tube having a length of 3 - Feet has been kept inverted into the reservoir or bowl of mercury, called a cistern.

The construction of Mercury barometers is given below with schematic diagram.

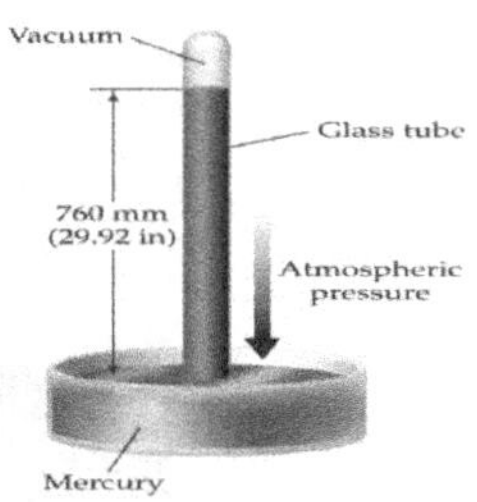

Diagram 3.17: Mercury Barometers

The change in the mercury level in the glass tube will indicating the sign in the variation of air or atmospheric pressure above the reservoir which can be measured by using the scale marked on the glass tube. Changes in atmospheric pressure occur prior to weather changes.

- Sudden drops in atmospheric pressure indicate weather will be stormy
- Sudden rises in pressure predicate that weather will be fair for short period of time.
- Gradual increases or decreases in atmospheric pressure point out that weather pattern will be sustained.

Measurement of pressure is only possible when the mercury level variation takes places if

- Whenever atmospheric pressure is increased on the reservoir, mercury level inside the glass tube rises.
- If atmospheric pressure decreases the mercury moves down inside the tube a vacuum is created, into the reservoir.
- The respected reading may be observed on scale which has been marked on glass tube.

For example, 1 atmospheric pressure causes a height of 760 mm of mercury in the tube and hence, 1 atmospheric pressure is also equivalent to 760 mm or 76 cm of mercury head. Since atmospheric pressure varies with altitude, the mercury barometer can also be used to measure the altitude of a place. Primarily, though, it is used for measurement of atmospheric pressure.

(ii) Aneroid Barometer: The purpose of aneroid barometer is same as that of Mercury barometers except the utilization of working fluid. In Mercury barometers the working fluid is Mercury while in aneroid barometer air will be a working medium. following are the basic constructional features of aneroid barometer.

- Operation of such type of barometers based on metal cell having small amount of air , metal cell may be one or more connected in series.
- One end terminal of metal cell is connected to the instruments while other terminal is connected to the pointer, the connection between system and pointer has been made by levers and pulleys.
- Whenever air pressure is increased , then the metals cells comes in contact due to the air pressure, levers and pulleys operated hence pointers movement takes place on dial gauge.

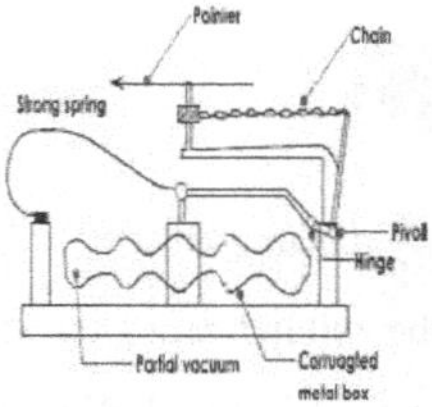

Diagram 3.18: Aneroid Barometer

(iii) Liquid Barometer: This is also a atmospheric pressure measuring device, the constructional features are described below with schematic diagram.

- It consists of two containers one is at top and another is at bottom, top container and bottom container are connected through a glass tube.
- Top container is completely air tight while the bottom container consists a liquid having an air gap or open around the tubes for atmospheric pressure.
- The liquid may be water or mineral oil (lighter) because it gives a greater change with smallest variation of pressure.
- The liquid barometer uses back-pressure in the upper chamber instead of a vacuum to keep the column to a reasonable length.
- But the variation of temperature may be give us inaccurate measurement.

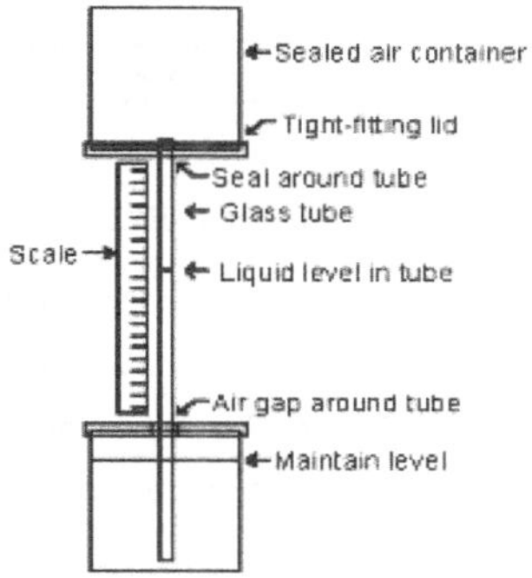

Diagram 3.19: Basic Liquid Barometer

3.9.1. Application of Pressure Measuring Instruments

In general pressure measurement is performed not only in mechanical industries but also in other industries. following are some important field of pressure measurement , that is given below .

- The pressure of steam in a boiler is measured for ensuring safe operating condition of the boiler.
- Pressure measurement is done in continuous processing industries such as manufacturing and chemical industries.
- Pressure measurement helps in determining the liquid level in tanks and containers.
- In many flow meter (such as venture-meter, orifice meter, flow nozzle, etc.,) pressure measurement serves as an indication of flow rate.

- Measurement of pressure change becomes an indication of temperature (as used in pressure thermometers-fluid expansion type).
- Apart from this, pressure measurement is also required in day-to-day situations such as maintaining optimal pressure in tubes of vehicle tyres.

In each & every application of pressure measurement, the device used in pressure measurement depends upon the intensity of pressure, here some important classification is given below.

Intensity of pressure	Applied instruments
Low pressure	Manometer
High and low pressure range	Bourdon tube pressure gauge Diaphragm gauge. Bellows Gauges.
Low vaccum pressure	McLeod vacuum gauge Thermal conductivity gauges. Ionisation gauges.
Very high pressure	Bourdon tube pressure gauge. Diaphragm gauge. Bulk modulus pressure gauge.

3.10. Pressure Measurement by Manometers

Manometer is very popular and simpler device that are generally used for measuring the pressure at a particular point in a fluid system , by balancing the column of the fluid by the same or another column of fluid as a reference .The liquid may be water , mercury , mineral oil (light oil) , methanol , gas etc. **The equation of manometer for calculating the pressure is given below.**

$$P = P_2 - P_1 = \rho g h_w \quad \text{.. 8}$$

Where ,

P= Differential pressure

P_1= Pressure applied to the low-pressure connection

P_2= Pressure applied to the high-pressure connection

h_w= The height differential of the liquid columns between the two legs of the manometer

ρ = Mass density of the fluid within the columns

g = Acceleration of gravity

3.11. Classification of Manometer

In general manometer is classified on the basis of intensity pressure and the field of application, following are some important type of manometers that are generally applied in industrial application.

(i) U Tube Manometers: According to the name U-Tube, it implies that a pressure measuring device having the form of English letter **"U"** . hence it is a simple glasses tube bent in the shape of letter **"U"** partially filled with a liquid according to the application, when there is no pressure is applied then atmospheric pressure is applied on both the limbs representing the same level of liquid. when pressure will applied any one of the limbs then liquid level rises on the other limb in similar manner if pressure drop occurs then liquid level fall in the tube . thus we can measure the height of column with the help of marked scale we can measure the height of rise and fall of liquid .

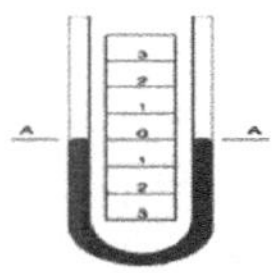

Diagram 3.20: U-Tube Manometers with Zero Pressure

The construction and procedure of another type of U-Tube simple Manometers is given as It consist a glass tube of U–shaped whose one end is attached to the gauge point **'A'.** The gauge point is closed end of the tube has a vacuum and other end is open to the atmosphere or the open end is attached to the item whose pressure is to be measured. It can measure both positive and negative pressures. It contains liquid of specific gravity greater than that of a liquid of which the pressure is to be measured. It has simple construction, easily operable and no need of mechanical items like gears, levers, etc.

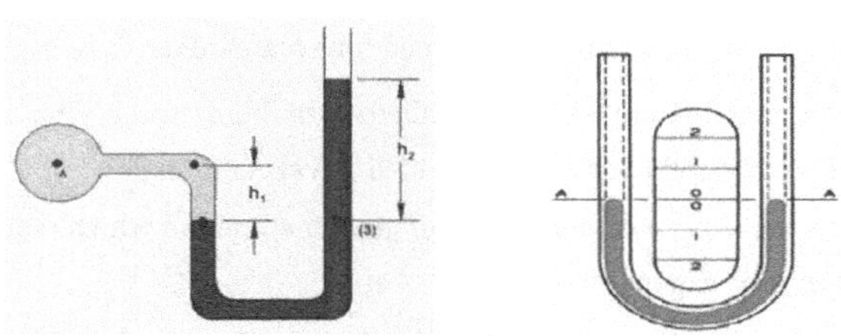

Diagram 3.21: U-Tube Manometers

One limb is closed while another is open to atmosphere or connected to device

Mathematically the Pressure at point A can be calculated as

$$P = W_2h_2 - W_1h_1 \quad \text{.................................. 9}$$

Where ,

W = Specific weight of liquid ,

P = Pressure measurement at point "**A**",

$\mathbf{h_2}$ **&** $\mathbf{h_1}$ are rise and fall of liquid in the glass tube.

Advantages

- Simple in construction
- Low cost hence easy to buy
- Very accurate and sensitive
- It can be used to measure other process variables

Disadvantages

- Fragile in construction
- Very sensitive to temperature changes
- Error can happen while measuring the height

(ii) Inclined tube Manometer: Whenever our requirement is to measure the low pressure with accuracy we will generally applied Inclined tube manometer, the construction and others details are given below.

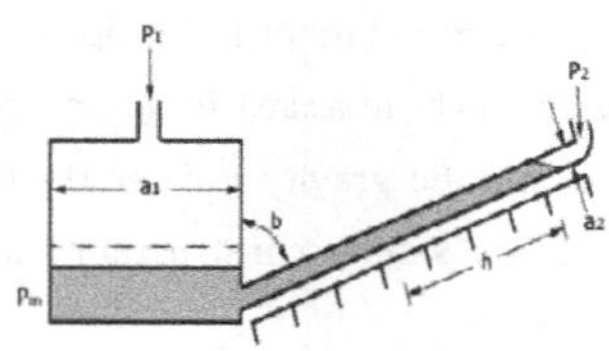

Diagram 3.22: Inclined tube Manometer

- It has two section one section is filled with liqid/water whose top is open to atmosphere while other section is inclined at a specific angle .
- The inclined tube is an enlarged limb of manometer upon which reading is impressed. The scale may be expanded scale.
- Scale reading 0.01 inches of liquid height give us an equivalent pressure of 0.000360 PSI per division
- Generally applied in air and gas installations.

The pressure may be calculated as

$$\mathbf{P_1 - P_2 = P_m\, h\, Cos\, b\, (1 + a_2 / a_1)}$$

$$P_A - P_B = \gamma_2 l_2 \sin\theta + \gamma_3 h_3 - \gamma_1 h_1 \quad \text{.............................. } \mathbf{10}$$

Advantages

- One need only compare the height of one liquid column, not the difference in height between two liquid columns.
- The cross-sectional area of the liquid column in the well is so much greater than that within the transparent manometer tube that the change in height with in the well is usually negligible.

Disadvantages

- A small change in a fluid height causes larger displacement in the transparent tube when it is inclined to the angle and filled by oil
- High sensitivity

(iii) Well type manometer: A well-type manometer is a common form of manometer which is similar to the U-tube manometer. the constructional featured of well type manometer is given as

- It consists of a well type container connected with vertical glass tube according to the diagram.

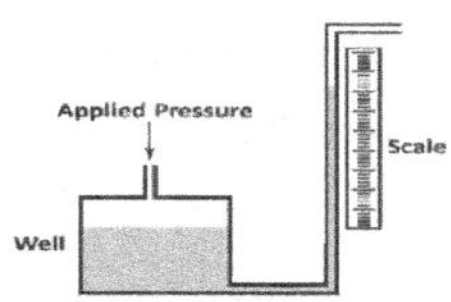

Diagram 3.25: Well type Manometer

- A well type container or large reservoir is filled with liquid/water and open at top to atmospheric pressure or open to connecting device whose pressure is to be measured.
- Whenever the pressure variation is achieved on the well top opening, the liquid rise /fall is takes place whose value can be read out directly from the scale.

Advantages

- It does not require the observer to make a calculation by looking at both sides of the tube.
- It is manufactured with the high degree of accuracy.
- The well type manometer provides us easier means of pressure measurement.

Disadvantages

- It places certain operational requirements not found with the U-type

(iv) Dual –tube Manometer: A dual tube manometer is a manometer that is designed to read out very high pressures. A high pressure causes the need for a longer indicating tube, which is very inconvenient to the person reading the manometer. A dual-tube manometer solves this problem by having two tubes to read the pressure, a standard well-type manometer and a well-type manometer with the well at the 100-inch reading on the indicating tube.

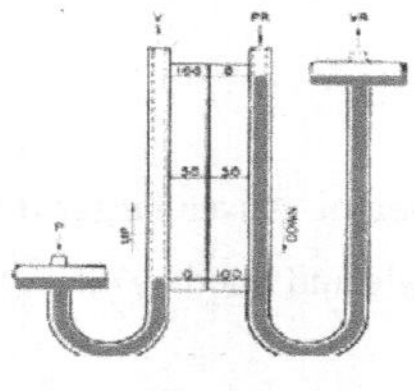

Diagram 3.26: Dual–tube Manometer

Advantages

- High range pressure is more accurate with this instrument.

Disadvantage

- Complicated operation
- Inaccurate for the exact pressure value due to large–scale calibration

(v) Differential U-Tube Manometer: The constructional featured of Differential U-Tube Manometer is given as.

- It is applied for measuring the pressure difference in a pipe between two points ,
- Differential manometers consist of an u-tube in which U-Tube mano-metric liquid is heavier than the liquid for which the pressure difference is to be measured and should not immiscible with it.

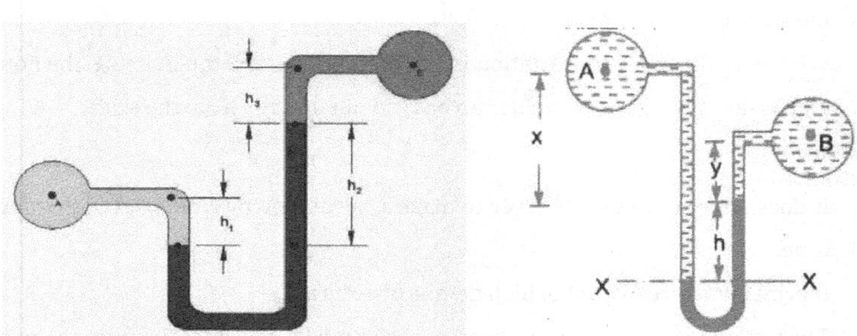

Diagram 3.27: Differential U-Tube Manometer

For Differential U-Tube Manometer - 1 pressure is calculate as

Pressure difference between A and B is given by equation

$$P_A - P_B = w_2h_2 + w_3h_3 - w_1h_1 \quad \text{.......} \quad 11$$

For Differential U-Tube Manometer – 2 pressure is calculate as

Difference of pressure at A and B

$$= P_A - P_B \quad \text{.......} \quad 12$$

$$= gh\rho_2 - \rho_1 + \rho_1 g y - \rho_2 g x \quad \text{.......} \quad 13$$

(vi) Inverted U-Tube Manometer: Inverted U-Tube manometer consists of an inverted U – Tube containing a light liquid. It is used to measure the differences of low pressures between two points where better accuracy is required. It generally consists of an air cock at top of manometric fluid type.

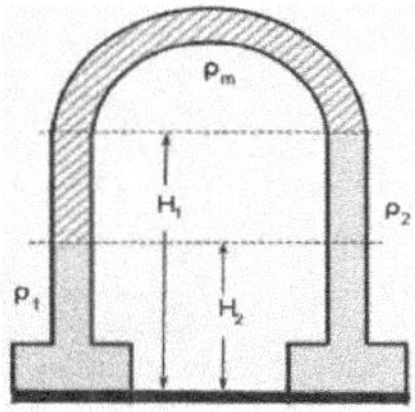

Diagram 3.28: Inverted U-Tube Manometer

$h1$ = height of the liquid in the left limb

$h2$ = height of the liquid in the right limb

h = difference of light liquid

ρ_1 = density of liquid in the left limb

ρ_2 = density of the liquid in the right limb

ρ_L = density of light liquid.

Pressure difference can be calculated from the following equation

$$P_1 - P_2 = \{ \rho_1 g h_1 - \rho_2 g h_2 - \rho_L g \, (h_2 - h_1) \} \quad \text{.......} \quad 14$$

Advantages

- Easy to fabricate and inexpensive
- Accurate and very sensitivity
- Not affected by vibration

Disadvantage

- Usually bulk and large in size
- Fragile in nature
- Reading is affected by change in temperature and gravity.

(vii) Micro-manometer : The micro-manometer is another variation of liquid column manometers that is based on the principle of inclined tube manometer and is used for the measurement of extremely small differences of pressure. The meniscus of the inclined tube is at a reference level as shown in the figure below, viewing through a magnifier provided with cross hair line. This is done for the condition, $P_1 = P_2$. The adjustment is done by moving the well up and down a micrometer. For the condition p_1 not equal to p_2, The shift in the meniscus position is restored to zero by raising or lowering the well as before and the difference between these two readings gives the pressure difference in terms of height . The schematic diagram is given below

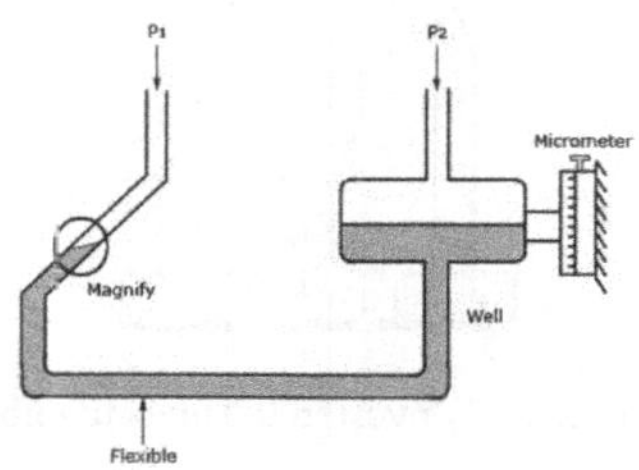

Diagram 3.29: Micro-manometer

$$P_1 = (\rho_m - \rho_1)gh_3 + \rho_m gh_2 - \rho_1 gh_1 \quad \ldots\ldots\ldots\ldots\ldots \mathbf{15}$$

where ,

h_3 = Falling liquid level reservoir,

h_2 = Rise of the liquid in the tube,

A = area of the reservoir,

'a' = area of the tube,

(viii) Enlarged-Leg Manometer : In the enlarged-leg manometer , It has a float in the enlarged-leg which is utilized for indication or recording. The two legs can be changed for changing the measurement span. The schematic diagram is given below.

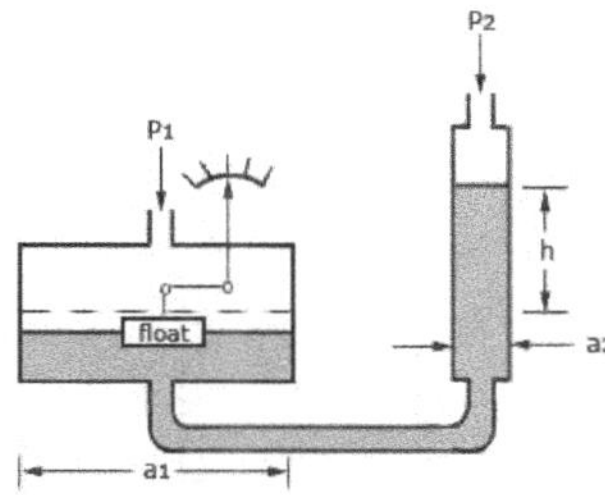

Diagram 3.28: Enlarged-Leg Manometer

(ix) Ring-Balance Manometer : Ring-Balance Manometer tube is made up of polythene or other light and transparent material. This tube is bent in the form of a ring and is supported at the centre by a suitable pivot. The tubular chamber is divided in to two parts by spilling, sealing, and filling with a suitable light liquid like kerosene or paraffin oil for isolating the two pressures. Pressure taps are made with two flexible tubing's. Pressures P_1 and P_2 act against the sealed walls as shown in the figure below, and rotate the ring which is balanced by the counter weight w. in this type of manometer mercury is working as a mano-metric fluids due to low vapour pressure , non-sticky nature , wide temperature range , but high density and installation cost is high.

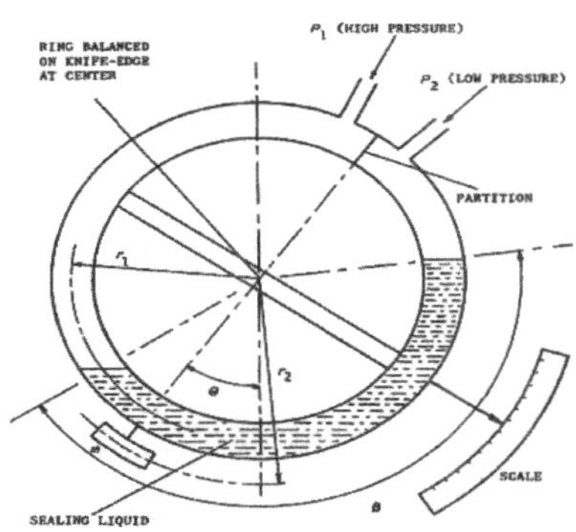

Diagram 3.29: Ring-Balance Manometer

(x) Advanced U-Tube Manometer : An enhanced version of a manometer is shown below with a seal liquid over the manometer liquid to separate the process fluid from the manometer fluid for any probable source of trouble like absorption, mixing or explosion and so on. Seal pots with large diameters are also placed for increasing the range.

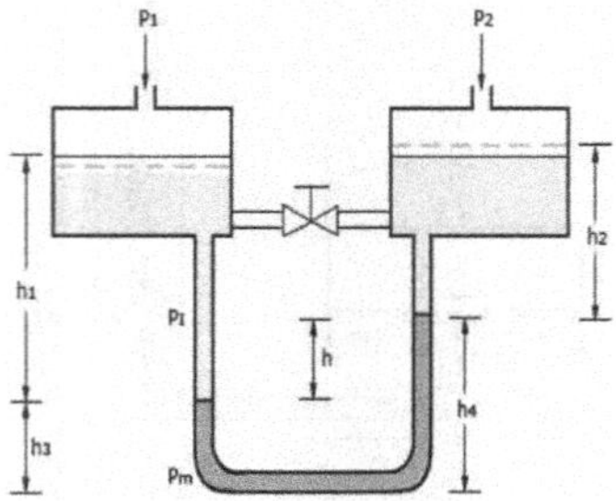

Diagram 3.30: Manometer With Large Seal Pots

3.12. Dead Weight Tester

The basic purpose of dead weight tester is to measuring a pressure that is generated by weight, on a given calibrated pressure gauge. The schematic diagram of dead weight tester is given below.

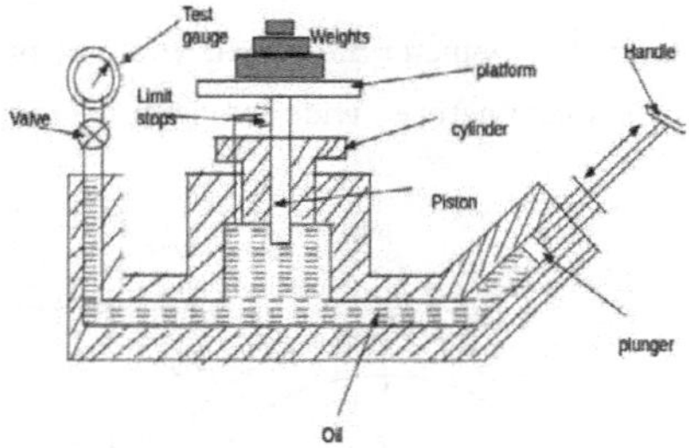

Diagram 3.31: Deadweight Tester

Components of Deadweight tester

Plateform	Cylinder
Piston	Plunger
Operating valve	Pressure gauge
Oil	Handle

It consists of following steps of operation

- It consists of chamber filled with oil which should be free from any kind of impurities, chamber also have a cylinder inside which piston is fitted that is free to movements, as clear from the diagram.

- On the top of the piston a platform is attached upon which weight is kept on the other hand plunger is provided in which handle maintaining and creating a pressure .
- Kept any specific weight on the platform.
- Operating the plunger in such a way that fluid pressure is applied to the other side of the piston create enough force to lift the piston-weight combination.
- We have got the piston weight combination floats freely within the cylinder between limiting stops.
- Crate equilibrium of Piston-Weight Combination with Applied pressure by plunger handle.
- When equilibrium reached we have a particular reading on test gauge, with the help of calibrated scale measurement can be performed.

For equlibrium condition

Pressure Force = Weights on Platform + Friction Drag.

P = Mg + F

$$P = Mg + F / A \quad \ldots\ldots\ldots\ldots\ldots\ldots\ldots\ldots\ldots\ldots\ldots\ldots\ldots\ 16$$

Where ,

P = Pressure applied due to weight platform

M = Mass in **Kg**

g = Acceleration due to gravity , $\mathbf{m / s^2}$

F = Friction drag in **Newton**

A = Equivalent area of piston – cylinder combination in $\mathbf{m^2}$

After the plunger has been released "**Gauge Pressure must be equal to the known Applied Pressure**" but the accuracy of the dead weight tester is generally depends upon the friction force which is acted between piston and cylinder as well as variation in gravitational constant **'g'**.

At static equilibrium position, chamber pressure can be given by the following equation,

$$p = \frac{F}{A_e} + \sum \text{error corrections}$$

Where ,

A_e = Equivalent area of the piston

F = External force (weight of the piston + the additional weight of calibrated masses)

P = Chamber Pressure

An indicated pressure (Pi), can be calculated as ,

$$P = Pi\,(1 + E_b + E_g)$$

Where ,

E_g = Gravity Effects correction,

$E_g = -\,(\lambda_{air} / \lambda_{masses})$

E_b = Air Buoyancy Effects correction,

$E_b = -\,\{2.637 \times 10^{-3}\cos 2\phi + 9.6 \times 10^{-8}z + 5 \times 10^{-5}\}$

Pi = Indicated Pressure

3.13. Pressure Gauges and Transducers

The function of pressure gauge & pressure transducer is to be performed measurement of pressure in order to achieve effective &safe mode of operation in an industrial application . Sometimes these gauges may or may not be gives us direct reading of measurement but in case of pressure transducer it converts the pressure into measurable output of electrical or mechanical form for any operational activities called **pressure transducer or pressure transmitter**.

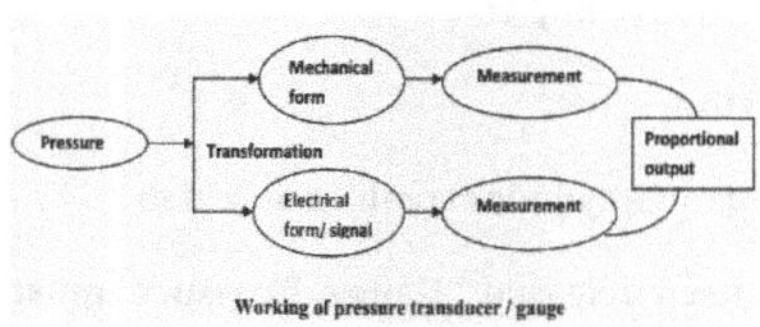

Working of pressure transducer / gauge

The classification of such type of transducer/gauge are given below

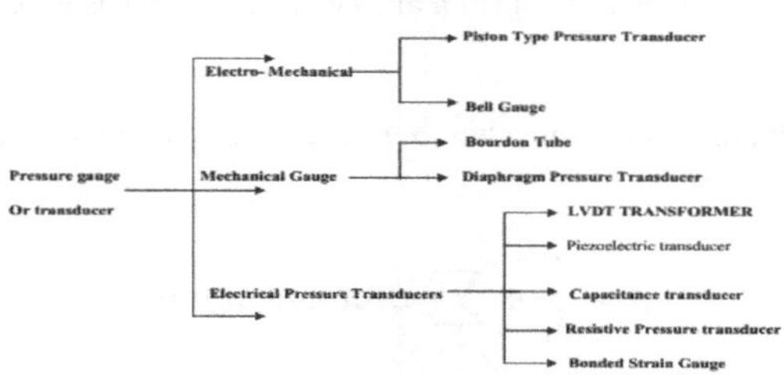

Type of pressure transducer

3.13.1. Electro- Mechanical Pressure Gauge or Transducer

Electro- Mechanical transducer is nothing but it is a combination of both technique in which some portion of measurement is performed by electrically manner & remaining portion of transformation is done mechanically then we have got final result of measurement. Following type of devices are described below.

(i) Bell Gauge : The schematic diagram of such type of gauge is given below .

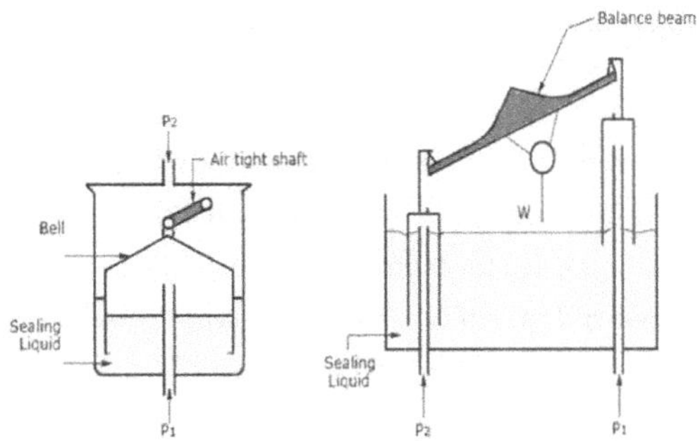

Diagram 3.31: Single Bell and Double Bell Gauge

Elements of Bell gauge

(i) Sealing liquid (oil)

(ii) Bell

(iii) Air tight shaft

(iv) Hanging weight /counter weight

(v) Balance beam

(vi) Pressure entry and exit portion

Working procedure of Bell gauge : Bell type gauge is used for measuring low static pressure or vacuums pressures.

- It consists of a container filled with sealing liquid (oil) in which bell shaped container is immersed as shown in figure, Inside the container bell can free for to and fro motion which can be achieved by mechanical arrangement operated with the help of link and lever mechanism, when pressure is applied , the bell rises up hence closing the inlet of atmospheric pressure on the other hand bell downward with decreasing pressure intensity also closes the inlet of applied pressure. the end of mechanical shaft is connected to pointer which has movement over calibrated scale whereby we can measure the applied pressure .

- On the other hand in double bell type , a counter weight is suspended at the centre of the beam which must be balanced when pressure is applied, the two pressures to be compared are led into the inside of the bells , and the pressure differential is indicated by the pointer which moves with the balance beam.

With the help of bell type gauge we can measure the pressure range between 0.06 Pa and 4 K Pa. The static pressure may be as high as 4 to 6 M Pa or 145 psig.

Advantage

- The gauge is very sensitive to small changes of pressure.
- Such type of gauge is unaffected by the temperature.
- Bell type gauge is applied for controlling furnace pressure.

(ii) Piston Type Pressure Transducer: The Piston Type Pressure Transducer is as shown in figure ,

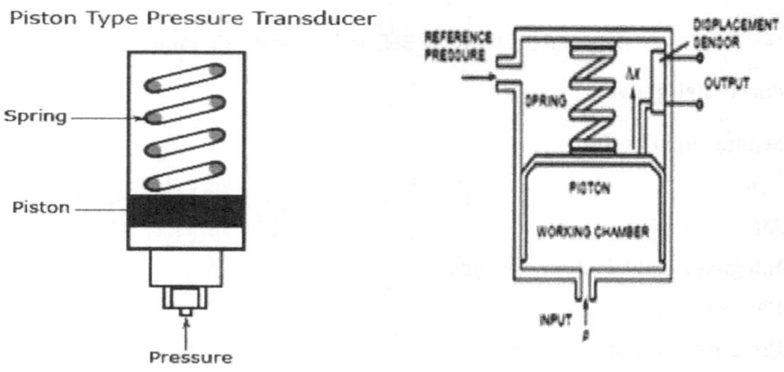

Diagram 3.32: Piston Type Pressure Transducer

Elements of Piston Type Pressure Transducer

- Pressure Chamber
- Compression Spring
- Piston
- Displacement sensor
- Output
- Cylinder

In such type of transducer pressure is applied through the inlet of measuring device , this applied pressure energy comes in contact with piston creates displacement of the piston , inside the upper portion of the cylinder a compression spring is installed due to the effects of pressure its get compressed. hence the displacement of compression spring is directly proportional to the applied pressure, which has been sensed by displacement sensor and finally converted into measurable electrical signal. it is commonly used in hydraulic application. this kind of device is applied for high dynamic loads& used for pressures measurement of between 15 PSI and 15,000 PSI. it is free from shock , sudden pressure changes , and vibrations

3.13.2. Mechanical Pressure Gauge

In this type of pressuring measuring device only mechanically operated mechanism is applied to full fill the desired purpose. Some important devices is described below.

(i) Bourdon Tube pressure gauge: The bourdon type pressure gauge is generally applied for measuring the high pressure range about 700 MPa, the constructional featured are as shown in figure .

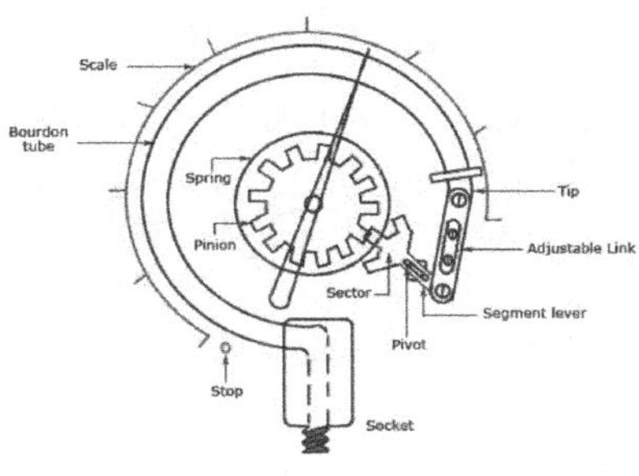

Diagram 3.33: Bourdon Tube Pressure Gauge

The basic elements of Bourdon Tube pressure gauge

- Bourdon tube
- Scale
- Spring
- Pinion
- Adjustable link socket
- Segment lever
- Socket

Bourdon tube is a elastic type pressure transducer, It consists of a elastic tube having a shape of circular cross section area of thickness between 0.01 and 0.05 inches whose one end is attached to socket of device with the help of soldering and other free end is closed , called tip . The circular cross-section of the tubing is flattened and then the flattened tubing is formed into a circular or helical shape. When a pressure is applied to the inside of the tube, the flattened section tries to regain its former circular cross-section. Tip, Adjustable link, Lever connected to sector which has been finally meshing with the pinion gear teeth. When the fluid pressure enters into the bourdon tube , it has been creating the motion in Adjustable link that is transmitting to pinion gear through lever and sector. In the center of the tube arrangement of the device, the pointer is attached which can be freely moved over a graduated scale. Hence we can measure the desired proportional pressure.

For best result tube must have best elastic properties and corrosion resistance , the following material/metal has been recommended for tube manufacturing **phosphor-bronze, silicon-bronze, beryllium-copper, inconel, and bronze or nickel steel.** Following shapes of bourdon tubes are mainly used in an industrial application.

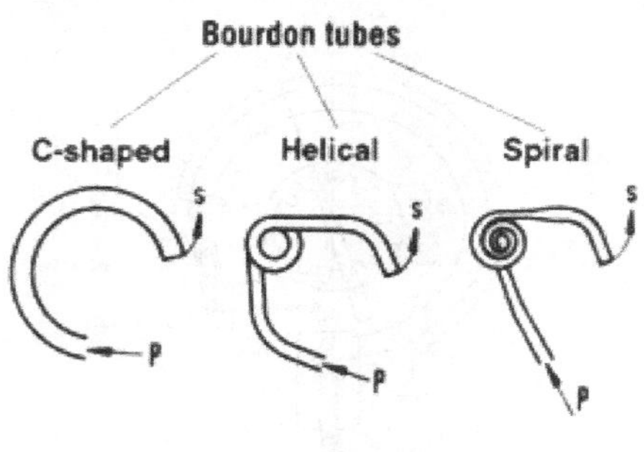

Diagram 3.34: Types of Bourdon Tubes

Advantages

- Low cost with simple construction.
- High ratio accuracy.
- High pressure measuring range.

(ii) Diaphragm Pressure Transducer : Such type of pressure transducer is generally applied for low intensity of **pressure measurement**. Diaphragm Measuring Elements like flat diaphragms and capsules diaphragms are most commonly used for low pressures measurement.

Diaphragm Pressure Transducer

Flat Diaphragm Transducer **Capsules Diaphragm Transducer**

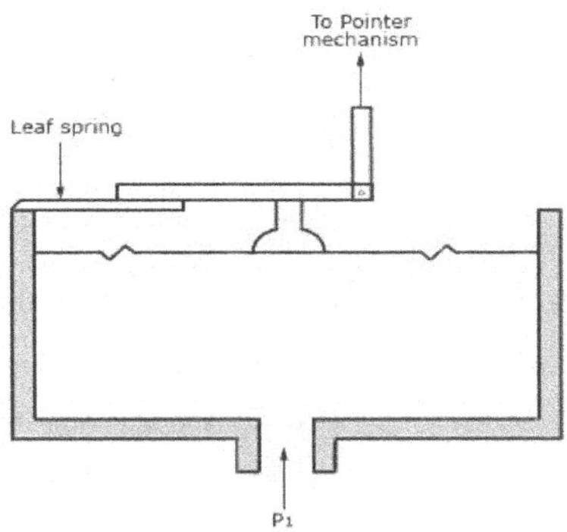

Diagram 3.35: Flat Diaphragm Transducer/Gauge

(iii) Flat Diaphragm Transducer : In such type of pressure measuring device , it consists of a diaphragm made up of metal which behave like a spring or elastic membrane , the edges of diaphragm are fixed inside the device as shown in diagram due to their shape and position, diaphragms have high mechanical resistance and are less shock sensitive . whenever a pressure force is applied , the elastic deformation of Diaphragm is achieved and deflection in upward direction takes place which will transmit it to rack and pinion arrangement upon which a moving pointer is installed which can be moved over a graduated scale indicating the proportional pressure measured value. They are commonly made up of steel or high-alloy steels of stainless steel. For measuring fluids pressure, special corrosion-resistant materials like **tantalum, titanium** , Hastelloy are applied.

(iv) Capsules Diaphragm Transducer: But in order to maintain the elastic limit condition of Diaphragm spring must be controlled if not damage of Diaphragm is possible due its fragility. Hence capsules of such kind Diaphragm is manufactured and arrange all the capsules according to the diagram. Whenever pressure is entered inside the capsules, the deflection will be the algebraic sum of individuals. The measuring system is same as that of flat diaphragm transducer. Following are the variety of diaphragm.

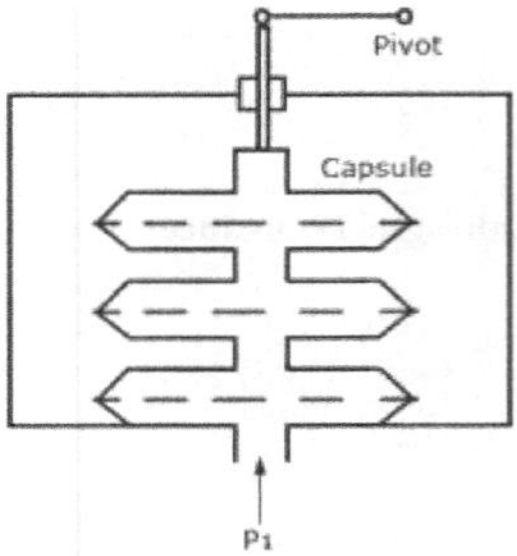

Diaphragm spring	Diagram
Flat Diaphragm	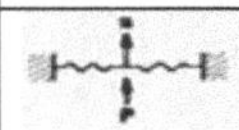
Capsules Diaphragm	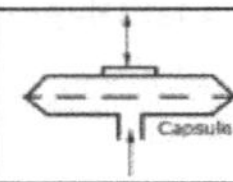
Corrugated diaphragm	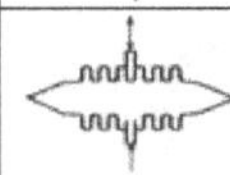

Diagram 3.36: Capsules Diaphragm Transducer

(v) Bellows: Bellows are also generally applied for pressure measurement of range between **0.2 to 1 kg/cm²**. It consists of a cylindrical shaped container having a corrugations at their periphery inside which supporting compression/expansion spring is provided depending upon the application as shown in figure. When pressure force is applied the length variation of bellows takes place means bellows may expands or contract. Other ends of the bellows connected to the mechanical linkage having a moving pointer if expansion or contraction created linkage activated and pointer moves on graduated scale, readinng on scale indicating proportional pressre of mechanism. bellows are generally manufacturing by **Phospher bronze, brass, stainless steel etc.**

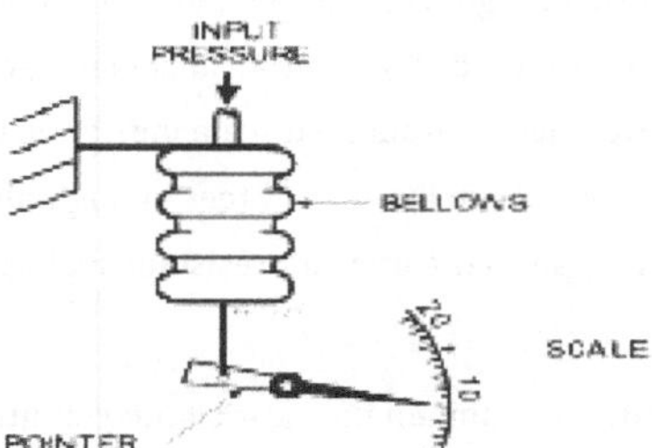

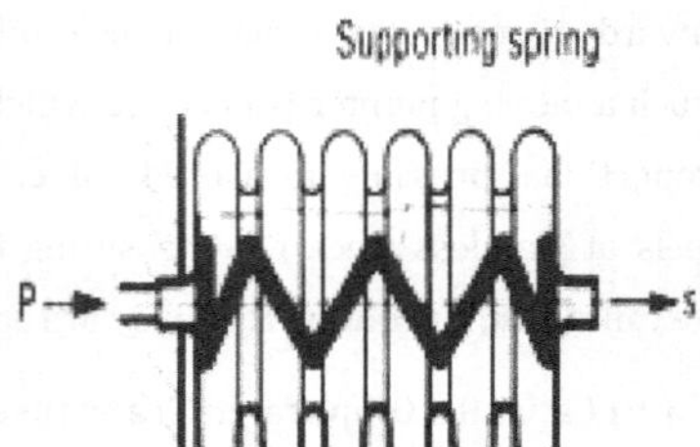

Diagram – 3.37 Types of Bellow

There may be various possible type of bellows depending upon the arrangement of supporting springs, which can be understand by following diagram.

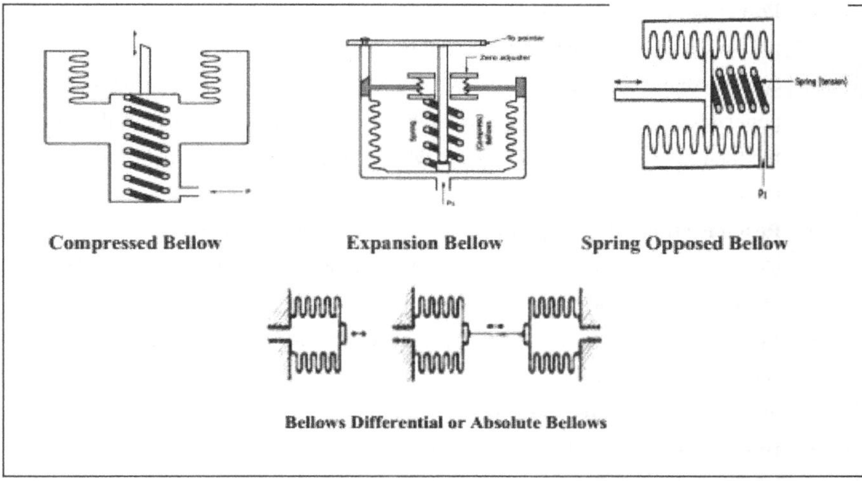

Type of bellows

3.13.3. Electrical Pressure Transducers

The function of electrical pressure transducer is same as that of other pressuring measuring devices but electrical pressure transducers is new advance form of such kind of transducer. In previous version of pressure transducer it can be sensed and measured it if measurement is possible. With the help of EPT the measurement process is simplified. It can sense the pressure and change in to the physical variable after that processed electrically

Following are some important terms related to pressure transducers:

- **Output signal :** The output signal is always proportional to the supply voltage.
- **Temperature error :** Every pressure transducer has a specific output signal in association with temperature error .
- **Measuring range/measuring span :** The measuring range is the span between the start and the end of the measurements.
- **Maximum pressure range :** The maximum pressure defines the pressures at which a pressure measuring converter can be exposed without it suffering permanent damage, i.e. zero-point offset.
- **Burst pressure:** Burst pressure indicates the pressure below or up to if it exceeds can lead to the total destruction of the pressure measuring converter .

- **Power supply :** An power supply is required to operate pressure transducers and pressure measuring converters. The most common is 24 V direct current.
- **Output signal :** Current and voltage signals have become established as the most popular for output signals. Exactly which output signal is chosen depends mainly on the indicating and control devices to which the pressure measuring converters send their output signal.
- **Response time :** The response time defines the period of time needed by the output signal when there is a sudden change of pressure. It is usually quoted with in a range of 10% to 90% of the measuring range.
- **Accuracy :** The accuracy is stated as a percent of the full scale deflection and usually specifies the maximum deviation of the measured characteristic curve from the ideal line between the start and end of the measurement.
- **Hysteresis :** Hysteresis is the difference in the output signals at a given pressure between when the applied pressure is rising and falling .

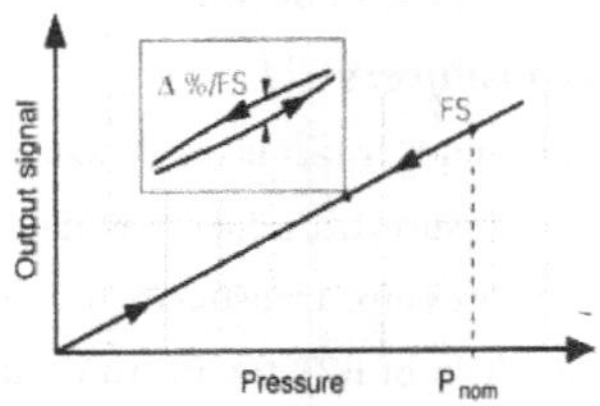

Hysteresis

Following are the some important type of electrical pressure transducers which is very important from subject point of view

3.14. Linear Voltage Differential Transducer (LVDT)

It is also an important type of electrical pressure transducers in which a linear displacement or motion is converted into electrical signal LVDT works on the principle of mutual induction and measurement indicating the proportional displacement. The constructional features of LVDT is given below

Elements of LVDT

- Displacement arm connected to linear motion device which is free for to and fro motion inside the former
- Soft iron magnetic core former

- Primary winding (p)
- Secondary winding (s_1 & s_2)
- AC voltage supply
- Output voltage terminal

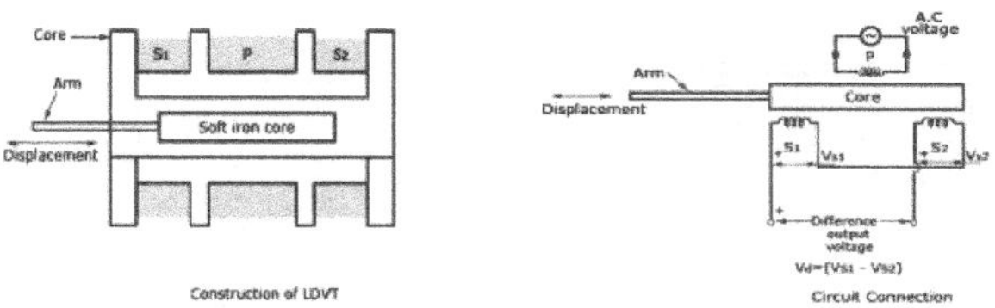

LVDT constructional featured **LVDT connection**

Diagram – 3.38 Linear Voltage Differential Transducer

Operational Features of LVDT

It consists of a one primary winding (P) and two secondary windings S_1 and S_2which is connected in series. All the three elements are bounded on cylindrical former, each & every windings have a same no of equal turns .When AC voltage is supplied to the primary winding, it generate two voltages V_1 and V_2 at the two secondary windings S_1 and S_2 respectively ,The AC excitation voltage is 5 to 12 V and the operating frequency is about 50 to 400 HZ. hence output voltage will be ($V_1 - V_2$) , the intensity of output voltage will be depends upon the soft iron core .

- **Null Position** – It is a position when soft iron core is at the center of the former so therefore linkage of magnetic flux is equal on both secondary windings hence voltage induced will be equal. Measured output will be equal to $V_1 - V_2 = 0$
- **Right from Null Position** – In this position, linkage value of magnetic flux is in secondary winding (s_2) is higher than that of primary (s_1). Hence ($v_2 - v_1$) > 0 will be in phase in V_2 .
- **Right from Null Position** – In this position, linkage value of magnetic flux is in secondary winding (s_2) is smaller than that of primary (s_1). Hence ($v_1 - v_2$) > 0 will be in phase in V_1 .

The position of the moving arms indicating the variation of output voltage which will be proportional to the displacement, A linear displacement typically has units of **millimeters (mm)** or **inches (in.)** and a negative or positive direction associated with it. it can also be applied for measuring force, weight, pressure or velocity of moving mechanism.

3.15. Rotary Variable-Differential Transformer

The function of Rotary Variable-Differential Transformer is same as that of LVDT except of motion of arm. The constructional featured are given as in diagram.

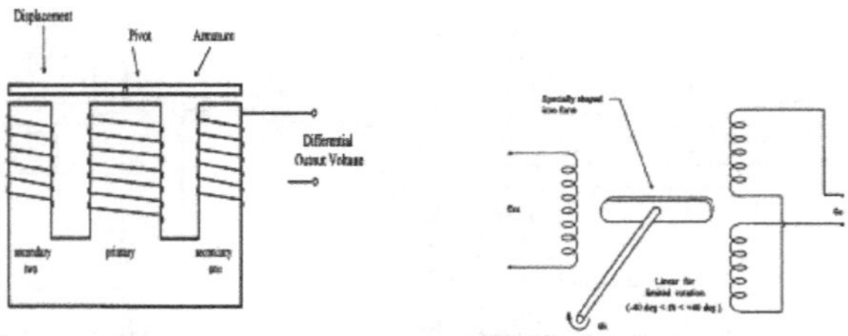

Diagram - 3.39 Rotary Variable-Differential Transformer

Operation of RVDT

The primary winding is wound on the center leg of an iron core, and the secondary windings are wound on the outer legs of the core. The armature is rotated by an externally applied force about a pivot point above the center leg of the core. When the armature is displaced from its reference or balance position, the reluctance of the magnetic circuit through one secondary coil decreases, simultaneously increasing the reluctance through the other coil. The induced EMFs in the secondary windings, which are equal in the reference position of the armature, are now different in magnitude and phase as a result of the applied displacement. The induced EMFs in the secondary coils are made to oppose each other and the transformer operates in the same manner as an LVDT. The rotating variable transformers may be sensitive to vibrations. If a dc output is required, a demodulator network can be used, as in the case of LVDT.

3.16. Piezoelectric Pressure Transducer

Piezoelectric pressure transducer is very important device used to measuring the pressure intensity/force generated by the mechanical mechanism or other process .

The governing principle of such type of transducer is that whenever any kind of force/ pressure is applied on special kind of natural crystal then crystal generates electrical potential on the surfaces. The natural crystal is recognize as Quartz cryastal. The generation of charge represent the rate that will be proportional to the rate of change of force/pressure applied as input.

Electrical Potential α Applied Pressure

The piezoelectric transducer is as shown in figure in which a quartz crystal /quartz plate is kept between a solid base and the force summing member. If a force is applied through the pressure inlet, then same force will be acted on to the crystal and hence a potential difference will be generated on the crystal due to its natural property. The voltage produced will be proportional to the magnitude of the applied force/pressure. This device is best suitable for dynamic measurement.

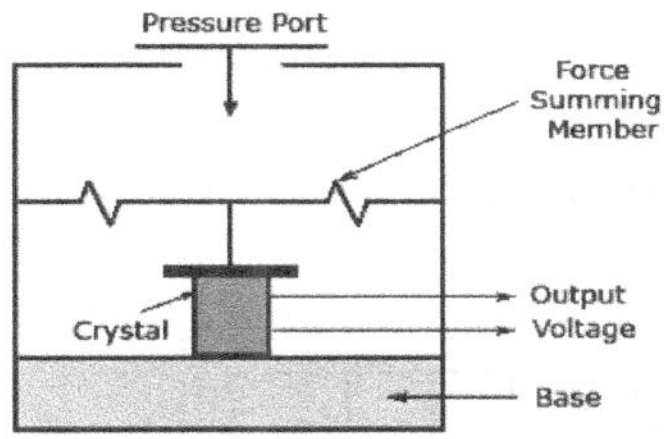

Diagram 3.40: Piezoelectric Pressure Transducer

Excepts the quartz some others materials also representing the Piezo-electric properties like tourmaline, barium titanate, and Lead zirconate, lithium niobate (LiNbO3), lithium tantalate (LiTaO3) , Rochelle salt (KNaC4H4O6.H2O).

(i) Capacitive pressure Transducer: It consists of a pair of parallel plates with the middle plate moving with pressure and producing a differential capacitor system. If the value of applied pressure $\mathbf{P_1}$&$\mathbf{P_2}$ are equal then pressure gauge shows zero deflection but When a differential pressure exists, the thin steel diaphragm moves towards the low pressure side and the output voltage $\mathbf{e_0}$measured as the difference of voltage $\mathbf{e_1}$ and $\mathbf{e_2}$ across the two capacitors formed with this movable plate is given by the equation

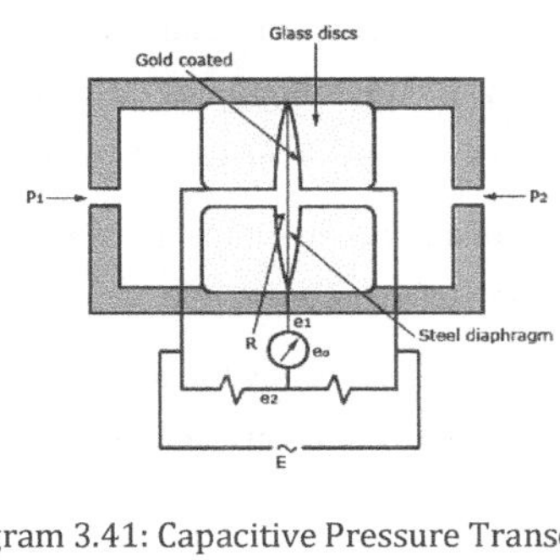

Diagram 3.41: Capacitive Pressure Transducer

$$e_0 = e_1 - e_2 = \mathbf{E\,L / D}$$

L – Displacement of the diaphragm

D – Diameter

The thickness of the diaphragm is carefully controlled, the thickness determines the pressure range of the sensor. When the sensor is made of silicon the plates of the capacitor are heavily doped regions within the substrate. A ceramic sensor usually has gold plates screen printed on the substrate.

(ii) Capacitance Transducer: Such kind of pressure transducer is based on the phenomenon of dielectric constant. solids ,liquids and gases have a definite value of dielectric constant but when it comes in any kind of pressure the value of dielectric constant has been changed .

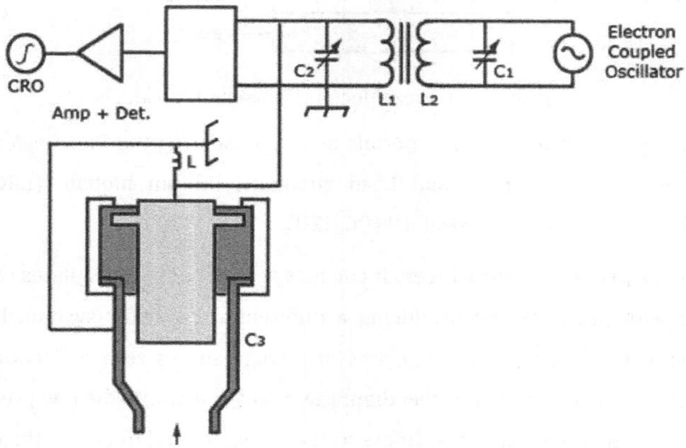

Diagram 3.42: Capacitance Transducer

The figure below shows an arrangement of a cylindrical capacitor applied for high pressure measurements. The measurement of this capacitance is done by a resonance circuit, The oscillogram giving the variation of the output voltage with capacitance.

(iii) Resistive Pressure transducer: Resistive transducer is used to measuring the pressure, the following element assists the process.

- At very initial state pressure is sensed by the primary sensor, primary sensor may be **bourdon tube** pressure gauge, bellows or manometer.

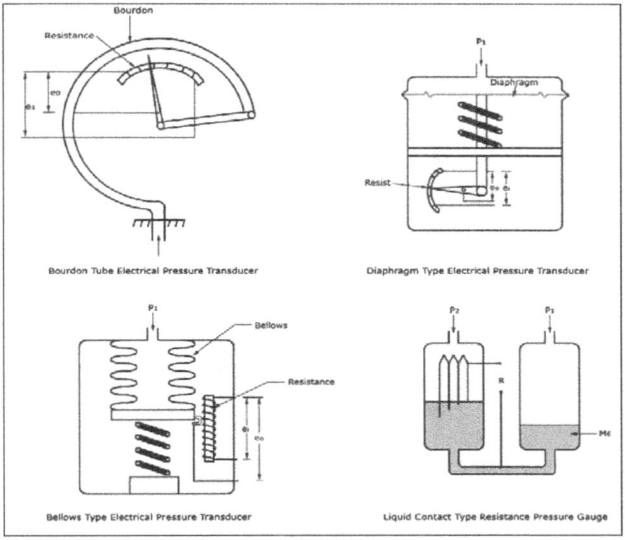

Diagram 3.43: Primary Sensor for Resistive Pressure Transducer

- The primary sensor output is then converted into electrical signals, the intensity of resistance variation helps in measuring the pressure.
- The intensity of pressure can be measured with the help of current meter.

(iv) Resistive Pressure Transducer (Strain Gauge) : A resistive pressure transducer has strain gauges bonded to the surface of the non-media contacted side of the diaphragm so that any change in pressure will cause a momentary deformation of the elastic material resulting in a change of the resistance of the strain gauge that is converted into a useable electrical signal.

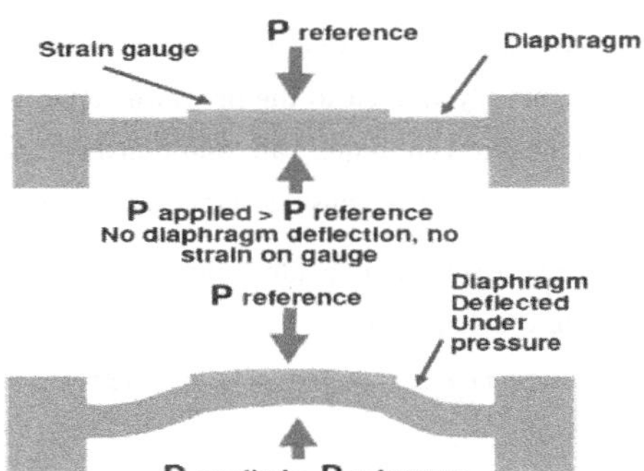

Diagram 3.44: Resistive Pressure Transducer (Strain Gauge)

(v) Thin Film Strain Gauge: Strain gauge is built upon the backside of the pressure sensing diaphragm. It uses what's called a **"sputtering"** process to place an atomic-level bond onto the surface of the diaphragm. This produces a highly stable output.

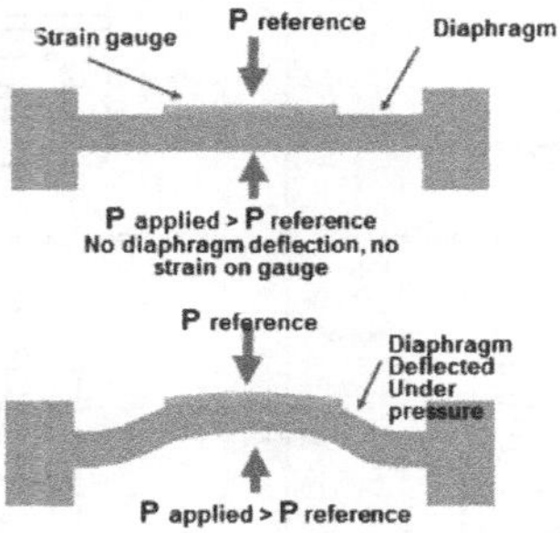

Diagram 3.45: Thin Film Strain Gauge

- Sensing element is formed by machining the inside of a stainless-steel button until the remaining steel is very thin
- Strain gauge material is deposited onto steel diaphragm forming a Wheatstone-bridge resistor network
- Strain transfers to the resistors as pressure is applied to the back of the element unbalancing bridge and thus causing a resistance change that is multiplied to provide a high level output (volts, mill volts, or current)

Advantages

- Over pressure protection
- Better effective resolution
- Excellent resistance to shock, vibration, and dynamic pressure changes.
- Pressure sensor is usually welded to the pressure connection = hermetically sealed measurement cell that does not require any additional sealing materials

(vi) Capacitive Pressure Sensors: It consists of following elements

- Two capacitive pressure sensor are made in the form of plates, one is permanently fixe another plate is work as diaphragm.
- Around the circumference it is clamped and applied pressure for bending .as shown in figure

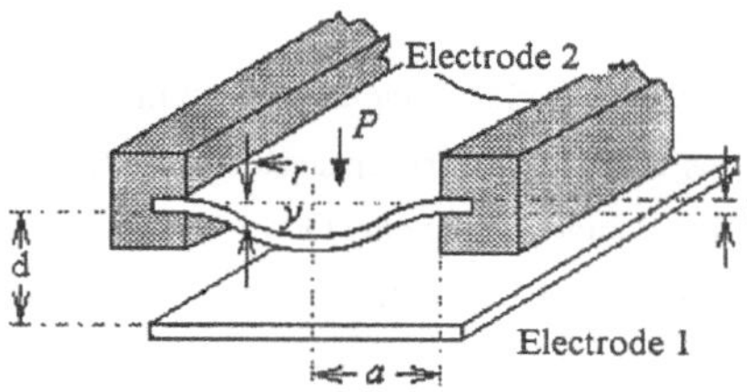

Diagram 3.46: Capacitive Pressure Sensors

$$y = 3\left(1 - v^2\right)\left(a^2 - r^2\right)\ P/16\ Et^3 \qquad \text{.................................. 17}$$

where ,

a = The radius of diaphragm

t = The thickness of diaphragm

E = Young's modulus

v = Poisson's ratio

Whenever diaphragm deforms, the distance between fixed plate and diaphragm reduces so therefore capacitance ΔC increases that can be calculated by following relationship.

$$\Delta C/C = \left(1 - v^2\right) a^4 P/16\ Et^3 \qquad \text{.................................. 18}$$

3.17. Bonded Strain Gauge

The bonded strain gauge is very popular technology applied for higher pressure measurement . Bonded strain gauge transducers may be classified

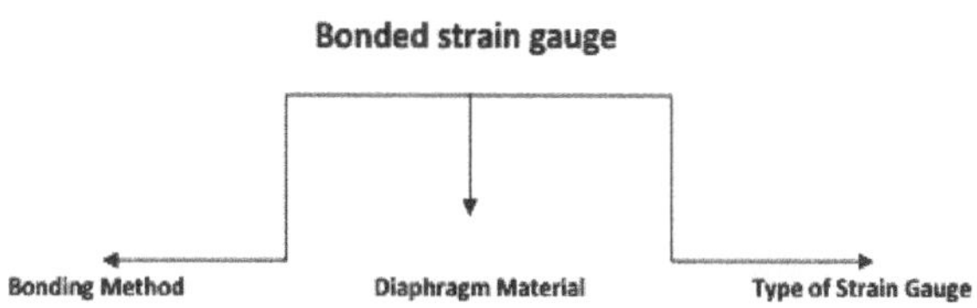

- A diaphragm is attached / fitted to the components of pressure is being measured and bonding strain gauges to one side. A pressure input causes the diaphragm to bend, stressing the strain gauges, creating an electrical output. The strain gauges are made of many different materials, metals and semiconductors being the most common. Different strain gauges are characterized by a number called the gauge factor. The gauge factor is the proportion of resistance change over applied, the gauge factor for metals can be anywhere from 2 to 4.5. A semiconductors gauge factor will vary from 100 to 175.

- Four strain gauges are bonded to each diaphragm in the areas of greatest stress, two of the strain gauges in an area of compression, the other two in an area of tension. Electrically, the strain gauges are connected in a Wheatstone bridge configuration. As the diaphragm bends the arms of the bridge are unbalanced causing the output voltage to increase.

(i) Bonded semiconductor strain gauge : Bonded semiconductor strain gauge , A silicon bar may be bonded to a diaphragm to form a sensor with relatively high output. Making the diaphragm from a chemically inert material allows this sensor to interface with a wide variety of media .

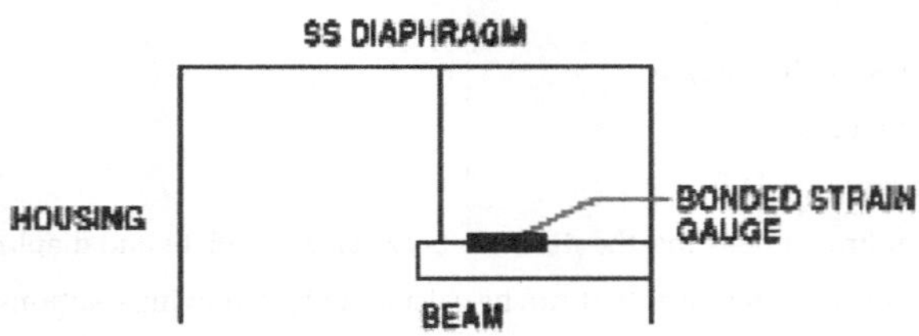

Diagram 3.47: Bonded Semiconductor Strain Gauge

(ii) Strain Gauge Pressure Sensors: Strain gauge sensors originally used a metal diaphragm with strain gauges bonded to it. A strain gauge measures the strain in a material subjected to applied stress. Consider a strip of metallic material with electrical resistance given by:

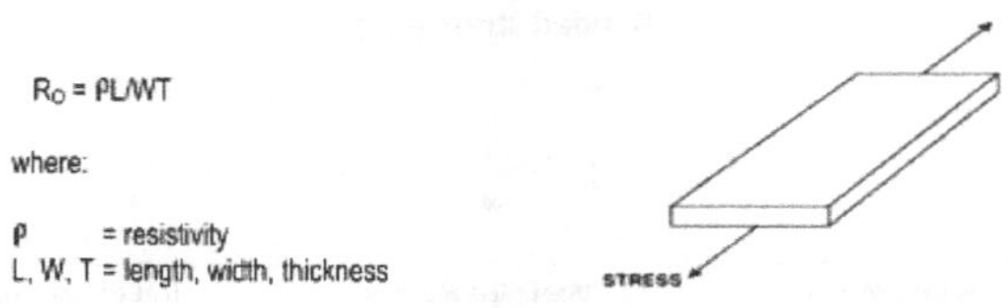

Diagram – **3.48** Strain Gauge Pressure Sensors

Metallic strain gauges depend only on dimensional changes to produce a change in resistance. A stress applied to the strip causes it to become slightly longer, narrower, and thinner , resulting in a resistance of ,

$$R = \rho(L + \Delta L) / (W - \Delta W)(T - \Delta T), \text{ or}$$
$$R \cong R_O(1 + 3\Delta) \quad \text{..........} \quad \mathbf{19}$$

3.18. Pressure Tube with Bonded Strain Gage

The principle of a pressure tube with **bonded** strain gage may be under stood by referring to

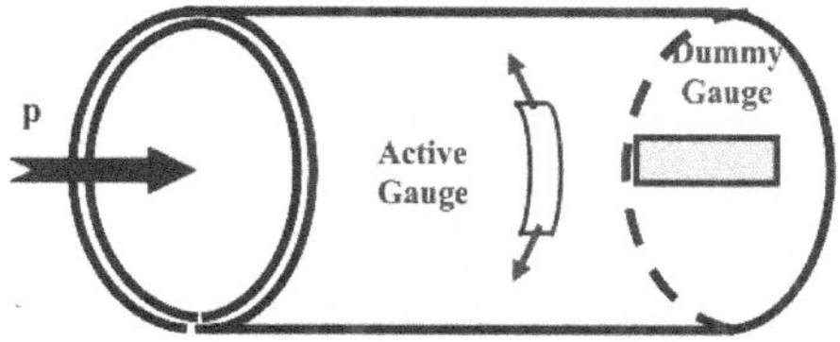

Diagram 3.49: Bonded Strain Gage

Active gauge : Active gauge responds to hoop strain due to pressure

Dummy gauge: Dummy gauge helps in temperature compensation

Strain gage theory: Strain gage consists of a resistance element whose resistance is a function of its size.

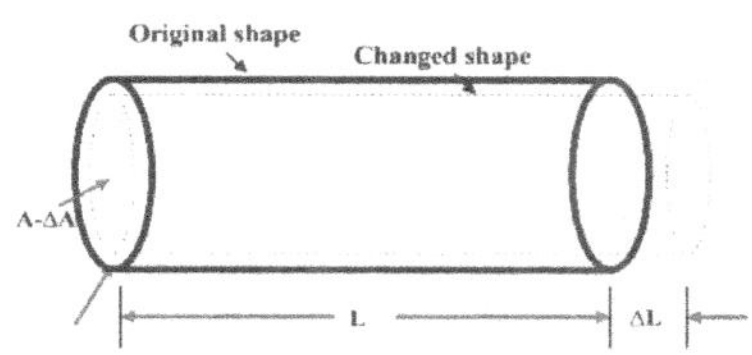

Diagram 3.50: Deformation of a Wire Element

Electrical resistance of a wire of Length L, area of cross section A and made of a material of specific resistance ρ is given by

$$R = \frac{\rho L}{A}$$

Change in dimension

$$\frac{dR}{R} = \frac{d\rho}{\rho} + \frac{dL}{L} - \frac{dA}{A} \quad \text{.......... 20}$$

The ratio of fractional resistance change of the wire to the longitudinal strain is called the gage factor (GF) or sensitivity of a strain gage and is thus given by

3.19. Total Static Pressure Measurement in Moving Fluids

(i) Static pressure measurement : When the fluid in static condition it realize static pressure which is distributed in equal intensity but

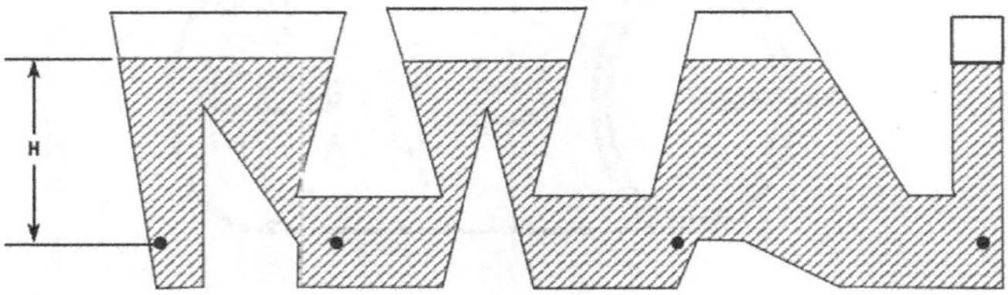

Diagram 3.51: Fluid System

for example consider the fluid system diagram, in which static fluid is distributed over the given area , pressure will equal in horizontal direction and independent of the shape of the container while only pressure variation is achieved between vertical portion . Pressure is increases with depth due to its weight. Consider another container

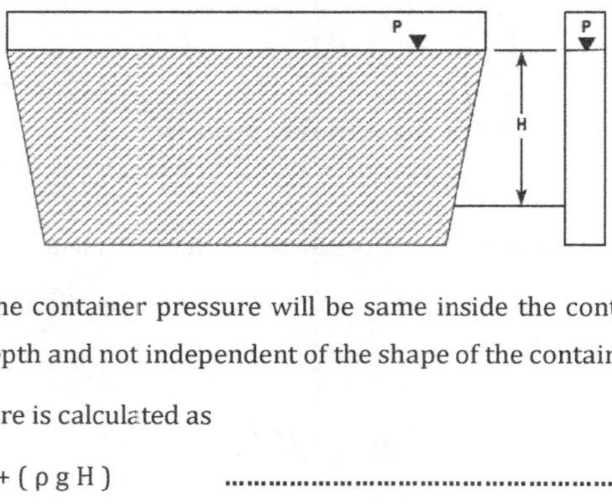

At any depth of the container pressure will be same inside the container; the pressure is dependent only on depth and not independent of the shape of the container.

Thus static pressure is calculated as

$$= P + (\rho g H) \quad \text{...} \quad \mathbf{21}$$

Where ,

P_{abs} = Absolute Pressure at depth H

P = atmospheric pressure

ρ = the density of the fluid

g = the acceleration due to gravity

3.20. Dynamic Pressure Measurement

In static pressure measurement, pressure can be calculated with the help of mathematical relationship but in case of dynamic pressure measurement it is more complicated and difficult task.

Dynamic pressure measurement has two terms

- Initially we have to measure the static pressure of fluid system.
- In second terms the velocity of the flowing fluid has to be considered.
- The third terms is total pressure which is equal to static pressure plus the dynamic pressure

Pressure Measurement In Steady-State Dynamic Condition: In this state the pressure measurement is performed in following steps.

- Static pressures Must be measured by tapping into the fluid stream perpendicular to the fluid flow.
- A static pressure tap is located in the duct wall at point A.

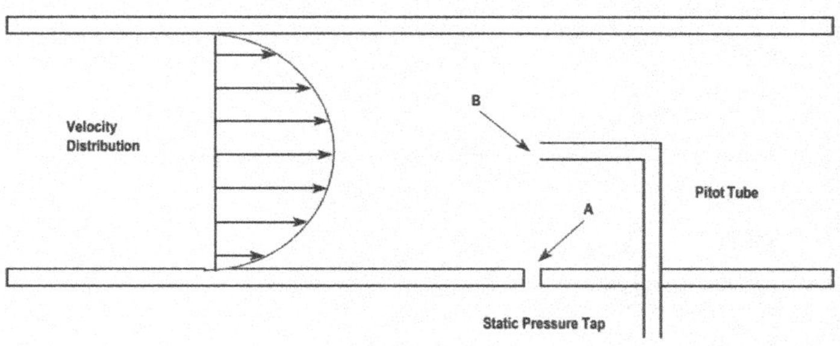

- The Pitot tube inserted besides static pressure tap in the stream line, at the point B we can measured the total pressure of the system called stagnation pressure which has frictionless zero velocity . This is a summation of static pressure plus the dynamic pressure.
- To obtain accurate pressure measurements the location of the pressure taps and probes must be correct, the pressure tap holes and probes must have minimum friction, and free from any obstructions for smooth flow.
- The types of probes may be of following three types. The diagrammatic structured is given below.

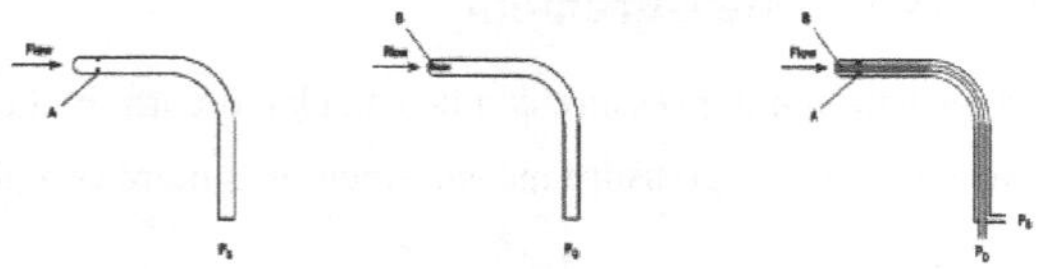

Static Pressure Probe

Total Pressure Pitot tube Static & Total Pressure Pitot Tube

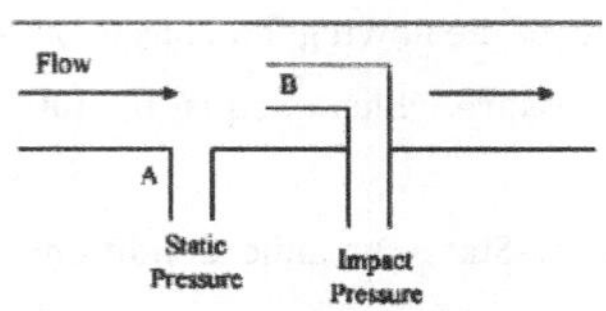

- Static pressure is the pressure of fluids that are stationary.
- Dynamic pressure is the pressure exerted by a fluid when it impacts on a surface or an object due to its motion or flow. the dynamic pressure will be **(P_B - P_A).**
- Impact pressure is the sum of the **Static** and **Dynamic Pressures** on a surface.

$$P_I = \rho V_O^2/2$$

where:

ρ = fluid density
V_O = fluid velocity

Bernoulli's theorem states that for horizontal flow the following relation holds:

$$P_S = P_O + P_I$$

where:

P_S = stagnation (or total) pressure
P_O = static pressure

This relationship is useful for determining the flow velocity in a variety of applications
Rearranging the equation gives:

$$V_O = \sqrt{2(P_S - P_O)/\rho}$$

3.21. Flow Measurement

Flow rate measurement is quite very essential parts of any industrial application, these may be related to so many kind of liquid like water, oil, gas etc . Flow rate measurement is not only applied for safe & effective operation but also maintaining the quality of industrial processes. These measurements determine the requirement of fluid rate needed exactly passing through a point at any instance. Flow meters measured the flow rate, how much capital is needed. Following are some important type of flow rate measuring instruments.

(i) Differential Pressure Flow Meters : The basic concept of Differential Pressure Flow Meters is that whenever we put any kind of restriction in fluid flow area , there is a creation of pressure drop due to the effect of change in flowing fluid velocity . The overall structure of the of flow can be examined with the help of Bernoulli equation and continuity equation. Hence we can measure the flow rate and pressure drop.

According to Bernoulli's equation

$$P_1 + \frac{\rho V_1^{\ 2}}{2} + \gamma Z_1 = P_2 + \frac{\rho V_2^{\ 2}}{2} + \gamma Z_2 \quad \text{.................................... 22}$$

Where ,

p_1 = flow velocity at section -1

V_1 = flow velocity at section -1

Z_1 = datum head at section -1

ρ = Density of fluid

Similarly for section - 2

P_2 = Flow velocity at section -1

V_2 = Flow velocity at section -1

Z_2 = Datum head at section -1

consider if pipe is horizontal then $z_1 = z_2$

$$P_1 + \frac{\rho V_1^{\ 2}}{2} = P_2 + \frac{\rho V_2^{\ 2}}{2} \quad \text{...23}$$

According to continuity equation

$$Q = A_1 V_1 = A_2 V_2 \quad \text{.. 24}$$

A_1 = Area at section **-1 -1**

V_1 = Flow velocity at section **-1-1** And

A_2 = Area at section **-2-2**

V_2= Flow velocity at section **-2-2**

Combining eqation 1 & 2

$$Q = A_2 \sqrt{\frac{2(P_1 - P_2)}{\rho\left[1 - \left(\frac{A_2}{A_1}\right)^2\right]}} \quad \text{.. 25}$$

Thus Eqation (3) representing the flow rate in a pipe

(ii) Venturi-meter: The venture-meter has a converging cone shape inlet section, a cylindrical throat and a diverging cone outlet section. The diagram-3.56 below shows the venturi meter with uniform cylindrical section before converging entrance, a throat and divergent outlet.

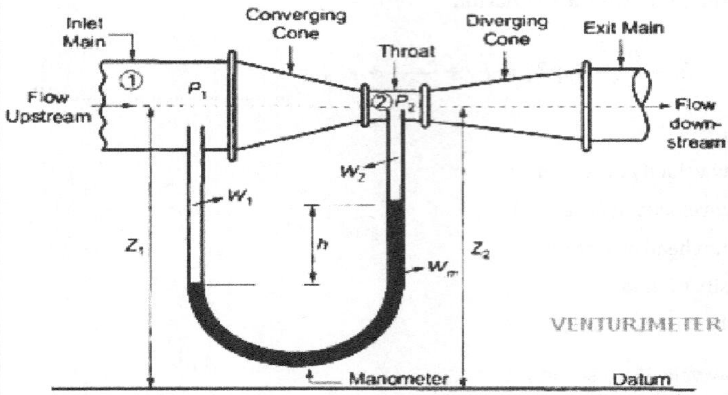

Diagram 3.56: Venturi-meter

The convergent cone angle – **21⁰**

The Divergent cone angle – between **15 ⁰ to 7⁰**

Components of Venture-meter

- Inlet section
- Convergent cone section
- Divergent cone section
- Outlet section

Working

In the given diagram, the inlet section of device is fitted with the pipe, after that the cross-section area is gradually decreasing, due to the decreased cross section area pressure energy of flowing fluid is converted into the kinetic energy or velocity, the minimum cross section area of the device at which pressure energy is fully converted into the kinetic energy or having lowest value of pressure energy is called throat. When the fluid enters into the divergent section again kinetic energy or velocity decreased and pressure energy is in increasing order.

Venturi tubes are very common differential pressure flow meters because of their simplicity and high pressure and energy recovery. Venturi tubes are very accurate over wide flow ranges and the gradual contraction and expansion reduces the drag and allows for a low pressure loss.

If V_1 and V_2 are the average upstream and downstream velocities and ρ is the density of the fluid, then using Bernoulli's equation we get,

$$\alpha_2 V_2^2 - \alpha_1 V_1^2 = \frac{2g\,(P_a - P_b)}{\rho} \qquad \text{.......................... 26}$$

where $\alpha_1$and α_2 are kinetic energy correction factors at two pressure tap positions. Assuming density of fluid to be constant, the equation of continuity can be written as:

$$V_1 = \left(\frac{D_2}{D_1}\right)^2 V_2 \qquad \text{....................................... 27}$$

where D1 and D2 are diameter of pipe and throat in meters respectively. Eliminating V1from equation (1) and equation (2) we get,-

$$V_2 = \frac{1}{\sqrt{\alpha_2 - \alpha_1 \beta^4}} \sqrt{\frac{2(P_1 - P_2)}{\rho}} \qquad \text{...................... 28}$$

where β is the ratio of the diameter of throat to that of diameter of pipe. If we assume a small friction lose between two pressure taps, the above equation (3) can be corrected by introducing empirical factor C_v and written as,

$$V_2 = \frac{C_V}{\sqrt{1 - \beta^4}} \sqrt{\frac{2(P_1 - P_2)}{\rho}} \qquad \text{................................ 29}$$

The small effect of the kinetic energy factors α1and α2are also taken into account in the definition of Cv. Volumetric flow rate Qa can be calculated as:

$$Q_a = V_2 S_2 = \frac{C_V S_2}{\sqrt{1 - \beta^4}} \sqrt{\frac{2(P_1 - P_2)}{\rho}} \qquad \text{........................... 30}$$

Where,S_2 is the cross sectional area of throat in m^2. Substituting $(P_1 - P_2) = \rho g H$ in above equation (5) we get,

$$Q_a = V_2 S_2 = \frac{C_V S_2}{\sqrt{1-\beta^4}} \sqrt{2g\Delta H} \quad \text{.......................... 31}$$

where ΔH is the manometric height difference

ΔH = (Specific gravity of mano-metric fluid –specific gravity of mano-metric fluid of water)

Orifice Meter: Orifice meter is an important device for calculating the discharge / flow rate or flow velocity in pipe. Actually it is a channel introduced between the pipe line for interrupting the flow due to that we can achieved pressure drop between two section as shown in figure, in the figure we have two section, section having a diameter equal to the pipe (D) on the other hand another section having a minimum diameter (D_e) which is a minimum cross sectional area of the jet is known as the vena contracta. And installed manometer for measuring the pressure drop.

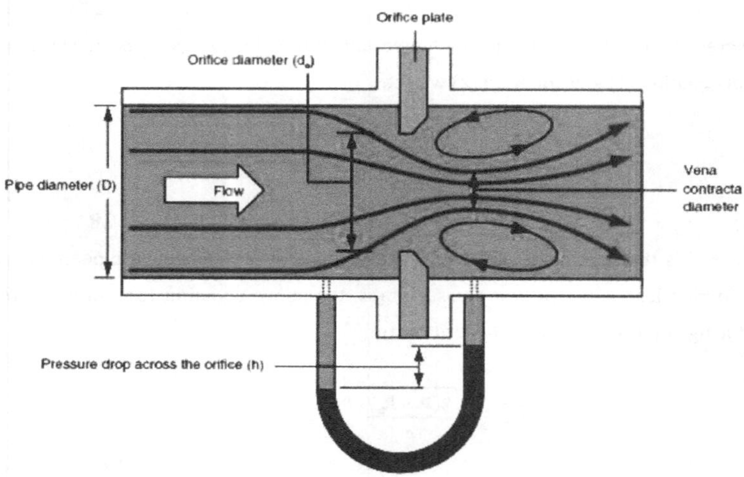

Diagram 3.57: Orifice Meter

Whenever fluid is passing through the pipe, the pressure energy is suddenly increased slightly and drop up to minimum cross section area called vena contracta at which we have achieved too much pressure drop. during this energy conversion portion pressure energy is decreased at the same time velocity is increasing after that the value of pressure gradually increases up to a maximum value then flow will be normal. All of the pressure loss is not recovered because of friction and turbulence losses in the stream. hence If all other factors remain constant, then the differential is proportional to the square of the rate of flow. The Orifice plates in the Orifice meter, is made up of stainless steel of varying grades, phosper bronze, nickel and monel metal .

$$Q = A_2 \sqrt{\frac{2(P_1 - P_2)}{\rho\left[1 - \left(\frac{A_2}{A_1}\right)^2\right]}} \quad \text{.........................} \mathbf{32}$$

Where ,

D = Diameters of Pipe

De = Diameters of orifice opening

A_1 = Area of Pipe

Ae = Area of orifice

P1 - P2 = Pressure Difference

Q = Discharge

Tap Location

Following tapes are generally located on device for reliable measurement.

- **Flange Taps:** These taps are located one inch from the upstream face of the orifice plate and one inch from the downstream face with a **+ 1 / 64 to +1 / 32** tolerance.
- **Pipe Taps:** These taps are located 2½ pipe diameters upstream and 8 pipe diameters downstream.
- **Vena-Contracta Taps**: These taps are located one pipe diameter upstream and at the point of minimum pressure downstream, this point is called the vena-contracta. This point, however, varies with the Beta ratio.
- **Corner Taps:** These taps are located immediately adjacent to the plate faces, upstream and downstream.
- **Orifice Flanges:** The Primary function of Orifice Flanges is to hold an orifice plate in a pipe with help of the orifice flange union.

Advantage

- It is simple in construction, reliable and inexpensive device.
- Flow rate measurement can be performed accurately without actual calibration.
- It has occupies less space.
- It may be applied for flow rates measurement in large size pipes.

Disadvantage

- At downstream side the pressure recovery performance is poor .
- The orifice plate may be affected by corrosion hence inaccurate reading may be possible.
- The orifice plate has low physical strength.
- The coefficient of discharge is low.

Applications

- The orifice plate is standardized device applied for flow rates measurement of pure fluid like concentric orifice.
- In general the flow rate measurement is performed for liquid like water but whenever necessary flow rate measurement may be possible in which fluid containing suspended particle like solids, oil mixed with water or wet steam .Eccentric or Segmental orifice plates may be applied.

Design and construction of orifice plate

- The upstream edge of the orifice must be sharp and square.
- Minimum plate thickness based on pipe I.D.
- Orifice bore size must be standardized.
- The orifice to pipe diameter ratio **d/D** called Beta ratio, must not exceed from recommended limits.

3.22. Type of Orifice Plate

Following type of Orifice Plates is to be used according to the standardization.

(i) Concentric Orifice plate: The thin plate, concentric orifice is the most commonly used orifice plate. In the design and use of orifice plates, several basic factors must be followed to assure accurate, reliable measurement. The plate should not depart from flatness along any diameter by more than 0.25mm per mm or 0.01 inch per inch of the dam height **(D-d) / 2**. To ensure conformance with recommended practices, the beta ratio must not exceed recommended limits.

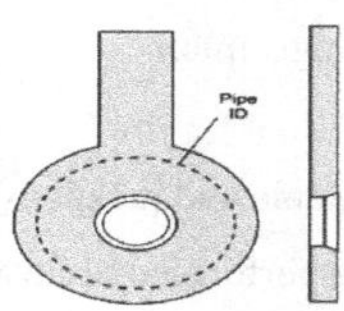

Concentric Orifice Plate

(ii) Eccentric Orifice Plates : The eccentric plate has a round opening (bore) tangent to the inside wall of the pipe. This type of plate is most commonly used to measure fluids which carry a small amount of non-abrasive solids, or gases with small amounts of liquid, since with the opening at the bottom of the pipe, the solids and liquids will carry through, rather than collect at the orifice plate.

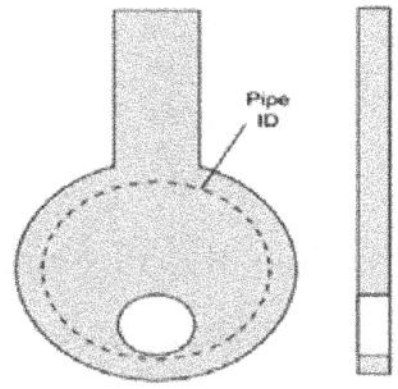

Eccentric Orifice Plates

(iii) Segmental Orifice plate: Plate The opening in a segmental orifice plate is comparable to a partially opened gate valve. This plate is generally used for measuring liquids or gases which carry non-abrasive impurities, including light slurries or exceptionally dirty gases. The predictable accuracy of both eccentric and segmental plates is not as high or reliable as the concentric plate.

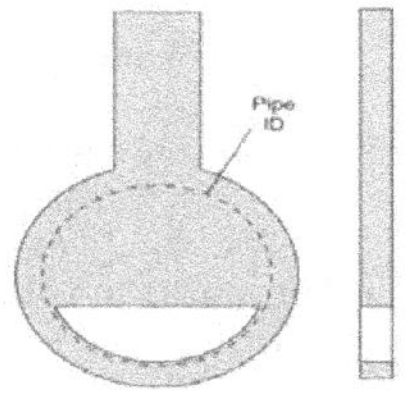

Segmental Orifice Plate

(iv) Quadrant Edge Plate: The quarter-circle or quadrant edge orifice is used for high viscosity fluids with low Reynolds numbers. The orifice incorporates a rounded edge of definite radius which is a particular function of the orifice diameter.

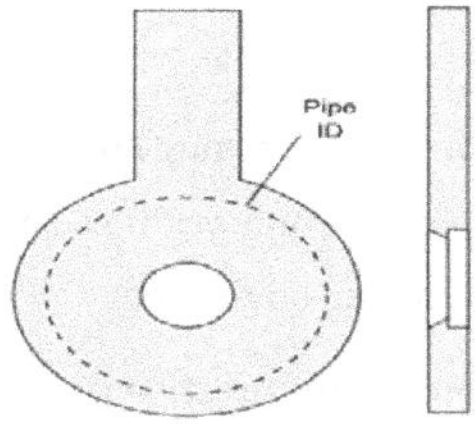

(v) Conic Edge Plate: The conic edge plate has a 45° bevel facing upstream into the flowing stream. It is useful for measuring fluids that have even lower Reynolds numbers than the quadrant edge.

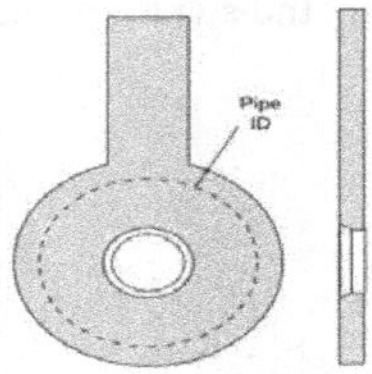

(iv) Rota meter: The working principle of orifice meter & Venturi-meter is based on the constant area of device and pressure drop that has been imposed by obstruction , and the pressure drop is depends upon the flow rate variation but in case of Rota-meter it works on the principle of constant pressure drop variable area meter. The constructional feature of Rota meter is given as

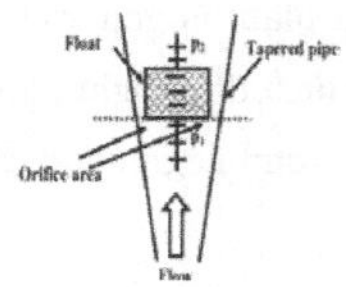

Rota Meter

- It consists of a transparent vertical glass tubes of variable area. Tapered from bottom to top.
- The fluid flows upward from bottom to top through a vertical conical tube whose diameter increases in the upward direction.
- A metallic float is inserted inside the tube which is free for to and fro motion.
- Whenever float moves ups and down, the annular area has been changed.
- When Fluid flow is takes place an upwards thrust is applied on float and it rises .

Achieve an equilibrium position means

(Upwards Thrust = Gravity Force)

OR

(Upward Pressure Force + Buoyancy Force) = Gravitational Force

Taking a respective reading of equilibrium position on glass tubes that will represent the flow rate. There are three forces are acting on the float

Gravitational Force Acting in Downward Direction (F_G) = $F_G = V_s \cdot \rho_s g$

Force of buoyancy Acting in upward Direction (F_b) = $F_b = V_s \rho_m g$

Resistance Force to flow (F_f) = $F_f = c_w A_s \rho_m / (v_2)$

Vs : Volume of the float

m_s : Mass of the float

ρ_s : Density of the float

ρ_m : Density of the measuring medium

c_w : Resistance coefficient

A_s : Cross-sectional area of the float at the reading edge

V : Flow velocity of the measuring medium

D_K : Inside diameter of the cone at the reading point

D_S : Diameter of the float at the reading edge

Volume flow rate:

$$q_v = \frac{\alpha}{\rho_m} D_s \sqrt{g \cdot m_s \cdot \rho_m \left(1 - \frac{\rho_m}{\rho_s}\right)}$$

Mass flow rate:

$$q_m = \alpha \cdot D_s \sqrt{g \cdot m_s \cdot \rho_m \left(1 - \frac{\rho_m}{\rho_s}\right)} \qquad \text{.......................... 33}$$

where **α** is

$$\alpha = \sqrt{\frac{1}{C_w}} \qquad \text{.. 34}$$

Type of Float

Following are important type of float available for Rota-meter, type of float depends upon the applied pressure, weight, shapes of float and material used.

Type of float	Name of float
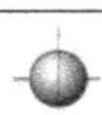	**Ball float**
	**Viscosity-immune float**
	**Viscosity-dependent float**
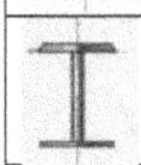	**Float for low pressure drop**

Advantage

- Rota meter is simple in construction with easy installation
- Flow rate can be measured directly on calibrated scale, there is no need of any other helping device.

(v) Pitot Tube : Pitot tube is widely used for velocity measurement in aircraft. Its basic principle can be understood from **Figure –1 & Figure –2** , If a blunt object is placed in the flow channel, the velocity of fluid at the point just before it, Will be zero. Then considering the fluid to be incompressible.

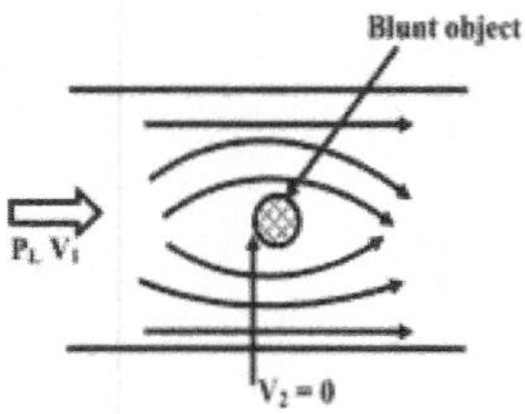

Figure – 1 Pitot Tube : Basic Principle

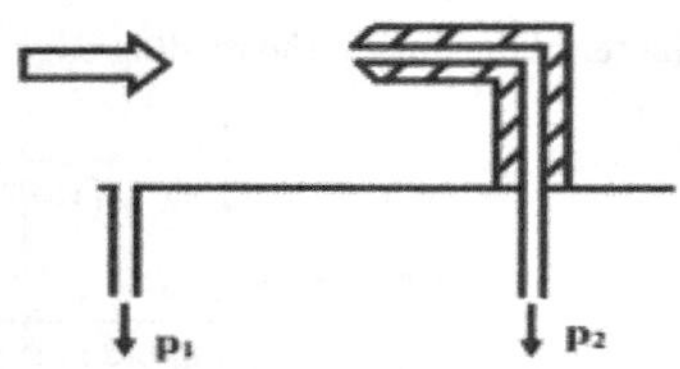

Figure – 2 Pitot Tube: Construction

$$\frac{p_1}{\gamma}+\frac{v_1^2}{2g}=\frac{p_2}{\gamma}+\frac{v_2^2}{2g} \qquad \text{.................................... 35}$$

Now $V_2 = 0$

$$\frac{v_1^2}{2g}=\frac{p_2-p_1}{\gamma}$$

$$v_1=\sqrt{\frac{2g}{\gamma}(p_2-p_1)} \qquad \text{.. 36}$$

Miscellaneous Measuring Devices

3.23. McLeod Gauge

McLeod Gauge is a vaccum gauge, it is nothing but it is a modification in a manometer by which we can measure the absolute pressure of gases with accuracy and also applicable for calibrating the other measuring gauges. The schematic diagram is as shown in figure. the operating principle of McLeod Gauge is based on the Boyle"s Law .

$$P_1V_1 = P_2V_2 \qquad \text{...................................... 37}$$

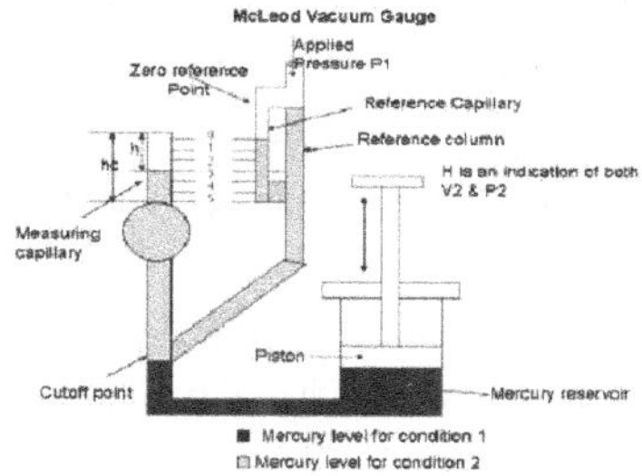

Diagram 3.58a: Mcleod Gauge

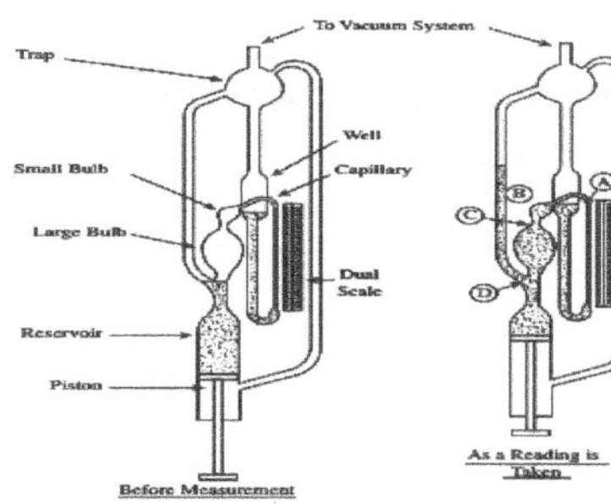

Diagram 3.58b: Mcleod Gauge Indicating Reading

Following are the basic Components of McLeod Gauge

- It consists of reference capillary tube working as a reference column, having a zero reference point. This reference column is connected to a bulb and measuring capillary and the place of connection of the bulb with reference column is called as cut off point.
- The device also consists of mercury reservoir which is operated by a piston for maintaining the intensity of applied pressure ,whenever the mercury level is raised above cut off point it will cut off the pressure to the bulb and measuring capillary.

Working of McLeod Gauge

- The intensity of applied pressure which is being measured initially applied on to the reference column of McLeod Gauge.
- After that piston is operated there by the mercury level in the gauge is raised to fill the volume (v_1) hence applied pressure (p_1) fills the bulb and the capillary.
- Again the piston is operated then mercury level in the gauge increases. When the mercury level reaches the cutoff point, a known volume of gas is occupied by bulb and measuring capillary tube.

- Again the piston is operated to raise the mercury level in the gauge increases. so the trapped gas in the bulb and measuring capillary tube are compressed.
- This is done until the mercury level reaches the " **Zero reference Point** "
- Hence the volume of the gas in the measuring capillary tube is read directly by a scale besides it.
- It will be the difference in height "H" of the measuring capillary and the reference capillary becomes a measure of the volume **(V_2)** and pressure **(P2)** of the occupied gas.
- With the help of this gauge we can measure the pressure from 10 Torr to 10^{-6} Torr

Applying Boyle's Law

If the value of V_1,$V_2$and $P_2$are known, then applied pressure P_1can be calculated

$$P_1 V_1 = P_2 V_2 \quad \text{.......} \quad 38$$

Therefore ,

$P_1V_1 = (P_1 + h) ah$

$P_1 V_1= P_1 ah + ah^2$

$P_1 V_1 - P_1 ah = ah^2$

$P_1= ah^2/ (V_1- ah)$

Since **ah << V_1**hence it can be neglected.

Therefore, $\mathbf{P_1 = ah^2 / V_1}$

P_2 = Pressure of gas at final condition, that is, after compression = **P_1+h**

Where ,

a = area of cross – section of the measuring capillary tube

h_c = height of measuring capillary tube

thus $\mathbf{V_1 = V + a\, h_c}$

3.24. Pirani Vacuum Gauge

The Pirani gauge is also a vaccum gauge used for measuring the pressure in vaccum. sometimes it is also known as thermal conductivity gauge or thermalvacuum gauge.

Working principle : The basic principle is based on the fact that when ever an electric wire having a cetrain resistance carrying an electric current it is heated up to the equlibrium temperature . this is invented by joule.

Hence

$$H = I^2 R T \quad \text{.......} \quad 39$$

Where ,

H = Heat Generated

I = Current passing through electrical wire

T = Temperature

R = Resistance

The electrical wire increases its temperature at the state when an equal amount of heat is removed from the electrical wire to ambient. The heat transfer may be possible of three modes like conduction, convection and radiation. Resistance of the wire depends upon the temperature, thus we can say that

Temperature Rise or fall **α** change in Resistance

If either temperature or Resistance indicating the proportional change in the base input.

With the help of Pirani gauge we can measure the resistance that wills proportional to the applied pressure. The schematic diagram of Pirani gauge is given as,

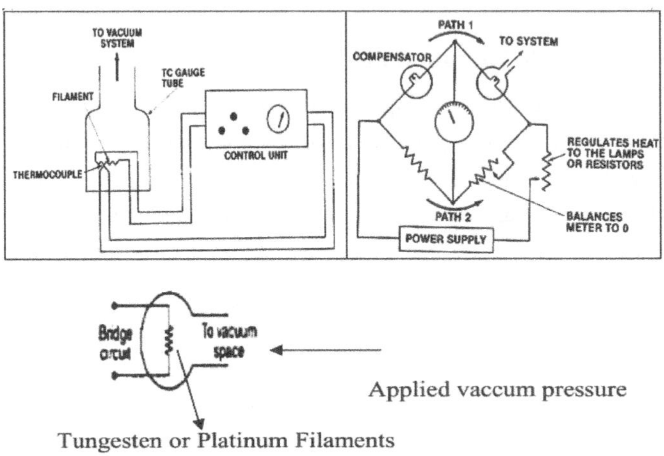

Diagram 3.59: Pirani Vacuum Gauge

The arrangement of pirani gauge is explained with the help of wheat-stone bridge circuit, in which one arm consists of a vacuum chamber in side which platinum or tungsten filament is installed , the other arm of the bridge having a compensate or which is enclosed in a reference vacuum whose function is to control the variation of process temperature including atmospheric temperature.

When electrical current is passing through the platinum filaments, it's get heated due to its self resistance then pressure is applied to pirani gauge chamber whose intensity is being measured. due to the effect of pressure not only density altered but also changing the conductivity of filaments , finally we have got a change of the electric wire resistance . readout the value of resistance change which will be proportional to applied pressure .The readout device also detects the amount of current or voltage that is necessary to return the bridge circuit to balance after a change in the vacuum being measured.

This gauge is operated between the pressure range of 100 Torr or 0.1 m Torr.

Advantage

- It is inexpensive and convenient to use.
- Applicable for pressure measurement of vapors as well as gases.
- This gauge represents the linear relationship between pressure and resistance

Disadvantage

- It is not applied for reliable pressure measurements higher than 1.0 torr .
- Pirani gauges is not highly accurate

3.25. Thermocouple Vaccum Gauge

The working principle of Thermocouple vaccum gauge is same as that of the pirani gauge , In pirani gauge we are measuring the resistance which will be proportional to the applied pressure but in case of thermocouple gauge we will measure the temperature only that will be proportional to the applied pressure . The diagrammatic arrangement is given below.

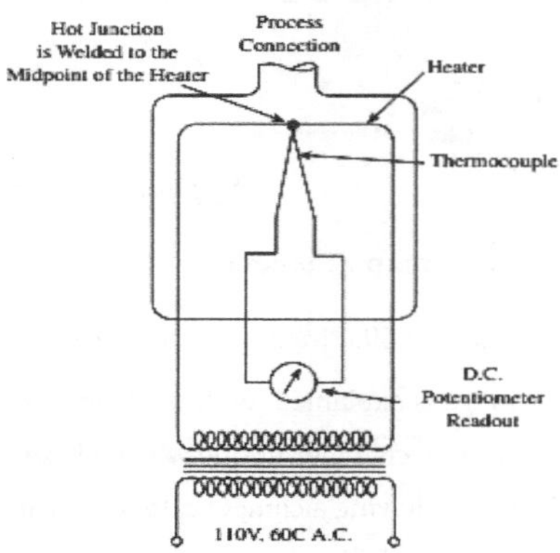

Diagram 3.60: Thermocouple Vaccum Gauge

It consists of filament and thermocouple which is enclosed in a glass tube, the wire filaments is heated by supplying the constant AC or DC current. Thermo couple is installed with the help of welding at the centre of the filaments to obtain the reading of temperature of the filament directly. The temperature is depends upon the amount of heat lost to the surroundings by conduction and convection. Thus temperature of filaments is proportional to the pressure of surrounding gas that will also be the proportional to output voltage of the device.

Temperature of Filaments **α** Pressure **α** Output Voltage

An opening in the tube is provided for connection to the vacuum system being measured. In operation, the constant current passing through the heater wire is in the order of 20 to 200 mA, and the **thermocouple** sensor develops a full-scale output in the order of 20 mV DC. For any constant value of current through the filament, the temperature increases as the pressure in the tube is reduced.

The received **thermocouple** output signals can be measured with the help of Milli-volt meters and Potentiometers or moving coil meter. Hence The moving coil instrument may be directly calibrated to read the pressure. In this device temperature limitation is on prime consideration because at higher temperature the gas may be decomposed or deposited on to the inner surface of gauge tube elements, to achieve accurate and safe operation the temperature must be less than or equal to 250°C. If temperature increases above this range, compensation for temperature variation is provided by the reference gauge that has been fully evacuated and sealed.

Advantage

- Thermocouple Vacuum gage is applied for pressure measurement approximately 0.001 to 1 torr absolute pressure.

3.26. The Bridgman Gauge

The another type of pressure gauge is Bridgman gauge which works on the principle of resistance change of wire with pressure variation, The resistance of fine wires changes with pressure according to the following linear relationship.

$$R = R_1 (1 + \alpha P)$$

the schematic diagram of gauge is given as.

Where R_1 = Resistance at 1 atmosphere in ohms

α = Pressure coefficient

P = Gage pressure in KN / m^2

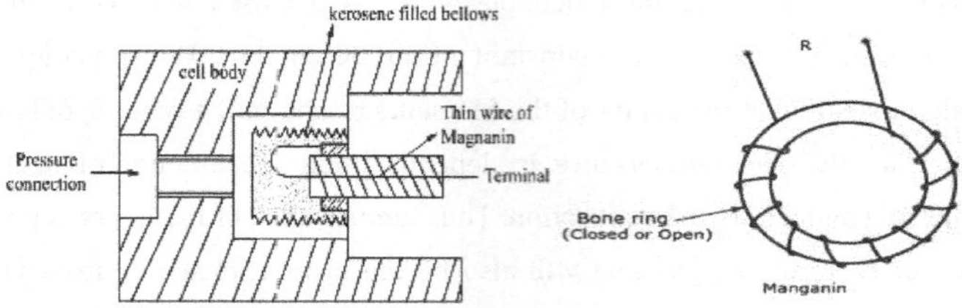

Diagram 3.61: The Bridgman Gauge

In such kind of gauge fine wire of Magnanin having a composition of **(84% Cu, 12% Mn, 4% Ni)** or may be of gold chrome is enclosed in a flexible bellows or pressure container which is filled with kerosene oil working as a sensing elements of pressure gauge having a resistance of about 100 Ω. whenever the device is connecting to the pressure measurement of any mechanism, the applied intensity of pressure, compress to flexible bellows thereby the resistance of the sensing wire has been changed thus change in the resistance will be proportional to the applied pressure. the resistance changed can be calculated with the help of Wheatstone bridge circuits and the formula – 1 applied to find out the pressure. The above said resistance change may be used for measurement of pressures as high as 100,000 Atm, 10.00KN/m^2. A pressure transducer based on this principle is called a Bridgman gage.

3.27. Penning Vacuum Gauge

Penning gauge is a cold cathode type vaccum gauge, sometimes it is also known as ionization gauge which is used for measuring the pressure intensity of the gas .The schematic diagram of gauge is given as,

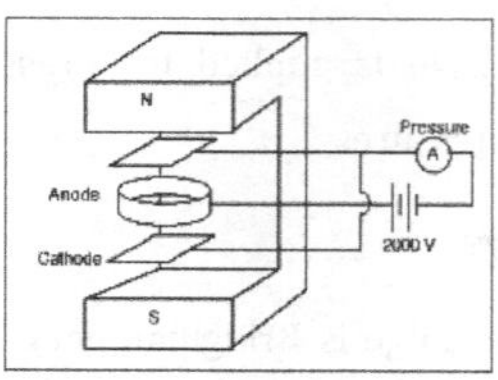

Diagram 3.62: Penning Vacuum Gauge

Constructional Features & Working of Penning Gauge

- It has two electrodes ie. Anode and Cathode both are connected to power supply of 2KV, The intensity of current is controlled by the variation of the resisting unit.
- The arrangement of Anode & Cathode is kept inside the magnetic field whose intensity is about between 700 to 300Gauss which is used for increasing the ionization current.

- The anode consists of a tungsten wire mounted in the center of the tube.
- When power is supplied, the electrons emitted from the Cathode which is deflected by means of magnetic field toward the Anode.
- The space between cathode & anode where collision is takes place so that electrons starting ionize the gas.
- The ionization creates the secondary electrons which will take part in increasing the rate of ionization.
- The electrons are captured by the anode and setup an equilibrium if

No of Electrons create /second = Positive ion current to the Cathode + Electron current to the Anode.

- This amount of current will be proportional to the applied pressure of the gas.

Example 1: Determine the time constant for a bellows type pressure gauge, If the pressure gauge consists of following data,

Bellows Area – **0.4 m^2**

Spring constant – **100 N/m**

Internal Diameter of pipe - **2.5 mm** = 2.5 x 10^{-3} m^2

Length of pipe - **10 m**

Inside volume of bellows or **Air Volume – 1.5 m^3**

Temperature – **50 °c**

Pressure – **6 bar**

Dynamic viscosity **(μ)= 1.846 x 10^{-5} kg / ms**

Assume that – **287 J / kg-K**

Solution

Given that

$$A = 0.8 \text{ m}^2, k = 100 \text{ N/m}, D = 2.5 \text{ mm}, P = 6 \text{ bar} = 6 \times 10^5 \text{ Pa}$$

$$L = 10 \text{ m}, \; V = 1.5 \text{ m3}, \; T = 50\,^{\circ}c = 323^{\circ}c, \; R_g = 287 \text{ J / kg-K}$$

I. **Calculate the air density**

$\rho = P / R_gT = 6 \times 10^5 / 287 \times 323 = 6.47 \text{ kg / m}^3$

II. **kinematic viscosity**

$(\mu / \rho) = (1.846 \times 10^{-5} / 6.47) = 2.85 \times 10^{-6}$

III. **Capacity of bellows**

$$C = \frac{A^2\rho}{K} + \frac{V}{nR_gT}$$

= [{(0.8)^{2}x 6.47 / 100} + { 1.5 / 1.4 x 287 x323 }] = 4 x 10^{-6} m^2 / s

IV. **Fluid or Air Resistance**

$$R = \frac{8\nu L}{\pi r^4}$$

= [{ (8x 2.85x10^{-6} x10) } + { 3.14 x (2.5 x 10^{-3} / 2)^4}] = 4 x 10^5 m^2 / s

V. **The time constant for pressure transducer**

$\tau = RC$ = 4 x 10^5 x 4 x 10^{-6} m^2 / s = 0.16 s

Example 2: Determine the time constant for a bellows type pressure gauge , If the pressure gauge consists of following data ,

Bellows Area – **2 m^2**

Spring constant – 250 **N/m**

Internal Diameter of pipe - **1.5 mm** = 1.5 x 10^{-3} m^2

Length of pipe - **8 m**

Inside volume of bellows or **Air Volume – 1.5 m^3**

Density of air – 995.7 kg / m^3

Kinematic viscosity (υ)= **0.846 x 10^{-6} kg / ms**

Radius of connecting tube (r) = (**ID / 2**) = 1.5 / 2 = 0.75 mm = 0.00075 m

Solution

(i) Calculate , Flow resistance imposed by tubes

$$R = \frac{8\nu L}{\pi r^4}$$

= {(8 x 0.846 x 10 $^{-6}$ x 8) / 3.14 x (75 x 10 $^{-5}$)^4}

= { 54.144 x 10^9 / 99351562.5 } = **0.5444 m^2 / s**

(ii) Calculate , Capacitance of Pressure Gauge

$$C = \frac{A^2\rho}{K}$$

= (2^2 x 995.7) / 250 = 15.95 m^2 / s

(iii) Calculation for Time Constant ,

$$\tau = RC$$

= 0.544 x 15.95 = 9 sec. Approximately

Example 3 : A steel diaphragm is of 25 mm diameter is applied in pressure measuring device which a maximum pressure carrying capacity of 20 MPa, The transducer is required to give an output of 5 *mV/V* at the maximum pressure. the concern value of Young's modulus (E), Poisson ratio (ν) & material bulk modulus(E) is about 225 G Pa , 0.325 , 3.29 x 10^9 respectively. find out following,

(i) What will be the thickness of the diaphragm?

(ii) How much maximum deflection will achieved in the diaphragm ?

(iii) Natural frequency of the diaphragm ?

Solution

Given that

Radius of diaphragm '**a**' = (25 / 2) = 12.5 mm = 0.0125 m

Pressure difference **(P_1 – P_2)** = 20 MPa = 2 x 10^7 Pa

Voltage ratio **(V_O / V_I)** = **5** mV / V

(i) What will be the thickness of the diaphragm

$$t = a\sqrt{\frac{0.75(p_1 - p_2)(1-v^2)\times 10^3}{E}}\frac{1}{\sqrt{\frac{V_o}{V_i}}}$$

t = (0.0125 x 1 / $\sqrt{5}$) [{0.75 (2 x 10^7) (1- 0.325^2) X10^3 } / 3.29 X 10^9 $]^{1/2}$

= 0.0125 x 0.4472 x 2.019330 = 0.0011288 m = 1.12 mm

(ii) How much deflection is achieved in the diaphragm ?

$$y_{max} = \frac{3(p_1 - p_2)\left(1 - \upsilon_p^2\right)r^4}{16E_m t^3}$$

y_{max} = δy_{max} at r = 0

δ(o) = {(3 x 2 x 10^7) (1 – 0.325^2) ($0.0125)^4$} / (16 x 3.29 x 10^9) x (0.00112 $)^3$

= (1.4648 / 73.9554) = 0.01980 m = 19.8 mm

(iii)Natural frequency of the diaphragm ?

$$\omega_n = 10.21\sqrt{\frac{E_m t^2}{12\left(1 - \upsilon_p^2\right)\rho r^4}}$$

= 10.21 [(3.29 x 10^9) ($0.01112)^2$/12(1-0.325^2) x 7830 x ($0.0125)^4$]=10.21 x 22 = 225.4 Hz

Example 4 : Calculate the temperature of the RTD . the Wheatstone bridge circuit is given below with suitable diagram & concerning data of bridge circuit is given below.

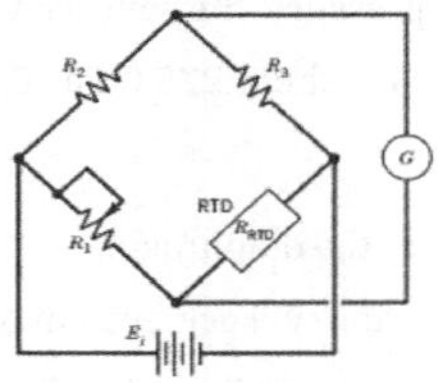

Resistance value R2 and R3 are equal to **30 V,**

Resistance of RTD is 22 V at a temperature of 0C ,

RTD Coefficient of resistance is 0.003925 C^{-1}

The value of R1 required to balance the bridge is 40 Ω

The resistance equation of the RTD can be given as ,

$$R_{RTD} = R_0[1 + \alpha(T - T_0)]$$

Solution

Given that

R(0°C) = 22 V ,

α = 0.002325 C^{-1} ,

R_1 = 40 Ω .

For balanced bridge condition , we have a following relationship

$$\mathbf{R_{RTD} : R_1 = R_3 : R_2}$$

Put the value R_{RTD} = 40 Ω , $\mathbf{R_0}$ = 22 Ω , $\boldsymbol{\alpha}$ = 0.002325 C^{-1} , **T** = 0°C , then the equation becomes

$$R_{RTD} = R_0[1 + \alpha(T - T_0)]$$

40 Ω = 22 {(1 + 0.002325 (T - 0) } Ω

Thus, the temperature of RTD will be **350 °C**

Example 5 : Find the value of constant " **β** " **of thermistor when** it is operated between temperatures ranging from 100 to 150C. consider following data regarding the problem ,

Thermistor resistance (R_T) is 60 kV at 25C ,

Reference resistor (R_1) = 130.5 KV,

Dissipation Constant δ = 0.09 m W /C ,

Voltage supplied = 1.564 V.

Thermistor resistance is given by the following equation,

$$R_T = R_1\left(\frac{E_i}{E_1} - 1\right)$$

Solution

The results of these measurements are as follows:

Temperature (°C)	R_1 Voltage (V)	R_T (Ω)
100	1.501	5477.4
125	1.531	2812.9
150	1.545	1604.9

The linear form of equation for thermistor can be represented as ,

$$\ln\frac{R_T}{R_0} = \beta\left(\frac{1}{T} - \frac{1}{T_0}\right)$$

Applying this equation to the measured data, with R_0 = 60 V , we have got three value of β which are 3546.7 , 3629.9 , 3650.2 .

$\ln(R_T/R_0)$	$(1/T - 1/T_0)[\text{K}^{-1}]$	β(K)
−2.394	-6.75×10^{-4}	3546.7
−3.060	-8.43×10^{-4}	3629.9
−3.621	-9.92×10^{-4}	3650.2

Example 6 : A J-type thermocouple measures a temperature of 500C and is referenced to 0C. The length of the thermocouple wire is 15 ft in order to run from the measurement point to the ice bath and to a potentiometer. The resolution of the potentiometer is 0.09 mV. If the thermocouple wire has a resistance per unit length, as specified by the manufacturer, of 7.6 Ω/f. Find The residual current in the thermocouple circuit.

Solution

Given that,

Length of thermocouple = 15 ft,

Potentiometer resolution = 0.09 mV,

Total resistance of the thermocouple = 76 Ω

Applying Ohm's law, we have following relationship between Current, Voltage & Resistance

I = (V / R)

I = (0.09 mV / 76 Ω)

I = 11.8 x 10^{-9} Ampere

Example 7 : Consider the pond having a water at 33°C , find the value of absolute and gauge pressures at a depth of 15 m below if density of water is 998. 207 kg / m³ and specific gravity of mercury is 11.21, consider atmospheric pressure is 1.0132 x 10⁵ pa (absolute)

Solution

Given that,

Height of water (H) = 15 m,

Position of free liquid surface (H_O) = 0

Temperature of water (T) = 33°C,

Density of water (ρ) = 998. 207 kg / m³,

S.G. of mercury = 11.21 .

Solution

(i) Calculation of absolute pressure

$P_{ab = Atm\ pressure + (}\rho\ g\ H / W)$

Absolute pressure = $(1.0132 \times 10^5) + \{(998.207 \times 9.8 \times 15) \div 1\}$

= 1.486×10^5 N / m² = **1.486 kPa abs**

≈ 1.486 atm abs

(ii) Calculation of Gauge pressure

$P_{Gauge\ =\ Pabs} - P_o = 1.486 \times 10^5 - 1.0132 \times 10^5 =$ **4.63×10^4 N / m²** ≈ 0.5 bar

Example 8 : A dead weight tester indicates 150 lb / inch² or 150 psi , at 73°C where $\phi = 74$, z = 640 ft . Take g_{air} = 0.069 lb/ft³ and g_{mass}= 692 lb/ft³. Correct the indicated reading for known errors.

Solution

Given that

Pi = 150 psi , z = 640 ft , $\phi = 74^0$

Apply formula

$$P = Pi\ (1 + e_B + e_G)$$

(i) Correction for buoyancy effects (e_b) =

$$e_2 = -\gamma_{air}/\gamma_{masses}$$

$e_b = -(0.069 / 692)$

e_b = - 0.00009971

(ii) Correction for Gravity effects (e_g) =

$$e_l = -(2.637 \times 10^{-3} \cos 2\phi + 9.6 \times 10^{-8} z + 5 \times 10^{-5})$$

$= - \{ 2.637 \times 10^{-3} \cos 148 + 9.6 \times 10^{-8} \times 640 \times 5 \times 10^{-5} \} = - 0.390276$

Hence correct pressure will be given as ,

$P = Pi (1 + e_b + e_g)$

= 150 (1 - 0.00009971- 0.390276)

= 150 (1- 0.390375713072)

= 150 (0.609624286928)

= 91.44 lb / inch^2 .... **Answer**

Example 9 : Find the sensitivity of capacitive sensor , the related data is given below ,

Area of plate (A) = 20 mm^2 ,

Thickness of plates = 5 mm ,

Capacitance (c) = 1 ,

Constant (ε) = 8.85×10^{-12} F / mm

Solution

C = (c ε A / t) = $\{1 \times 8.85 \times 10^{-12} \times 20 \} / 5 = 35.4 \times 10^{-2}$ F

Sensitivity of sensor

Ks = $(\frac{\partial C}{\partial t})$ = 35.4×10^{-2} F

= ($35.4 \times 10^{-2} \times 1 \times 20$) / 25 **= 28.32×10^{-12} F / mm**

Example 10 : Find out the pipe flow rate , for square edge orifice plate or orifice meter , the data concerning the problem is given below ,

Diameter of pipe(d_1) = 20 cm ,

Diameter of orifice(do) = 10 cm ,

Pressure drop(H) = 50 cm Hg ,

Specific gravity of mercury = 13.5 ,

Density of water (ρ) = 1000 kg / m^3 ,

Viscosity (μ) = 1.08 x 10^{-3} Ns / m^2 ,

Consider the value $\mathbf{k_o = 0.625}$ & $\boldsymbol{\gamma = 1}$

Solution

(i) Calculate the velocity approach factor

E = 1 / (1- β^4)$^{1/2}$

Here β = (d_o / d_1) = (10 / 20) = 0.5

E = 1 / (1- 0.5^4) ^{½}= **1.032**

Hence flow rate (Q) can be calculate as ,

= $K_0 A_0 Y \sqrt{2gH[(S_m/S)-1]}$ = {0.625 x 3.14 x $(0.10)^2$} [(2 x 9.8 x 05) (13.5 /1) – 1]$^{1/2}$

= 0.054 m^3/s

Example 1 : Determine the time constant for a bellows type pressure gauge , If the pressure gauge consists of following data ,

Bellows Area – **0.4 m^2**

Spring constant – **100 N/m**

Internal Diameter of pipe - **2.5 mm** = 2.5 x 10^{-3} m^2

Length of pipe - **10 m**

Inside volume of bellows or **Air Volume – 1.5 m^3**

Temperature – **50 ^{o}c**

Pressure – **6 bar**

Dynamic viscosity (μ)= **1.846 x 10^{-5} kg / ms**

Assume that – **287 J / kg-K**

Solution

Given that

A = 0.8 m^2, k = 100 N/m , D = 2.5 mm , P = 6 bar = 6 x 10^5 Pa

L =10 m, V =1.5 m3, T = 50 ^{o}c = 323°c , R_g= 287 J / kg-K

Calculate the air density

$\rho = P / R_g T$

$= 6 \times 10^5 / 287 \times 323 = 6.47$ kg / m^3

(i) kinematic viscosity

(μ / ρ) = (1.846 x 10^{-5} / 6.47) = 2.85 x 10^{-6}

(ii) Capacity of bellows

$$C = \frac{A^2 \rho}{K} + \frac{V}{nR_g T}$$

= [{(0.8)^{2}x 6.47 / 100} + { 1.5 / 1.4 x 287 x323 }] = 4 x 10^{-6}m^2 / s

(iii)Fluid or Air Resistance

$$R = \frac{8\nu L}{\pi r^4}$$

= [{ (8x 2.85x10^{-6} x10) } + { 3.14 x (2.5 x 10^{-3} / 2)^4}] = 4 x 10^5 m^2 / s

(iv) The time constant for pressure transducer

$$\tau = RC$$

= 4 x 10^{5}x 4 x 10^{-6} m^2 / s = 0.16 s

Example 2 : Determine the time constant for a bellows type pressure gauge , If the pressure gauge consists of following data ,

Bellows Area – **2 m^2**

Spring constant – 250 **N/m**

Internal Diameter of pipe - **1.5 mm** = 1.5 x 10^{-3} m^2

Length of pipe - **8 m**

Inside volume of bellows or **Air Volume – 1.5 m^3**

Density of air – 995.7 kg / m^3

Kinematic viscosity (υ)= **0.846 x 10^{-6} kg / ms**

Radius of connecting tube (r) = (**ID / 2**) = 1.5 / 2 = 0.75 mm = 0.00075 m

Solution

Given that,

(i) Calculate , Flow resistance imposed by tubes

$$R = \frac{8\nu L}{\pi r^4}$$

= {(8 x 0.846 x 10 $^{-6}$ x 8) / 3.14 x (75 x 10 $^{-5}$) ^{4}}

= { 54.144 x 10^9 / 99351562.5 } = **0.5444 m² / s**

(ii) Calculate , Capacitance of Pressure Gauge

$$C = \frac{A^2 \rho}{K}$$

= (2^2 x 995.7) / 250 = 15.95 m² / s

(iii) Calculation for Time Constant ,

$$\tau = RC$$

= 0.544 x 15.95 = 9 sec. Approximately

Example 3 : A steel diaphragm is of 25 mm diameter is applied in pressure measuring device which a maximum pressure carrying capacity of 20 MPa, The transducer is required to give an output of 5 *mV/V* at the maximum pressure. the concern value of Young's modulus (E), Poisson ratio (v) & material bulk modulus(E) is about 225 G Pa , 0.325 , 3.29 x 10^9 respectively . find out following,

(i) What will be the thickness of the diaphragm?

(ii) How much maximum deflection will achieved in the diaphragm ?

(iii) Natural frequency of the diaphragm ?

Solution

Given that ,

Radius of diaphragm '**a**' = (25 / 2) = 12.5 mm = 0.0125 m

Pressure difference **(P_1 – P_2)** = 20 M Pa = 2 x 10^7 Pa

Voltage ratio **(V_O / V_I) = 5** mV / V

(i) What will be the thickness of the diaphragm

$$t = a\sqrt{\frac{0.75(p_1 - p_2)(1-\nu^2)\times 10^3}{E}}\frac{1}{\sqrt{\frac{V_o}{V_s}}}$$

t = (0.0125 x 1 / $\sqrt{5}$) [{0.75 (2 x 10^7) (1- 0.325^2) X10^3 } / 3.29 X 10^9]$^{1/2}$

= 0.0125 x 0.4472 x 2.019330 = 0.0011288 m = 1.12 mm

(ii) How much deflection is achieved in the diaphragm ?

$$y_{max} = \frac{3(p_1 - p_2)\left(1 - \upsilon_p^2\right)r^4}{16E_m t^3}$$

y_{max}= δy_{max}at r = 0

$\delta(o) = \{(3 \times 2 \times 10^7)(1 - 0.325^2)(0.0125)^4\} / (16 \times 3.29 \times 10^9) \times (0.00112)^3$

= $(1.4648 / 73.9554) = 0.01980$ m = 19.8 mm

(iii) Natural frequency of the diaphragm ?

$$\omega_n = 10.21\sqrt{\frac{E_m t^2}{12\left(1 - \upsilon_p^2\right)\rho r^4}}$$

= $10.21 [(3.29 \times 10^9)(0.01112)^2 / 12(1-0.325^2) \times 7830 \times (0.0125)^4] = 10.21 \times 22 = 225.4$ Hz

Example 4 : Calculate the temperature of the RTD. the Wheatstone bridge circuit is given below with suitable diagram & concerning data of bridge circuit is given below.

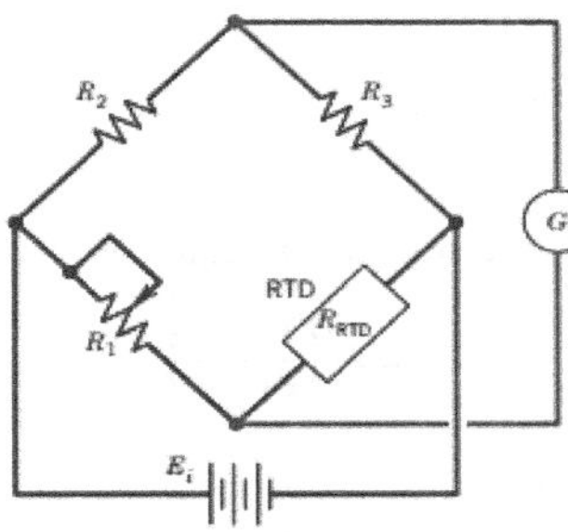

Resistance value $\mathbf{R_2}$and $\mathbf{R_3}$ are equal to **30 V,**

Resistance of RTD is 22 V at a temperature of 0C ,

RTD Coefficient of resistance is 0.003925 C^{-1}

The value of $\mathbf{R_1}$ required to balance the bridge is 40 Ω

The resistance equation of the RTD can be given as ,

$$R_{RTD} = R_0[1 + \alpha(T - T_0)]$$

Solution

given that ,

R(0°C) = 22 V ,

α = 0.002325 C^{-1} ,

R_1 = 40 Ω .

For balanced bridge condition , we have a following relationship

$\mathbf{R_{RTD} : R_1 = R_3 : R_2}$

Put the value R_{RTD} = 40 Ω , $\mathbf{R_O}$ = 22 Ω , $\boldsymbol{\alpha}$ = 0.002325 C^{-1} , **T** = 0°C , then the equation becomes

$$R_{RTD} = R_0[1 + \alpha(T - T_0)]$$

40 Ω = 22 {(1 + 0.002325 (T - 0) } Ω

Thus, the temperature of RTD will be **350 °C**

Example 5 : Find the value of constant " **β** " **of thermistor when** it is operated between temperatures ranging from 100 to 150C. consider following data regarding the problem ,

Thermistor resistance (R_T) is 60 kV at 25C ,

Reference resistor (R_1) = 130.5 KV ,

Dissipation Constant δ = 0.09 m W /C ,

Voltage supplied = 1.564 V.

Thermistor resistance is given by the following equation,

$$R_T = R_1\left(\frac{E_i}{E_1} - 1\right)$$

Solution

The results of these measurements are as follows:

Temperature (°C)	R_1 Voltage (V)	R_T (Ω)
100	1.501	5477.4
125	1.531	2812.9
150	1.545	1604.9

The linear form of equation for thermistor can be represented as ,

$$\ln\frac{R_T}{R_0} = \beta\left(\frac{1}{T} - \frac{1}{T_0}\right)$$

Applying this equation to the measured data, with R_0 = 60 V , we have got three value of β which are 3546.7 , 3629.9 , 3650.2 .

$\ln(R_T/R_0)$	$(1/T - 1/T_0)[K^{-1}]$	β(K)
−2.394	-6.75×10^{-4}	3546.7
−3.060	-8.43×10^{-4}	3629.9
−3.621	-9.92×10^{-4}	3650.2

Example 6 : A J-type thermocouple measures a temperature of 500C and is referenced to 0C. The length of the thermocouple wire is 15 ft in order to run from the measurement point to the ice bath and to a potentiometer. The resolution of the potentiometer is 0.09 mV. If the thermocouple wire has a resistance per unit length, as specified by the manufacturer, of 7.6 Ω/f. Find The residual current in the thermocouple circuit.

Solution

Given that ,

Length of thermocouple = 15 ft,

Potentiometer resolution = 0.09 mV,

Total resistance of the thermocouple = 76 Ω

Applying Ohm's law, we have following relationship between Current, Voltage & Resistance,

I = (V / R)

I = (0.09 mV / 76 Ω)

I = 11.8×10^{-9} Ampere

Example 7 : Consider the pond having a water at 33^oC , find the value of absolute and gauge pressures at a depth of 15 m below if density of water is 998. 207 kg / m^3 and specific gravity of mercury is 11.21 , consider atmospheric pressure is 1.0132×10^5 pa (absolute)

Solution

Given that ,

Height of water (H) = 15 m ,

Position of free liquid surface (H_0) = 0

Temperature of water (T) = 33^oC ,

Density of water (ρ) = 998. 207 kg / m^3 ,

S.G. of mercury = 11.21 .

Solution

(i) Calculation of absolute pressure

$P_{ab = \text{Atm pressure} + (}$ ρ g H / W)

Absolute pressure = (1.0132×10^5) + { (998. 207 x 9.8 x 15) ÷ 1}

= 1.486 x 10 ^{5}N / m ^{2} = **1.486 kPa abs**

≈ 1.486 atm abs

(ii) Calculation of Gauge pressure

$P_{Gauge} = P_{abs} - P_o$= 1.486 x 10 ^{5} - 1.0132 x 10^5 = **4.63 x 10 ^{4} N / m ^{2}**≈ 0.5 bar

Example 8 : A dead weight tester indicates 150 lb / inch^2 or 150 psi , at 73^{0}C where ϕ = 74 , z = 640 ft . Take g_{air} = 0.069 lb/ft^3 and g_{mass}= 692 lb/ft^3. Correct the indicated reading for known errors.

Solution

Given that ,

Pi = 150 psi , z = 640 ft , ϕ = 74^0

Apply formula :

P = Pi (1 + e_B+ e_G)

$$p = p_i(1 + e_1 + e_2)$$

(i) Correction for buoyancy effects (e_b) =

$$e_2 = -\gamma_{air}/\gamma_{masses}$$

e_2 = - (0.069 / 692)

e_2 = - 0.00009971

(ii) Correction for Gravity effects (e_g) =

$$e_1 = -(2.637 \times 10^{-3} \cos 2\phi + 9.6 \times 10^{-8} z + 5 \times 10^{-5})$$

= - { 2.637 x 10^{-3} cos 148 + 9.6 x 10^{-8} x 640 x 5 x 10^{-5} }

= - 0.390276

Hence correct pressure can be given as ,

P = Pi (1 + e_b + e_g)

= 150 (1 - 0.00009971- 0.390276)

= 150 (1- 0.390375713072)

= 150 (0.609624286928)

= 91.44 lb / inch^2 .... **Answer**

Example 9 : Find out the flow rate & power consumption through the horizontal venturi meter . the following related data is given below,

Diameter of pipe = 12cm

Diameter of throat = 4 cm

U-tube manometer reading = 50 cm

Coefficient of discharge (C_v) = 0.98

Loss of pressure head = 10 %

Solution

Given that

Diameter of pipe (D_1) = 12 cm = 0.12 m

Diameter of throat (D_2) = 4 cm = 0.04 m

Manometer reading (h_m) = 30 cm of Hg

Coefficient of discharge (C_v) = 0.98 ,

Pressure loss = 10 % of Pressure Difference

Density of Manometric Liquid (ρ_m) = 13600 kg/m^3

Density of water (ρ) = 1000 kg/m^3

(i) Calculate Pressure difference

$$P_a - P_b = (\rho_m - \rho) g\, h_m$$

$$= (13600 - 1000) \times 9.812 \times 0.5$$

$$= 61740 \text{ N / M}^2$$

$$\beta = D_2 / D_1$$

$$= 0.04 / 0.12 = 0.33$$

(ii) Others calculations

V_b = (0.98/(1-0.44)0.5) x (2 x 37089.4 / 1000)0.5 = 0.993 x 8.613 = 8.553 m/sec

Cross sectional area of throat = ($\Pi D_2^2 / 4$) = 3.14x 0.04^2 / 4 = 0.00126 m^2

Volumetric flow rate = 8.553 x 0.00126 = 0.01078 m^3/sec = **10.78 lit/sec**

Permanent pressure loss = 0.1 x61740 = 6174 N/m^2

Power consumption of the meter = 0.01078 x 6174 = **66 .5 watt**

Example 10 : A rotameter calibrated for metering has a scale ranging from 0.058m³/min to 0.585 m³/min. It is intended to use this meter for metering a gas of density 4.3 kg/m³ with in a flow range of 0.085 m³/min to 0.85 m³/min. What should be the density of the new float if the original one has a density of 2300 kg/m³?Both the floats can be assumed to have the same volume and shape.

Solution

Given that

$Q_{new} = 0.85\ m^3/min$

$Q_{old} = 0.58\ m^3/min$

Density of gas (ρ) = 4.3 kg/m³

Density of float (ρ_f)= 1900 kg/m³

For the same volume and shaped of the float

$$\mathbf{Q_{new} / Q_{old} = \{(\rho_{fnew} - \rho) / (\rho_{fold} - \rho)\}^{0.5}}$$

$$0.58 / 0.85 = (\rho_{fnew} - 4.3) / (1900 - 4.3)^{1/2}$$

$$1.46 = \{(\rho_{fnew} - 4.3) / 43.539\}^{1/2}$$

$$2.147 = (\rho_{fnew} - 4.3) / 43.539$$

$$0.0427 = \rho_{fnew} - 4.3$$

$$\rho_{fnew} = \mathbf{4.347\ kg / m^3}$$

Example 11 : Calculate the value of coefficient of velocity (**C_V**)for venturi meter, if following data is observed ,

Flow rate (Q) = 5.6 kg/sec ,

Pipe diameter = 200 mm ,

Throat diameter = 50 mm

Pressure drop in converging section = 100 mm of water .

$(P_a - P_b)$ = 100 mm of water = (0.1/10.33) x 101325 N/m² = 980.9 N/m² ,

Solution

Velocity at the throat is given by ,

$$V_b = \frac{C_v}{\sqrt{1-\beta^4}}\sqrt{\frac{2(P_a - P_b)}{\rho}}$$

$\beta = D_2 / D_1 = 50 / 200 = 0.25$

(i) **Volumetric flow rate** = 5.6 / 1000 = 0.0056 m^3/sec

(ii) **Velocity at the throat** = Volumetric flow rate / cross sectional area of throat

= 5.6 / $(\pi D_2^2/4)$ = 5.6 / $(\pi \times 0.05^2/4)$ =0.01099 m/sec

Hence the value of

C_v = **$0.01099 \times \{(1 - (0.05/0.20)^4)^{0.5}\} / \{(2 \times 100 /1000)^{0.5}$**

= 0.01099x{**(0.998 / 0.4472**} = 0.976 = 0.2227

Example 12 : Brine of specific gravity 1.7 is flowing through a 12 cm I.D. pipeline at a maximum flow rate of 1500 litres/min. A sharp-edged orifice connected to a simple U-tube mercury manometer is to be installed for the purpose of measurements. The maximum reading of the manometer is limited to 35 cm. Assuming the orifice coefficient to be 0.82, calculate the size of the orifice required.

Solution

Given that

Density of brine (ρ) = 1.7 x 1000 kg/m^3 = 1700 kg/m^3

Diameter of pipe (D_1) = 12 cm = 0.12 m

Maximum flow rate (Q) = 1500 liters/min = 1.5 m^3/min

Maximum manometer reading (h_m) = 35 cm = 0.35 m of Hg

Density of manometric fluid (ρ_m) = 13600 kg/m^3

Orifice coefficient (C_o) = 0.82

Solution

Given that

$(P_a - P_b)$ $=\{\rho_m - \rho\}\ g\ h_m$

$(- P_b)$ = (13600 - 1700) x 9.812 x 0.35 = 40817 N/m^2

Assume that ,

$(1 - \beta^4) \approx 1$

Therefore ,

$V_b = 0.82 \times (2 \times 40817 / 1500)^{0.5}$ = 12..09 m/sec ,

(i) Cross sectional area of orifice = volumetric flow rate / velocity at the orifice

= 0.15/12.09 = 0.01239 m^2

Diameter of orifice (D_2) = (C.01239 x 4/3.14)$^{0.5}$ = 0.0400 m = 4 cm

Example 13 : Find out the coefficient of discharge (C_d) with considering the following data

Flow rate is 5.2 m^3/min.

Size of venturimeter = 200 mm x 55 mm ,

Specific gravity of mercury = 0.82 ,

Reading of u-tube manometers = 85 mm of Hg column ,

Solution

Given below

Diameter of pipe or inlet of venturi meter (D_1) = 200 mm

Diameter of throat (D_2) = 55 mm

Density of mercury (ρ_m) = 13.6

Density of water (ρ) = 0.98

Respective formula

Velocity $$V_2 = \frac{C_d A_1}{\sqrt{A_1^2 - A_2^2}} \sqrt{2gh_m\left(\frac{\rho_m}{\rho} - 1\right)} \text{ and } Q = V_2 \times A_2$$

Flow rate $$Q = \frac{C_d A_1 A_2}{\sqrt{A_1^2 - A_2^2}} \sqrt{2gh_m\left(\frac{\rho_m}{\rho} - 1\right)}$$

Calculations

(i) Inlet Area (A_1) = (Π / 4) $D^2{}_1$

= (3.14 / 4) x (0.2)^2 = 0.0314m^2

(ii) Throat Area (A_2) =(Π / 4) $D^2{}_2$

= (3.14 / 4) x (0.55)^2 = 0.238 m^2

(iii) Flow rate (Q) = A_2V_2

V2 = (4.2 / 60) = 0.07 m^3 / s

(iv) Coefficient of discharge (C_d)

$$Q = \frac{C_d A_1 A_2}{\sqrt{A_1^2 - A_2^2}} \sqrt{2gh_m\left(\frac{\rho_m}{\rho} - 1\right)}$$

$2.50 = [(c_d \times 0.0314 \times 0.238)/\{(0.0314)^2 - (0.238)^2\}^{1/2}][2x9.81x0.85\{(13.6/0.9) - 1\}^{1/2}]$

$2.50 = [c_d\ 0.0075 / (0.0557)^{1/2}]\ [16.677 \times 14.12]^{1/2}$

$0.07 = [c_d 0.0318](235.47924)^{1/2}$

$c_d = 0.07 / (15.32 \times 0.0318)$

$c_d = \{ 0.07 / 0.477 \} = 0.1467$

Example 14 : A venturi-meter is used to measure liquid flow rate of 750 litres per second. The difference in pressure across the venturi-meter is equivalent to 6 m of the flowing liquid. The pipe diameter is 25 cm. Calculate the throat diameter of the venturi-meter. Assume the coefficient of discharge for the venturi-meter as 0.82.

Solution

Given that

Flow rate (Q) = 750 L /sec

Diameter of pipe (D_1) = 25 cm

Coefficient of discharge = 0.82

Manometric height (h) = 6 m

Solution

(i) Calculate, Area of pipe or inlet area of venturi meter (A_1) = $(\Pi / 4)\ D^2{}_1$

$= (3.14) \times (0.25)^2 = 0.0490\ m^2/s$

(ii) Area of throat

$Q = C_d \frac{A_1 A_2 \sqrt{2gh}}{\sqrt{A_1^2 - A_2^2}}$ then, $\frac{A_2}{\sqrt{A_1^2 - A_2^2}}$ will be

$= \mathbf{750 /[(2 \times 9.8 \times 6)^{1/2} \times 0.82 \times 0.049.]}$

$\mathbf{(A^2{}_1 / A^2{}_2) = 1.0000121}$

$\mathbf{A^2{}_2 = (0.0490)^2 / 1.000012}$

$\mathbf{A_2 = 0.049\ M^2}$

(iii) Diameter of Throat

$\mathbf{(A_2)} = (\Pi / 4)\ D^2{}_2$

$\mathbf{D^2{}_2 = (0.049 \times 4) / 3.14}$

$\mathbf{D_2 = 6.2\ cm}$

Example 15 : A venturi-meter is fitted in a pipe of 30 cm diameter inclined at 40° to the horizontal to measure the flow rate of oil having a specific gravity of0.8. The ratio of areas of main pipe and throat is 5 and the throat is at 1 m from the inlet along its length. The difference in manometer head is 40mm of mercury. Assuming the coefficient of discharge as 0.96. Find out the discharge or flow rate through the venturi-meter, the schematic diagram of venturi meter is given below .

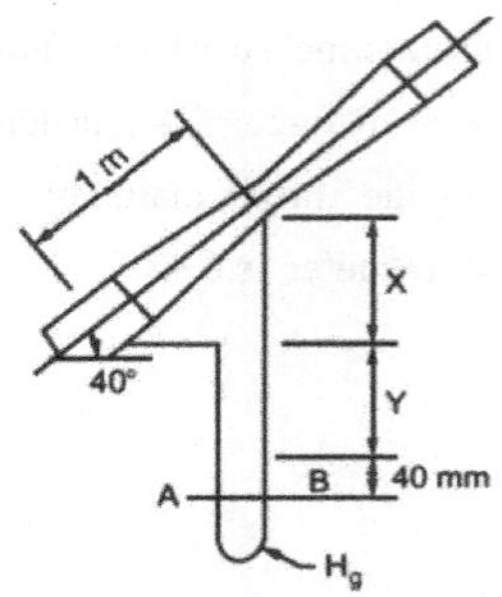

$$Q = C_d \frac{A_1 A_2}{\sqrt{A_1^2 - A_2^2}} \sqrt{2gh_m\left(\frac{\rho_m}{\rho} - 1\right)}. \ A_1 = \frac{\pi}{4} \times 0.3^2 = 0.0707 \text{ m}^2$$

$A_1/A_2 = 5$

$A_2 = 0.0707/5 = 0.0141$ m^2, **Substituting,**

$$Q = \frac{0.96 \times 0.0141 \times 0.0707}{\sqrt{0.0707^2 - 0.0141^2}} \sqrt{2 \times 9.81 \times 0.04 \times \left(\frac{13.6}{0.8} - 1\right)}$$

$= 0.0486$ m^3/s

Example 16 : find out the value of coefficient of velocity, coefficient of contraction and the theoretical discharge through the orifice by using following data .

Actual velocity of a liquid (v) = 5 m /s ,

Head of liquid (h) = 4 m ,

Discharge (Q) = 0.025 m3/s,

Orifice diameters (D_2) = 7cm = 0.07 m

Solution

(i) Flow velocity of orifice meter is given as , then coefficient of velocity will be ,

$$V = C_v \sqrt{2gh}$$

$5 = C_V (2 \times 9.81 \times 4)^{½}$

$C_V = 0.5859$

(ii) Coefficient of discharge ,

$$\text{Actual discharge } Q = C_d \sqrt{2gh} \times A_2$$

$0.025 = C_d\ (2 \times 9.81 \times 4)^{1/2} \times (3.14 / 4)\ (0.07)^2$

$C_d = \mathbf{0.0784}$

(iii) Coefficient of contraction ,

$$C_c = \frac{C_d}{C_v}$$

= (0.0784 / 0.5859) = 0.1323

Example 17 : Determine the volumetric flow of water in m^3/hr flowing in a pipe having a diameter of 40 mm through an orifice having a diameter of 20 mm. Manometer reading across the tapping is 15 cm of Hg. Assume correction factor = 1 and density of water is 1000 kg/m3.

Solution

Given that

Diameter of pipe D = 40 mm

Diameter of orifice d = 20 mm

Manometer head or differential pressure = 15 cm of Hg.

Differential pressure is calculated in terms of water column as

h = (13.6 – 1) × (15 × 10) mm of water = 1890 mm of water.

β = d / D = 20 mm/40 mm = 1/2

Velocity approach factor

$$M_{v.a} = \frac{1}{\sqrt{1-(d/D)^4}} = \frac{1}{\sqrt{1-(1/2)^4}} = 1.032$$

$$Q = C_d M_{v.a} A_2 \cdot \sqrt{2gh} \ \text{m}^3/\text{sec}$$

$$Q = 0.5999 \times 1.032 \times \pi/4\ (20 \times 10^{-3})^2 \cdot \sqrt{2 \times 9.81 \times 1890 \times 10^{-3}}$$

$$= 1.184 \times 10^{-3} \text{ m}^3/\text{sec} = 1.184 \times 10^{-3} \times 3600 = 4.262 \text{ m}^3/\text{hr}$$

Example 18 : Determine the nominal flow velocity V2 at the orifice having a diameter of 20 mm kept in a pipe of 40 mm diameter. Reynolds number R_a is 10^5. Assume density of water = 1000 kg/m3 and kinematic viscosity is 10–2 stokes. (Cm^2/sec).

Solution

Given that ,

$R_e = \rho V_1 D/\mu$ **then , the value of V_1 will be given as ,**

$$V_1 = R_e \mu/\rho D = R_e \nu/D$$

Kinematic viscosity (ν) = 10^{-2} stokes = 10^{-6} m^2/sec

$R_e = 10^5$

$D = 40 \times 10^{-3}$ m

$V_1 = 2.5$ m/sec

Then, from the continuity equation, we have

$A_1V_1 = A_2V_2$

Then V_2 will be given as,

$$= V_1 \times A_1/A_2 = 2.5 \times \frac{\pi/4(40 \times 10^{-3})^2}{\pi/4(20 \times 10^{-3})^2}$$

$$= 2.5 \times (4^2/2^2) = 2.5 \times (16/4) = 10.0 \text{ m/sec.}$$

Example 19 : Determine the volume flow rate of water through a pipe of 110 mm diameter when measured by

(i) An orifice plate of size 55 mm dia and

(ii) Venturi tube of throat size 55 mm dia.

Assume that, Pressure head is (250 / 1000) **pa**, Density of water is **1000 kg / m³**, Cd for orifice plate 0.6, Cd for venture-meter 0.99

Solution

Following equation will be useful in calculating the flow rate measurement,

$$Q = \frac{1}{\sqrt{1-(d/D)^4}} \times Cd \,.\, \frac{\pi d^2}{4} \times \sqrt{2gh}$$

(i) For orifice plate, the flow rate measurment will be given as,

$$Q = \frac{1}{\sqrt{1-(1/2)^4}} \times 0.6 \times \frac{\pi(75 \times 10^{-3})^2}{4} \times \sqrt{(2 \times 9.81 \times 250)/1000}$$

$$= 1.032 \times 0.6 \times 4.417 \times 10^{-3} \times 2.214$$

$$= 0.006057 \text{ m}^3\text{/sec.} = 6.057 \text{ litres/sec.}$$

(ii) For venturi- meter, the flow rate measurement will be given as,

$$Q = 1.032 \times 0.99 \times 4.417 \times 10^{-3} \times 2.214$$

$$= 9.99 \times 10^{-3} \text{ m}^3\text{/sec.} = 9.99 \text{ litres/sec.}$$

Example 20: An incompressible fluid is flowing through an orifice plate with a flow coefficient of 0.8 causing a pressure drop of 250 mm of water column. Calculate the fluid velocity.

Solution

Given that ,

Coefficient of discharge (C_d)= 0.8

$g = 9.8\ m/sec^2$

Pressure head (h)= 250 mm = 250×10^{-3} m

Applying the formula,

$$V = C_d \sqrt{2gh}$$

$V = 0.6 (2 \times 9.8 \times 250 \times 10^{-3})^{1/2}$

V = 1.77 m / sec

Example 21 : A pitot tube with a coefficient of 0.68 is used to measure the velocity of water in a pipe. The differential pressure head is 600 mm. find out the flow velocity ?

Solution

Given that ,

Coefficient of discharge (C_d)= 0.68

$g = 9.8\ m / sec^2$

Pressure head (h) = 600 mm = 600×10^{-3} m

Applying the formula,

$$V = C_d \sqrt{2gh}$$

$V = 0.68 (2 \times 9.8 \times 600 \times 10^{-3})^{1/2}$

V = 2.34 m / sec .

Example 22 : calculate the velocity of flow in an electromagnetic flow meter if ,

Flux density of liquid = 0.58 Weber/m^2 ,

Pipe diameter = 20 cm = 20×10^{-2} m

Induced voltage = 0.5 mV = 0.5×10^{-3}

Solution

Following formula is applied in this case ,

e = B v l

v = (e / B l)

$= 0.5 \times 10^{-3} /\{ 0.58 \times (20 \times 10^{-2})\} = 0.000058\ m/sec = 5.8 \times 10^{-3}\ cm/sec$

Example 23 : Find out , the induced emf in an electromagnetic flow meter due to the flow of a conductive fluid in a pipe with inner diameter of 1.75 cm. The flux density B = 4 mV. sec/cm^2 and volume flow rate Q = 450cm^3/min.

Solution

Given that ,

Flow rate (Q) = 450 cm^3/min ,

ID of pipe (d) = 1.75 cm = 1.75 x 10^{-2} m

Flux density (B) = 4 mV = 4 x 10^{-3} V sec = 40 v sec/m^2 = 40 weber / m^2

(i) Area (A) = (∏ / 4) d^2

= (3.14 / 4) x (1.75)^2 = 2 . 40 m^2

(ii) Velocity = 450 / (60 x 2.40) = 3.11 cm / sec

(iii) Induced Emf (e) = *B v l* = *40 x* 1.75 x 10^{-2} x 3.11 x 10^{-2}

= 0.02177 V = 21.77 mV

Example 24 : calculate the speed or velocity for following two condition ,

(i) When the pitot tube is placed at submarine , pressure head or reading of manometers 27 kN/m^2 and density of sea water is 992 kg/m3

(ii) When the pitot tube is mounted at aircraft, pressure head or reading of pressure gauge 9 kN/m^2 and density of air is 1000 kg/m3 .

Solution

Given that ,

(i) For submarine

Pressure head or pressure difference (ΔP) = 27 kN/m^2

Density of sea water (ρ) = 992 kg/m3 ,

Then the velocity or speed of submarine may be calculated as ,

$$V = \sqrt{\frac{2\Delta P}{\rho}}$$

= [(2 x 27000) / 992] ^{½}**= 7.4 m / sec .**

(ii) For aircraft ,

Pressure head or pressure difference (ΔP) = 9 kN/m^2

Density of air (ρ) = 1000 kg/m3 ,

$$V = \sqrt{\frac{2\Delta P}{\rho}}$$

= { (2 x 9000) / 1000 }$^{1/2}$ = **4.3 m / sec**

Example 25 : A rota meter uses a cylindrical float of 3.5 cm height, 2.5 cm diameter and density of 2800 kg/m3. The maximum inside diameter of the metering tube is 4.5 cm. Determine the maximum flow rate handling capacity of the rota meter if the fluid is water.

Assume that ,

Float volume (V_f) = 2.15 x 10^{-5} m^3

C_d = 0.82

Float diameter (D_f) = 2.5 x 10^{-2} m

Pipe diameter (D_p) = 4.5 x 10^{-2} m

Solution

The volume flow rate (Q) will be expressed by following relationship ,

$$Q = Cd \cdot \frac{D_p^{\,2} - D_f^{\,2}}{D_f} \sqrt{\frac{\pi g V_f (\rho_f - \rho)}{2\rho}}$$

Q = [0.82 x {(0.045^2 – 0.025^2) / 0.025} **]** [3.14 x 9.8 x 2.15 x 10^{-5} {(2.8 x 10^3) – 1000 **}]**

= [0.82 x 0.056] [0.04592] = **2.52 x 10^{-2} m^3/ sec**

Question for Discussion

1. Describe the construction and working of resistance temperature Detector (RTD) with its advantages and disadvantages
2. Describe with sketch bimetal strip thermometer stating salient features.
3. Sketch and describe the construction and working of Bellows pressure gauge.
4. Explain the construction and working of LVDT with its advantage and disadvantages
5. Explain the construction and working principle of Rotameter.
6. Explain the working principle of Ionization gauge with advantages and disadvantages
7. Explain the construction and working of bourdon tube pressure gauge with neat sketch.
8. Explain the various types of Bimetallic thermometer also state their application.
9. Discuss the construction and working of thermocouple with neat Sketch.
10. Discuss briefly about the temperature measurements using bimetallic strip and pressure thermometer.
11. Discuss briefly about the flow measurements using orifice meter and venturimeter.
12. What are the different types of bi- metallic sensors?
13. Mention the advantages and disadvantages of orifice meter.
14. With a neat sketch explain the constructional feature and working of (i) piezoelectric accelerometer, (ii) Ionization gauge for pressure measurement.
15. With neat sketch explain the construction and working principle of vapor pressure thermometer.
16. Explain with neat diagram the purpose and operating principle of a venturimeter.
17. Explain the working principle of an electrical resistance thermometer.
18. List the advantages of temperature measurement by using the resistance thermometer
19. What are rotometers? State its applications.
20. What are thermo couples? State its applications.
21. Explain the construction and working of Venturimeter and Rotameter
22. Explain the construction and working of bimetallic strip and Thermocouple
23. Explain with neat diagram the purpose and operating principle of a venturimeter.
24. Explain the working principle of an electrical resistance thermometer.
25. What are the differences between orifice and Venturimeter?
26. Explain working of Pressure thermometer and resistance thermometer.

27. List the advantages of temperature measurement by using the resistance thermometer.
28. Explain the construction & working of a Venturimeter.
29. Describe a method of orifice flow measurement using a suitable instrument.
30. Describe with neat sketches of Thermocouples
31. Explain in detail about the characteristics of fluid flow and measurement of fluid velocity
32. List out different types of thermocouples and the materials used. State and explain the laws governing thermocouples and explain the temperature measuring techniques with thermocouples
33. Describe the function of a bourdon tube pressure gauge in detail.
34. Describe any two types of flow measurement equipments.
35. Rotometer – Explain in detail with sketch.
36. Explain the working of bimetallic strip type temperature measurement system.
37. A simple thermocouple circuit is formed using two wires of different alloy: one end of the wires is twisted together to form the measuring junction, while the other ends are connected to a voltmeter and form the reference junction. A voltage is set up by the difference in temperature between the two junctions. For a given pair of alloy material and reference junction temperature, the temperature of the measuring junction is inferred from the measured voltage difference. For a measurement, what variables need to be controlled? What are the dependent and independent variables?
38. Calculate the resistance of a platinum wire that is 2 m in length and has a diameter of 0.1 cm. The resistivity of platinum at 25C is 9.83 x 10-6 V-cm. What implications does this result have for the construction of a resistance thermometer using platinum?
39. An RTD forms one arm of a Wheatstone bridge, as shown in Figure 8.38. The RTD is used to measure a constant temperature, with the bridge operated in a balanced mode. The RTD has a resistance of 25 V at a temperature of 0C, and a thermal coefficient of resistance, a = 0.003925C1. The value of the variable resistance R1 must be set to 37 V to balance the bridge circuit. Determine the temperature of the RTD.
40. An RTD forms one arm R4 of a Wheatstone. The RTD is used to measure a constant temperature, with the bridge operated in deflection mode. The RTD has a resistance of 25 V at a temperature of 0C, and a thermal coefficient of resistance, a = 0.003925 o C-1. If the RTD is subjected to a temperature of 100C and the input voltage to the bridge is 5V, what is the output voltage?

41. A thermistor is placed in a 100C environment, and its resistance measured as 20,000 V. The material constant, b, for this thermistor is 3650C. If the thermistor is then used to measure a particular temperature, and its resistance is measured as 500 V, determine the thermistor temperature

42. Define and discuss the following terms related to thermocouple circuits:

    a. thermocouple junction

    b. thermocouple laws

    c. reference junction

    d. Peltier effect

    e. Seebeck coefficient

43. The thermocouple circuit in represents a J-type thermocouple with the reference junction having T2 = 0C. The output emf is 13.777 mV. What is the temperature of the measuring junction, T1

44. The thermocouple circuit is composed of copper and constantan and has an output voltage of 6 mV for T1 = 200C. What is T2?

45. Estimate the required level of uncertainty in the measurement of resistance for a platinum RTD if it is to serve as a local standard for the calibration of a temperature measurement system for an uncertainty of 0.005C. Assume R(0C) = 100 V.

46. An iron-constantan thermocouple is placed in a moving air stream in a duct, The thermocouple reference junction is maintained at 212F. The wall temperature, Tw, is 500F, and the air velocity is 200 ft/s. The emf output from the thermocouple is 14.143 mV. Determine the thermocouple junction temperature

47. A temperature measurement system consists of a digital voltmeter and a T-type thermocouple. The thermocouple leads are connected directly to the voltmeter, which is placed in an air-conditioned space at 25C. The output emf from the thermocouple is 10 mV. What is the measuring junction temperature?

48. Flexivity, k, is a property of bimetallic strips that if known allows the radius of curvature to be calculated directly as

$$r_c = \frac{d}{2k(T_2 - T_1)}$$

where T_2 is the temperature of interest, T_1 is the assembly temperature of the bimetallic strip, and r_c is infinite. If a 3-cm strip is assembled at 25°C, plot the shape of the strip at temperatures of 50°C, 75°C, and 100°C for a k value of 7.3×10^{-5} K^{-1}.

49. Briefly explained the principle of employed in manometers used for measurement of pressure

    One end of U-tube manometer is connected to a piezometer opening in an air duct and the other end is left open to atmosphere. if the mercury level in the manometer on the duct side stands 0 .2 m than on the end open to atmosphere , calculate the absolute pressure of the air inside the duct. take atmospheric pressure Pa = 1.01325 N/m^2

50. Explain the mechanism of bourden tube pressure gauge and explain how the tube functions and why the curved tube tends to straighten out when pressure is applied inside ? mention the material used for the tube and the operating range of particular materials.

51. Explain with the suitable sketch, the difference between bellows gage and a diaphragm gauge that is applied for pressure measurement.

52. Explain with the suitable sketch, the working of metallic diaphragm , capsules and bellow gauge . also described the material or metal required for it manufacturing .

53. Explain the working principle with suitable sketch of McLeod gauge and Pirani gauge

54. Explain the operational principle of dead weight tester with suitable diagram used for force measurement .

    A U – tube mercury manometer is used to measure the presuure, above atmospheric pressure, of water in a pipe. the right limb of manometer contains mercury and is open to atmosphere. the mercury is in contact with water in the left limb. explain its working

55. A liquid filled thermometer has a bulb volume 5 times that of the capilliary plus bourdon tube. estimate the error due to increase of ambient temperature of 10^0 from the temperature of calibration.

56. Determine the true gas temperature, when hot gas flowing through a large pipe with small thermocouple inserted at the centre. the following data is recorded under steady state conditions.

    Temperature of tube wall = 340^0C, thermocouple temperature = 150^0C , emissivity of thermocouple = 0.75, convective film coefficient = 46.5 w/m^2, radiation for black body = 5.67×10^{-8} w/m^2

57. Explain various factors which affects the response of various temperature sensing devices.

58. A bimetallic strip elements is made up of strips of nickel – chrome – iron alloy and invar bonded together at 250C. each strip has a thickness of 1 mm , the length of cantilever is 50 mm .

   i. calculate radius of curvature at 1500C

   ii. free ends movements when if composition having thermal expansion 11 x 10^{-6}

59. In a thermistor, the characteristic equation is given as **Rt = Roa** $\exp^{b/T}$. where R_0 = resistance at ice temperature, R_t=resistance at absolute temperature, a & b are constant. calculate the thermistor temperature if R_t=1000 Ω , R_o = 4000 Ω , a = 250 x 10^{-6} , b = 2200

60. Differentiate between following,

   (i) Resistance thermometer and thermistors,

   (ii) Resistance thermometer and thermocouple,

61. Determine the diameter of throat of venturi meter, the pipe diameter = 10 cm, reading of differential U-tube manometer=60 cm, discharge = 20 lit/sec, coefficient if discharge = 0.95, distance between throat and inlet =20 cm .

62. Why rotameter is called variable area flow meter, explain with suitable sketch.

63. Described the advantage and disadvantage of venturi meter, rotameter and orifice meter.

64. Plot the resistance of a platinum wire that is 5 m long as a function of wire diameter ranging from 0.1 mm to 2 mm. Estimate the allowable tensile load for the 0.1 mm wire. (Hint: The modulus of elasticity of platinum is 145 GPa.)

65. Explain the sensing elements commonly used in pressure gauges.

   A liquid is flowing past a restriction in a process pipe with a volumetric flow of 10 ft3 /sec, causing a pressure drop of 2 in H_2O across the restriction. Calculate the volumetric flow if the pressure drop across the restriction increases to 5inH_2O.

66. An incompressible fluid is flowing through an orifice plate with a flow coefficient of 0.9, causing a pressure drop of 10 in H_2O. Calculate the fluid velocity

67. Explain the advantages and disadvantages of the three main types of orifice plates.

References

1. T.D. McGee, Principles and Methods of Temperature Measurement, Wiley-Interscience, New York, 1988.
2. R.P. Benedict, Fundamentals of Temperature, Pressure, and Flow Measurements, 3rd ed., Wiley, New York, 1984.
3. R.J. Moffat, Gas temperature measurements, Temperature-Its Measurement and Controlin Science and Industry, Vol. 3, Part 2, Reinhold, New York, 1962.
4. B. Munson, D. Young, T. Okishi, Fundamentals of Fluid Mechanics, 6th ed., Wiley, New York, 2009.
5. E.O. Doebelin, Measurement Systems: Application and Design, 5th ed., McGraw-Hill Science/Engineering/Math, New York, 2003
6. R.J. Goldstein, ed., Fluid Mechanics Measurements, 2nd ed., CRC Press, New York, 1996
7. Instrument Society of America (ISA), A Guide for the Dynamic Calibration of Pressure Transducers, ISA-37.16.01-2002, Instrument Society of America, 2002
8. H.G. McLeod, Vacuum Gauge, Philosophical Magazine, 1874. Reprinted in History of Vacuum Science and Technology, edited by T.E. Madey and W.C. Brown, American VacuumSociety, New York, 1984, pp. 102–105
9. W.G. Brombacher, D.P. Johnson, J.L. Cross, Mercury Barometers and Manometers, National Bureau Standards Monograph, 1960.
10. H. Rouse, S. Ince, History of Hydraulics, Dover, New York, 1957.
11. International Organization for Standardization (ISO), Measurement of fluid flow by means ofpressure differential devices inserted in circular cross-section conduits: Parts 1 through 4, ISO 5167, ISO, Geneva, 2003
12. American Society of Mechanical Engineers, Measurement of Fluid Flow in Pipes Using Orifice, Nozzle and Venturi, ASME Standard MFC-3M-1985, ASME International, New York, 1985
13. American Society of Mechanical Engineers, PTC 19.5-Flow Measurements, ASME International, New York, 2005.
14. J. Hall, (ed.). Sensor/Transducer Course, Instruments & Control Systems, 1990.
15. Hewlett-Packard Company, Practical Temperature Measurement-Applications, Note 290, 1987, Hewlett Packard Company.
16. Omega Engineering, Temperature Measurement Handbook 1986, Stanford, CT: Omega Engineering
17. Engineering Metrology, I.C. Gupta, Dhanpat Rai Publications, Delhi.

18. Mechanical Measurements, R.K. Jain Khanna Publishers, 1994.
19. Mechanical Measurements, Beckwith Marangoni and Lienhard, Pearson Education, 6th Ed., 2006.
20. Crane Co. Flow of Fluids through Valves, Fittings, and Pipe. Joliet, IL: Crane Co., 1996.
21. Giles, R. V. Theory and Problems of Fluid Mechanics and Hydraulics, Shaum's Outline Series, 2d ed., New York: McGraw-Hill, 1962.
22. D.W. Spitzer, Industrial Flow Measurement, Research Triangle Park, NC: ISA, 1990.
23. TC Ltd 2001 Guide to Thermocouple and Resistance Thermometry, issue 6.0.
24. J. Fraden, Handbook of Modern Sensors: Physics, Designs, and Applications, Springer-Verlag, 2003.
25. D. Johnson, Pressure Sensing: It's Everywhere, Control Engineering, 2001.

CHAPTER 4

STRAIN MEASUREMENT

4.0. Stress

Whenever any external force is applied on a body it resists the applied force for deformation this force per unit area is called stress.

Stress = Resisting force/Unit area

Strain: If any kind of deformation is achieved due to external load then this deformation per unit length is called strain. or in brief it is ratio of change in length to original length

Strain = Change in length/Original length

The Strain Gauge : The strain gauge is a device which is used for measuring the change in strain, the basic principle of such a device is that if electrical resistance varies with load application then it will be proportional to the amount of strain in that device. Electrical conductors possess resistance that is based on the following relationship,

$$R = \rho \frac{L}{A}$$

Where,

- R is resistance in ohms,
- ρ is resistivity constant, which is a property of the specific conductor material of the wire.
- L is the length of the wire in cm and
- A is the cross-sectional area in cm2.

If the change in the length of conductor or area of the wire is created then it will cause a change in the resistance. This phenomenon is called piezo-resistivity. There are various type of strain gauges available to fulfill the purpose, some of them are given below.

4.1. Type of Strain Gauge

Following are some important type of strain gauge which has been applied in measuring device.

(i) Bonded Strain Gauge: Bonded strain gauge made up of very fine wire or metallic foil arranged in a grid form. the cross section of fine wire or metallic foil is minimized in order to reduce or eliminate the effect of shear strain. The grid is bonded to a thin backing, called the carrier. then the strain experienced by the test specimen is transferred directly to the strain gauge, which responds with a linear change in electrical resistance. The range of such strain gauge varies from 30 to 3000 Ω, with 120, 350, and 1000 Ω. The schematic diagram of such strain gauge is given below.

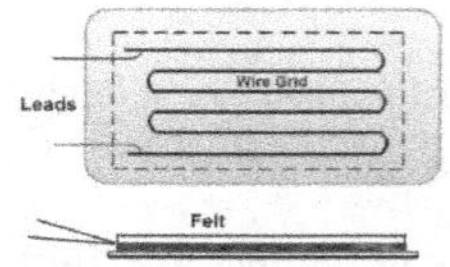

Diagram 4.1: Bonded Strain Gauge

(ii) Foil type Strain Gauge: The foil strain gage has metal foil on the electric insulator of the thin resin, and gauge leads attached, as shown in Fig.

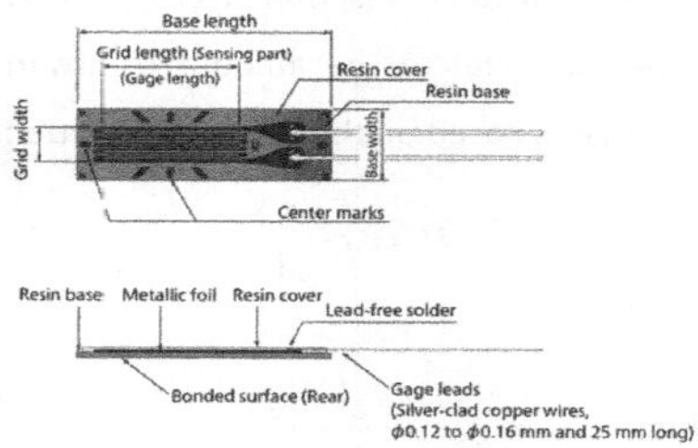

Diagram 4.2: Foil Strain Gauge

The strain gage is bonded to the measuring object with a dedicated adhesive. Strain occurring on the measuring site is transferred to the strain sensing element via adhesive and the resin base. For accurate measurement, the strain gage and adhesive should be compatible with the measuring material and operating conditions such as temperature, etc.

(iii) Thin-film Strain Gauges : Thin-film strain gage is more advanced form of strain gauge because it doesn't require any kind of adhesive bonding. A thin film strain gauge is constructed by first depositing an electrical insulation, usually a ceramic onto the stressed metal surface, and then depositing the strain gage onto that insulation layer. Techniques used to bond the materials molecularly are:

- Vacuum deposition
- Sputtering method

(iv) Semiconductor Strain Gages : Semiconductor strain gauges make use of the piezo-resistive effect in certain semiconductor materials such as silicon and germanium in order to obtain greater sensitivity and higher-level output. Semiconductor gauges can be produced to have either positive or negative changes when strained. They can be made physically small while still maintaining a high nominal resistance. Semiconductor strain gage bridges may have 30 times the sensitivity of bridges employing metal films, but are temperature sensitive and difficult to compensate. Their change in resistance with strain is also nonlinear. They are not in as widespread use as the more stable metal-film devices for precision work; however, where sensitivity is important and temperature variations are small, they may have some advantage. **The one of the most important factor of the strain gauge is its sensitivity to strain which is defined by gauge factors.**

4.1.1. Gauge Factor

Gauge factor is defined as the ratio of fractional change in electrical resistance to the fractional change in length (strain). The value of Gauge Factor for bonded metallic strain gauges is considered about 2.

Strain gauge arrangement & respective output : Since we know that , resistance change in electrical circuit can be easly detected & measured with the help of Wheatstone bridge circuit. In similar manner we can also apply the same principle to measure resistance changes of a strain gage. The following possible arrangement of strain gage may be tabulated below.

Bridge circuit	Resultant output
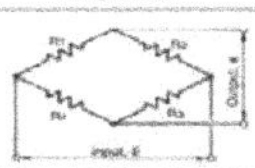 **Basic bridge circuit (Wheatstone Bridge)**	$R_1 = R_2 = R_3 = R_4$, or $R_1 \times R_3 = R_2 \times R_4$ If voltage is supplied and output is zero then it recognize as a balanced bridge condition .
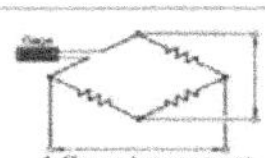 **1-Gauge Arrangement**	R_1 is repliced by strain gauge then **output** will be $e = \frac{1}{4} \cdot \frac{\Delta R_1}{R_1} \cdot E$ or, $e = \frac{1}{4} \cdot K \cdot \varepsilon_1 \cdot E$
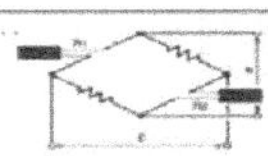 **2-Gauge Arrangement**	$e = \frac{1}{4}\left(\frac{\Delta R_1}{R_1} - \frac{\Delta R_2}{R_2}\right)E$ or, $e = \frac{1}{4} K(\varepsilon_1 - \varepsilon_2)E$
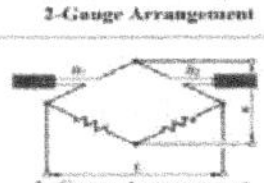 **2 -Gauge Arrangement**	$e = \frac{1}{4}\left(\frac{\Delta R_1}{R_1} + \frac{\Delta R_3}{R_3}\right)E$ or, $e = \frac{1}{4} K(\varepsilon_1 + \varepsilon_3)E$
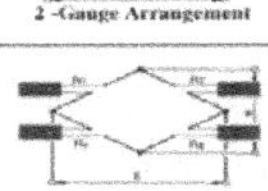 **4-Gauge Arrangement**	$e = \frac{1}{4}\left(\frac{\Delta R_1}{R_1} - \frac{\Delta R_2}{R_2} + \frac{\Delta R_3}{R_3} - \frac{\Delta R_4}{R_4}\right)E$ OR $e = \frac{1}{4} \cdot K(\varepsilon_1 - \varepsilon_2 + \varepsilon_3 - \varepsilon_4)E$ Where K is called Gauge Factor & ε_1 , ε_2 , ε_3 , ε_4 is called strain value

4.2. Measurement of Strain

Strain measurement are performed by following method.

(i) Wheatstone bridge : The Wheatstone bridge, consists of four resistive arms with an excitation voltage, VEX , that is applied across the bridge.

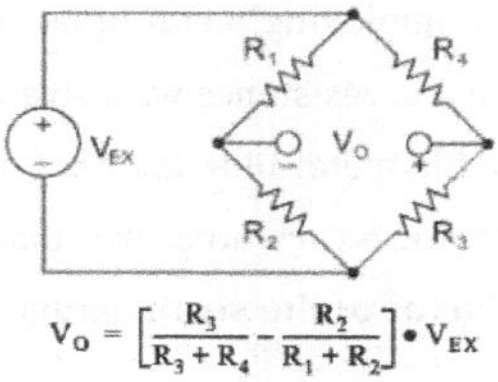

Diagram 4.3a: Wheatstone Bridge Circuit

The output voltage of the bridge, VO, will be equal to:

$$V_O = \left[\frac{R_3}{R_3 + R_4} - \frac{R_2}{R_1 + R_2}\right] \bullet V_{EX}$$

when $R_1 / R_2 = R_{G1} / R_{G2}$, the voltage output V_O will be zero. Hence, the bridge is said to be balanced. Any change in resistance in any arm of the bridge will result in a non zero output voltage .

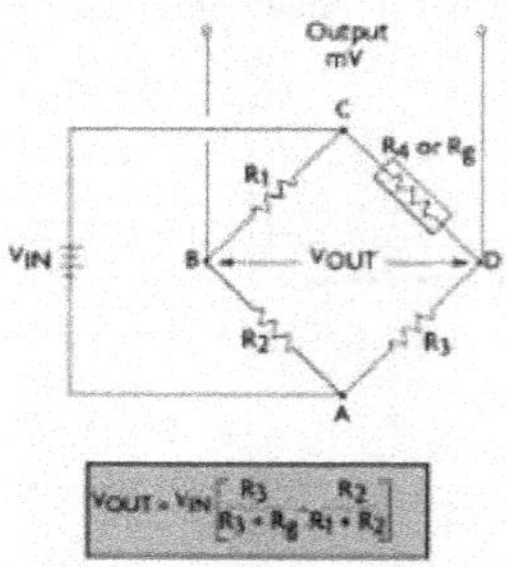

Diagram 4.3b: Wheatstone Bridge Circuit

Procedure for strain measurement : If we want to measure the strain by strain gauge , replacing any one of arm resistance by strain gauge .

- Replace resistance R4 with an active strain gauge, any changes in the strain gauge resistance will unbalance the bridge and having a certain output voltage.
- If R_G is a new strain induced due change in resistance.

- If ΔR is change in resistance then it is given by the following relationship

[$\Delta R = R_G \cdot GF \cdot \varepsilon$]

If R1, R2, R3 and RG is resistance then output voltage will be given as **1 / (1 + GF• ε/2)**

$$\frac{V_O}{V_{EX}} = -\frac{GF \bullet \varepsilon}{4}\left(\frac{1}{1 + GF \bullet \frac{\varepsilon}{2}}\right)$$

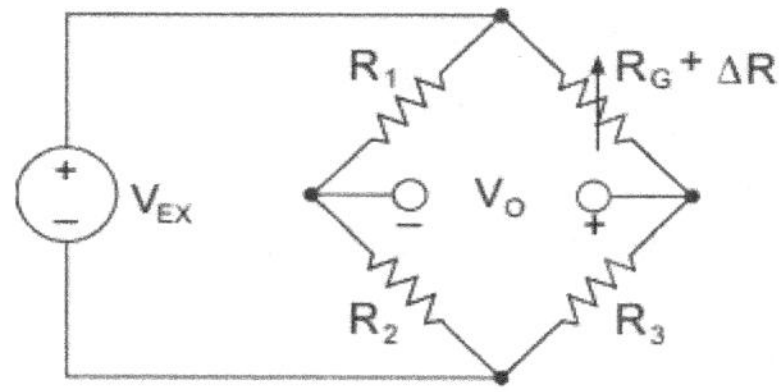

Diagram 4.4: Quarter-Bridge Circuit

In order to eliminate the effect of temperature, we also introduce the new strain gauge in the circuit, called dummy gauge. Thus

- One strain gauge is active strain gauge **(RG + ΔR),**
- Second strain gauge is the dummy gauge.

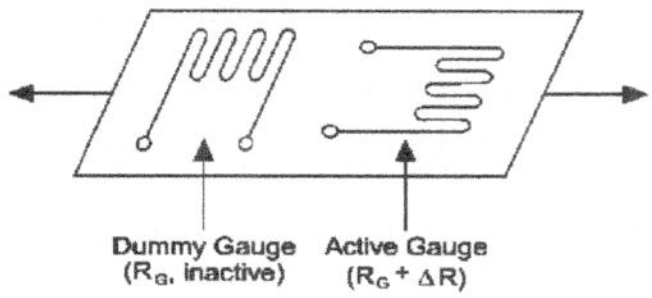

Diagram 4.5: Dummy Gauge

- In a bridge one resistance strain gauge is in tension (RG+ΔR) and the other in compression (R_G–ΔR). the schematic diagram is given below .

$$\frac{V_O}{V_{EX}} = -\frac{GF \bullet \varepsilon}{2}$$

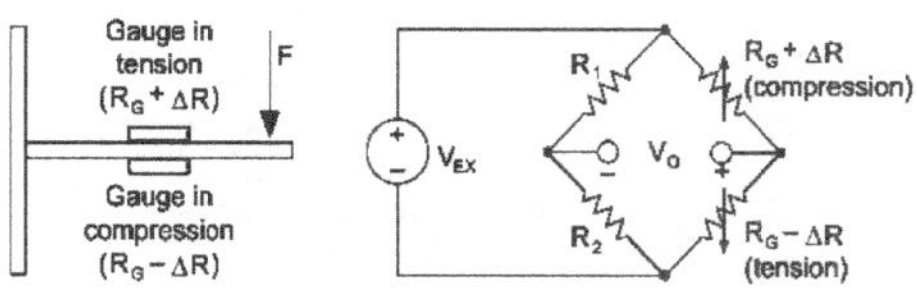

Diagram 4.6: Half-Bridge Circuit

We can further replace the active resistance by the resistance strain gauge , in which two strain gauges in tension and two gauges in compression. The full-bridge circuits is given below

$$\frac{V_O}{V_{EX}} = -GF \bullet \varepsilon$$

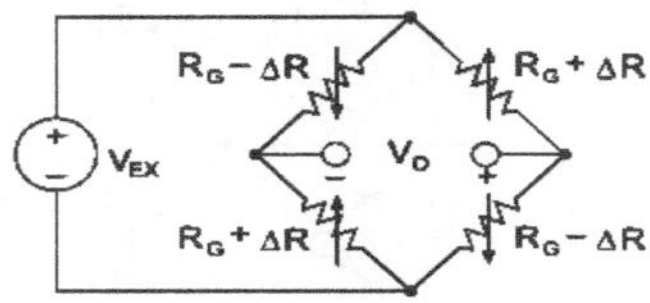

Diagram 4.7: Full-Bridge Circuit

4.3. Apparent Strain

Apparent strain sometimes also known as no load strain hence the question is arises here that how it developed , actually it is an expansion of gauge due to thermal effects , This expansion is described by the expansion coefficient of the material . Material expansion, completely depend upon changes in temperature, thus we can say that apparent strain is the result of the interaction of the thermal coefficient of the strain gauge and the difference in expansion between the gauge and the test specimen. The variation in the apparent strain of various strain-gage materials as a function of operating temperature is shown in Figure . In addition to the temperature effects, apparent strain also can change because of aging and instability of the metal and the bonding agent. Compensation for apparent strain is necessary if the temperature varies while the strain is being measured. In most applications, the amount of error depends on the alloy used, the accuracy required, and the amount of the temperature variation. If the operating temperature of the gage and the apparent strain characteristics are known, compensation is possible.

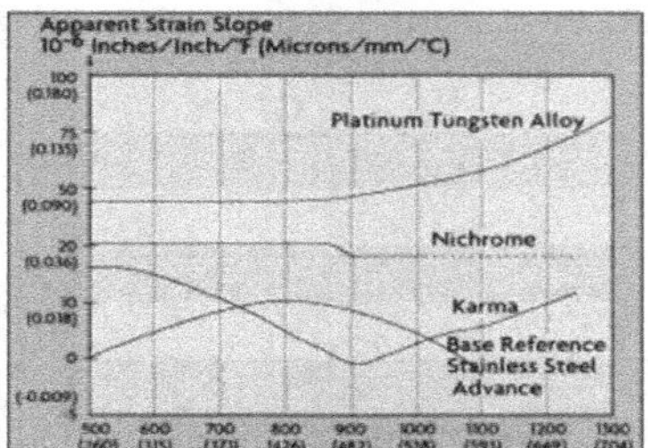

Diagram 4.8: Apparent Strain

4.4. Self-Temperature-Compensation Gauges

Self-temperature-compensation strain gauges are designed to adjust resistive temperature coefficient of their resistive elements to linear expansion coefficient of measuring objects in order to get apparent strain close to zero. When it bonded to a suitable material. The graph below shows apparent strain output of 3-wire strain gauge.

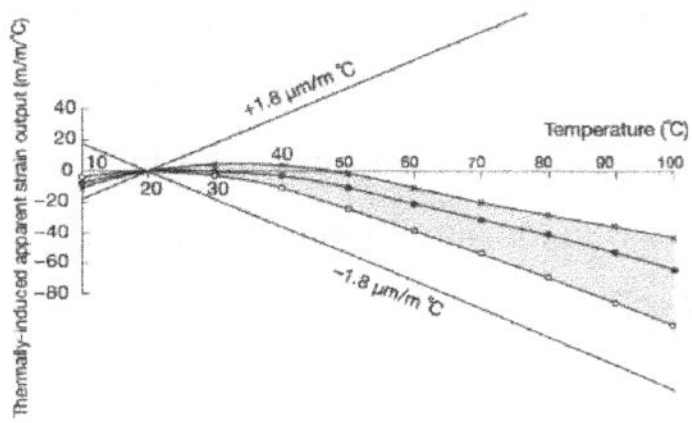

Diagram 4.9: Temperature Compensation of Foil Gauge

The measuring object and the resistive element of the strain gauge have linear expansion coefficients **β_S** and **βg**, respectively. Then, the strain gauge bonded on the surface of the object provides a thermally-induced apparent strain **ε_T** per **1°C** that is expressed with the following equation:

$$\varepsilon_T = (a / k_s) + (\beta_S - \beta g)$$

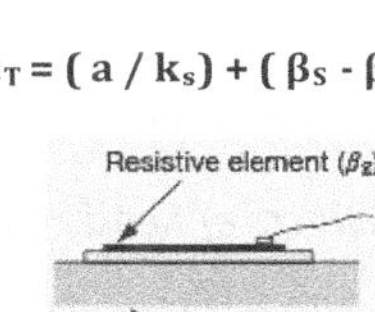

Diagram 4.10: Foil Gauge Attachment

where,

a is a resistance temperature coefficient of resistive elements

k_s is a gauge factor of strain gauge

- **Method of Compensating Gauge Factors :** If the gauge factor of the strain gauge is different from that (2.00) of the strain amplifier, the real strain **ε** can be obtained by the following equation

$$\varepsilon = (2 / k_s) \times \varepsilon_0$$

where ,

ε_0 is measured strain

k_s is a gauge factor of strain gauge

4.4.1. *Temperature Compensation Due to the Effect of Lead Wire*

If we have 2-lead wire system, strain output from the bridge is affected by temperature effect of the lead wire. To prevent the temperature effect, we will applied 3-lead wire system. If 3 leads are connected to the strain gauge as shown below, one half the lead wire resistance is applied to the adjacent side of the bridge to compensate the resistive components of the 2 leads affected by a similar temperature change, and thus the bridge output is free from any temperature effect of the lead wire. The temperature effect of a third lead connected directly to the amplifier can be ignored since the amplifier provides high input impedance. As precautions in using the 3-wire system, the 3 leads should be the same in type, length and cross-section to receive the same temperature effect.

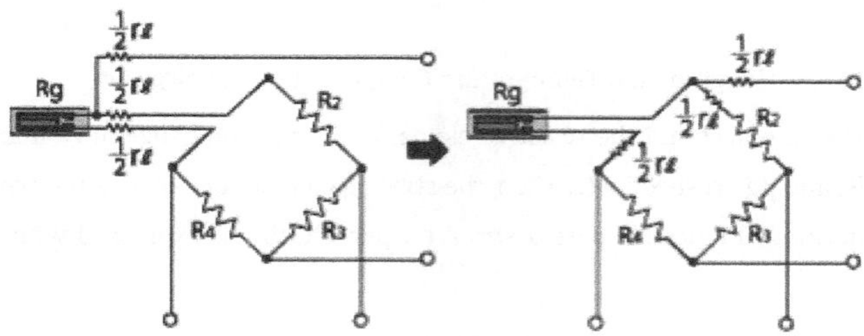

Diagram 4.11: Lead Wire Arrangement for Temperature Compensation

4.4.2. *Compensation Method of Effects of Lead Wire Extension*

If the lead wire or cable is extended with the quarter or half bridge system, additional resistance is initiated in series to the strain gauge, thereby decreasing the apparent gage factor. For example, if a 10m long lead wire with 0.3 mm^2 conductors is used, the gage factor decreases by **1%.** In these cases, the real strain **ε** can be obtained by the following equation.

$$\varepsilon = (1 + r_e / R_g) \times \varepsilon_i$$

where

ε_I is a measured strain

R_g is a resistance of strain gauge

r_e total resistance of Lead wire

4.5. Strain Gage Bridges Circuits

There are so many type of bridge circuit are implemented in the strain measurement technique, following are the some important bridge circuit with their specific characteristics.

	Gage method Connection system	Application	Circuit	Output	Remarks
1	**1-active-gage 2-wire** Nr. of gage:	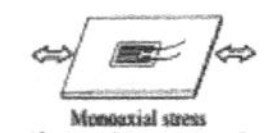 Monoaxial stress (uniform tension or compression)	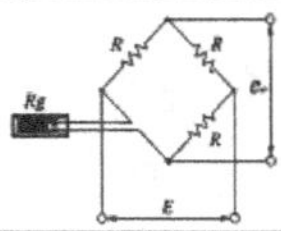	$e_0 = \frac{E}{4} K_s \cdot \varepsilon_0$ K_s : Gage factor ε_0 : Strain E : Bridge voltage e_0 : Output voltage Rg : Gage resistor R : Fixed resistor	No temp. compensation; x1 output; nonlinearity correction
2	**1-active-gage 3-wire** Nr. of gage:	 Monoaxial stress (uniform tension or compression)	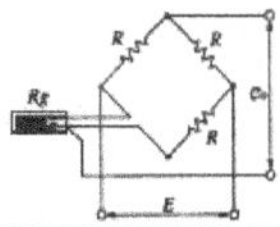	$e_0 = \frac{E}{4} K_s \cdot \varepsilon_0$	needed for large strainNo temp. compensation; thermal effect on leadwires cancelled; x1 output; non- linearity correction needed for large strain
3	**1-active-gage (2 in series) 2-wire (cancelling bending strain)** Nr. of	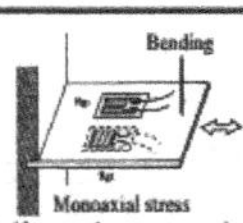 Monoaxial stress (uniform tension or compression)	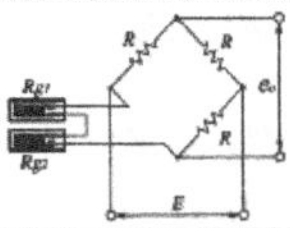	$e_0 = \frac{E}{4} K_s \cdot \varepsilon_0$ Rg_1 ······ Strain : ε_1 Rg_2 ······ Stra : ε_2 $\varepsilon_0 = \frac{\varepsilon_1 + \varepsilon_2}{2}$ R : Fixed resistor $Rg_1 + Rg_2 = R$	No temperature compensation; bending strain cancelled; x1 output
4	**1-active-gage (2 in series) 3-wire (cancelling bending strain)** Nr. of gages: 2	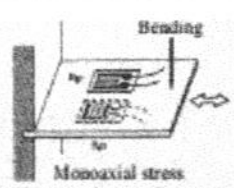 Monoaxial stress (uniform tension or compression)	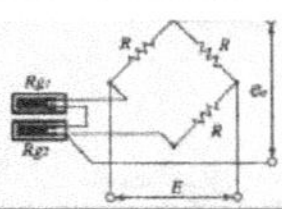	$e_0 = \frac{E}{4} K_s \cdot \varepsilon_0$ Rg_1 ······ Strain : ε_1 Rg_2 ······ Strain : ε_2 $\varepsilon_0 = \frac{\varepsilon_1 + \varepsilon_2}{2}$ R : Fixed $Rg_1 + Rg_2 = R$	No temperature compensation; thermal effect on leadwires cancelled; bending strain cancelled; x1 output
5	**Active-dummy (2-gage method)**	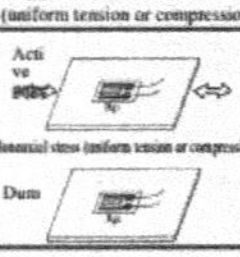 Monoaxial stress (uniform tension or compression)	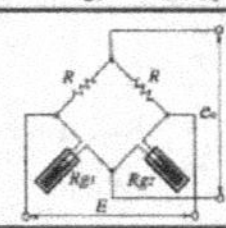	$e_0 = \frac{E}{4} K_s \cdot \varepsilon_0$ K_s : Gage factor ε_0 : Strain E : Bridge voltage e_0 : Output voltage Rg_1 : Strain : ε_0 R : Fixed Rg_2 : 0	Temperature compensation; thermal effect on leadwires cancelled; x1outppt
6	**2-active-gage (at right angles)**	 Monoaxial stress (uniform tension or compression)	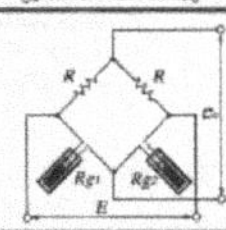	$e_0 = \frac{(1+\nu)E}{4} K_s \cdot \varepsilon_0$ ν : Poison's ratio Rg_1, Rg_2 : Gage resistor Rg_1 ······ Strain : ε_0 Rg_2 ··· Strain : $-\nu\varepsilon_0$ R : Fixed resistor	Temperature compensation; thermal effect on leadwires cancelled; ε (1+ν) output
7	**2-active-gage (cancelling bending strain)** Nr. of gages: 2	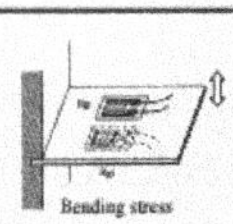 Bending stress	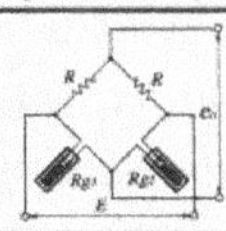	$e_0 = \frac{E}{2} K_s \cdot \varepsilon_0$ Rg_1 ······ Strain : ε_0 Rg_2 ··· Strain : $-\varepsilon_0$ R : Fixed	Temp. compensation; thermal effect on leadwires cancelled; comp.(tension) strain cancelled; x2 output
8	**2-wire** Nr. of gages:	 Monoaxial stress (uniform tension or compression)	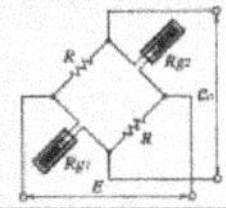	$e_0 = \frac{E}{2} K_s \cdot \varepsilon_0$ Rg_1 ······ Strain : ε_0 Rg_2 ······ Strain : ε_0 R : Fixed resistor	No temp. com- pensation; bending strain cancelled by gages cemented on front and back; nonlinearity compensation needed for large strain; x2 output
9	**Opposite 2-active-gage 3-wire** Nr. of gages: 2	 Monoaxial stress (uniform tension or compression)	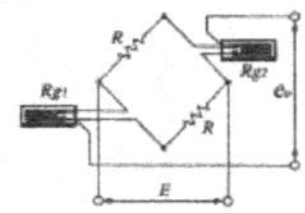	$e_0 = \frac{E}{2} K_s \cdot \varepsilon_0$ Rg_1 ······ Strain : ε_0 Rg_2 ······ Strain : ε_0 R : Fixed resistor	No temp. compensation; thermal effect on leadwires cancelled; bending strain cancelled by gages cemented on front and back; x2 output; correction of nonlinearity needed Temp. compensation; thermal effect on leadwires cancelled; compression (tension) strain cancelled; x4 output
10	**4-active-gage (bending strain measuring)**	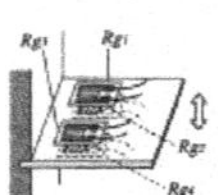 Bending stress	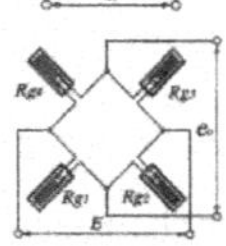	$e_0 = K_s \cdot \varepsilon_0 \cdot E$ Rg_1, Rg_3 ······ Bending strain : ε_0 Rg_2, Rg_4 ······ Bending strain : $-\varepsilon_0$	

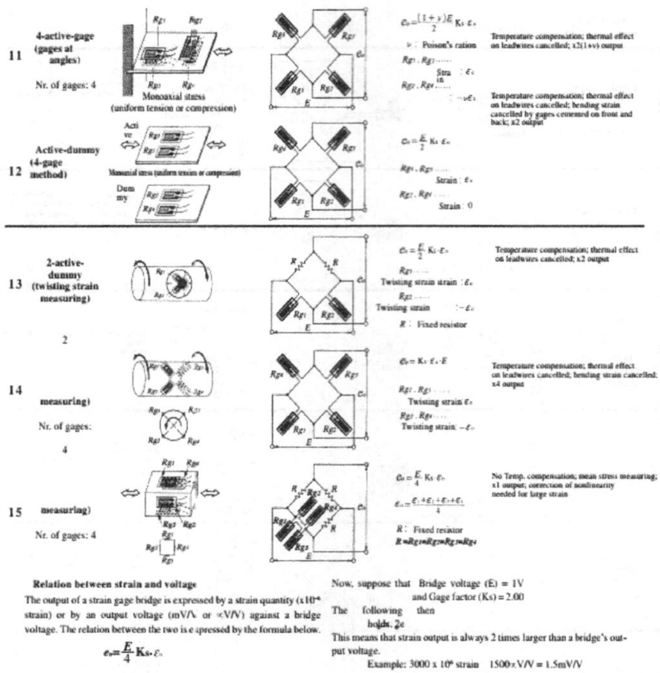

4.6. Gage Factor (GF)

Gage factor is the term used to define the fractional change in resistance of a gauge with applied strain. The larger the gage factor, the higher the resistance change and the resulting output, resolution, etc. Metal wire and foil gages have gage factors between 2 and 4.

Thus, gage factor (GF) is given by equation

$$GF = \frac{\Delta R}{R_0 \varepsilon} = 1 + 2\upsilon + \Pi_L Y$$

Where ,

υ is Poisson's ratio,

Y is Young's modulus for the semiconductor material,

ε is strain,

ΔR is the strain induced gage resistance change and

R₀ is the zero strain gage resistance value.

4.6.1. *Effect of Gauge Factors on Resistance*

(i) Increase in linearity : The single element nonlinearity at 1000 micro-strain may be improved over 10 times. This is also illustrated in Figure 4.12, The lower the resistivity the higher the inherent linearity.

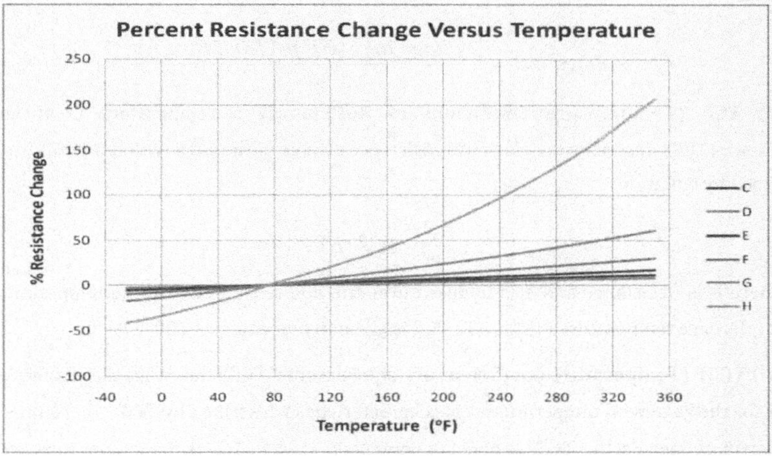

Diagram 4.12: Curve Temperature Vs Resistance

(ii) Reduction in temperature coefficient of resistance (TCR) : The temperature coefficient of resistance may be reduced from 50% / 100°F to about 2% / 100°F. The decrease of temperature coefficient of resistance with decreasing resistivity is illustrated in above Figure .

(iii) Reduction in temperature coefficient of gage factor (TCGF) : The temperature coefficient of gage factor may be decreased from -23% / 100°F to 6% / 100°F by decreasing resistivity.

(iv) Reduction in radiation sensitivity: Along with reducing resistivity, increased impurity concentrations markedly reduce radiation's effect on resistivity, gage factor, and temperature characteristics. Piezo-resistive strain gages are majority carrier devices without depletion layer isolation, which also reduces the sensitivity to radiation.

increase in operating temperature range : In general, the lower the resistivity, the greater the operating temperature range. Gages with impurity concentrations greater than 2 x 10^{19} impurities/$cm^3$are suitable for cryogenic applications.

4.6.2. Some Important Terms Related to Strain Gauge Measurement

(i) Gauge Output : Consider an equation in which 4 arm Wheatstone bridge circuit is used with input voltage . then the total strain is the arithmetic sum of strain magnitudes in all four arms of a Wheatstone bridge assuming tension and compression in adjacent arms of a four gauge bridge

$$Voltage\ out = \frac{(Voltage\ in)\times(GF)\times(Total\ Strain)}{4}$$

(ii) TCR (Temperature Coefficient of Resistance) : Temperature Coefficient of Resistance (TCR) describes the electrical resistance change of the gage with temperature. TCR , is defined by equation

$$\frac{\Delta R}{R} = TCR \cdot \Delta T$$

where R is Resistance and T is temperature. The difference between the temperature (T) and a reference temperature (T_0) is ΔT , TCR is given in percentage / 100°F .

(iii) TCGF (Temperature Coefficient of Gage Factor) : TCGF describes the change in gage factor for the gage with temperature. This characteristic is described by TCGF, or Temperature Coefficient of Gage Factor. TCGF is given in percentage/100°F. **ΔGF** is the temperature induced gage factor changes.

Analogous to TCR , TCGF is defined by the following equation

$$\frac{\Delta GF}{GF} = TCGF \cdot \Delta T$$

(iv) Gage Factor-Sensitivity : Gage factor is a measurement of the sensitivity of the strain gage and is defined as the resistance change per unit of initial resistance divided by the applied strain.

$$\mathbf{\Delta R = GF \times R \times \varepsilon}$$

Where ,

ΔR , Change in resistance ,

G_F,is a Gage Factor ,

R ,is Initial resistance of the gage

ε ,is a Strain

Foil gauges have a gage factor of 1 to 4, for silicon gages it has between 30 to over 200.

4.7. Displacement Measurement

Displacement Measurement is also a very important part of industrial measurement, following method are generally applied for displacement measurement.

(i) Linear Variable-Differential Transformer (LVDT) : The linear variable-differential transformer, LVDT, is a passive inductive transducer finding many applications. It consists of a single primary winding positioned between two identical secondary windings wound on a tubular ferromagnetic former, as shown in Diagram 4.13. The primary winding is energized by a high-frequency 50 Hz to 20 kHz ac voltage. The two secondary windings are made identical by having an equal number of turns and similar geometry. They are connected in series opposition so that the induced output voltages oppose each other. The output voltages of individual secondaries v_1 and v_2 at null position will be zero if any displacement in the core position indicates by the deflection from the null position. The output waveform v_o depends upon the core position, When the core is positioned in the middle, there is an equal coupling between the primary and secondary windings, thus giving a null point or reference point of the sensor.

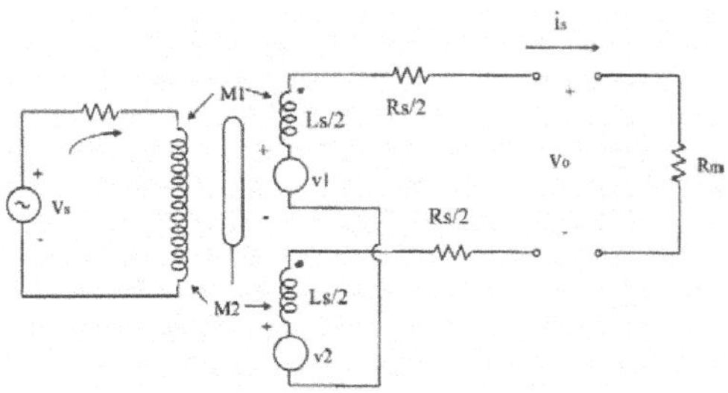

Diagram 4.13: Linear Variable-Differential Transformer

As long as the core remains near the center of the coil arrangement, output is very linear. The linear ranges of commercial differential transformers are clearly specified, and the devices are seldom used outside this linear range. The ferromagnetic core or plunger moves freely inside the former, thus altering the mutual inductance between the primary and secondaries. With the core in the center, or at the reference position, the induced emfs in the secondaries are equal; and since they oppose each other, the output voltage is zero. When the core moves, say to the left, from the center, more magnetic flux links with the left-hand coil than with the right-hand coil. The voltage induced in the left-hand coil is therefore larger than the induced voltage on the right-hand coil. The magnitude of the output voltage is then larger than at the null position and is equal to the difference between the two secondary voltages. The net output voltage is in phase with the voltage of the left-hand coil. The output of the device is then an indication of the displacement of the core. Similarly, movement in the opposite direction to the right from the center reverses this effect, and the output voltage is now in phase with the emf of the right-hand coil .

(ii) Rotary Variable-Differential Transformer: A variation from the linear-variable differential transformer is the rotary core differential transformer shown in **Diagram 4.14**. Here, the primary winding is wound on the center leg of an E core, and the secondary windings are wound on the outer legs of the core. The armature is rotated by an externally applied force about a pivot point above the center leg of the core. When the armature is displaced from its reference or balance position , the reluctance of the magnetic circuit through one secondary coil decreases , simultaneously increasing the reluctance through the other coil. The induced emf 's in the secondary windings , which are equal in the reference position of the armature, are now different in magnitude and phase as a result of the applied displacement. The induced emf's in the secondary coils are made to oppose each other and the transformer operates in the same manner as an LVDT. The rotating variable transformers may be sensitive to vibrations. If a dc output is required, a demodulator network can be used, as in the case of LVDT.

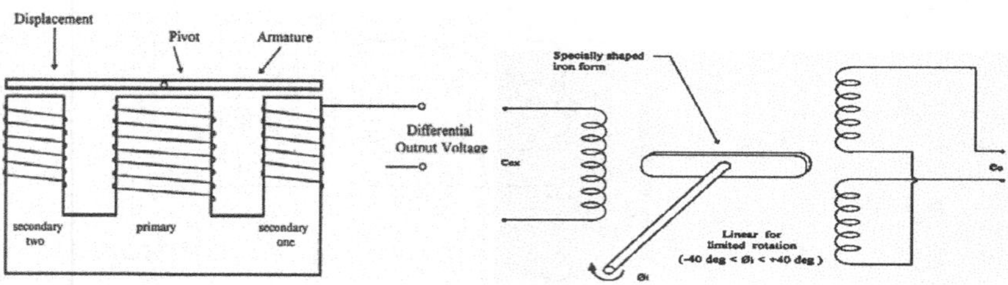

Diagram 4.14: Rotary Variable-Differential Transformer

(iii) Capacitive Displacement Sensors : The measurement of distances or displacements is an important aspect of many industrial, scientific, and engineering systems. The displacement is basically the vector representing a change in position of a body or point with respect to a reference point. Capacitive displacement sensors satisfy the requirements of applications where high linearity and wide ranges (from a few centimeters to a couple of nanometers) are needed. The basic element of Capacitive Displacement Sensors are given below,

(i) Two simple electrodes with capacitance C

(ii) The capacitance is a function of the distance d (cm) between the electrodes of a structure, the surface area A (cm^2) of the electrodes, and the permittivity ε

$$C = f(d, A, \varepsilon)$$

(iv) Variable Distance Displacement Sensors : A capacitor displacement sensor, made from two flat coplanar plates with a variable distance *x* apart , is illustrated in **Diagram 4.15.** Ignoring fringe effects, the capacitance of this arrangement can be expressed by

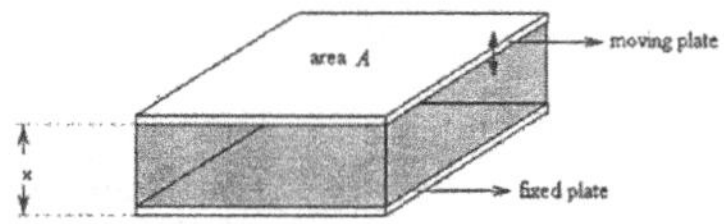

Diagram 4.15: Variable Distance Displacement Sensors

$$C(x) = \varepsilon A/x = \varepsilon_r \varepsilon_0 A/x$$

Where ,

ε is the dielectric constant or permittivity

ε_r is the relative dielectric constant

ε_0 is 8.854188 × 10–12 F/m , the dielectric constant of vacuum

x is the distance of the plates in m

A is the effective area of the plates in mm

The capacitance of this transducer is nonlinear with respect to distance *x*, having a hyperbolic transfer function characteristic. The sensitivity of capacitance to changes in plate separation is

$$dC/dx = -\varepsilon_r \varepsilon_0 A/x^2$$

Equation indicates that the sensitivity increases as xdecreases. This type of sensor is often used for measuring small incremental displacements without making contact with the object.

(v) Variable Area Displacement Sensors: Alternatively, the displacements may be sensed by varying the surface area of the electrodes of a flat plate capacitor, as illustrated in **Diagram 4.16.** In this case, the capacitance would be :

$$C = \varepsilon_r \varepsilon_0 (A - wx)/d$$

where ,

w = The width

wx = The reduction in the area due to movement of the plate

Then, the transducer output is linear with displacement *x*. This type of sensor is normally implemented as a rotating capacitor for measuring angular displacement. The rotating capacitor structures are also used as an output transducer for measuring electric voltages as capacitive voltmeters.

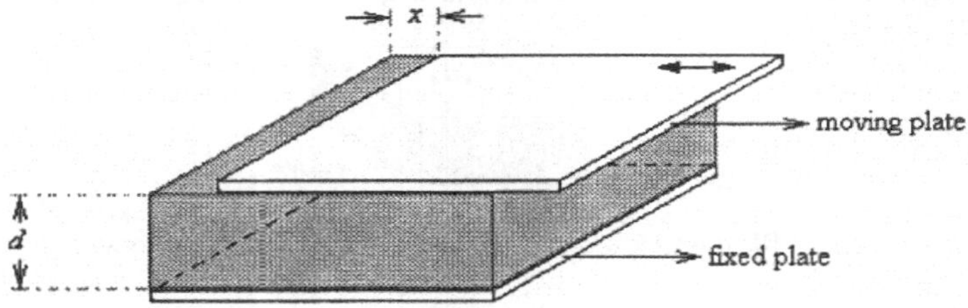

Diagram 4.16: Variable Area Displacement Sensors

(vi) Variable Dielectric Displacement Sensors :- In some cases , the displacement may be sensed by the relative movement of the dielectric material between the plates , as shown in Diagram 4.17. The corresponding equations would be:

$$C=\varepsilon_0\, w\left[\varepsilon_2 l-\left(\varepsilon_2-\varepsilon_1\right)x\right]$$

Diagram 4.17: Variable Dielectric Displacement Sensors

Where,

ε_1 = The relative permittivity of the dielectric material

ε_2= The permittivity of the displacing material (e.g., liquid)

In this case, the output of the transducer is also linear. This type of transducer is predominantly used in the form of two concentric cylinders for measuring the level of fluids in tanks. A non conducting fluid forms the dielectric material.

4.8. Angular Displacement Measurement

In this category we will about following method,

(i) Capacitive Pick Up : Capacitance is well known to be function of effective area of the conductors, separation between them, and the dielectric strength of the material in separation. Any one of these three can be the varying factor causing a change in capacitance. The capacitive transducer along with this equipment is based on the principle of variation of effective area of the conductors, other parameters i.e. separation, distance and dielectric strength being kept constant. A two gauged condenser normally used in transistor radio receivers, is used to demonstrate the measurement of angular displacement measurement.

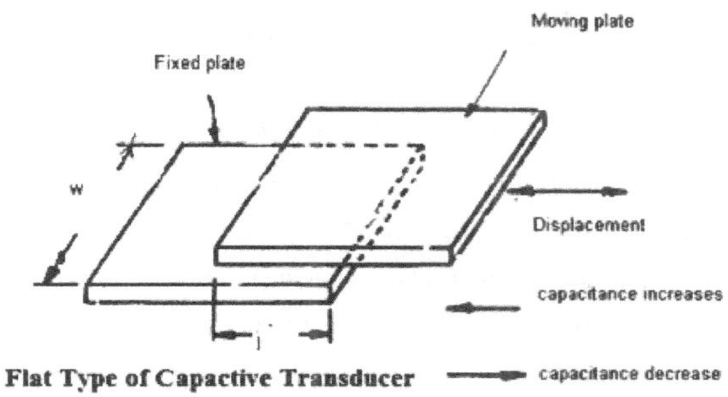

Diagram 4.18: Flat Type Capactive Transducer

Working & operation : Connect the capacitive pick up cable to the input socket on the main unit. Set the input angular displacement & DPM at zero indication value if it does not you can used to potential meter marked **MIN** and maintain zero value representing zero angular displacement of the rotating shaft. The indication on the DPM representing the angular displacement with respect to input.

(ii) Potentiometers: Resistive displacement sensors are commonly termed as potentiometers or **"pots."** A pot is an electro-mechanically device containing an electrically conductive wiper that slides against a fixed resistive element according to the position or angle of an external shaft. See **Diagram 4.19**, the resistive element is "divided " at the point of wiper contact. To measure displacement, a pot is typically wired in a "voltage divider", the circuit is as shown in **Diagram 4.19**. The circuit's output, a function of the wiper's position, is an analog voltage available for direct use or digitization. Calibration maps the output voltage to units of displacement.

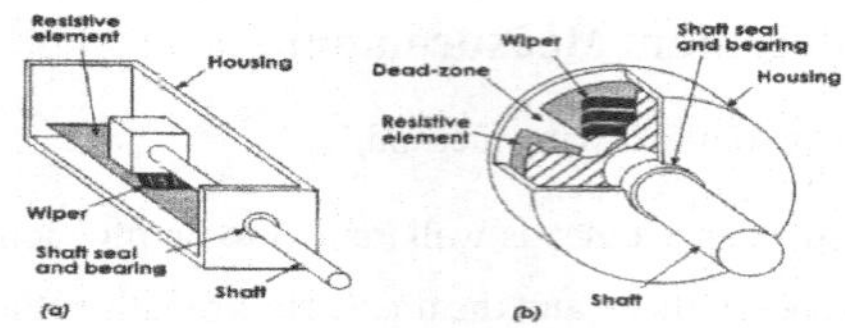

Linear-Motion Potentiometers Rotary Potentiometers

Diagram 4.19: Potentiometers

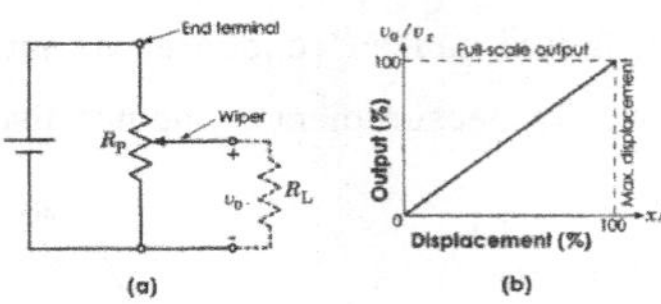

Diagram 4.20: Potentiometers

a. Shows potentiometer as a resistor with an arrow representing the wiper.

b. Shows an ideal linear output function where *x* represents the wiper position.

(iii) Linear Potentiometer Transducer : A linear potentiometer **transducer** consists of a potentiometer, which is short circuited by a slider. The other end of the slider is connected to a slider arm. The force summing device on the slider arm causes linear displacement of the slider causing the short circuit of a certain portion of the resistance in the potentiometer. Let the whole resistance positions on the potentiometer be ABC. Let the resistance position caused by the slider movement be BC. As the movement of the slider moves further to the right, the amount of resistance increases. This increase in resistance value can be noted according to the corresponding change in the linear displacement of the slider. The change in resistance can be calculated with the help of a Wheatstone bridge.

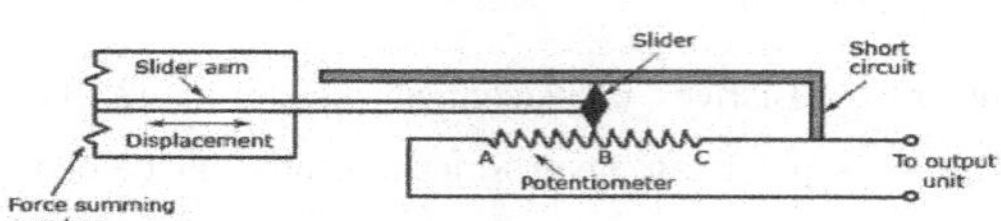

Linear potentiometer construction

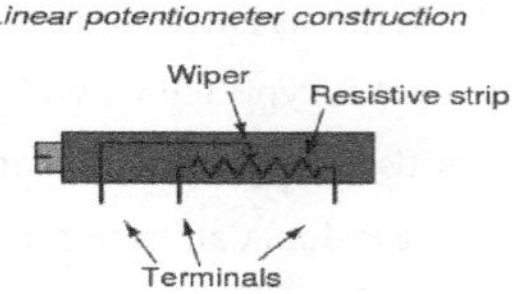

Diagram 4.21: Linear Potentiometer

(iv) The Rotary Potentiometer: In terms of operating principles, the rotary potentiometer is the same as the linear potentiometer. The wiper moves over a fixed resistance to provide a varying output signal proportional to the movement. The difference is in the design and manufacture so that the resistance forms into an arc and that the wiper rotates about a central pivot. The rotary potentiometer has been included to illustrate the conversion of linear motion into rotary motion in this example by the use of a worm and wheel arrangement, though any similar arrangement producing the same effect could be used. In addition, although the shaft rotates by the movement of the Linear Assembly the final motion delivered to the potentiometer is rotary. The main objective of this device is to use of the rotary potentiometer for measuring linear motion. While moving the Linear Assembly over its full range of movement visually inspect the effect this has on the shaft of the rotary potentiometer in terms of the angle moved through.

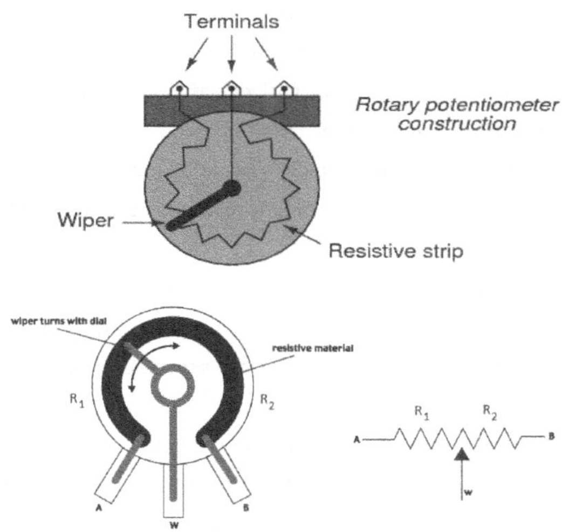

Diagram 4.22: Rotary Potentiometer

4.9. Type of Potentiometer

Except above potentiometers, the following type also generally used in an industrial application , industries may be of mechanical , electronic or electrical etc .

(i) On the basis of Resistive Element : Resistive element can be classified as either wire wound, or non wire wound.

- **Wire wound** elements offer excellent temperature stability and high power dissipation abilities. The coils quantize measurement according to wire size and spacing. Providing the resolution limits are acceptable, wire wound elements can be a satisfactory choice for precision measurement, **Wire wound** elements contain tight coils of resistive wire that quantize measurement in step-like increments. The most commonly used resistance elements in this potentiometer are nickel, chromium or nickel copper because it has low temperature coefficient of resistance, very cost effective and applied for handling large currents capacity. The winding of the resistance wire will depend on the different types of resistance changes due to the slider motion.
- **non-wire wound** elements present a continuous sheet of resistive material capable of essentially unlimited measurement resolution.

(ii) Cermets Potentiometer: Cermets Potentiometer, offer unlimited resolution and reasonable noise levels. Their advantages include high power dissipation abilities and excellent stability in adverse conditions ,very cost effective , low temperature coefficient of resistance at high temperatures .This type of potentiometer is made from a material called Cermets composed of ceramic-metal alloy which is made by mixing a paste of precious metal particles and a ceramic. Some of the most common mixtures used are palladium silver glass and palladium oxide glass.

(iii) Carbon Potentiometers: It is made by depositing a thin film of carbon and a thermosetting plastic binder. This device is mostly used for alternating current (ac) purposes. They offer unlimited resolution and low noise, but are sensitive to environmental stresses (e.g., temperature, humidity) and are subject to wear.

(iv) Conductive plastic : It has a smooth film with unlimited resolution , low friction , low noise , and long operational life. They are sensitive to temperature and other environmental factors and their power dissipation abilities are low however, they are an excellent choice for most precision measurement applications.

(v) Hybrid Potentiometers : Hybrid elements have a wire wound core with a conductive plastic coating , which combines wire wound and conductive plastic elements .They are applied for precision measurement. In this type of device conductive plastic elements imparts low noise, long life, and unlimited resolution, excellent temperature stability as well as limit the power dissipation abilities also.

(vi) Thin Metal Film Potentiometer : This device is in the form of a thin vapour deposited layer of metal on glass or ceramic base. This is also used for ac applications.

4.10. Velocity Measurement

Velocity may be defined as it is a displacement created by the particles due to the effect of the application of forces with respect to time. It is a vector representation of the displacement. The velocity is expressed in meters per second **(M/S)**, centimeter per second **(CM/S)** and angular velocity is expressed as radians per second **(RAD/S)** or revolutions per minute **(RPM)**

On the basis of particle motion, the velocity measurement can be broadly classified in the following two categories.

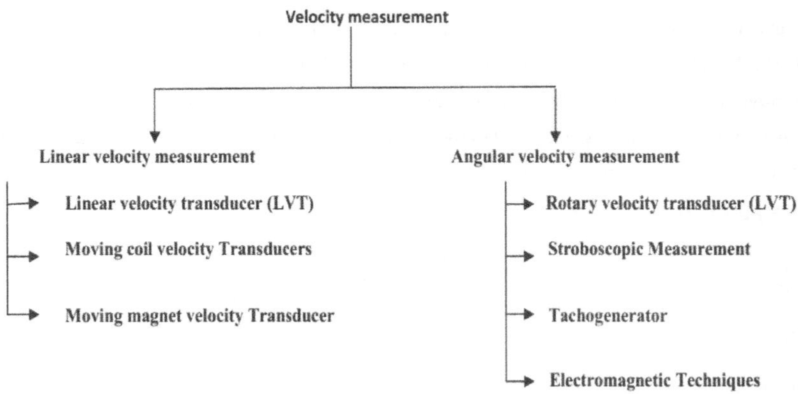

(i) Linear Velocity transducer : A linear velocity transducer (LVT) is a inductive type transducer that's works on the principle of Linear Variable – Differential Transformer (LVDT) . The differentiate between LVDT & LVT is given as,

- In case of **LVDT,** it consists of two secondary winding and one primary winding supplied to high-frequency of about **50 Hz** to **20 kHz AC** voltage to primary winding.
- But **LVT** consists of only two coils/winding, single permanent magnet, and output DC voltage terminal.
- **LVDT** employed for displacement measurement while **LVT** applied for speed or velocity measurement.

Operational & Constructional Features

LVT consists of two electrical winding or coils are wrapped with opposite polarity, having a permanent magnet core looking like a metal rod which is free for to and fro motion inside a hollow cylindrical tube called a bobbin , the schematic diagram of the **LVT** is given as ,

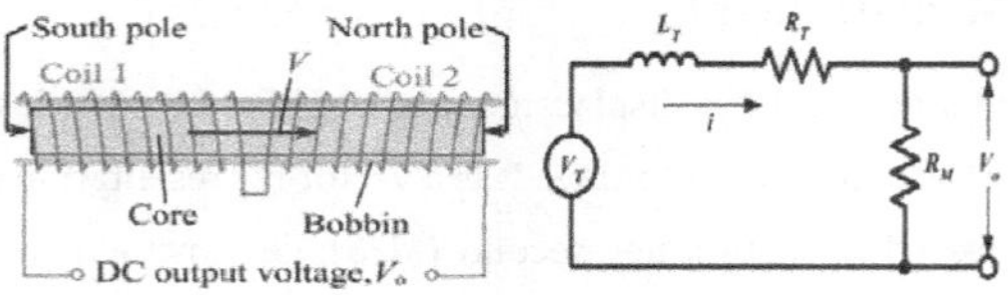

Linear Velocity Transducer Equivalent Circuit

Diagram 4.23: Linear Velocity Transducer

When permanent magnet slide inside the bobbin, north and south poles of magnetic rod has been activated thereby the South Pole induces a voltage in primary **winding-1** on the other hand the North Pole induced in **winding-2.** Hence DC voltage is generated due to magnetic core movements. The intensity of voltage generated is proportional to the speed or velocity of any mechanism. According to Lorentz law if magnetic core is moving near about or across the electrical wire or coils, then current flows through the wire. The intensity of induced voltage can be expressed as

$$\mathbf{V_T = B\,L\,v}$$

Where

V_T = The voltage generated by the transducer,

B = Flux density

L = Length of the conductor

v = Velocity of conductor

(ii) Moving Coil Velocity Transducers : The basic purpose of such kind of transducer have to measure the velocity of moving components or mechanism. The foundation of such kind of transducer is based on that whenever any electrical coil taking movement in a permanent magnetic field then EMF will be induced in the moving coil. The intensity of induced EMF will be proportional to the applied velocity. The mathematical expression indicating the relationship between various operational elements.

$$\mathbf{V = N\,(d\emptyset/dt)}$$

Where ,

V = intensity of induced voltage

N = Number of turns in a the coil

$d\emptyset/dt$ = Rate of change of flux in the coil

Working Operation :- In any industrial or manufacturing process our primary object is to perform the measurement with closed accuracy or having minimum error , to achieve the goal the device is kept in the antimagnetic case .the device consists of an arm whose one end is pivoted and other end attached to the mass of definite quantity , upon which coil is wrapped .All the such kind of arrangement is made inside the permanent magnetic field , the schematic arrangement is given as ,

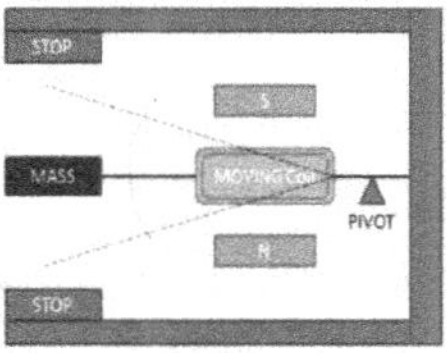

Diagram 4.24: Moving Coil Velocity Transducers

The mechanism / process whose velocity we want to measure is initially applied to the arm, due to the input velocity , coil motion takes place in the magnetic field thereby we have got an induced voltage in the coil . The intensity of induced voltage will be proportional to the applied velocity.

Output Voltage α Velocity

Advantages

- The antimagnetic case reduces the effects of stray magnetic field.
- There is high stability under varying temperature conditions.

(iii) Moving magnet type velocity Transducer : The basic principle of Moving magnet type velocity transducer is same as that of Moving Coil Velocity Transducers instead, It consists of movement of only magnetic bar. When a permanent magnet occupied to and fro motion inside a coil, the reluctance variation takes place hence output voltage is directly proportional to the applied velocity. the schematic diagram is given as,

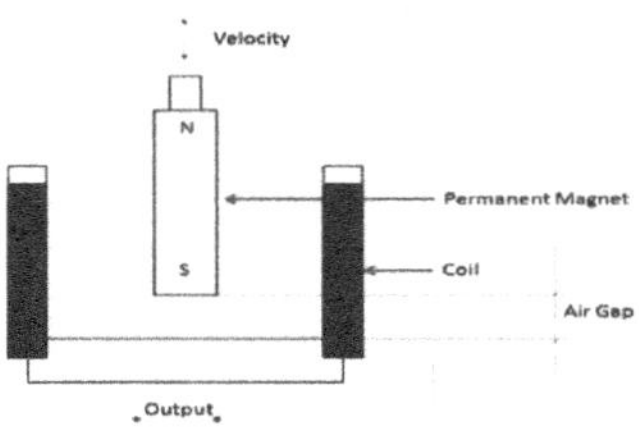

Diagram 4.25: Moving Magnet Type Velocity Transducer

Working operation : The instrument is attached to the mechanism or process whose velocity is being measured. The applied motion generates the motion of the permanent magnet between the fixed coils, an output terminal of the device is provided for measuring the output voltage variation. When the magnetic bar is in motion, we can read out the output voltage of device. Finally we have got a proportional output voltage with respect to the applied velocity. Thus ,

Output Voltage **α** Applied Velocity

Advantages

- It requires a minimum maintenance cost.
- The output voltage has linear relationship to velocity means **Output Voltage α Applied Velocity.**
- Minimum cost of manufacturing .

(iv) Angular velocity measurement : There are so many techniques are available for measuring the Angular velocity measurement of rotational components or mechanism. Amongst those **rotary velocity** transducer is one. The basic concept of such kind of transducer is that whenever any electrical coil rotated in the permanent magnetic field then **EMF** will be induced in the moving coil. The schematic diagram of **Rotary Velocity Transducer** is given as,

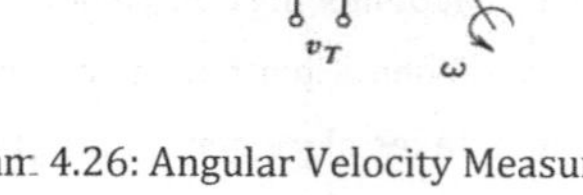

Diagram 4.26: Angular Velocity Measurement

The measured value of induced **EMF** will helps in determining the angular velocity of rotating components. The following mathematical expression is used for measuring the angular velocity

$$\mathbf{E = N\,B\,A\,\omega}$$

Where ,

N = No of turns in coil

B = Intensity of magnetic field in tesla ,

A = Rotational area of coils ,

For rectangular shape **A = LW** ,

For cylindrical shape coils we will consider the lateral surface area = $\mathbf{2\pi rl}$

ω = **Angular velocity** , E = **Induced EMF**

4.11. Tachogenerator

Tachometer generators or tachometers are device which has two components one is mechanical components behave as input & other is electrical components behave as output . The complete unit is working as an electro - mechanical devices which is used for following two purpose

- For measuring the rotational speed / velocity of motors, engines, conveyors, machine tools, speedo meters of automobiles / vehicle , steam turbines etc.
- An electro-mechanical device may also be used for generation of electrical power from mechanical energy; hence generators will generate voltage proportional to shaft speed.

Thus we can say that Tacho-generator

A tachometer or Tacho-generator is an instrument used to measuring the rotational speed of a shaft / disk or angular measurements are made with the help of device called tachometer, as in case of motor or other machine tool. Rotational Speed is measured in the revolutions per minute (RPM).

OR

An instrument used to measure angular velocity as of shaft , either by registering the number of rotations during the period of contact , or by indicating directly the number of rotations per minutes.

OR

An instrument which either continuously indicates the value of rotary speed , or continuously displays a reading of average speed over rapidly operated short intervals of time .

Working principle : The tacho-generator type of velocity transducer in which , its input is a rotational speed of shaft and its output is a voltage $\mathbf{V_o}$ proportional to the input angular velocity $\boldsymbol{\omega_s}$.

$$V_o = K_t \cdot \omega_s$$

Where $\mathbf{K_t}$ is a constant of proportionality.

The principle of operation of a tacho-generator, is that whenever any rotational components. Rotated inside the permanents Magnetic field, the electromotive force (voltage) is induced in the armature coil .The magnitude of this voltage is directly proportional to the rotational velocity / speed of armature. The armature output terminals are connected to a voltmeter circuit, which converts the voltage into speed .the diagrammatic arrangement is given as,

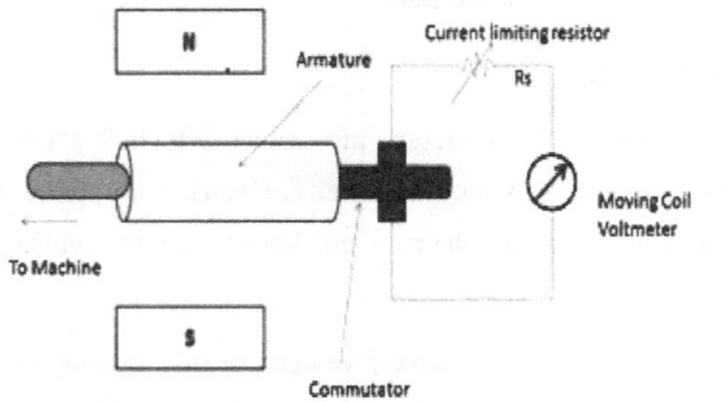

Diagram 4.27: Functional Diagram of Techo-generator

Tacho-generator is classified in two category

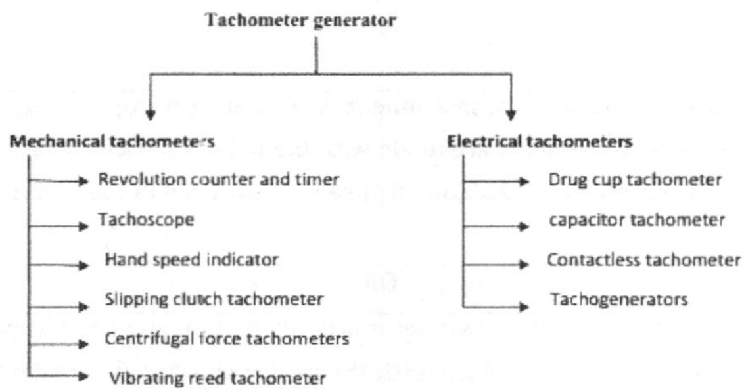

Mechanical tachometers : Following are some important type of Mechanical tachometers which can be described as

(i) Tacho-scope : Tacho-scope is used to measuring the rotational speed of the shaft which consists of following components .

- Tacho-scope consists of two parts, one is revolution counter and other is stopwatch/ timer, the revolution counter, counted a no of revolution on the other hand stopwatch / timers indicated a time in second. the schematic arrangement is given as,

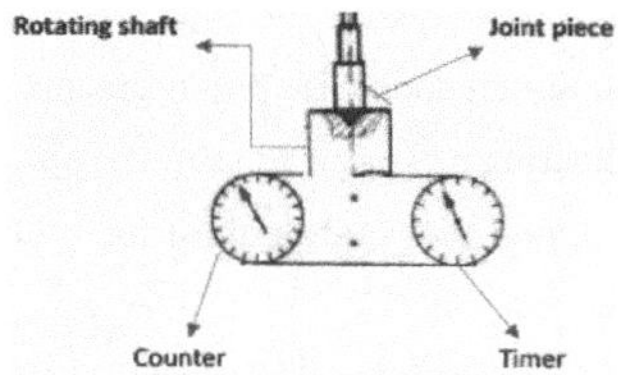

Diagram 4.28: Tachoscope

- Couple the device with the rotating shafts, adjust zero to both revolution counter as well as stopwatch or timer.
- We will start the rotation and pressing the key whose function is to operate the start and stop of revolution counter and stopwatch simultaneously.
- After certain length of time we will stop by pressing the same key. The length of time may be about 30 seconds.
- From the revolution counter we can determine the No of revolution & time from stopwatch which is in second.
- Tachoscopes have been used to measure speeds up to 5000 rpm

Since rotating shaft &device joint are connected with or without the help of joint piece , the joint piece may be following type

Joint piece	Name of joint piece	Description of shaft
	Conical rubber joint piece	Bore at the end of the shaft
	Pyramid shaped metal peak	Rectangular swage at the end of the shaft
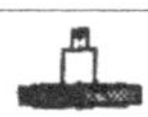	The shaft-end adhesion disc	No bore at the end of the shaft

And the angular velocity may be calculated as

$$\omega = 2 \cdot \pi \cdot n \text{ (Rad / Sec)}$$

Where , the revolution number of the rotating shaft is the number of revolutions per second.

(ii) Drag cup tachometer : Drag cup tachometer sometimes known as eddy current or drag type tachometer because it is based on the phenomenon of eddy current generation by which we can measure the speed of rotating shaft . its construction & working is given below.

- It consists of an aluminum cup , inside the cup permanent iron magnetic rotated which is coupled to the rotational mechanism . When this process takes place it generated an eddy currents which is acting on to the aluminum cup.
- This eddy current developed the rotational torque which rotate the cup in opposite to the torque of the spring.
- When the value of the developed torque is equal to the value of a spiral spring . then the installed pointer with the aluminum cup moves over the calibrated scale by which we can measure the requires speed of rotating shaft . as shown in figure .

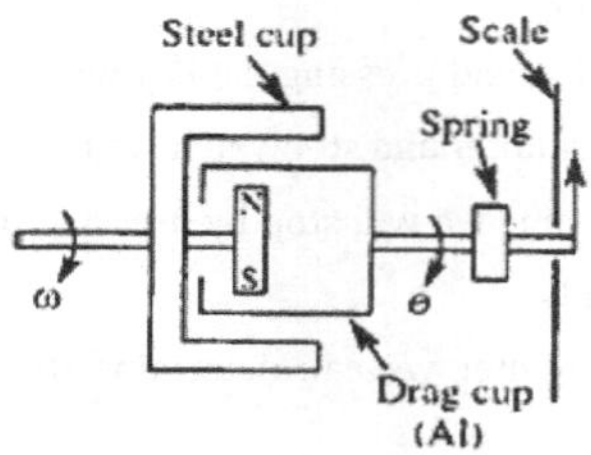

Diagram 4.29: Drag Cup Tachometer

- The maximum possible rotational speeds of such kind of device is about 12000 rpm. a very practical example of such kind of tachometer is automobile speedo meters applying for angular speed of the wheels.

(iii) Centrifugal force tachometers : The operating principle of such kind of tachometers is based on the principle of centrifugal force. It consists of two fly balls or masses attached with the spindle shaft through the pivoted arm, as shown in figure.

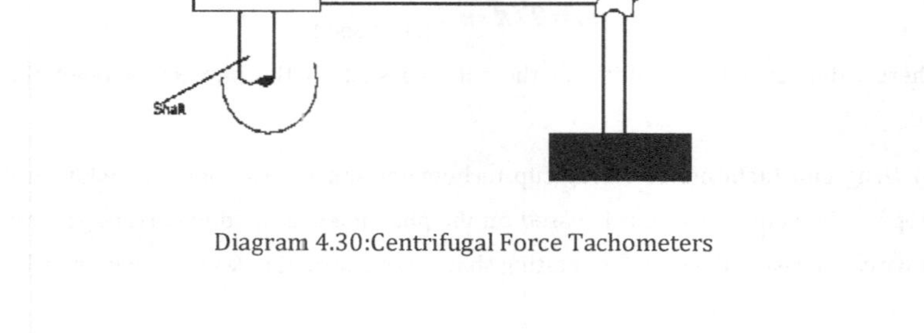

Diagram 4.30:Centrifugal Force Tachometers

when the device is connected to the rotating shaft whose speed is to be measured . the spindle rotation takes place there by fly balls have to be moved in outward direction from the center of spindle axis simultaneously compress the spring that is fitted on to the spindle, due to the effect of centrifugal force . The one end of the compression spring attached with the sleeve while grooved portion of the sleeve engaged to the pinion through the lever segments . when rotational shaft is connected to the spindle , the To &Fro motion of sleeve transmitting the motion to pointer which is moving over the calibrated scale , hence we can read out the speed of rotating shaft .

(iv) Speedo meter (Hand speed Tachometer): It is circular in shape and has a graduated dial and an indicator. On the dial is attached a starting and reset button. As shown in figure ,

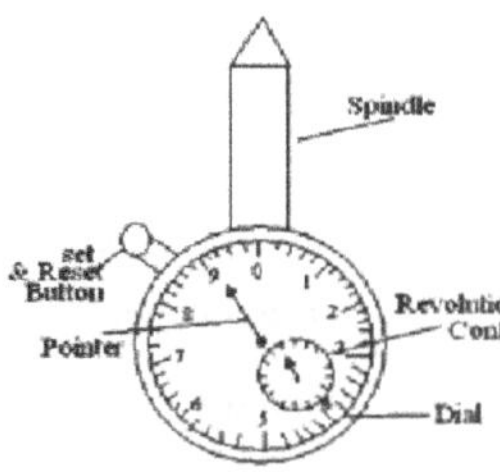

Diagram 4.31: Hand Speed Tachometer

It has a spindle which can be attached to the shaft whose speed is to be measured. The spindle is connected to mechanical counters which display the count. This meter has an inbuilt stop watch and a mechanical counter with automatic disconnect. These can be used to indicate speed of **20,000 to 30,000 R.P.M.** with an accuracy of **1** percent.

(v) Revolution counter : The revolution counter is also very important device generally applied for low speed measurement. The constructional featured & operation is given below.

- It consists of warm gear which is attached with the rotating shaft whose speed is to be measured. The schematic diagram is given below.

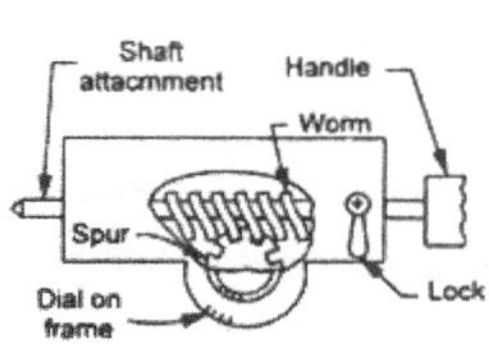

Diagram 4.32: Revolution Counter

Two measuring dial has been provided with the device that is inner dial & outer dial ,The inner dials represent one revolution of the outer dials and the outer dials represent on revolution of the rotating shaft .The worm gear & spur gear always engaged to each other by which pointer is operated and moves over the calibrated dial.

- Such kind of tachometer has a separate stop watch which is provided with the revolution counter for time measurement.
- With the help of revolution counter we can determine the average speed of rotating shaft only.

Advantage

- It is Simple & easy for operation.

Limitation

- Only applied for low speed measurement.
- Operational error may be possible.

(v) Slipping clutch tachometer : In the slipping clutch tachometer the rotating shaft is attached with the mechanism whose speed is to be measured . As shown in figure,

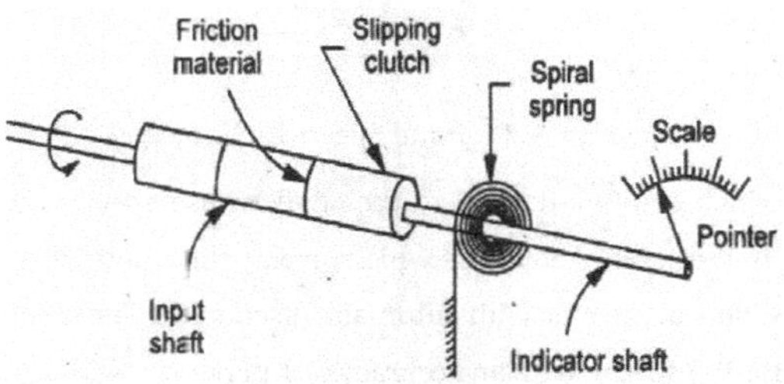

Diagram 4.33: Slipping Clutch Tachometer

When speed is applied indicator shaft is also rotated in association with the input speed shaft. a indicator shaft consists of a pointer which rotate against the torque of spiral spring over the calibrated scale , indicating the speed of rotating shaft .

(vi) Inductive type tachometer : In the Inductive type tachometer the rotating shaft is connected to the rotor having a desired no of teeth on its periphery comes in near about to permanent magnate upon which coil is mounted , then emf is induced in that coil . The output pulse is then modified with the help of amplifier which can be supplied to digital counter. Hence we can readout the speed of rotating shaft .

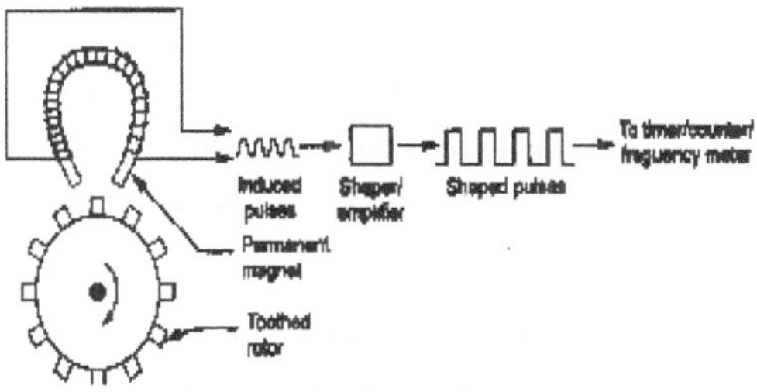

Diagram 4.34: Inductive Pick-up Tachometer

(vii) Capacitive type tachometer :- In capacitive type tachometer the rotating vane passes between the capacitor plates a capacitance change occurs in the form of a pulse. These pulses is amplified and then fed to the digital counter.

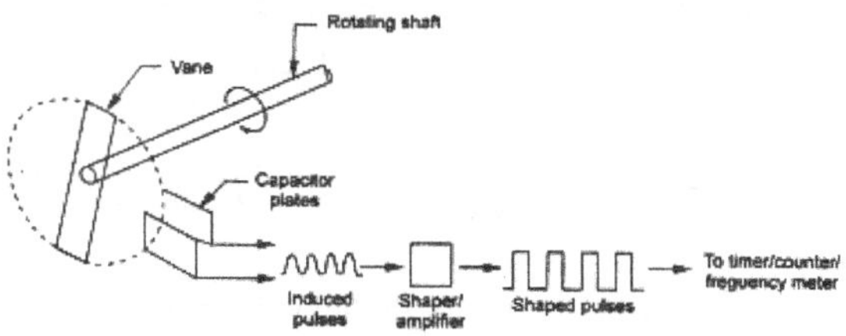

Diagram 4.35: Capacitive Pick-up Tachometer

(viii) Photo electric pickup : This is method of speed measurement based on photo electric effect. An opaque disc with evenly spaced holes on its periphery is to be attached to the shaft whose speed to be measured. A light source is placed on one side and a light sensitive transducer on the other side of the disc. Both are in alignment with the holes in disc. As the disc rotates the intermittent light falling on the photo cell produces voltage pulses whose frequency is a measure of the speed of the shaft.

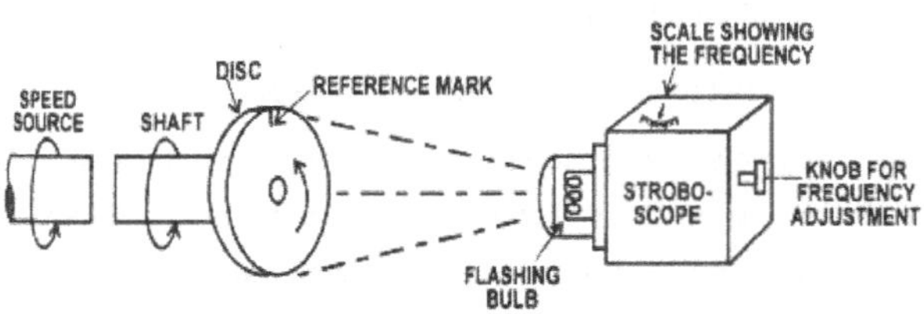

Diagram 4.36: Photo Electric Pickup

4.12. AC Tachometer Generators

In general the tachometers consists of small magnet which convert the rotational speeds into DC or AC voltage signal .The armature may be either the rotor (rotating part) or the stator (stationary parts of mechanism) upon which coil is mounted .in case of AC tacho-generators we have following basic elements .

- It consists a stator which behave like a armature in AC tacho-generators upon which two coils are mounted one is primary and other is secondary.
- Rotar of the device having a permanents magnets which is coupled to the mechanism whose speed is being measured as shown in figure.

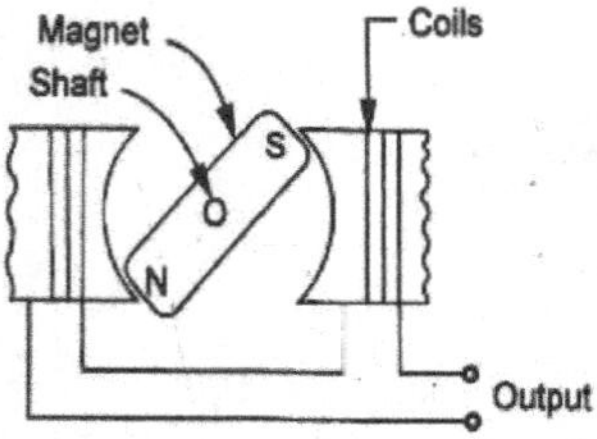

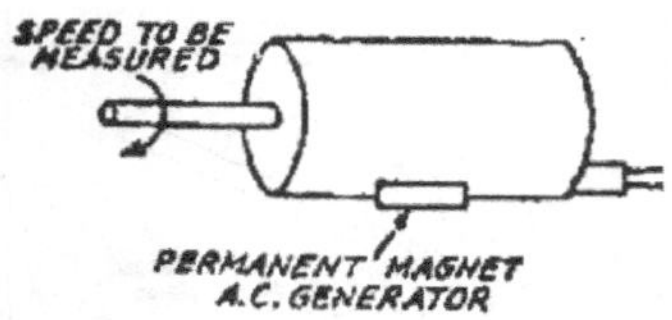

(a) Front View of AC Tachometer Generators (b) Rotor View of AC Tachometer Generators

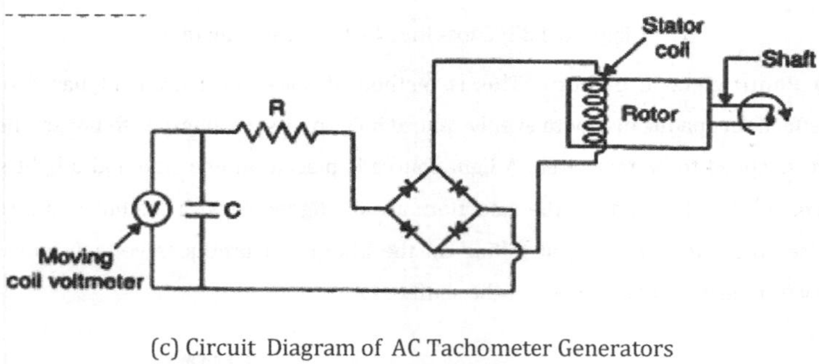

(c) Circuit Diagram of AC Tachometer Generators

Diagram 4.37: AC Tachometer Generators

- When rotar is stationary, no reading is achieved in the moving coil galvanometer while primary voltage is rectified before then it will be measured by moving coil galvanometer, indicating the proportional value of rotor speed.
- It is an accurate device for speed measurement, the maximum permissible value may be about 500 RPM .

4.13. D.C. Tacho-meter Generator

The D.C Tachometer generators is also applicable for speed measurement of a rotating mechanism .it consists of following components.

- It has a small permanent magnets whose structure is just looking like a horse-shoe.
- The armature is kept inside the, permanents magnetic field having coils or winding. Called the rotor of the D. C. tachometer generator which is coupled with the rotational elements of any mechanism whose speed is to be measured.
- An output terminals is extracted from the windings in order to achieve required measurement of voltage . as shown in diagram,

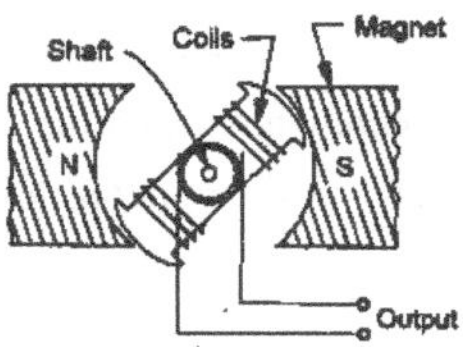

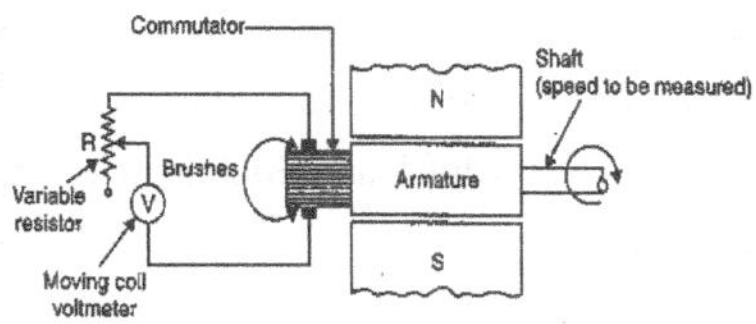

Front View of DC Tachometer Generators Circuit Diagram of DC Tachometer Generator

Diagram 4.38: DC Tachometer Generators

Working operation: When the shaft of the machine revolves, the armature of the tachometer generator revolves inside the permanent magnetic field due to the effects of rotation E.M.F. is induced , since the magnetic field is fixed permanent magnetic field hence the **E.M.F** generated is proportional to the speed directly which can be measured with the help of moving coil instruments indicating the direct reading of speed on calibrated scale. The instrument is also requires some form of commutation which presents the problem of brush maintenance. In order to achieve greater accuracy the air gap between magnet & armature must be uniform.

Note : Compared to other velocity transducers the main advantage of a tachometer generator is that it provides an output without the need of any power supply. Its main drawbacks, on the other hand, are that it has an analogue output that can suffer from large fluctuations, and it has a shorter operational life. Both of the later two drawbacks are caused by the presence of the commutator, in the D.C. type tachometer generator and can therefore be reduced if an A.C. type tachometer generator is used since this one employs slip rings to extract the power from the armature windings. The A.C. output would however require rectification and smoothing before interfacing.

Advantages

- The intensity of output voltage is small so we can measure by conventional d.c voltmeters.
- The polarity of output voltage directly indicates the direction of rotation.

Disadvantages

- The maintenance cost is high due to of commutator and brushes.
- Due to the distortions in the permanent magnetic field, error full measurement may be possible.

4.14. Variable-Reluctance Tachometer Generators

These sensors are based on Faraday's law of electrical EMF in the sensor depends upon the angular velocity of the motion. The variable-reluctance tachometer generator consists of a ferromagnetic , toothed wheel attached to a rotating shaft, and a winding on permanent magnet having **N & S** pole.

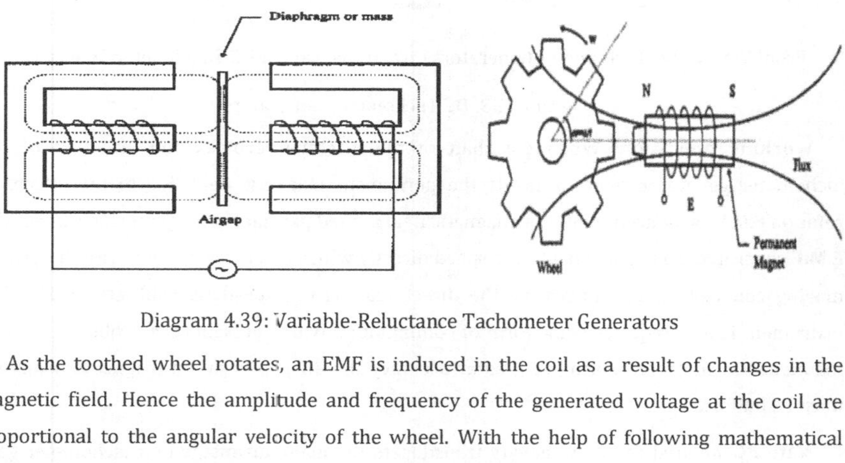

Diagram 4.39: Variable-Reluctance Tachometer Generators

As the toothed wheel rotates, an EMF is induced in the coil as a result of changes in the magnetic field. Hence the amplitude and frequency of the generated voltage at the coil are proportional to the angular velocity of the wheel. With the help of following mathematical relationship we can measure the angular velocity of the rotating shaft,

$$V_{\mathbf{indduced}} = KN\omega \sin N\omega$$

Where ,

$V_{\mathbf{indduced}}$ = Output induced voltage

N = Number of tooth on rotating wheel

ω = Angular velocity of rotating wheel

K = constant of proportionality

Note

- Whenever angular velocity measurement is performed on the basis of voltage amplitude output signal, the respective measurement will have certain error due to sensitivity toward noise and loading errors .
- Whenever angular velocity measurement is performed on the basis of frequency out signal, the respective reading will have lesser error than that of voltage amplitude output signal because digital device is applied to transmitting the output signal for recording purpose.

Applications of Tacho-generator

- Measuring shaft speed
- Measuring conveyor belt speed
- Calibration (comparing speeds to verify that panel mounts are reading correctly)
- Measure linear speed
- Measure distance (length)

4.15. Stroboscopic

The stroboscope is a device employed for those machine components, which are moving / rotating at high rate speed. Following are the basic heading of stroboscope application.

- It is an instrument to measuring rotational speed of moving shaft.
- To observe the condition of an object.
- To capture images with a camera.
- To provide a light source for automatic image inspections.

Constructional & Operational Feature of Stroboscope

The stroboscope consists of source of flashing light whose frequency and controlling is managed by stobotron. Thestroboscope is a device applied for measuring the, Rotary motion / Periodic motion/Speed of shaft. it consists of flashing light source whose controlling & flashing frequency is performed by frequency oscillator .The speed is measured by a adjusting the frequency so that the moving objects are visible only at specific period of time. If a strong light is flashed on moving objects, we have a position at which the objects will appear to be stationary. As shown in **Diagram 4.40,**

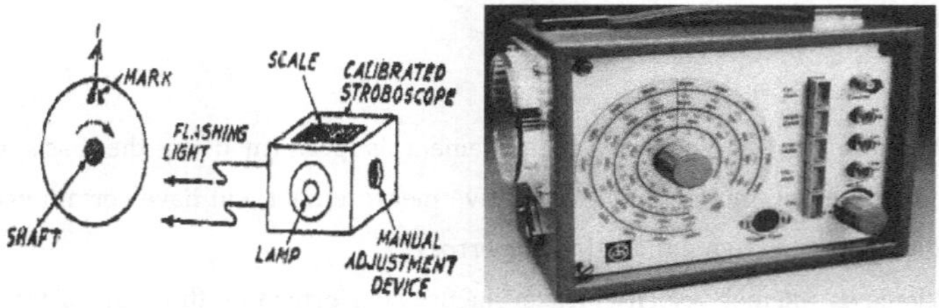

Diagram 4.40: Shaft Speed Measurement by Stroboscope

In case of speed measurement of rotating shaft , a metallic disc has attached to the rotating shaft upon which marked is impressed, with the help of stroboscope , flashed light has been fallen upon the impressed mark and adjust in such a manner that the mark becomes stationary, hence the flashing light frequency will be proportional to the speed of rotating shaft that can be read out with the help of scale calibrated in terms of speed .

For high speeds measurement, synchronization can be achieved N times, with ω_1 representing the maximum achievable synchronization speed, and ω_2 , ω_3 , . . . representing successively lower synchronization speeds. The measured rotational speed is then calculated as

$$\omega = \frac{\omega_1 \omega_N (N-1)}{\omega_1 - \omega_N}$$

4.16. Electrical Stroboscope

An Electrical stroboscope consists of a flash that is generated by the neon gas. The flashing rate is controlled by a variable frequency electronic oscillator, the frequency of which can be either read off a dial attached to the oscillator tuning co-directed on to a rotating a reciprocating or an oscillating member, the rate of flashing is so adjusted that the member appears stationary. Thus the speed is proportional to the frequency.

Frequency of Flashed α Speed of shaft

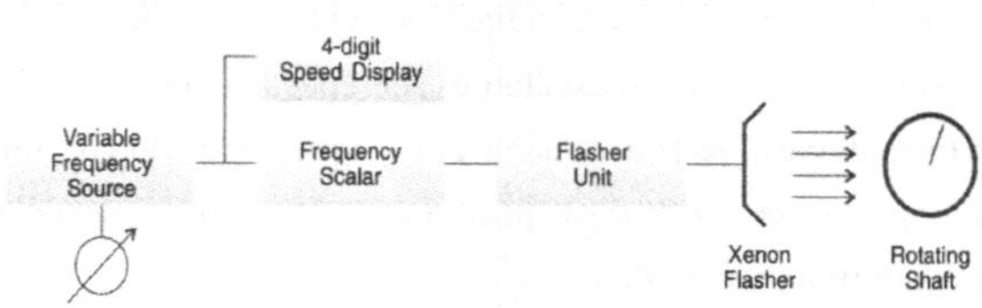

Diagram 4.41: Electrical Stroboscope

(i) Mechanical disc type Stroboscope : The schematic diagram of mechanical disc type stroboscope is as shown in figure ,

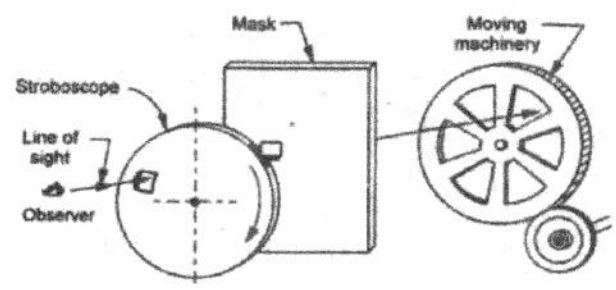

Diagram 4.42: Mechanical Disc Type Stroboscope

The stroboscope is based on the phenomenon of vision when an object is viewed intermittently. The human sense of vision is so slow to react to light stimuli that it is unable to separate two different light impulses reaching the eye within a very short Period of time (less than 0.1second). A succession of impulses following one another at brief intervals are observed by the eye as a continuous unbroken sequence. A mechanical disk type stroboscope consists essentially of a whirling disk attached to motor whose speed can be varied and measured. A reference mark on the rotating shaft on the shaft appears to be stationary. For this condition , the shaft speed equals that of rotating disk , or some even multiple of this speed and is given by

$$\text{Shaft speed} = \frac{(\text{disk speed}) \times (\text{number of openings in the disk})}{\text{number of images}}$$

Advantage of stroboscope

- It can be measured rotating / angular speed between the ranges of **600 to 2000RPM.**
- It is non-contact speed measurement method.
- It can generate high intensity flashing light.
- Speed can be read out directly in **RPM.**
- Speed measurement is possible without shaft modification.

Applications

The stroboscope is applied for purpose like ,**Designed for quality control , motion analysis and maintenance, production line checks , gear movement and engagement , propellers , turbines , monitoring vibrating parts , printing presses , oil leak checks etc .**

(ii) Electromagnetic velocity meters : Electromagnetic velocity meter is a device which used for measuring the flow velocity with the help of conducting fluids , the basic concept of such kind of device is that when a magnetic field is applied through an electrically conducting fluid, motion of the fluid induces a voltage across two electrodes placed in the flow field . The schematic diagram of such kind of device is given as,

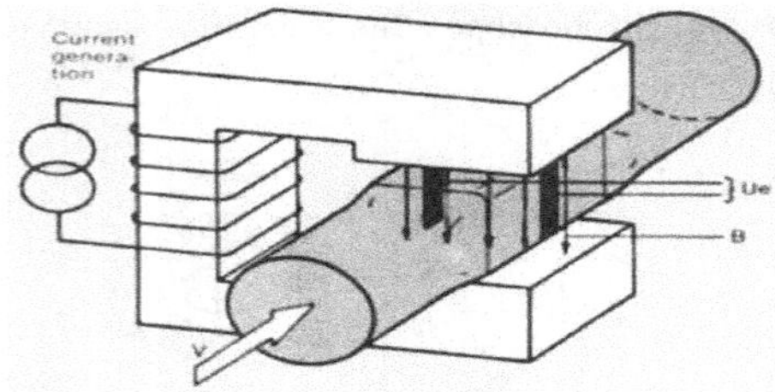

Diagram 4.43: Electromagnetic Velocity Meters

A preferred form of the velocity meter is circular in cross section and has a pair of electrodes for sensing the electric signal, the electrodes being disposed on the flow meter in a manner tending to minimize misalignment error. The strength of the output voltage increases with speed, and if properly calibrated, can infer fluid velocity. The output voltage is depends upon conductance of conducting fluid. Seawater is generally applied to fulfill the desired purpose. The velocity meter is particularly useful in a ship's log, ship's hull and oceanography experiments.

4.17. Force

Force may be defined as a push or a pull type action that can change the position of the objects , the position either rest condition or uniform motion condition . The effect of applied force will also change the direction or shape of an object. A force is a vector quantity. A vector quantity is a quantity which has both magnitude and direction. The SI unit of force measurement is **'Newton' (N).**

$$\textbf{Newton} = \textbf{1 kg m/s}^2$$

4.17.1. Force Measurement Systems

A force measuring system is a system consists of transducer and other associated devices . A transducer is a device that receives a one form of the energy and converts it into another measurable form of the energy. The secondary devices may be powered the transducer or process the output from the transducer before it is readout. A dial gauge, a computer with analogue to digital converters and excitation circuitry are generally work as another associated devices and the output may be representing the intensity of force. For example

- Whenever the force is an act upon a metal cylinder which is compressed by the force, the change in size alters the electrical resistance of a strain gauge bonded to the surface of the cylinder, and the associated device is used to measures the change in resistance.

Method of force measurement : Following are the basic method by which we can perform the force measurement.

1. In the first method balancing is performed with respect to standard known mass through system lever mechanism.
2. By determining the acceleration of a known mass then we can find out the intensity of force applied.
3. Establishing equilibrium of magnetic force that is generated by interaction between magnate and current carrying conductor/coil.
4. Force is applied on to a particular surface area and then measuring it as a pressure.
5. Force is applied on to elastic elements by which we can measure the deformation as a intensity of forces.

4.17.2. Force Measurement

A Load Cell or Load Sensor/Load transducers is a passive transducer which is defined as a transducer that converts an input mechanical force into an measurable output signal like , electrical signal or an electronic signal . Load cell designs can be based on according to the type of output signal generated (pneumatic, hydraulic, electric, and electronic) or on the basis of weight detection like bending , shear , compression , tension etc.

(i) Hydraulic load cells: Hydraulic load cell is a force balance devices applied for measuring force generated by the weight which can be converted into change in pressure of the internal filling fluid. The schematic diagram is given as,

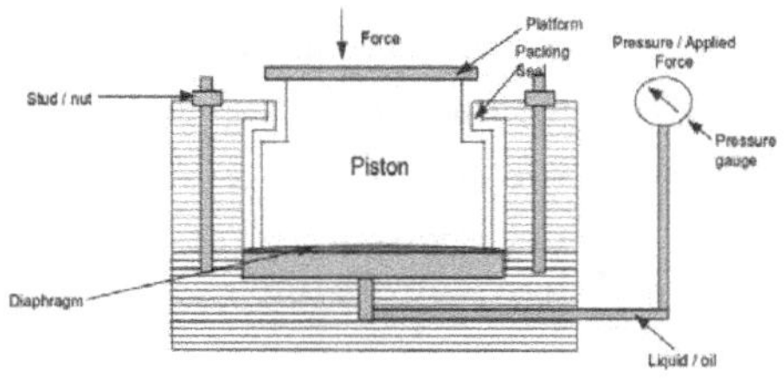

Diagram 4.44: Hydraulic Load Cells

It consists of following elements,

- A container filled with a liquid/oil
- Platform upon which weight/load is applied for creating force which is attached to the piston. At the bottom end a diaphragm is installed whose function is to create deformation when force is applied.

- At the bottom of the cylindrical shell a liquid pipe is connected to pressure gauge.
- **Diaphragm:** The working principle of diaphragm is based on the concept of deflection of thin sheet whose function is to create distortion when force is applied . It is a thin plates may be a shaped of circular, rectangular or square form depends upon the application ; in this case it has a circular form. Applied for determining the intensity of force/weight as a pressure.

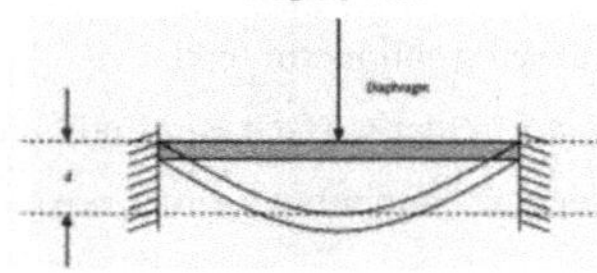

Diaphragm

Working : In such type of load cell , a load is applied on to the platform whose intensity is transferred to the piston creating the compression to filling fluid confined within an elastomeric diaphragm chamber, As force increases, the pressure of the hydraulic fluid is also increases thus the indicated value can be measured with the help of pressure gauge which will be proportional to the applied load, It is a most common type of load cell which has no electrical elements .

(ii) Pneumatic load cells : Pneumatic load cells is works on the same operating principle as that of hydraulic load cells which is force-balance principle. It has multiple dampener chambers in respect of hydraulic load cell device. When the weight/force is applied to diaphragm which is made up of thin metallic materials or flexible material, it is deformed. After the deformation we will applied a pneumatic air pressure for balancing the deformation. The block diagram of pneumatic load cells is given as,

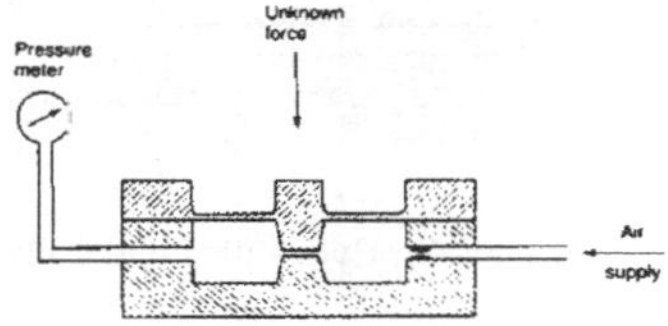

Diagram 4.45: Pneumatic Load Cells

Hence the applied pressure for maintaining the deformation by applied load / weight or force, will be proportional to force applied up on diaphragm. That can be measured with the help of pressure gauge.

Advantage

- It can used for measuring low value of weight/force in the form of pressure with accuracy.
- It is explosion proof and insensitive to temperature variations.

Disadvantage

- It has slow speed of response.
- There is a requirement of clean dry air or nitrogen.

(iii) Strain Gauge load cell : The Strain Gauge load cell is based on the principle that if length of the electrical conductor variation is achieved due to the effect of load or stress application then proportional resistance change takes place or we can say that a device which convert the load acting on strain gauge into electrical signals. The construction & working is given as,

- Strain gauge load cell consists of cylindrical tube to which strain gauges are attached. A load applied on the top collar of the cylinder compress the strain gauge element which changes its electrical resistance.
- The gauges are bonded onto a structural member by suitable adhesive that deforms when force / load is applied.
- Four strain gages are used to achieve higher accuracy and temperature compensation. Two of the gauges are usually in tension , and two in compression.
- When weight is applied , the strain will change due to the effect of weight / force application thus electrical resistance of the gauges will proportional to the applied load which can be measured with the help of Wheatstone bridge circuit, as shown in figure.

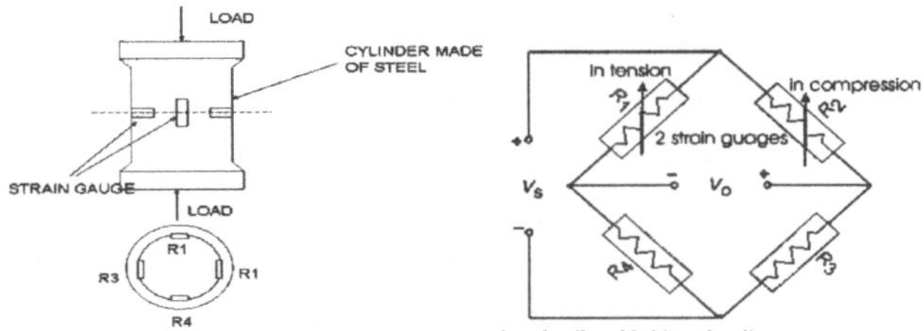

Diagram 4.46: Strain Gauge Load Cell & Bridge Circuit

- Since, The electrical signal output is very small therefore It is amplified with the help of an amplifier before sending it to the measurement system.

- The strain gauges are used to measure forces up to 10 MN.
- **Output voltage is calculated as**

$$V_0 = \left(\frac{R_3}{R_3 + R_4} - \frac{R_2}{R_1 + R_2}\right) \times V_{EX}$$

- Here are some important type of strain gauge which are generally applied for load application.

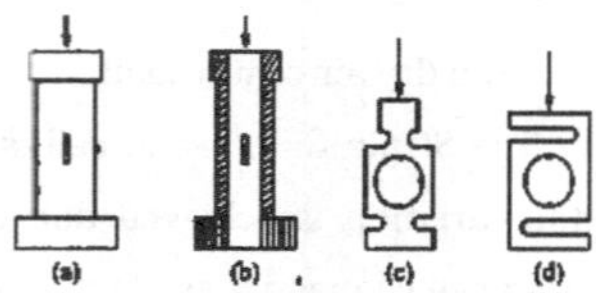

(iv) Inductive Sensing : Inductive weight sensors use the change in inductance of a solenoid coil with changing position of an iron core. Illustrated as in Figure.

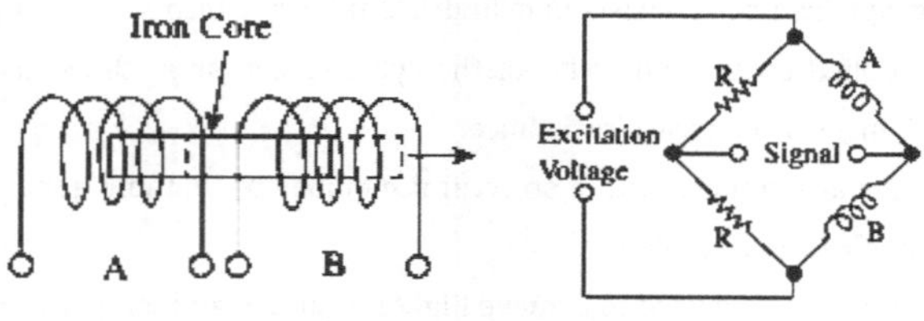

Diagram 4.48: Inductive Sensing

If iron core moves to the right, the inductance of coil B increases and the inductance of coil A is reduced. Arranging the two coils in a Wheatstone bridge with resistors completing the bridge network provides a means for developing a voltage signal proportional to the core position.

(v) Capacitive Load **Cells:** A capacitive load cells, is based on the phenomenon of capacitance variation is used to measure the intensity of applied force. The schematic diagram of such kind of load cell is given below,

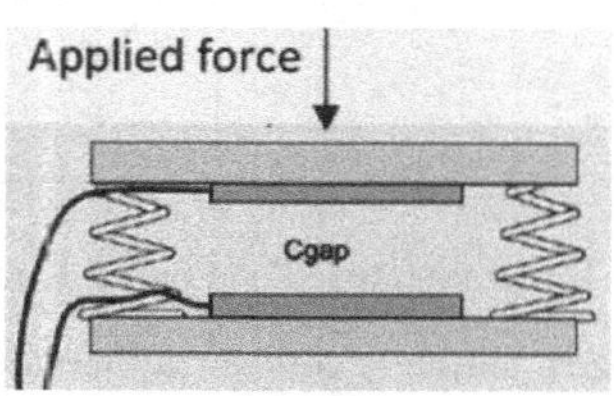

Diagram 4.49: Capacitive Load Cells

It consists of two metal plates separated by an air gap, Whenever a force or load or pressure is applied , the air gap between the plates has been changed so that the intensity of charge storage capacity is also changed , hence the charge storing capacity of capacitor plate is directly proportional to the area between the two plates and reciprocal to the gap between the plates , the plot between the capacitance & force is given below .

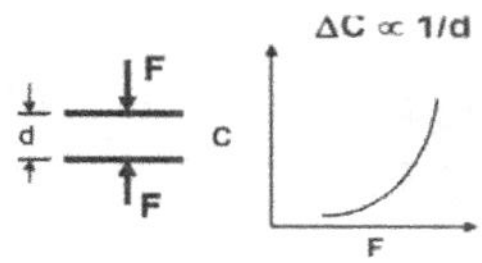

Diagram 4.50: Curve between Capacitance & Distance

The capacitance **C** between terminals is given by the following expression

$$C = K\,\varepsilon_0 \frac{A}{d}$$

Where

C = Capacitance in farads

A = Area between two plates

d = Thickness of the gap between the two plates

ε_0 = Dielectric constant of free space

4.18. Piezoelectric Electric Load Cell/Force Transducers

Piezoelectric electric load cell is nothing but it is a transducer whose function is to converts one form of energy into another form. In some practical industrial application we are unable to measure the mechanical forces or load directly so therefore piezoelectric transducer can be applied for determining the required intensity of input .The maximum input application may be about 100MHz .According to principle whenever force or pressure is applied on such kind of specific material it converts it into electric voltage or vice-versa and this voltage is function of the force or pressure . Such kind of material is also known as piezoelectric sensor for example Quartz crystal which is made from silicon and oxygen arranged in crystalline structure (SiO_2) but Barium Titanate , Lead zircon- natetitanate (PZT), Rochelle salt, tourmaline are artificial crystal.

Working & operation :- whenever force / weight / load is applied on certain crystalline material it generate electrical charge on its faces which will be proportional to the intensity of applied force , but in actual practice the intensity of generated charge is small hence there is a requirement of amplifier , called charge amplifier whose function is to modified consistency of the weak electrical charge signal there after it can be measured , indicating the proportional applied force , the schematic diagram of such kind of device is given as ,

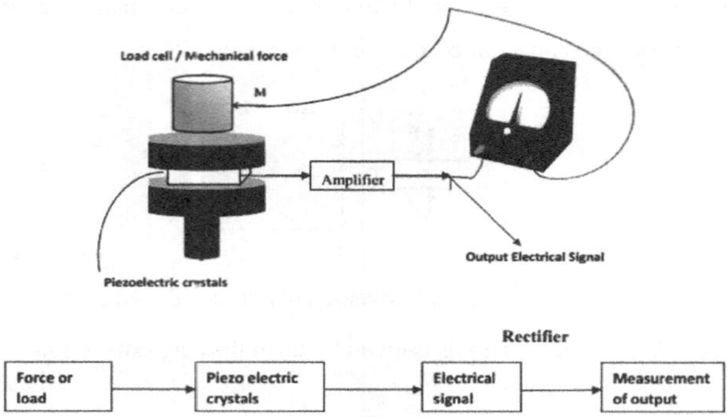

Diagram 4.51: Piezoelectric Effects

Such kind of crystal is natural quartz crystal but in present practice artificial crystal has been also used to fulfill the requirement like Barium Titanate, Lead zircon-natetitanate (PZT), Rochelle salt etc. which is recognized as piezoelectric crystal or Active sensing elements . In general we have to considered two condition of applied forces compressive force & tensile force, which is briefly demonstrated under here with suitable diagram.

- When there is no stressed applied on quartz crystals no charges will be induced on it.

- When compressive force is applied, positive charges are induced in one side and negative charges are induced in opposite side. The crystal size gets thinner and longer due to compressive stress.

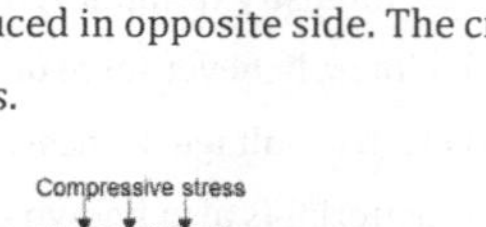

- In case of tensile stress, charges are induced in reverse as compare to compressive stress and quartz crystal gets shorter and fatter.

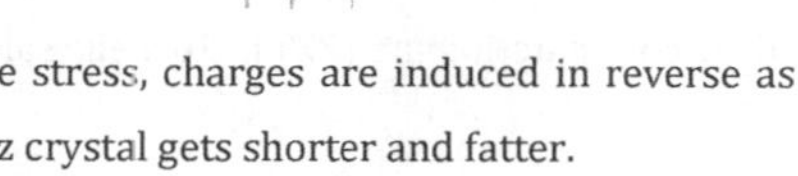

Multiple -component piezoelectric force transducers :- Such kind of device is also based on the same principle , measure the forces in three orthogonal axes . Force **F** acts upon the transducer and is transmitted to each of three discs with the same magnitude and direction. As shown in figure,

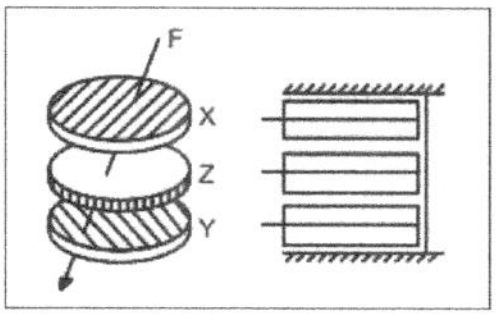

Multiple-component Piezoelectric Force Transducers

Each piezoelectric crystal ring has been cut along a specific axis and the orientation of the sensitive axis coincides with the axis of the force component to be measured. Each of them produces a charge proportional to the force component specific to that disc. When charge has been generated there is a necessity of charge amplification here after we can collect it through the electrodes inserted into the stack.

4.19. Magneto-Elastic Force Transducer

The magneto-elastic force transducer is also used to measuring the intensity of mechanical force or load , the constructional feature & operation has been discussed below .

- It consists of two electromagnetic coils (primary coil & secondary coil) which is installed perpendicular to each other and fixed upon the transducer main body.
- AC supply is connected to the primary coil while secondary coil are bound through holes in a column having a two output terminal for measurement , as shown in figure.

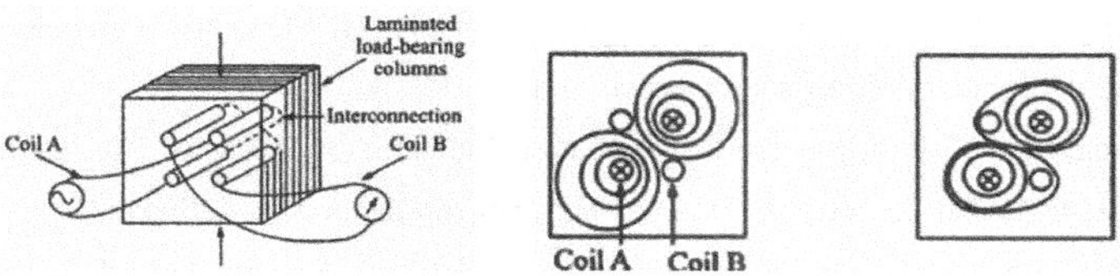

Unload Position Loaded Position

Diagram 4.52: Magneto-Elastic Force Transducer

- When the load is applied upon transducer body column, the magnetic field alteration takes place thus the permeability of steel is decreased in the direction of the force and increases perpendicular to it.

- Due to change in permeability, the magnetic flux intensity change achieved, so therefore voltage is induced in the secondary winding hence the induced voltage will be proportional to the applied force.
- The maximum loading/force application range may be order of 2 kN to 5 MN, it is generally applied for rolling mills.

Advantages of Load cell

- Rugged and compact construction
- No moving parts
- Can be used for static and dynamic loading
- Highly Accurate
- Wide range of measurement
- Can be used for static and dynamic loading

Disadvantages of Load cell

- Mounting is difficult
- Calibration of load cell is a very time consuming procedure

Miscellaneous Method for Force or Load Measurement

4.20. Analytical Balance or Direct Method

(i) Equal arm balance : It is a direct comparison method , in which Direct comparison is to be taken between unknown force and the gravitational force . The direction of force is always parallel to the gravitational force, and only magnitude determination takes place. this instruments consists of following elements.

- It has balance arm which rotates about the point **"O"** and two forces or load are hanged i.e. $\mathbf{W_1}$ and $\mathbf{W_2}$ at the ends of the balanced arm. Where $\mathbf{W_1}$ is an unknown force and $\mathbf{W_2}$ is standard mass or reference mass.
- The centre of gravity of the instruments is indicated by the point " **G** " of the balance arm, and **WB** is the weight of the balance arm and the pointer acting at **G.**
- When the unknown force and the gravitational force are in balanced position means **W1 = W2 ,** therefore angular deflection θ will be zero.
- If unknown force and the gravitational force are in unbalanced position so that the force/load **W1&W2** , are unequal then this is indicated by the angle θ which the pointer makes with the vertical. As shown in figure.

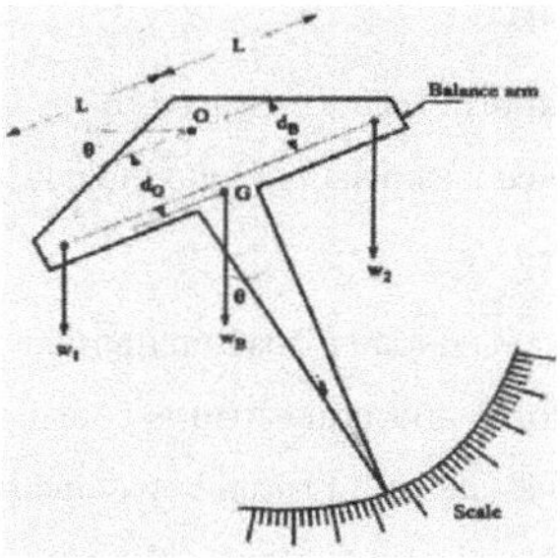

Diagram 4.53: Analytical Balance or Equal Arm Balance

- The sensitivity S of such a is given by the following relationship which is angular deflection per unit of unbalancing.

  Sensitivity

$$S = \frac{\theta}{w_1 - w_2} = \frac{\theta}{\Delta W}$$

Where, ΔW is the difference between W_1 and W_2. and

The sensitivity **S** can be calculated as

$$W_1 (L \cos\theta - dB \sin\theta) = W_2 (L \cos\theta + dB \sin\theta) + WB\ dG \sin\theta$$

For small deflection angles $\sin\theta = \theta$ and $\cos\theta = 1$ and the above equation becomes

$$W1 (L - dB\,\theta) = W2 (L + dB\,\theta) + WB\ dG\ \theta$$

Hence sensitivity will be

$$S = \frac{\theta}{w_1 - w_2} = \frac{L}{(w_1 + w_2)d_B + d_G W_B}$$

When $W_1 = W_2$

$$S = \frac{\theta}{\Delta w} = \frac{L}{2Wd_B + W_B\, d_G}$$

The sensitivity of the balance will be independent of the weight **W** provided it is designed such that **dB = 0** then

$$S = \frac{L}{W_B\, d_G}$$

An equal arm analytical balance suffers from a major disadvantage. It requires a set of weights which are at least as heavy as the maximum weight to be measured.

4.21. Unequal Arm Balance

The basic purpose of unequal arm balance is used to measure the heavier weight with the help of lighter weights , this device is consists of following parts , the constructional features & operation is given below .

- It has two arms , one is called load arms and another is called power arm , load arm is applicable for load or force and other arm is considered as power arm by which we will always try to establishing equilibrium between both the arms .
- Consider the unequal arm balance where Mass m acts as power on the beam and exerts a force of **Fg** due to gravity where **Fg = m g.** This force acts as counter posing force against the load applied test force **Ft.**
- The beam consists of knife edge **'q'**. The test force **Ft** is applied by a screw or a lever through a knife edge **'p'** until the pointer indicates that the beam is horizontal. For balance of moments, the schematic diagram is given as ,

Ft x a = Fgx bor Ft= Fgx b/a = m x g x b/a = constant x b

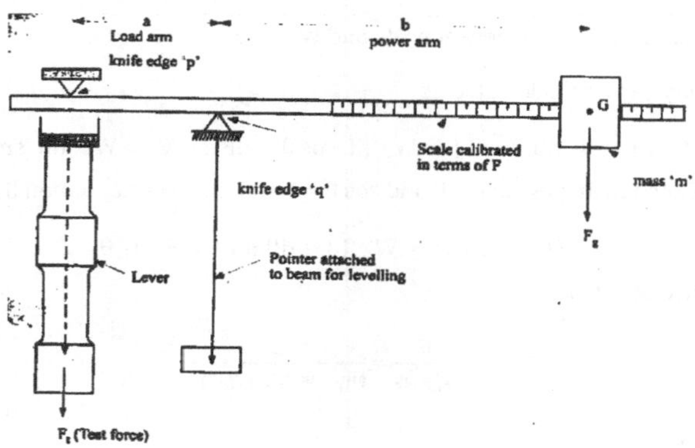

Diagram 4.54: Unequal Arm Balance

For null balance $W_1 b = W_2 a$

$$W_2 = W_1 \times \frac{b}{a}$$

- Hence the test force is proportional to the distance **' b'** of the mass from the pivot. Hence, if mass **' m'** is constant and the test force is applied at a fixed distance **' a '**from the knife edge **' q'**, the right hand of the beam that is the power arm may be calibrated in terms of force **Ft.**

- The device application is depends upon the value of "**g**" , since this device is only applicable for measuring the load/force on the earth surface . If other than earth surface we can use value of "**g**" according to the application .
- This instrument is only applied for tensile force.

Finally we have got the following relationship

$$\mathbf{F_t = m_t\, g}$$

For balancing ,

$$\mathbf{m_1\, g\, a = m\, g\, b}$$

or

$$\mathbf{m_1 = m \times b / a = a\ constant \times b}$$

4.22. Platform Balance (Multiple Lever System)

Both the above instruments are generally applied for measuring the load or force whose intensity is low. but when we want to measure the large weight/load multiple lever instruments is used .

- In this instruments large weight **W** is divided into two weight , **Wp** and **Wg** where, **Wp** = weight of poise and **Ws** = Weight of Pan .
- When the unknown load **W** is applied to the platform, the poise weight **Wp** is set to **zero** with respect to beam scale as well as counter piece is adjusted to obtain Initial zero balance.
- For simplification the weight **W** can be replaced by two arbitrary weights **W1**and **W2**.The instruments figure is given below,

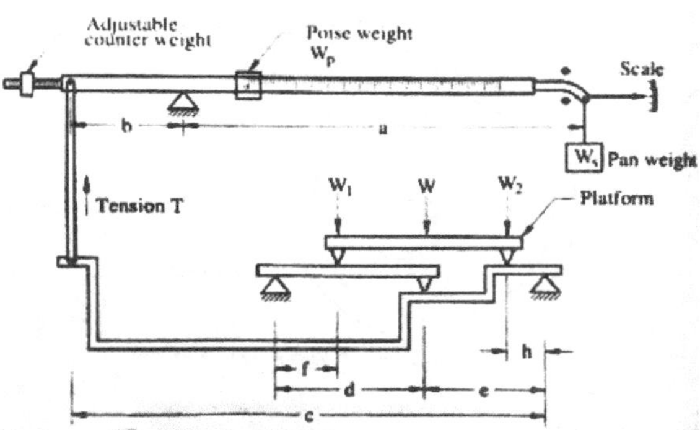

Diagram 4.55: Platform Balance (Multiple Lever System)

Hence the mathematically we can write as

$$T \times b = Ws \times a \qquad \text{.....(1)}$$

$$T \times c = W1\, f / (de + W2\, h) \qquad \text{.....(2)}$$

$$T \times c = h\,(W1 + W2)\, h\, W \qquad \text{.....(3)}$$

$Ws(a / b) = W h / d$ where unknown weight $\mathbf{W = \{ (a / b)(c / h) \} Ws}$

Where $\mathbf{M = \{a / b)(c / h \}}$ is multiplication ratio of scale which is indication of weight that should be put in the pan to balance the weight on the platform.

- If the beam scale is so divided that a movement of poise weight **Wp** by 1 scale division represents a force of **x**kg, then a poise movement of y scale divisions should produce the same result as a weight ***Wp*** placed on the pan at the end of the beam. Hence,

$$\mathbf{W_p = x\, y\, a \text{ Or } x = W_p / a}$$

4.23. Proving Ring

It is also a another device which has been applied for measuring the force/load application since long time, the force/load may be of compressive or tensile nature. The constructional & operational features is given as,

- A proving ring consists of circular ring of rectangular cross section area.
- When it is subjected to tensile or compressive forces across its diameter its gets deflected.
- Hence the deflection is directly proportional to the applied force/load. the schematic diagram is given as ,

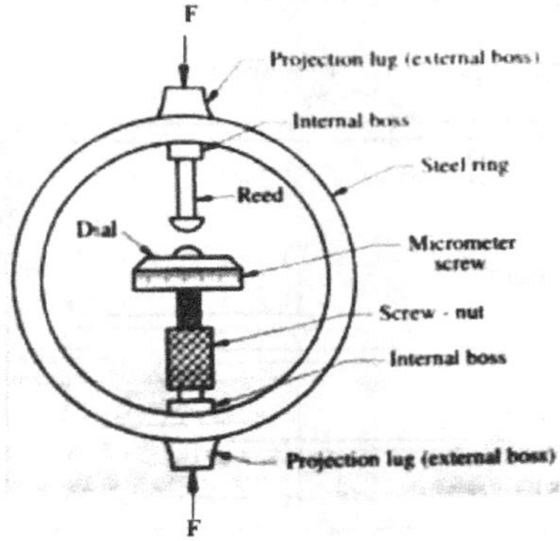

Diagram 4.56: Proving Ring

- The deflection/displacement has been measured with the help of **LVDT** and force is calculated by the following relationship ,

$$F = [16 / \{(\Pi/2) - (\Pi/4)\}] \times (EI / D^3)\, y$$

Where,

F is the force,

E is the young's modulus,

I is the moment of inertia of the section

D is the outside diagram of the ring, y is the deflection.

- The deflection & accuracy of proving rings is measured with the help of micrometer .If precise measurements is required , the micrometer is mounted on a vibrating read device which will detecting & observing the vibratory motion.
- Proving Ring is applied for static force measurement within the range of 1.5 KN to 1.5MN, but it has also applied for calibrating tensile testing machines.

Torque Measurement

Torque: The torque may be defined as the intensity of force which is acting at a certain distance from fixed point that will try to rotate the object about its axis is called the torque. The fixed point is called pivot point &The distance between the pivot point and force acting point is called the **moment arm** , and is denoted by ' **r** '.

Torque = Distance between fixed point & force point x applied force

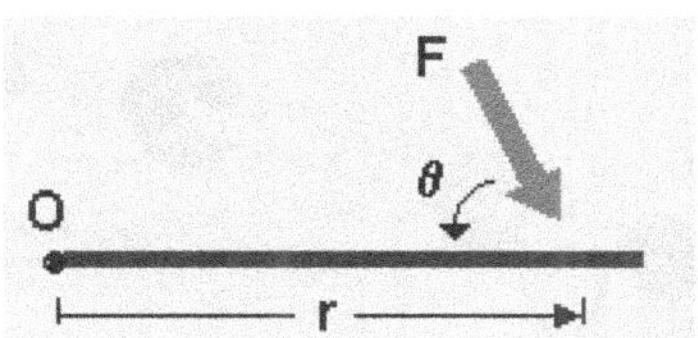

When Torque is applied on the particular objects it will deflect the object from its axis at an angle say **θ** then Torque may be defined as ,

$$\tau = \mathbf{r} \times \mathbf{F} = \mathbf{r}\,\mathbf{F} \sin \theta$$

SI units of torque is a **Newton-meter**

4.24. Torque Measurement Technique

There are so many method are available which is applied for many industrial operation in order to determining the intensity of applied torque. The classification of technique are given as,

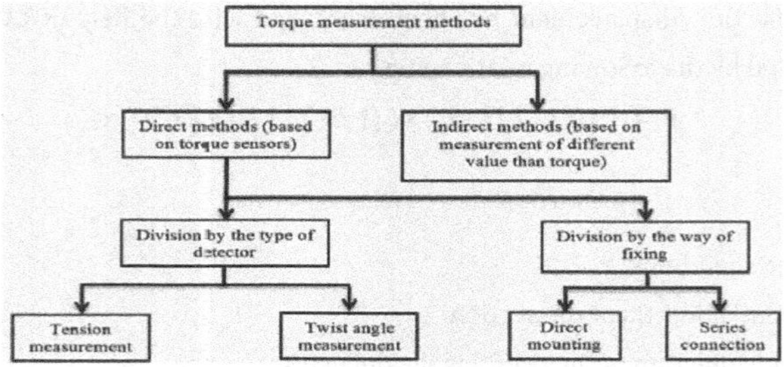

Torque Measurement Technique

4.24.1. Mechanical Torsion Meter

When a shaft is connected between a driving source and driven load, an angular displacement occurs on the shaft between its ends. This angle of twist is measured as a torque.

Principle Parts of Mechanical Torsion Meter

- A rotating shaft which has two drums and two flanges mounted on it as shown in figure
- One drum carries a pointer and other drum has a torque calibrated scale.
- A stroboscope is used to take readings on a rotating shaft.

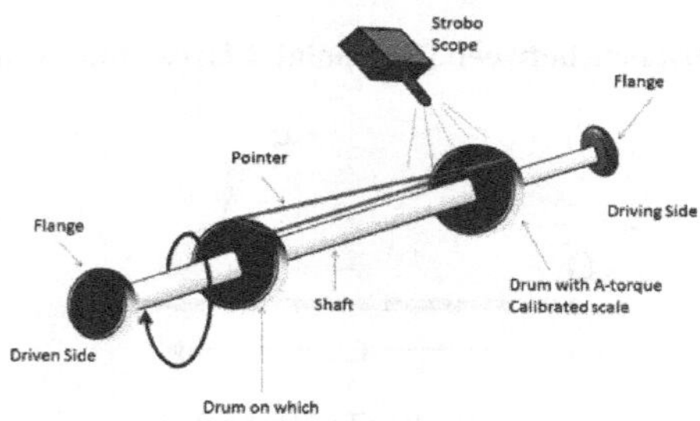

Diagram 4.57: Mechanical Torsion Meter

Operation of Mechanical Torsion Meter

- One end of the shaft of the torsion meter is connected to the driving source and its other end to the driven point.
- An angle of twist between the two flanges will be proportional to the torque applied to the shaft.
- The angle of twist will be proportional to the measure of torque.

- The angular twist will indicate the position of the pointer on calibrated scale corresponding to the torque.
- For measurement of the reading on calibrated scale, a stroboscope is applied, the stroboscope's flashing light is made to fall on the scale and the flashing frequency is adjusted till a stationary image is obtained.

4.24.2. Strain Gauge Torsion Meter

Strain gauge torsion meter is also a very practical method which is applied for measuring a torque of rotating shaft. It is based on the principle that whenever any electrical conductor subjected to the stress that may be of tensile or compressive nature then respective strain will be changed. Following are the constructional & operating features of Strain gauge torsion meter.

- The Four wire bonded strain gauges are mounted on 45° with the axis of rotation and are placed in opposite direction.

The Strain gauge is attached on to the rotating shaft with the help of some kind of adhesive.

When the twisting torque is applied on theshaft , gauge R2 and R4 will experience a tensile force due to that elongation takes place or action of pure shear stress on the diagonal axis , at the same time R1 and R3 is subjected to compression on the diagonal axis as shown in figure ,

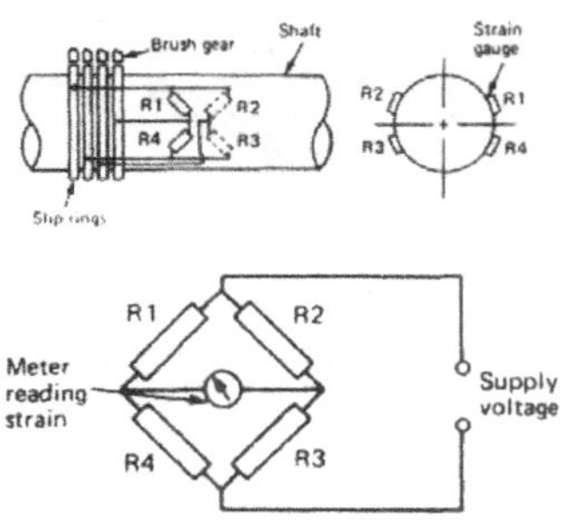

Diagram 4.58: Strain Gauge Torsion Meter

- For torque measurement it is desirable to have the sensors mounted 90°relative to each other since the principal strains, compression and tension, will be perpendicular to each other. Compression will cause the resistance to decrease and tension will cause it to increase. Thus the strain induced resistance ratio, between the sensors, will change more compared to bending or pulling, resulting in a stronger signal.

- Hence these induced tensile and compressive stresses are responsible for the change in strain causes the electrical resistance to change of the strain gauge system or bridge. And the shaft torque can be calculated as

$$\tau = E \ / \ (VK)\ JE \ / \ r\ (1\text{-}\mu)$$

Where ,

E = change in voltage , V = applied voltage , K = gauge factor

J = polar moment of inertia , E = young modulus , r = radius of shaft

Note: A problem is encountered in rotating shafts for making an electrical connections to the strain gauges this problem can be overcome by using of slip rings and brushes for the system but it may leads to increasing the cost.

4.25. Torque Reaction Method

In general, we know that whenever power is transmitted through the rotating shaft , It has two main parts one is power source and another is power absorber or sink , sometimes it is defined as a power dissipater , the power source and power absorber / sink connected through the power transmitter or shaft whose function is to transmit the generated torque . The intensity of torque can be measured by applying cradling method, as shown in diagram

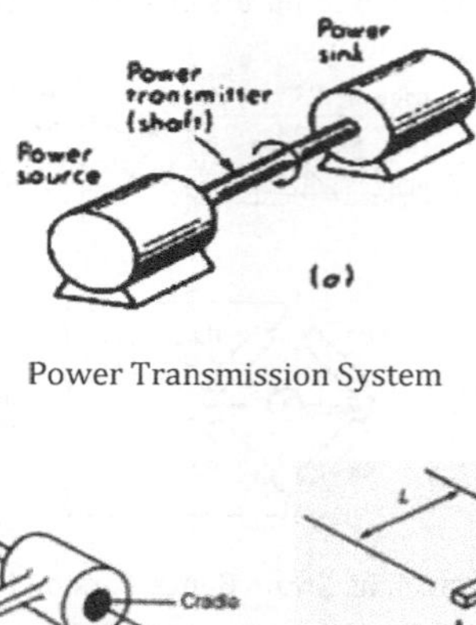

Power Transmission System

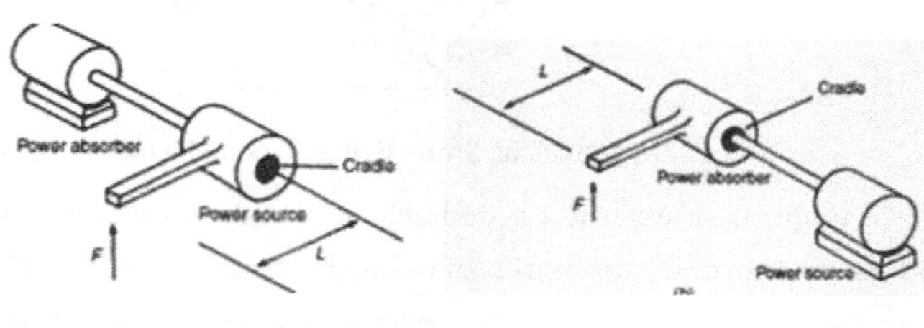

(b) (c)

Reaction Force Measurement by (b) Power Source and (c) Power Absorber Cradling

Diagram 4.59: Torque Reaction Method

cradling may be done either to power source or power absorber at the end of the shaft in bearings which will generated the reaction force due to the application of torque , which is measured on Pendulum scales and hence we know that the torque will be the multiplication of reaction force **F,** and the arm length **L** , that is given as ,

Torque = Reaction Force x Arm Length T = F X R

4.26. Electrical Torsion Meter

In Electrical Torsion Meter, two slotted disc are mounted on to the rotating shaft , as shown in figure ,

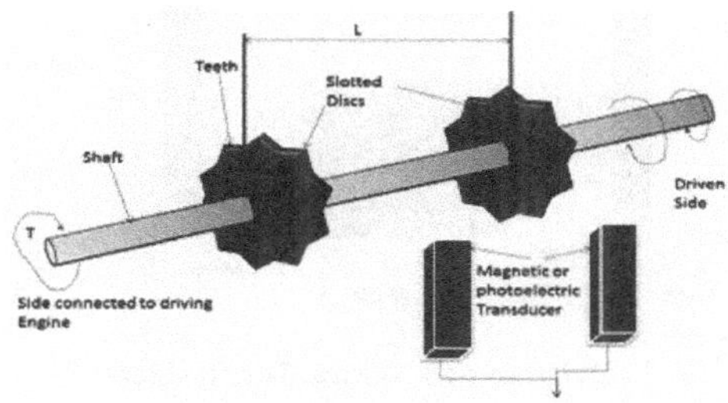

Output for Measurement

Diagram 4.60: Electrical Torsion Meter

Elements of Electrical Torsion Meter

- A shaft connected between a driving source and a driven load.
- Two slotted discs installed on either side of the shaft.
- A magnetic transducer or photo electric transducer is applied for counting pulses from the slotted disc.

Working : Initially , when shaft is in static condition , there will be zero value of voltage pulses & time difference but Whenever torque is applied on a shaft , the rotational motion is achieved . if we kept the magnetic transducer near about to the slotted discs then we have got the relative displacement between the two slotted discs that is mounted on the shaft and hence teeth produced voltage pulses in the transducers with the shaft rotation because when torque is applied on the shaft, there is a relative displacement of the slotted discs thereby twist experienced by the shaft and hence the teeth of both the discs will not align to each other therefore the voltage pulses produced in the transducer will have a time difference , Thus time lapse between the pulses of the two discs will be proportional to the twist or torque of the shaft.

4.27. Optical Torsion Meter

This device is based on the principle that when an angular twist is created by the application of torque on rotating shaft can be measured with the help of optical arrangement indicating the angular deflection of light rays which will proportional to twist or proportional to torque.

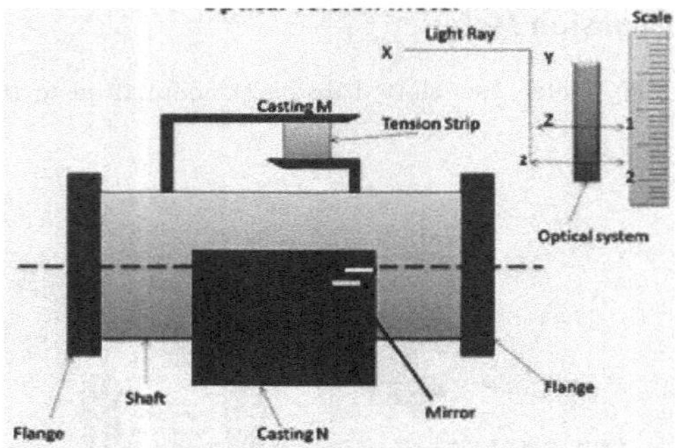

Diagram 4.61: Optical Torsion Meter

Working of device : It consists of two casting , tension strips , mirror and optical system. as shown in figure ,Under zero torque conditions, reflected light are in same phase with each other. If the shaft rotation starts torque transmission takes place as a result of that a relative movement achieved between both the castings **i.e. M** and **N** there by mirror attachment will also change their position. Hence mirrors attachment continuously reflect a light rays so that there will be an angular deflection of the light rays occurs which can be measured from the calibrated scale. This angular deflection of the light rays is proportional to the twist on the shaft or the torque of the shaft. The cost of such instruments is relatively low, and an additional advantage in many applications is their small physical size.

4.28. Torque Measurement by Prony Brake

The Prony brake is also another torque-measuring device. It consists of following elements.

- It has rotating shaft whose torque is to be measured and a rope or belt is wrapped around the brake drum which is attached to the rotating shaft.
- One end of the rope is attached to a spring balance and the other end is attached to rigid supports with spring balances also. The function of rigid support loading screw is to supplied the tightness or slackness effect of rope or belt when frictional torque applied on the break drum.

- When the shaft rotation takes place, the tension will be different at the both side of the rope. The difference in tension will be will be proportional to the applied torque. The schematic diagram is as shown in figure.

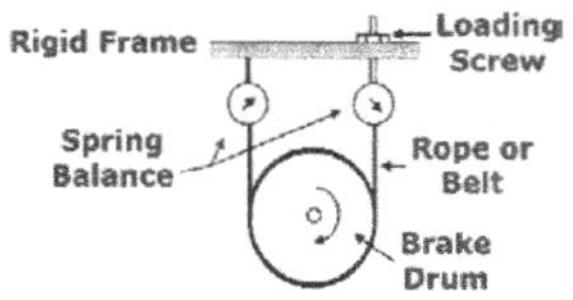

Diagram 4.62: Prony Brake

- Suppose the shaft radius is $\mathbf{R_s}$ and that of the rope is $\mathbf{R_r}$, then the effective radius ,$\mathbf{R_e}$ of the rope and drum with respect to the axis of rotation of the shaft is given by $\mathbf{R_e = R_s + R_r}$, Hence the torque in the shaft **T** will be $\mathbf{T = F_e \times R_e}$, where **Fe** effective force applied by the rope on the shaft .
- In this method of torque measurement, a large amount of frictional power is consumed between the rope and shaft which is dissipated as a heat. In order to overcome the heat effect we will continuously applied cold water for cooling to the system.

4.29. Differential Transformer Torsion Meter

It is also an another torque measuring device which consists of

- Two casting clamped with rotating shaft having air gap between magnetic circuits , both casting are joined with the help of thin steel strips.
- It has two differential transformer fitted to a casting, the two coils are wound one on the casting and another is on soft iron core. Another differential transformer is also attached to the magnetic indicating circuit. as shown in figure,

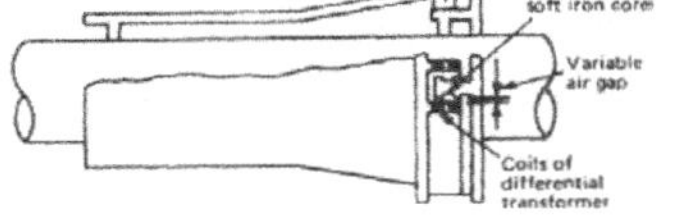

Diagram 4.63: Differential Transformer Torsion Meter

- The primary coils of the two transformers are connected in the series and A.C. supply is provided thereby **E.M.F.** is induced in secondary coils which are in opposition.

- When a torque is applied to the shaft the air gap of the shaft transformer will change, the function of micrometer screw is to achieving no current flows condition.
- When the air gaps between both the transformers being equal. The applied torque will be directly proportional to the width of the air gap.

4.30. Power Estimation from Rotational Shaft

The torque measurement technique is basically applied for estimating the following objectives

- How much mechanical power is being developed by the rotating shaft?
- How much mechanical power is required for operating a particular machinery ?
- To determine the speed of rotational shaft?

The amount of developed power is calculated from the following relationship.

$$\mathbf{P = 2\,\pi\,NT}$$

where ,

N is the angular speed in revolutions per second,

Torque measuring devices commonly used to fulfill the desired purpose is known as **dynamometers.** It is the special name given to the power-measuring apparatus that includes absorbing or driving means and where in torque is determined by the reaction forces on a stationary part (the stator). Dynamometers are generally classified into three categories.

- **Absorption dynamometers:** They absorb the mechanical energy as torque is measured, and hence are particularly useful for measuring power or torque developed by power sources such as engines or electric motors.
- **Driving dynamometers:** These dynamometers measure power or torque and as well provide energy to operate the devices to be tested. They are, therefore, useful in determining performance characteristics of devices such as pumps, compressors etc
- **Transmission dynamometers:** These are passive devices placed at an appropriate location within a machine or in between machines to sense the torque at that location. They neither add nor subtract the transmitted energy or power and are sometimes referred to as torque meters.

(i) Absorption Dynamometers : The basic principle for absorption dynamometers is that when ever mechanical power is applied, it converted into heat due to friction throughout the measurement. absorption dynamometers is used in various engine, turbines & electrical machinery, brakes like prony, hydraulic brake, fan brake etc. Absorption dynamometers , often called brakes because their operation depends on the creation of a controllable drag torque, convert mechanical work into heat. Temperature rise within a dynamometer is controlled by carrying away the heat energy, usually by transfer to a moving fluid, typically air or water. Drag torque is created by inherently dissipative processes such as: friction between rubbing surfaces, shear or turbulence of viscous liquids, the flow of electric current, or magnetic hysteresis .

4.31. Prony Brake Dynometer

In Prony brake dynometer, it consists of shaft which is connected to the driving engine & delivered it to driven load , this mechanical power is converted into heat due to frictional force between blocks and the break drum. as shown in figure .

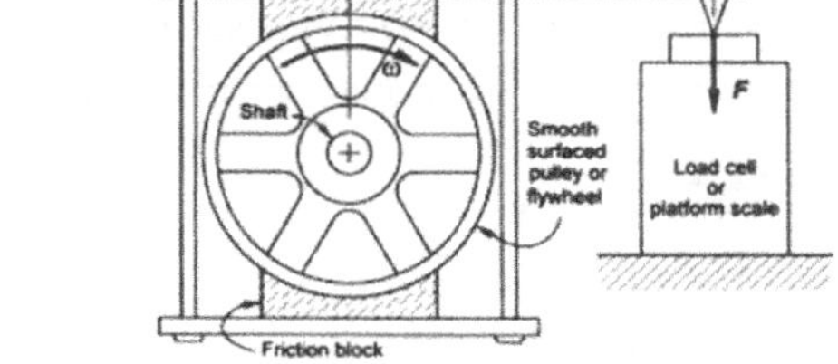

Diagram 4.64: Prony Brake Dynamometer

A loading screw is provided on the rigid support whose function is to supplied the tightness or slackness effect between blocks and the break drum, the generated heat is then removed by applying continuously flow of water or air .The torque exerted by the prony brake is given as .

$$\mathbf{T = F.\,L}$$

where force **F** is measured by conventional force measuring instruments, like balances or load cells etc. The power dissipated in the brake is calculated by the following equation.

$$P = \frac{2\pi\ NT}{60} = \frac{2\pi\ FLN}{60}\ Watts.$$

Where ,

F force is in Newton's,

L is the length of lever arm in meters,

N is the angular speed in revolution per minute,

P is power in watts.

Limitation

- It is inexpensive but depends upon the loading condition.
- During the operation of Prony brake , high temperature is generated that's result in decreasing the coefficient of friction which will be responsible for brake failure. the temperature rise is controlled by cooling. This is done by running water into the hollow channel of the flywheel.
- When the machine torque is varies means it is subjected to change with time , then coefficient of friction is also varies and hence the reading of force may be error full or difficult to measurement .

4.32. Eddy Currents Dynamometers

The constructional features of this type of dynamometer is given as, It consists of

- It consists of a toothed steel rotor having a rotational motion in a permanent magnetic field that is generated by direct current through coils in the stator, induces eddy currents in electrical circuit . the schematic diagram is as shown in figure ,

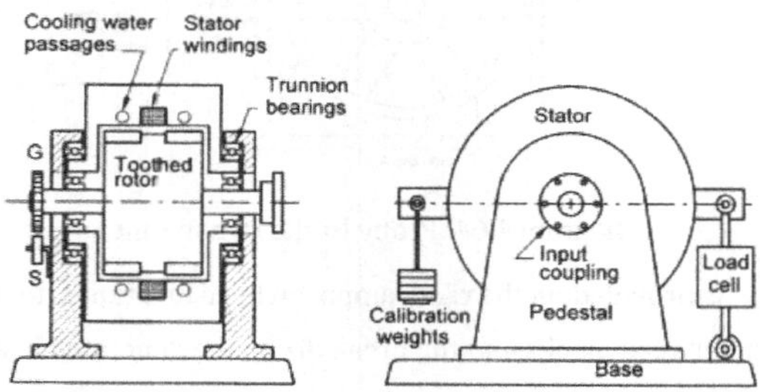

Diagram 4.65: Eddy Currents Dynamometers

- When Electromagnetic forces is generated by moving rotar and magnetized by the magnetic field then eddy currents formation takes places creates a drag torque, indicating then mechanical input power whose intensity can be controlled with variation of AC supply to stator coil .

4.33. Hydraulic Dynamometer

It is implies by the name that in this device water is used as working fluid in break operation. In hydraulic dynamometer, the power sink which uses fluid friction for dissipation of the input energy and thereby measures the input torque or power with the help of a reaction arm. As shown in figure,

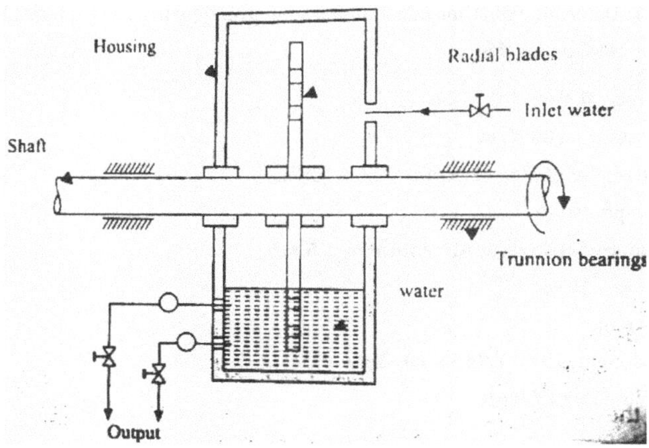

Diagram 4.66: Hydraulic Dynamometer

The power absorption capacity is managed by the water level in the housing at a particular speed. Before torque measurement, it is requirement of force measurement , to achieve it at the end of a reaction arm of radius **r** , a load cell or calibrated scale is provided , measurement of the force at the known radius, helps in calculating the torque **T** .

$$T = F X R$$

Work done & power calculation

Work done / minute = Force x distance = **F X R**

= Average torque x Angular displacement = (2 π N T / 60)

Then power (p) = **(2 π N T / 60) watts**

P = Tω watts

ω = (2 π N / 60) Rad / sec

Advantages

- In hydraulic dynamometer constant supply of water running through the breaking medium acts as a coolant.
- The brake power of very large and high-speed engines can be measured. The hydraulic dynamometer may be protected from hunting effects by means of a dashpot damper.
- In hydraulic dynamometer there is a flexibility in controlling the operation

Example 1: Determine the time constant for a bellows type pressure gauge , If the pressure gauge consists of following data ,

Bellows Area – **0.4 m²**

Spring constant – **100 N/m**

Internal Diameter of pipe - **2.5 mm** = 2.5 x 10^{-3} m²

Length of pipe - **10 m**

Inside volume of bellows or **Air Volume – 1.5 m³**

Temperature – **50 °c**

Pressure – **6 bar**

Dynamic viscosity (μ)= **1.846 x 10^{-5} kg / ms**

Assume that – **287 J / kg-K**

Solution

given that ,

A = 0.8 m², k = 100 N/m , D = 2.5 mm , P = 6 bar = 6 x 10^5 Pa

L =10 m, V =1.5 m3, T = 50 °c = 323° c , R_g= 287 J / kg-K

(i) Calculate the air density

ρ = P / R_gT = 6 x 10^5 / 287 x 323 = 6.47 kg / m³

(ii) kinematic viscosity

(μ / ρ) = (1.846 x 10^{-5} / 6.47) = 2.85 x 10^{-6}

(iii) Capacity of bellows

$$C = \frac{A^2 \rho}{K} + \frac{V}{nR_g T}$$

= [{(0.8)²x 6.47 / 100} + { 1.5 / 1.4 x 287 x323 }]= 4 x 10^{-6}m² / s

(iv) Fluid or Air Resistance

$$R = \frac{8 \nu L}{\pi r^4}$$

= [{ (8x 2.85x10^{-6} x10) } + { 3.14 x (2.5 x 10^{-3} / 2)⁴}]= 4 x 10^5 m² / s

(v) The time constant for pressure transducer

$$\tau = RC$$

= 4 x 10^5 x 4 x 10^{-6} m² / s = 0.16 s

Example 2 : Determine the time constant for a bellows type pressure gauge , If the pressure gauge consists of following data ,

Bellows Area – **2 m²**

Spring constant – 250 **N/m**

Internal Diameter of pipe - **1.5 mm** = 1.5 x 10^{-3} m²

Length of pipe - **8 m**

Inside volume of bellows or **Air Volume** – **1.5 m³**

Density of air – 995.7 kg / m³

Kinematic viscosity (υ)= **0.846 x 10^{-6} kg / ms**

Radius of connecting tube (r) = (**ID / 2**) = 1.5 / 2 = 0.75 mm = 0.00075 m

Solution

a. **Calculate , Flow resistance imposed by tubes**

$$R = \frac{8\nu L}{\pi r^4}$$

= {(8 x 0.846 x 10^{-6} x 8)/3.14 x (75 x 10^{-5})^4} = {54.144 x 10^9 / 99351562.5 }=0.5444 m² / s

b. **Calculate , Capacitance of Pressure Gauge**

$$C = \frac{A^2 \rho}{K}$$

= (2^2x 995.7) / 250 = 15.95 m² / s

c. **Calculation for Time Constant ,**

$$\tau = RC$$

= 0.544 x 15.95 = 9 sec. Approximately

Example 3 : A steel diaphragm is of 25 mm diameter is applied in pressure measuring device which a maximum pressure carrying capacity of 20 MPa , The transducer is required to give an output of 5 mV / V at the maximum pressure. the concern value of Young's modulus (E), Poisson ratio (ν) & material bulk modulus(E) is about 225 G Pa , 0.325 , 3.29 x 10^9 respectively . find out following,

(i) What will be the thickness of the diaphragm?

(ii) How much maximum deflection will achieved in the diaphragm ?

(iii) Natural frequency of the diaphragm ?

Solution

Given that ,

Radius of diaphragm '**a**' = (25 / 2) = 12.5 mm = 0.0125 m

Pressure difference **(** $P_1 - P_2$ **)** = 20 MPa = 2 x 10^7 Pa

Voltage ratio **(** V_o / V_I **)** = 5 mV / V

(i) What will be the thickness of the diaphragm

$$t = a\sqrt{\frac{0.75(p_1 - p_2)(1-\nu^2)\times 10^3}{E}}\frac{1}{\sqrt{\frac{V_o}{V_s}}}$$

t = (0.0125 x 1 / $\sqrt{5}$) [{0.75 (2 x 10^7) (1- 0.325^2) X10^3 } / 3.29 X 10^9 $]^{1/2}$

= 0.0125 x 0.4472 x 2.019330

= 0.0011288 m

= 1.12 mm

(iv) How much deflection is achieved in the diaphragm ?

$$y_{max} = \frac{3(p_1 - p_2)\left(1 - \upsilon_p^2\right)r^4}{16E_m t^3}$$

y_{max} = δy_{max} at r = 0

δ(o) = {(3 x 2 x 10^7) (1 – 0.325^2) ($0.0125)^4$} / (16 x 3.29 x 10^9) x ($0.00112)^3$

= (1.4648 / 73.9554)

= 0.01980 m = 19.8 mm

(v) Natural frequency of the diaphragm ?

$$\omega_n = 10.21\sqrt{\frac{E_m t^2}{12\left(1 - \upsilon_p^2\right)\rho r^4}}$$

= 10.21 [(3.29 x 10^9) ($0.01112)^2$ / 12(1-0.325^2) x 7830 x ($0.0125)^4$]

= 10.21 x 22

= 225.4 Hz

Example 4 : Calculate the temperature of the RTD . the Wheatstone bridge circuit is given below with suitable diagram & concerning data of bridge circuit is given below.

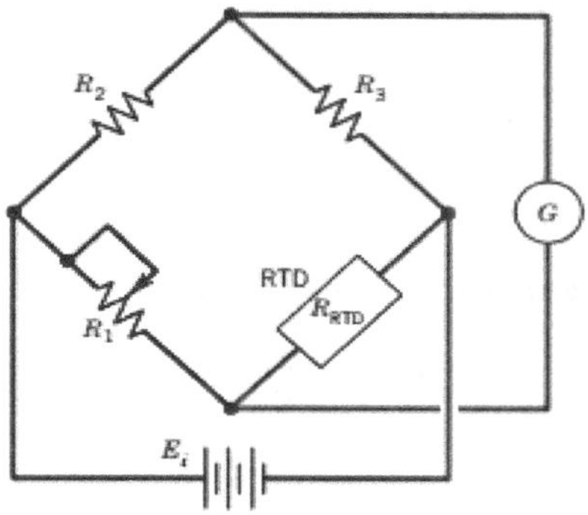

Resistance value $\mathbf{R_2}$and $\mathbf{R_3}$ are equal to **30 V,**

Resistance of RTD is 22 V at a temperature of 0C ,

RTD Coefficient of resistance is 0.003925 C^{-1}

The value of $\mathbf{R_1}$ required to balance the bridge is 40 Ω

The resistance equation of the RTD can be given as ,

$$R_{RTD} = R_0[1 + \alpha(T - T_0)]$$

Solution

given that ,

R(0°C) = 22 V ,

α = 0.002325 C^{-1} ,

R_1 = 40 Ω .

For balanced bridge condition , we have a following relationship

$\mathbf{R_{RTD} : R_1 = R_3 : R_2}$

Put the value R_{RTD}**=** 40 Ω , $\mathbf{R_O}$ = 22 Ω , $\boldsymbol{\alpha}$ = 0.002325 C^{-1} , **T** = 0°C , then the equation becomes

$$R_{RTD} = R_0[1 + \alpha(T - T_0)]$$

40 Ω = 22 {(1 + 0.002325 (T - 0) } Ω

Thus, the temperature of RTD will be **350 °C**

Example 5 : Find the value of constant " **β** " **of thermistor when** it is operated between temperatures ranging from 100 to 150C. consider following data regarding the problem ,

Thermistor resistance (R_T) is 60 kV at 25C ,

Reference resistor (R_1) = 130.5 KV ,

Dissipation Constant δ = 0.09 m W /C ,

Voltage supplied = 1.564 V.

Thermistor resistance is given by the following equation,

$$R_T = R_1\left(\frac{E_i}{E_1} - 1\right)$$

Solution

The results of these measurements are as follows:

Temperature (°C)	R_1 Voltage (V)	R_T (Ω)
100	1.501	5477.4
125	1.531	2812.9
150	1.545	1604.9

The linear form of equation for thermistor can be represented as ,

$$\ln\frac{R_T}{R_0} = \beta\left(\frac{1}{T} - \frac{1}{T_0}\right)$$

Applying this equation to the measured data, with R_0 = 60 V , we have got three value of β which are 3546.7 , 3629.9 , 3650.2 .

$\ln(R_T/R_0)$	$(1/T - 1/T_0)[K^{-1}]$	β(K)
−2.394	-6.75×10^{-4}	3546.7
−3.060	-8.43×10^{-4}	3629.9
−3.621	-9.92×10^{-4}	3650.2

Example 6 : A J-type thermocouple measures a temperature of 500C and is referenced to 0C. The length of the thermocouple wire is 15 ft in order to run from the measurement point to the ice bath and to a potentiometer. The resolution of the potentiometer is 0.09 mV. If the thermocouple wire has a resistance per unit length, as specified by the manufacturer, of 7.6 Ω/f. Find The residual current in the thermocouple circuit.

Solution

Given that

Length of thermocouple = 15 ft,

Potentiometer resolution = 0.09 mV,

Total resistance of the thermocouple = 76 Ω

Applying Ohm's law, we have following relationship between Current, Voltage & Resistance

I = (V / R)

I = (0.09 mV / 76 Ω)

I = 11.8×10^{-9} Ampere

Example 7 : Consider the pond having a water at 33°C , find the value of absolute and gauge pressures at a depth of 15 m below if density of water is 998. 207 kg / m^3 and specific gravity of mercury is 11.21 , consider atmospheric pressure is 1.0132×10^5 pa (absolute)

Solution

Given that

Height of water (H) = 15 m ,

Position of free liquid surface (H_0) = 0

Temperature of water (T) = 33°C ,

Density of water (ρ) = 998. 207 kg / m^3 ,

S.G. of mercury = 11.21 .

Solution

(ii) Calculation of absolute pressure

$$P_{ab} = \text{Atm pressure} + (\rho g H / W)$$

Absolute pressure = $(1.0132 \times 10^5) + \{(998.207 \times 9.8 \times 15) \div 1\}$

$= 1.486 \times 10^5 N / m^2 =$ **1.486 kPa abs** ≈ 1.486 atm abs

(iii)Calculation of Gauge pressure

$P_{Gauge} = P_{abs} - P_o$

$= 1.486 \times 10^5 - 1.0132 \times 10^5 =$ **4.63×10^4 N / m^2** ≈ 0.5 bar

Example 8 : A dead weight tester indicates 150 lb / inch^2 or 150 psi , at 73^{0}C where ϕ = 74 , z = 640 ft . Take g_{air} = 0.069 lb/ft^3 and g_{mass}= 692 lb/ft^3. Correct the indicated reading for known errors.

Solution

Given that ,

Pi = 150psi , z = 640 ft , ϕ = 74^0

Apply formula :-

P = Pi (1 + e_B+ e_G)

$$p = p_i(1 + e_1 + e_2)$$

(i) Correction for buoyancy effects (e_b) =

$$e_2 = -\gamma_{air}/\gamma_{masses}$$

e_2 = - (0.069 / 692)

e_2 = - 0.00009971

(ii) Correction for Gravity effects (e_g) =

$$e_1 = -(2.637 \times 10^{-3} \cos 2\phi + 9.6 \times 10^{-8} z + 5 \times 10^{-5})$$

= - { 2.637 x 10^{-3} cos 148 + 9.6 x 10^{-8} x 640 x 5 x 10^{-5} } = - 0.390276

Hence correct pressure can be given as ,

P = Pi (1 + e_b + e_g)

= 150 (1 - 0.00009971- 0.390276)

= 150 (1- 0.390375713072)

= 150 (0.609624286928)

= 91.44 lb / inch^2 .... **Answer**

Example 9 : Find out the flow rate& power consumption through the horizontal venturi meter . the following related data is given below,

Diameter of pipe = 12cm

Diameter of throat = 4 cm

U-tube manometer reading = 50 cm

Coefficient of discharge (C_v) = 0.98

Loss of pressure head = 10 %

Solution

Given that

Diameter of pipe (D_1) = 12 cm = 0.12 m

Diameter of throat (D_2) = 4 cm = 0.04 m

Manometer reading (h_m) = 30 cm of Hg

Coefficient of discharge (C_v) = 0.98 ,

Pressure loss = 10 % of Pressure Difference

Density of Manometric Liquid (ρ_m) = 13600 kg/m^3

Density of water (ρ) = 1000 kg/m^3

(i) Calculate Pressure difference

$\mathbf{P_a - P_b = (\rho_m - \rho) g\, h_m}$

= (13600 – 1000) x 9.812 x 0.5

= 61740 N / M^2

β = $\mathbf{D_2 / D_1}$

= 0.04 / 0.12 = 0.33

(ii) Others calculations

$\mathbf{V_b}$ **=** (0.98/(1-0.44)0.5) x (2 x 37089.4 / 1000)0.5 = 0.993 x 8.613 = 8.553 m/sec

Cross sectional area of throat = (Π D_2^2 / 4) = 3.14x 0.04^2 / 4 = 0.00126 m^2

Volumetric flow rate = 8.553 x 0.00126 = 0.01078 m^3/sec = **10.78 lit/sec**

Permanent pressure loss = 0.1 x61740 = 6174 N/m^2

Power consumption of the meter = 0.01078 x 6174 = **66 .5 watt**

Example 10 : A rotameter calibrated for metering has a scale ranging from 0.058m^3/min to 0.585 m^3/min. It is intended to use this meter for metering a gas of density 4.3 kg/m^3 with in a flow range of 0.085 m^3/min to 0.85 m^3/min. What should be the density of the new float if the original one has a density of 2300 kg/m^3?Both the floats can be assumed to have the same volume and shape.

Solution

Given that ,

Qnew = 0.85 m^3/min

Q_{old} = 0.58 m^3/min

Density of gas (ρ) = 4.3 kg/m^3

Density of float (ρ_f)= 1900 kg/m^3

For the same volume and shaped of the float

$\mathbf{Q_{new} / Q_{old} = \{(\rho_{fnew} - \rho) / (\rho_{fold} - \rho)\}^{0.5}}$

$0.58 / 0.85 = (\rho_{fnew} \; - 4.3) / (1900 - 4.3)^{½}$

$1.46 = \{(\rho_{fnew} - 4.3) \; / 43.539\}^{1/2}$

$2.147 = (\rho_{fnew} - 4.3) \; / 43.539$

$0.0427 = \rho_{fnew} - 4.3$

$\rho_{fnew} = \mathbf{4.347\ kg / m^3}$

Example 11 : Calculate the value of coefficient of velocity **(C_V)**for venturi meter, if following data is observed ,

Flow rate (Q) = 5.6 kg/sec

Pipe diameter = 200 mm

Throat diameter = 50 mm

Pressure drop in converging section = 100 mm of water .

$(P_a - P_b)$ = 100 mm of water = (0.1/10.33) x 101325 N/m^2 = 980.9 N/m^2 ,

Solution

Velocity at the throat is given by ,

$$V_b = \frac{C_v}{\sqrt{1-\beta^4}}\sqrt{\frac{2(P_a - P_b)}{\rho}}$$

$\beta = D_2 / D_1$ = 50 / 200= 0.25

(i) Volumetric flow rate = 5.6 / 1000 = 0.0056 m^3/sec

(ii) Velocity at the throat = Volumetric flow rate / cross sectional area of throat = 5.6 / $(\pi D_2^2/4)$ = 5.6 / $(\pi \times 0.05^2/4)$ =0.01099 m/sec

Hence the value of

C_v = **0.01099 x{(1 - (0.05/0.20)^4)$^{0.5}$} / {(2 x 100 /1000)$^{0.5}$**

= 0.01099x{**(0.998 / 0.4472**}

= 0.976

= 0.2227

Example 12 : Brine of specific gravity 1.7 is flowing through a 12 cm I.D. pipeline at a maximum flow rate of 1500 litres /min. A sharp-edged orifice connected to a simple U-tube mercury manometer is to be installed for the purpose of measurements. The maximum reading of the manometer is limited to 35 cm. Assuming the orifice coefficient to be 0.82, calculate the size of the orifice required.

Solution

Given that

Density of brine (ρ) = 1.7 x 1000 kg/m^3 = 1700 kg/m^3

Diameter of pipe (D_1) = 12 cm = 0.12 m

Maximum flow rate (Q) = 1500 liters/min = 1.5 m^3/min

Maximum manometer reading (h_m) = 35 cm = 0.35 m of Hg

Density of manometric fluid (ρ_m) = 13600 kg/m^3

Orifice coefficient (C_o) = 0.82

Solution

Given that

(i) $(P_a - P_b) = \{\rho_m - \rho\} g h_m$

$(P_a - P_b)$ = (13600 - 1700) x 9.812 x 0.35 = 40817 N/m^2

Assume that ,

$(1 - \beta^4) \approx 1$

Therefore ,

V_b = 0.82 x (2 x 40817 / 1500)$^{0.5}$ = 12..09 m/sec ,

(ii) Cross sectional area of orifice = volumetric flow rate / velocity at the orifice

= 0.15/12.09 = 0.01239 m^2

Diameter of orifice (D_2) = (0.01239 x 4/3.14)$^{0.5}$ = 0.0400 m = 4 cm

Example 13 : Find out the coefficient of discharge (C_d) with considering the following data

Flow rate is 5.2 m^3/min

Size of venturimeter = 200 mm x 55 mm

Specific gravity of mercury = 0.82

Reading of u-tube manometers = 85 mm of Hg column

Solution

Given below

Diameter of pipe or inlet of venturi meter (D_1) = 200 mm

Diameter of throat (D_2) = 55 mm ,

Density of mercury (ρ_m) = 13.6

Density of water (ρ) = 0.98

Respective formula:

Velocity $$V_2 = \frac{C_d A_1}{\sqrt{A_1^2 - A_2^2}} \sqrt{2gh_m\left(\frac{\rho_m}{\rho} - 1\right)} \text{ and } Q = V_2 \times A_2$$

Flow rate $$Q = \frac{C_d A_1 A_2}{\sqrt{A_1^2 - A_2^2}} \sqrt{2gh_m\left(\frac{\rho_m}{\rho} - 1\right)}$$

Calculations

(i) Inlet Area (A_1) = (Π / 4) $D^2{}_1$

= (3.14 / 4) x $(0.2)^2$ = $0.0314 m^2$

(ii) Throat Area (A_2) =(Π / 4) $D^2{}_2$

= (3.14 / 4) x $(0.55)^2$

= $0.238 m^2$

(iii)Flow rate (Q) = A_2V_2

V2 =(4.2 / 60) = $0.07 m^3 / s$

(iv) Coefficient of discharge (C_d)

$$Q = \frac{C_d A_1 A_2}{\sqrt{A_1^2 - A_2^2}} \sqrt{2gh_m\left(\frac{\rho_m}{\rho} - 1\right)}$$

2.50 = [(c_d x 0.0314 x 0.238)/{($0.0314)^2$– $(0.238)^2\}^{1/2}$][2x9.81x0.85{(13.6/0.9) – $1\}^{1/2}$]

2.50 = [c_d 0.0075 / $(0.0557)^{1/2}$] [16.677 x 14.12 $]^{1/2}$

0.07 = [c_d0.0318]$(235.47924)^{1/2}$

c_d = 0.07 / (15.32 x 0.0318)

c_d = { 0.07 / 0.477 } = 0.1467

Example 13 : A venturi-meter is used to measure liquid flow rate of 750 litres per second. The difference in pressure across the venturi-meter is equivalent to 6 m of the flowing liquid. The pipe diameter is 25 cm. Calculate the throat diameter of the venturi-meter. Assume the coefficient of discharge for the venturi-meter as 0.82 .

Solution

Given that ,

Flow rate (Q) = 750 L /sec ,

Diameter of pipe (D_1) = 25 cm ,

Coefficient of discharge = 0.82

Manometric height (h) = 6 m

Solution

(i) Calculate, Area of pipe or inlet area of venturi meter (A_1) =(Π / 4) $D^2{}_1$

= (3.14) x (0.25)^2 = 0.0490 m^2/s

(ii) Area of throat

$$Q = C_d \frac{A_1 A_2 \sqrt{2gh}}{\sqrt{A_1^2 - A_2^2}}$$ then , $$\frac{A_2}{\sqrt{A_1^2 - A_2^2}}$$ will be

= 750 /[(2 x 9.8 x 6)^{½} x 0.82 x 0.049.]

($A^2{}_1$ / $A^2{}_2$) = 1.0000121

$A^2{}_2$ = (0.0490)^2 / 1.000012

A_2 = 0.049 M^2

(iii) Diameter of Throat

(A_2) = (Π / 4) $D^2{}_2$

$D^2{}_2$ **= (0.049 x 4) / 3.14 =** D_2 **= 6.2 cm**

Example 14 : A venturi-meter is fitted in a pipe of 30 cm diameter inclined at 40° to the horizontal to measure the flow rate of oil having a specific gravity of0.8. The ratio of areas of main pipe and throat is 5 and the throat is at 1 m from the inlet along its length. The difference in manometer head is 40mm of mercury. Assuming the coefficient of discharge as 0.96. Find out the discharge or flow rate through the venturi-meter, the schematic diagram of venturi meter is given below.

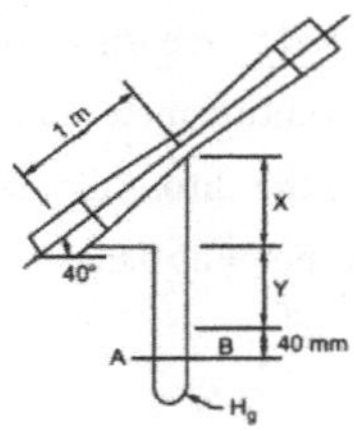

$$Q = C_d \frac{A_1 A_2}{\sqrt{A_1^2 - A_2^2}} \sqrt{2gh_m\left(\frac{\rho_m}{\rho} - 1\right)},\ A_1 = \frac{\pi}{4} \times 0.3^2 = 0.0707\ m^2$$

$A_1/A_2 = 5$

$A_2 = 0.0707/5 = 0.0141\ m^2$, **Substituting,**

$$Q = \frac{0.96 \times 0.0141 \times 0.0707}{\sqrt{0.0707^2 - 0.0141^2}} \sqrt{2 \times 9.81 \times 0.04 \times \left(\frac{13.6}{0.8} - 1\right)}$$

$= 0.0486\ m^3/s$

Example 15 : find out the value of coefficient of velocity, coefficient of contraction and the theoretical discharge through the orifice by using following data .

Actual velocity of a liquid (v) = 5 m /s ,

Head of liquid (h) = 4 m ,

Discharge (Q) = 0.025 m3/s,

Orifice diameters (D_2) = 7cm = 0.07 m

Solution

(i) Flow velocity of orifice meter is given as , then coefficient of velocity will be ,

$$V = C_v \sqrt{2gh}$$

5 = C_V (2 x 9.81 x 4) ½

C_V = 0.5859

(ii) Coefficient of discharge ,

$$\textbf{Actual discharge } Q = C_d \sqrt{2gh} \times A_2$$

0.025 = C_d(2 x 9.81 x 4)$^{1/2}$x (3.14 / 4) (0.07)^2

C_d = 0.0784

(iii) Coefficient of contraction ,

$$C_c = \frac{C_d}{C_v}$$

= (0.0784 / 0.5859) = 0.1323

Example 16 : Determine the volumetric flow of water in m^3/hr flowing in a pipe having a diameter of 40 mm through an orifice having a diameter of 20 mm. Manometer reading across the tapping is 15 cm of Hg. Assume correction factor = 1 and density of water is 1000 kg/m3.

Solution

Given that

Diameter of pipe D = 40 mm

Diameter of orifice d = 20 mm ,

Manometer head or differential pressure = 15 cm of Hg.

Differential pressure is calculated in terms of water column as

h = (13.6 – 1) × (15 × 10) mm of water = 1890 mm of water.

β = d / D = 20 mm/40 mm = 1/2

Velocity approach factor

$$M_{v.a} = \frac{1}{\sqrt{1-(d/D)^4}} = \frac{1}{\sqrt{1-(1/2)^4}} = 1.032$$

$$Q = C_d M_{v.a} A_2 \cdot \sqrt{2gh} \ \text{m}^3/\text{sec}$$

Example 17 : Determine the nominal flow velocity V2 at the orifice having a diameter of 20 mm kept in a pipe of 40 mm diameter. Reynolds number R_a is 10^5. Assume density of water = 1000 kg/m3 and kinematic viscosity is 10–2 stokes. (Cm^2/sec).

Solution

Given that ,

$R_e = \rho V_1 D/\mu$ **then , the value of V_1 will be given as ,**

$$V_1 = R_e \mu/\rho D = R_e \nu/D$$

Kinematic viscosity (v) = 10^{-2} stokes = 10^{-6} m^2/sec

$R_e = 10^5$

D = 40 x 10^{-3} m

V_1 = 2.5 m/sec

Then , from the continuity equation , we have

$$A_1V_1 = A_2V_2$$

Then V_2 will be given as ,

$$= V_1 \times A_1/A_2 = 2.5 \times \frac{\pi/4(40 \times 10^{-3})^2}{\pi/4(20 \times 10^{-3})^2}$$

$$= 2.5 \times (4^2/2^2) = 2.5 \times (16/4) = 10.0 \text{ m/sec.}$$

Example 18 : Determine the volume flow rate of water through a pipe of 110 mm diameter when measured by

(i) an orifice plate of size 55 mm dia and

(ii) Venturi tube of throat size 55 mm dia.

Assume that , Pressure head is (250 / 1000) **pa** , Density of water is **1000 kg / m³** , Cd for orifice plate 0.6 , Cd for venture-meter 0.99

Solution

Following equation will be useful in calculating the flow rate measurement ,

$$Q = \frac{1}{\sqrt{1-(d/D)^4}} \times Cd \cdot \frac{\pi d^2}{4} \times \sqrt{2gh}$$

(i) For orifice plate , the flow rate measurment will be given as ,

$$Q = \frac{1}{\sqrt{1-(1/2)^4}} \times 0.6 \times \frac{\pi(75 \times 10^{-3})^2}{4} \times \sqrt{(2 \times 9.81 \times 250)/1000}$$

$$= 1.032 \times 0.6 \times 4.417 \times 10^{-3} \times 2.214$$

$$= 0.006057 \text{ m}^3\text{/sec.} = 6.057 \text{ litres/sec.}$$

(ii) For venturi- meter , the flow rate measurement will be given as ,

$$Q = 1.032 \times 0.99 \times 4.417 \times 10^{-3} \times 2.214$$

$$= 9.99 \times 10^{-3} \text{ m}^3\text{/sec.} = 9.99 \text{ litres/sec.}$$

Example 19: An incompressible fluid is flowing through an orifice plate with a flow coefficient of 0.8 causing a pressure drop of 250 mm of water column. Calculate the fluid velocity.

Solution

Given that ,

Coefficient of discharge (C_d)= 0.8

$g = 9.8 \text{ m/sec}^2$

Pressure head (h)= 250 mm = 250 x 10^{-3} m

Applying the formula,

$$V = C_d \sqrt{2gh}$$

$V = 0.6 (2 \times 9.8 \times 250 \times 10^{-3})^{1/2}$

$V = 1.77$ m / sec

Example 20 : A pitot tube with a coefficient of 0.68 is used to measure the velocity of water in a pipe. The differential pressure head is 600 mm. find out the flow velocity ?

Solution

Given that ,

Coefficient of discharge (C_d)= 0.68

g = 9.8 m / sec^2

Pressure head (h) = 600 mm = 600×10^{-3} m

Applying the formula,

$$V = C_d \sqrt{2gh}$$

$V = 0.68 (2 \times 9.8 \times 600 \times 10^{-3})^{1/2}$

$V = 2.34$ m / sec .

Example 20 :calculate the velocity of flow in an electromagnetic flow meter if ,

Flux density of liquid = 0.58 Weber/m^2 ,

Pipe diameter = 20 cm = 20×10^{-2} m

Induced voltage = 0.5 mV = 0.5×10^{-3}

Solution

Following formula is applied in this case ,

$e = B\,v\,l$

$v = (e / B\,l)$

$= 0.5 \times 10^{-3} / \{ 0.58 \times (20 \times 10^{-2})\}$

= 0.000058 m/sec = 5.8×10^{-3} cm/sec

Example 21: Find out , the induced emf in an electromagnetic flow meter due to the flow of a conductive fluid in a pipe with inner diameter of 1.75 cm. The flux density B = 4 mV. sec/cm^2 and volume flow rate Q = 450cm^3/min.

Solution

Given that

Flow rate (Q) = 450 cm^3/min ,

ID of pipe (d) = 1.75 cm = 1.75 x 10^{-2} m

Flux density (B) = 4 mV = 4 x 10^{-3} V sec = 40 v sec/m^2 = 40 weber / m^2

1. Area (A) = (∏ / 4) d^2= (3.14 / 4) x ($1.75)^2$ = 2 . 40 m^2
2. Velocity = 450 / (60 x 2.40) = 3.11 cm / sec
3. Induced Emf (e) = *B v l = 40 x* 1.75 x 10^{-2} x 3.11 x 10^{-2} = 0.02177 V = 21.77 mV

Example 22 : calculate the speed or velocity for following two condition ,

(i) When the pitot tube is placed at submarine , pressure head or reading of manometers 27 kN/m^2 and density of sea water is 992 kg/m3

(ii) When the pitot tube is mounted at aircraft, pressure head or reading of pressure gauge 9 kN/m^2 and density of air is 1000 kg/m3 .

Solution

Given that

(i) For submarine

Pressure head or pressure difference (ΔP) = 27 kN/m^2

Density of sea water (ρ) = 992 kg/m3 ,

Then the velocity or speed of submarine may be calculated as ,

$$V = \sqrt{\frac{2\Delta P}{\rho}}$$

= [**(2 x 27000) / 992**] ½ **= 7.4 m / sec .**

(ii) **For aircraft ,**

Pressure head or pressure difference (ΔP) = 9 kN/m^2

Density of air (ρ) = 1000 kg/m3 ,

$$V = \sqrt{\frac{2\Delta P}{\rho}}$$

= { (2 x 9000) / 1000 $\}^{1/2}$ **= 4.3 m / sec**

Example 23 : A rota meter uses a cylindrical float of 3.5 cm height, 2.5 cm diameter and density of 2800 kg/m3. The maximum inside diameter of the metering tube is 4.5 cm. Determine the maximum flow rate handling capacity of the rota meter if the fluid is water.

Assume that ,

Float volume (V_f) = 2.15×10^{-5} m^3

$C_d = 0.82$

Float diameter (D_f) = 2.5×10^{-2} m

Pipe diameter (D_p) = 4.5×10^{-2} m

Solution

The volume flow rate (Q) will be expressed by following relationship ,

$$Q = Cd \cdot \frac{D_p^2 - D_f^2}{D_f} \sqrt{\frac{\pi g V_f (\rho_f - \rho)}{2\rho}}$$

Q = [0.82 x {(0.045² – 0.025²) / 0.025} **]** [3.14 x 9.8 x 2.15 x 10^{-5} {(2.8 x 10^3) – 1000 **}]**

= [0.82 x 0.056] [0.04592] = 2.52 x 10^{-2} m^3/ sec

Example 24 : Deduce an expression for Data acquisition and Reduction errors in case of Multiple temperature sensors .

Solution

Consider that we have Multiple temperature sensors or thermocouple say **M** and resultant of such a system is measured and stored through M-channel data acquisition system also considered that all the sensors are operated at normal temperature with same referenced junction and **20** reading for each **M** thermocouples are observed say it is **N** . then ,

N_i = 20 Readings Taken For Each Thermocouple,

Where, i = 1 , 2 , 3 , ... N

M_j = Total No of Thermocouple .... say **M**

Where, **j = 1 , 2 , 3 , ...N**

Suppose all the thermocouple is of T-type ,then **The mean value of all the reading (N) may be given by following relation,**

$$T_j = \frac{1}{N}\sum_{i=1}^{N} T_{ij}$$

And the pooled mean for **'M' thermocouples** is given by the following relationship ,

$$\langle \overline{T} \rangle = \frac{1}{M}\sum_{j-1}^{M} \overline{T}_j$$

Conclusion

- The Estimated Systematic Uncertainty during data acquisition cab be given as ,

Systematic Uncertainty = (Pooled Mean Temperature – Normal Known Temperature)

- The Estimated Stanard Random Uncertainty during data acquisition & reduction cab be given as

$$\textbf{Random Uncertainty} = \{ \overline{T}_j - \langle \overline{T} \rangle \}$$

$$= [\overline{T}_j = \frac{1}{N}\sum_{i=1}^{N} T_{ij} - \langle \overline{T} \rangle = \frac{1}{M}\sum_{i-1}^{M} \overline{T}_j]$$

The standard random uncertainty for the data acquisition and reduction instrumentation system can be deduced as ,

$$s_{\overline{T}} = \frac{\langle s_T \rangle}{\sqrt{M}}$$

Where ,

$$\langle s_T \rangle = \sqrt{\frac{\sum_{j=1}^{M}\sum_{i=1}^{N}\left(T_{ij} - \overline{T}_j\right)^2}{M(N-1)}}$$

And Degree of Freedom for such a system can be given as ,

Degree of Freedom = M (N - 1)

Question for Discussion

1. Explain the constructional & operational features of AC and DC. Techogenerator
2. What is the main difference between a moving coil and a moving iron instruments
3. What is tachometer ? Explain the d.c.tachogenerator.
4. Find the change in wire resistance for a strain gauge that has a nominal wire resistance of 150 Ω when it is subjected to a strain of 800 μm/m.
5. Explain in brief the following with neat sketch

   i) Prony brake dynamometer and

   ii) Rope brake dynamometer
6. Define LVDT. Explain its working with neat sketch. Also state its practical application.
7. Explain the working of R V D T with suitable sketch , also write down its merits & demerits
8. Explain followings briefly.

   (i) Stroboscope

   (ii) Resonance (vibrating reed) tachometer
9. Explain with the help of neat sketches the principle and construction of an auto-Collimator.
10. Give the detail classification of Tachometer and explain briefly photoelectric generator.
11. Give the classifications of Tachometers .
12. What is load cell ? Explain any one Method used for force measurement.
13. Explain about the torque measurements by using strain gauge method
14. Discuss briefly about the Torque measurements using torsion bars and proney brake.
15. Explain the working principle of electrical and mechanical torsion meter
16. Define the working of load cells ?
17. Discuss briefly about the dial type and reed type of mechanical comparator with a neat sketch.
18. Classify the various torque measuring method ? explain about strain gauges technique with suitable diagram .
19. Consider a single strain gauge of resistance 120Ω mounted along the axial direction of an axially loaded specimen of steel (E=200 GPa). If the percentage change in length of the rod due to loading is 3% and the corresponding change in resistivity of the strain gauge material is 0.3%, estimate the percentage change in the resistance of the strain gauge and its gauge factor; Poisson ratio=0.3. If the strain gauge is connected to a

measurement device capable of determine change in resistance with an accurate of ± 0.02 Ω, what is the uncertainty in stress that would result in using this resistance measurement device?

20. Write short notes on following
    (i) Mechanical comparators
    (ii) Electrical comparators
    (iii) Optical comparators
    (iv) Pneumatic comparators
21. With neat sketch explain the construction and working principle of differential pneumatic comparator.
22. How to measure the power by using rope brake dynamometer? Explain with a neat diagram
23. Explain how cup and vane type anemometers are used to measure air movement.
24. Write the advantages and disadvantages of the mechanical comparator?
25. Explain any one method used to measure very high pressure.
26. Write a detailed note on pressure measuring systems and pressure measuring transducers.
27. Explain the working and construction of Mcleod gauge with suitable sketch .
28. Explain briefly about
    (i) piezo-electric pressure transducer
    (ii) variable capacitance transducer
29. Explain the method of measurement of force by using a strain gauge load cell .
30. What are the different elastic transducers used for the pressure measurement. Illustrate the working principle of any one in detail.
31. A 100 Ω resistance strain gauge having gauge factor 1.5, mounted on shaft. and it forms one arm of a symmetrical Wheatstone bridge circuits . when shaft is subjected to tensile load , the galvanometer reading is about 5 mV when operating current is about 15 mA . calculate the strain
32. An electrical resistance strain gauge of resistance 100 Ω, and gauge factor 2.0 is bonded to steel having an elastic–limit stress of 400 MN/m2 and modulus of elasticity 200 GN / m2 . determine change in resistance due to
    (i) Change of stress equal to 10% of elastic range.
    (ii) An increase of 25Oc in temperature.

    Where,

α_R = 2 x 10-6 / OC (resistance temperature coefficient)

λg = 12 x 10-6 / oc (coefficient of linear expansion)

33. The torque transmitted by a propeller shaft is measured by pair of strain gauge bonded at 45 to axis of the shaft. each gauge has a nominal resistance of 100 Ω having a gauge factor of 2.0 and are connected to form half bridge circuits. the bridge supply voltage is 6 V, maximum strain is about 1000 micro strain, power transmitted 500 kw, rotational speed is about 1400 RPM and E = 200 x 109 N/m2. then determine
    (i) Diameter of shaft
    (ii) Output voltage at full power
34. Define gauge factor of resistance strain gauge ,
35. Show that in a wire resistance strain gauge, the value of gauge factor attributed to resistance change due to changes in gauge wire length , its cross sectional area and resistivity . under what condition, The relationship holds good F= 1 + 2μ

    Where,

    F = gauge factor

    μ = poission ratio
36. what is strain rosette? explain

    A rectangular rosette is placed on a steel plate , the strains are σ 1 = 500 μ cm / cm , σ 2 = 100 μ cm / cm , σ 3 = 200 μ cm / cm , calculate the principle stress and strain .
37. write short notes on following
    (i) Bonded and un bonded strain gauge
    (ii) Semi – conductor strain gauge
    (iii) Strain gauge material
    (iv) Electrical resistance strain gauge
    (v) Dummy gauge
38. Define the temperature compensation and how it is performed when strain measurement is performed with axial thrust , bending load and torque by using bonded strain gauge
39. Described, the basic working principle of stroboscope for speed measurement with suitable sketch

    A stroboscope light is fleshing 3600 time/min . the light is focused upon the end of the shaft with a single keyway in it and there appear to be 4 keyway , 90^0aparts . what will be the shaft speed .

40. A 45 teeth ferrous gear wheel was used with an electromagnetic pick – up to measure the angular velocity of hydraulic turbine. if the turbine shaft is rotating at 12Π rad / sec . then what will be the reading of frequency meter .
41. Explain briefly about speed counter, tacho-scope and tachometer
42. What is proving ring? how is it used to measure force.
43. What is dynamometer? classified it and explained any one in details.
44. Define the terms torque? classified about torque measuring technique, explain about torque reaction method with clear sketch.

References

1. L.J. Weymouth, J.E. Starr and J. Dorsey, Bonded resistance strain gauges, Experimental Mechanics 6(4): 19A, 1966.
2. Micro-Measurements Division, Measurements Group, Inc., Strain Gauge Selection: Criteria, Procedures, Recommendations, Technical Note 505-1, Raleigh, NC, 1989
3. D.S. Kumar, Mechanical measurement & control, Technical Publication, New Delhi, India.
4. E.E. Herceg, Schaevitz Handbook of Measurement and Control, Schaevitz Engineering, Pennsauken, NJ, 1976.
5. S.S. Gindy, Force and torque measurement, a technology overview, Experimental Techniques, 1985.
6. J.J., Jr. Elengo, Selecting and Applying Load Cells, Instruments and Control Systems, 1980.
7. Omegadyne, Inc., Omegadyne Pressure, Force, Load, Torque Databook, Sunbury, OH, 1996
8. Engineering Metrology, I.C. Gupta, Dhanpat Rai Publications, Delhi.
9. Mechanical Measurements, R.K. Jain Khanna Publishers, 1994.
10. Beckwith Marangoni and Lienhard, Mechanical Measurements, Pearson Education, 6th Ed., 2006.
11. A. Bray, G. Barbato and R. Levi, Theory and practice of force measurement, Academic Press, 1990.
12. Introduction to Capacitance Gages, ADE Technologies.
13. W. Du, S. Yelich, Resistive Sensors: Principles, Design, and Applications, Proceedings of the 2ndInternational Conference on Sensing Technology, 2007, 326-331.
14. W. Du, S. Yelich, Capacitive Sensors: Principles, Design, and Applications, Proceedings of the 2ndInternational Conference on Sensing Technology, Palmerston North, New Zealand, Pp. 332-337, 2007.
15. R.J. Melling, Ionization Vacuum Gauge Measures Absolute Pressure, Instruments and Control Systems, 1964.

CHAPTER 5

CONTROL SYSTEMS

5.0. Introduction

Control system is composed of two clause, i.e. control and another is system control is supplied where some kind of dynamic activity has been achieved with system where system representing a set of control devices.

The function of control system is to provide regulation, commanding and direction to any particular dynamic activity on the other hand a system is unit or collection of sub system of associated devices which helps in obtaining controllable response of a mechanism. Hence we can say that. A control system is an interconnection of various components forming a system configuration that will provide a desired system response. The basis for analysis of a system is the foundation provided by linear system theory. The input–output relationship represents the cause & effect relationship of the process, A **cause-effect relationship** is a relationship in which the processing of the input signal is behave as cause which create another event happening, a variation of output signal is an effect of any dynamic activity. In order to under stands the concept of control system, we have to two quantity–(i) input parameter & (ii) output parameter. Input may be stimulus or command applied to control system on the other hand output is a respective response of the input. In general Control engineering is based on the foundations of feedback theory and linear system analysis, related to network theory and communication theory.

But the field of application is not limited to network theory and communication theory but also applicable to aeronautical, chemical, mechanical, environmental, civil, and electrical engineering, traffic control systems, chemical processes, and robotic systems.

Historical background of control system: Here we are supplied some important historical view of control system.

- The first feedback system to be invented in modern Europe was the temperature regulator of Cornelis Drebbel (1572–1633) of Holland [1].
- Dennis Papin [1647–1712] invented the first pressure regulator for steam boilers in 1681. Papin's pressure regulator was a form of safety regulator similar to a pressure cooker valve.

- The first automatic feedback controller used in an industrial process is generally agreed to be James Watt's **fly ball governor**, developed in 1769 for controlling the speed of a steam engine [1,2]. It measured the speed of the output shaft and utilized the movement of the fly ball with speed to control the valve and therefore the amount of steam entering the engine. As the speed increases, the ball weights rise and move away from the shaft axis, thus closing the valve. The flyweights require power from the engine to turn and therefore cause the speed measurement to be less accurate.
- The first historical feedback system, claimed by Russia, is the water-level float regulator said to have been invented by I. Polzunov in 1765 [4]. The float detects the water level and controls the valve that covers the water inlet in the boiler.

5.1. Design of Control System

For designing of any control system, we must have to implement the following steps to find out the desirable response of process by the control system unit.

DESIGN PARAMETER OF CONTROL SYSTEM UNIT

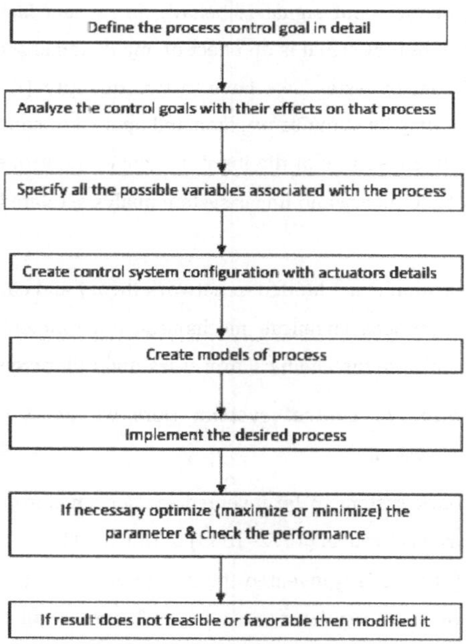

Digram 5.1: Design Parameter of Control System Units

Example of control system: The practical & common example of control system is steering control of the automobile , in which the direction of two front wheel can be regarded as the result or controlled output variable and the direction of steering wheels as the actuating signals or objectives. Hence the control system has combined two mechanism **(i) steering mechanism & (ii) dynamic system of automobile.** This can be under by following diagram.

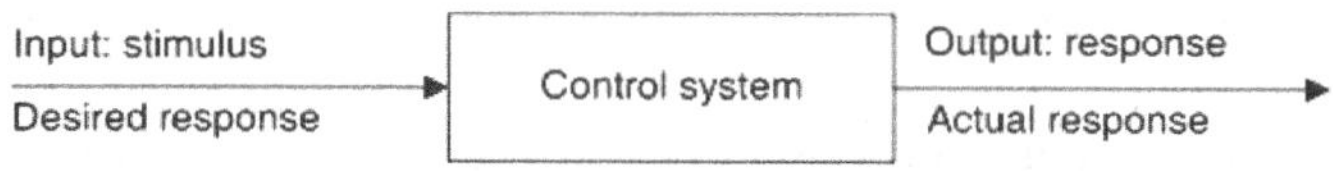

(a) Description of Control System

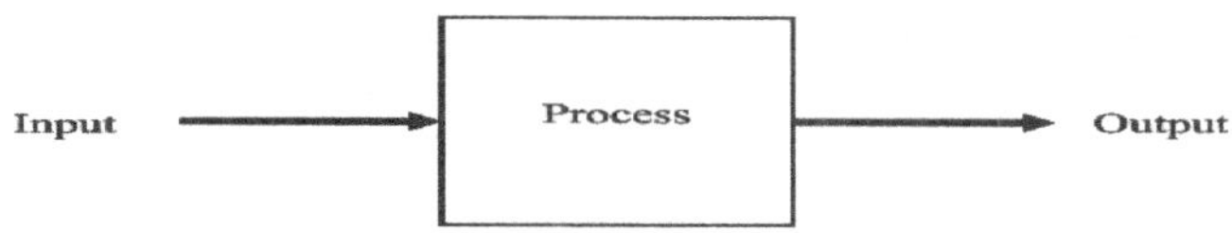

(b) Process Under Control

Diagram 5.2: Control System Analyses

Some other example of control system : Following are some other important example of control system.

- Heating and air conditioning system controlled by thermostat.
- Automobile steering & speed control.
- Automobile traffic control signal system of road traffic.
- Automatic light on & off system of room during night & day.
- Automatic hot water heater.
- Human heart using a pacemaker
- Automatic position control of high speed automated train system

5.2. Control System Concept

In any type of control system, we will have to considered the three type of variable, these are

- **Controlled variables :** Controlled variables are those which are independent variable and maintain the value at desired level throughout the process. These are also known as the set point or reference input. **For example –** voltage, temperature, level, position , pressure , pH value , density etc.

- **Manipulated variables : Manipulated variables are works as** input variables and adjusted dynamically to keep the controlled variables at their set-points & maintained it despite any disturbances. This manipulated variable is maintain & controlled by the process in such a manner to keep the controlled variables in the system at constant settings since **manipulated** variables are dependents upon on controlled variables. For example - Rpm of fan, Valve position, Motor speed, Damper position, or blade pitch.
- **Disturbance variables :** The disturbance variable is **an** unwanted input variables that will create the deviation in the controlled variable from their respective set points therefore it will affects the value of manipulated variable hence finally affects the process outputs that may be Internal factor , External factor , Environmental factors **.**

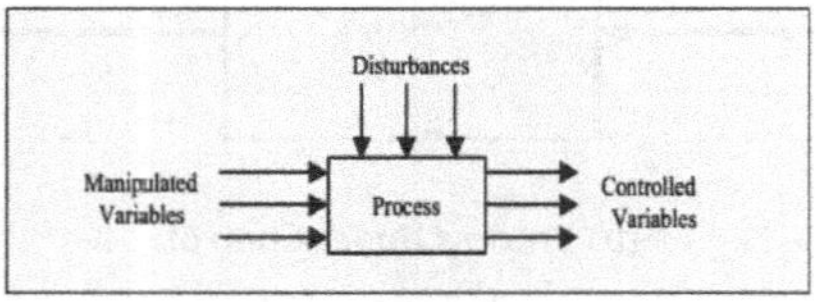

Diagram 5.3: Process Control Variable

The basic elements of control system consists of four elements, these elements is representing in the block diagram,

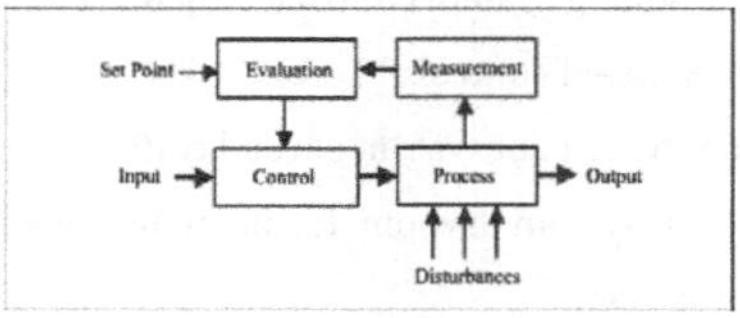

Diagram 5.4: Four Elements of Control System

(i) Process : Process is methodology by virtue of which we can tried to create any particular article/products. In general, a process consists of an assembly of equipment/device and material that is related to some manufacturing operation or process. the process always may be associated with the single variable or multivariable which has been controlled up to permissible limits.

(ii) Measurement: Measurement refers to the conversion of the process variable into an analog or digital signal that can be used by the control system. The device that performs the initial measurement is called a sensor or instrument. The result of any measurement is the conversion of a dynamic variable into some proportional information.

(iii) Evaluation : A device called a controller performs this evaluation. The controller can be a pneumatic, electronic, or mechanical device mounted in a control panel or on the process equipment. Throughout the evaluation process the measurement value is examined and compared with the desired value or set point. If it is deviate then corrective action will be performed.

(iv) Control : The control element in a control system is the device that exerts a direct influence on the process or manufacturing sequence. Devices such as electrical motors , pumps , and dampers are also used as control elements.

(v) System Error : It is defined as is the difference between the value of the set point value and the true value of the process variable maintained by the system. The system error is expressed in equation form by the following:

$$e(t) = PV(t) - SP(t)$$

Where ,

e (t) = System error as a function of time (t)

PV (t) = The process variable as a function of time

SP (t) = is the set point as a function of time

(vi) System Response : System response is the ability of a control loop to recover from a disturbance that causes a change in the controlled process variable.

5.3. Control System Basic Architecture

In any control system we have a particular input in the process and achieving a respective output but between the length of initial stage to final stage there may be involvement of so many other devices , which will help in accomplishing the final result with desired accuracy , therefore the overall structure of the control unit is known as Control system architecture , It is described in brief with suitable diagram.

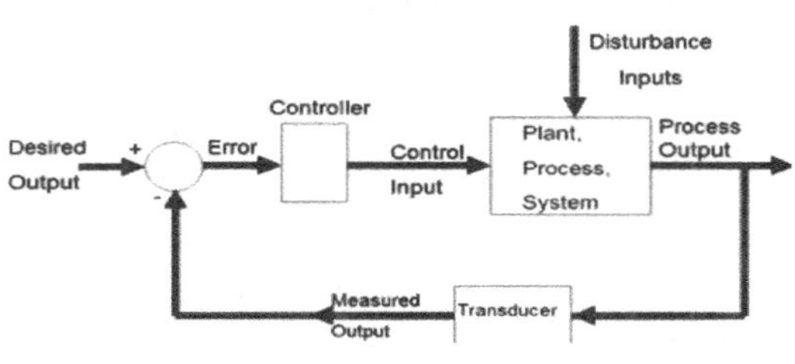

Diagram 5.5: Control System Architecture

(i) Desired Output / Reference Input : The input to the system is the reference input value, or set point, for the system output or It is the actual signal input to the control system which represents the desired operating value of the output. In a cruise control system like

- An electrical signal corresponding to the desired output speed.
- The automobile engine , the resultant output is the variable of the process that is being measured and compared to the input

(ii) Plant / Process / System : The device, process, or system that need to be controlled. the system control is nothing but it is a physical process in which we can either model or measure. The output of the system is may be of any type of the signal. In automobile control system, the output signal is the speed of the car.

- **Transducer** - The concept of the transducer is to convert the output signal into equivalent measurable value of other form in the system.
- **Error Signal** – The error signal may be of electrical type / mechanical type or may be any other depending upon the application , that is the difference between the desired output and true output signals.
- **Controller / actuating devices :** These mechanisms are usually designed especially for the system and applied to modify the behavior of the process in which changes in the process are accomplished to influence the output variable or we can say that **The actuating devices** is a power device that produces the input to the plant according to the control signal so that the output signal will approach the reference input signal.
- **Controlled variable (output) :** The quantity that must be maintained at prescribed value throughout the process.
- **Disturbance Input :** An unwanted input signal that affects the output signal. Disturbance may be an internal factor, External factor, environmental factors that affects the process behavior.

5.4. Feedback Loop of Control System

A feedback is a common and powerful tool when designing a control system. Feedback loop is the tool which takes the system output into consideration and enables the system to adjust its performance to meet a desired result of system. In any control system , output is affected due to change in environmental condition or any kind of disturbance. So one signal is taken from output and is fed back to the input. This signal is compared with reference input and then error signal is generated. This error signal is applied to controller and output is corrected. Such a system is called feedback system. Figure below shows the block diagram of feedback system.

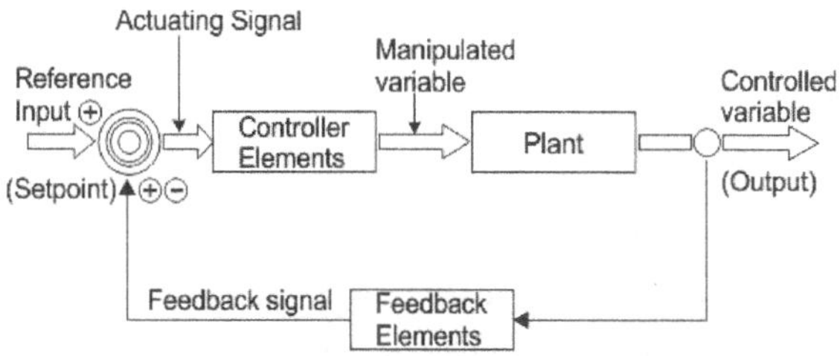

Diagram 5.6: Feedback Loop of Control System

When feedback signal is positive then system called positive feedback system. For positive feedback system, the error signal is the addition of reference input signal and feedback signal. When feedback signal is negative then system is called negative feedback system. For negative feedback system, the error signal is given by difference of reference input signal and feedback signal.

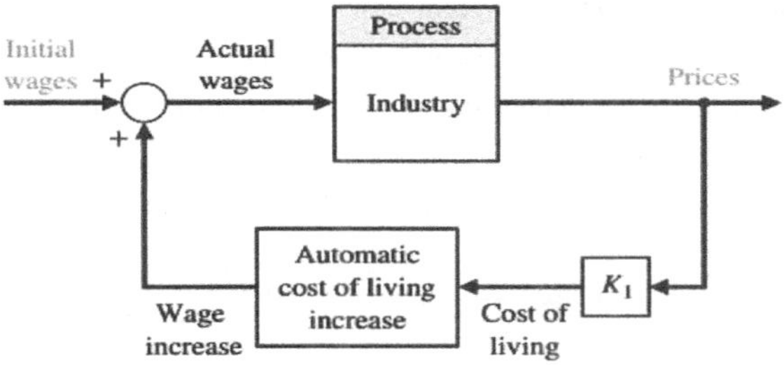

Diagram 5.7: Positive Feedback System

5.5. Effect of Feedback

Consider the following block diagram which indicating the feedback system ,

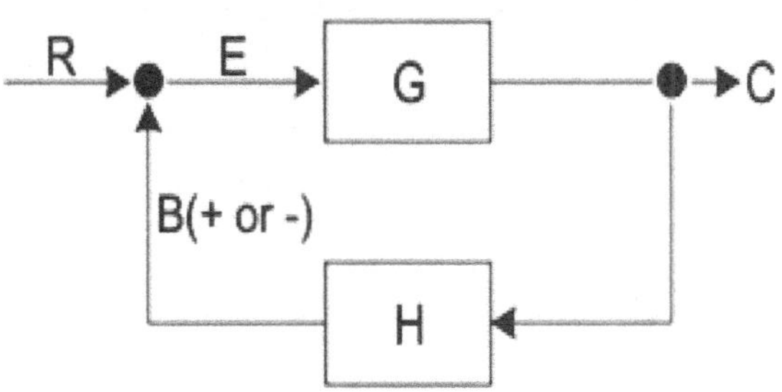

Effect of Feedback

Where ,

R = Input signal , E = Error signal

G = Forward path gain , H = Feedback

C = Output Signal , B = Feedback signal

- Error between system input and system output is reduced.
- System gain is reduced by a factor **1 / (1 ± GH).**
- Improvement in sensitivity.
- Stability may be affected.
- Improve the speed of response.

5.6. Control System Configuration

The control system configuration consists of following components .

(i) Block Diagrams of Control System : In control system, any kind of operation or process performed by system like physical, mechanical, electrical, chemical is represented with the help of pictorial block to simplify the control system process that's depends upon the cause-and-response relationship between input and output of the system and it is suitable way to derive transfer function of control element connected to the system .

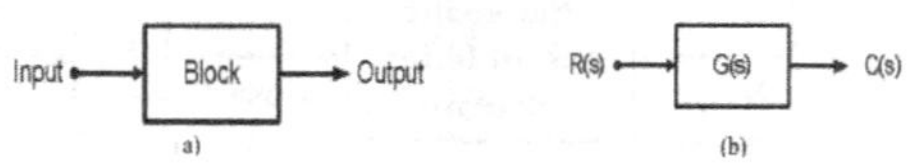

Diagram 5.8: Block Diagrams

(ii) Error signal : Consider the close loop control system in which following data is achieved as shown in figure ,

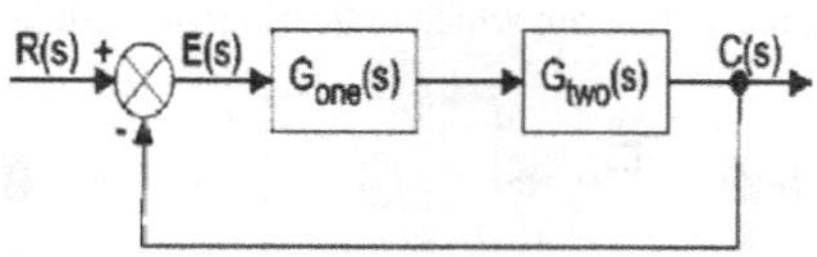

Diagram 5.9: Block Diagrams Representing Error

Where ,

C(s) is output signal or fed back of the system,

R(s) is input of the system

Then error signal is defined as the difference between input and output

$$\dot{E}(s) = R(s) - C(s)$$

(iii) Cascade Blocks : Whenever several control blocks are connected in cascaded manner, the transfer function of the entire system will be the product of transfer function of all individual blocks.

Diagram 5.10: Cascade Block

$$G(s) = \frac{C(s)}{R(s)} = \frac{C_1(s)}{R(s)} \cdot \frac{C_2(s)}{C_1(s)} \cdot \frac{C_3(s)}{C_2(s)} \cdot \cdot$$

Hence final block diagram and resultant transfer function can be represented as ,

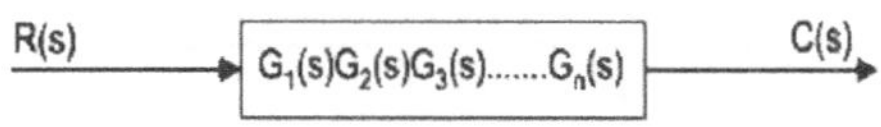

Resultant transfer function = $G_1(s)G_2(s)G_3(s) \cdot \cdots \cdot G_n(s)$

(iv) Parallel Blocks : whenever control block are arranged in parallel to each other and same input signal is provided then the resultant or over all transfer function will be equal to the algebraic sum of all individual transfer function of blocks. Such kind of system is as shown in figure .

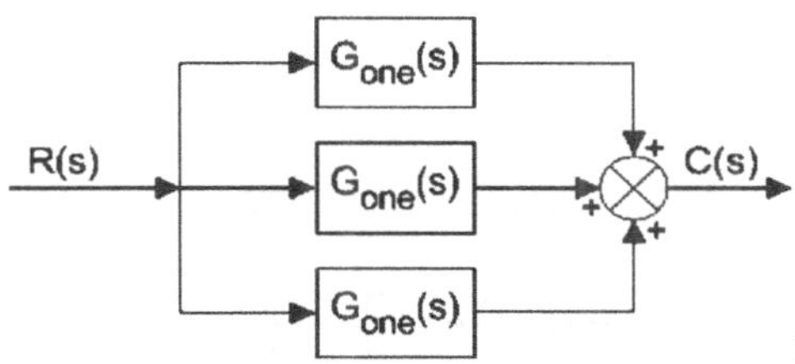

Diagram 5.11: Parallel Blocks Control System

$G(S) = [C_1(S) / R_1(S)] + [C_2(S) / R_2(S)] + [C_3(S) / R_3(S)] == G_1(S) + G_2(S) + G_3(S)$

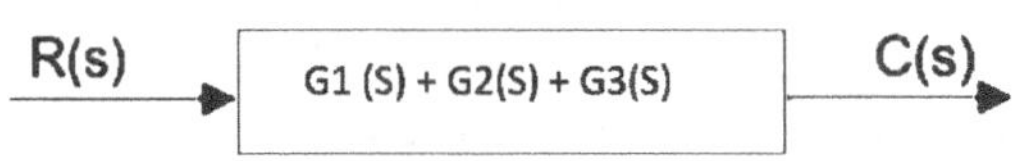

5.7. Transfer Function

The Transfer function of system is defined as it is a ratio between output to input . the block diagram is given as ,

Transfer function = Output / Input

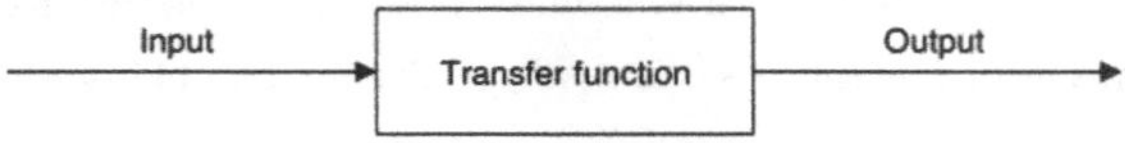

Diagram 5.12: Transfer Function

The transfer function is of two types & very important from subject point of view these are Open-Loop Transfer Function and Feed forward Transfer Function.

- **Open-Loop Transfer Function :** It is defined as the ratio between signal B(s) to the actuating error signal E(s) is called the open-loop transfer function . the block diagram is as shown as ,

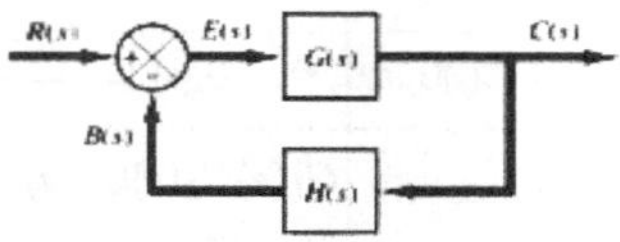

Open-Loop Transfer Function

$$Open-loop\ transfer\ function = \frac{B(s)}{E(s)} = G(s)H(s)$$

- **Feed forward Transfer Function :** It is defined as the ratio between output **C(s)** to the actuating error signal **E(s)** is known as feed forward transfer function , the block diagram is as shown as ,

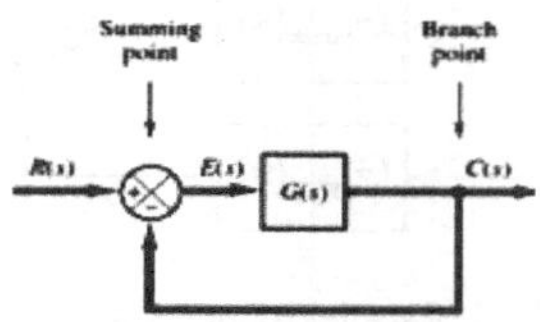

Feed forward Transfer Function

$$Feedforward\ transfer\ function = \frac{C(s)}{E(s)} = G(s)$$

- **Closed-Loop Transfer Function**: In Closed-Loop Transfer Function the output **C(s)** and input **R(s)** are related as ,

$$C(s) = G(s)E(s)$$

$$E(s) = R(s) - B(s)$$

$$= R(s) - H(s)C(s) \quad \text{..........................} \; 5.1$$

So that the value of

$$C(s) = G(s)\,[R(s) - H(s)C(s)]$$

$$\frac{C(s)}{R(s)} = \frac{G(s)}{1 + H(s)C(s)} \qquad \text{............................ 5.2}$$

Summing Points : A circle with a cross is the symbol that indicates assuming operation. With the help of summing point we can calculate the error between the actual output and the desired output and produce a positive or negative error signal value.

Error = Set point – Actual

- For **closed-loop systems** the summing point will be represented by circle with two crossed lines.

- Plus (+) symbol is used for **positive feedback control system** and sometimes known as "**summer**" as shown in figure , in which arrow represent the direction of signals .

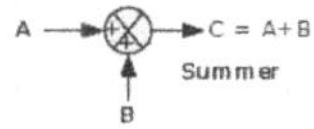

- Minus (–) symbol is used for **negative feedback control system** and sometimes known as "**comparator**" as shown in figure , in which arrow represent the direction of signals .

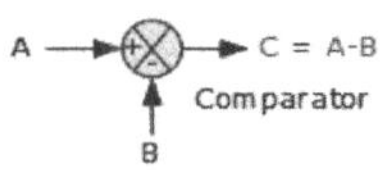

- **Open-Loop control system :** A system in which the output has no effect on the input action. In other words, in the open loop control system output & input both are independent to each other. Neither output measured nor error is considered for mitigation. A simple block diagram of such a system is given as ,

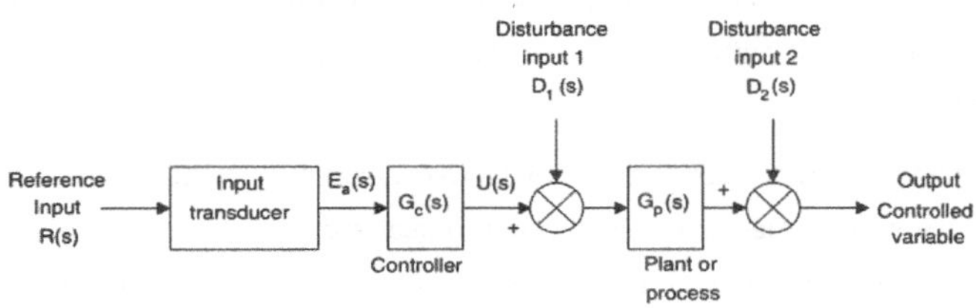

Diagram 5.13: A General Block Diagram of Open-Loop Control System

- **Closed-Loop Control System :** A system in which the output has an effect on the input quantity and always tried to maintain the desired output value. A simple block diagram of such a system is given as ,

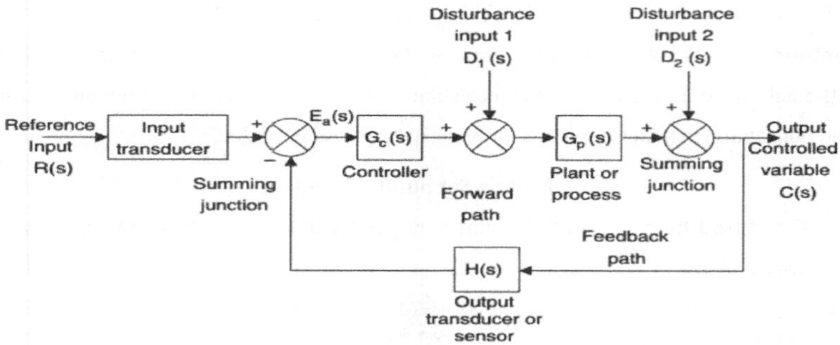

Diagram 5.14: A General Block Diagram of Closed-Loop Control System

Where ,

C(s) - Control output which is Transfer Function of **c(t)**

D(s) - Disturbance input which is Transfer Function of **d(t)**

$E_a(s)$ - Disturbance input which is Transfer Function of $\mathbf{e_a(s)}$

$G_a(s)$ - Transfer Function of the actuator

$G_c(s)$ - Transfer Function of the controller

$G_p(s)$ - Transfer Function for process

H(s) - Transfer Function for process

R(s) - Reference input which is Transfer Function of r(t)

5.8. Classification of Control System

In general control system is broadly classified into following categories but we will only consider here , open loop control system and close loop control system , the brief definition of remaining types is given as .

- **Natural control system :** It is defined as a control system which is created by nature , i.e. solar system, digestive system of any animal, etc.
- **Man-made control system :** It is defined as a control system that is created by humans being for its own comforts , i.e. automobile, power plants etc.

- **Automatic control system :** A control system that's governs to the basic concept of engineering & technology in association with mathematics , considered as Automatic control system , in this system sensors , actuators , transducer may be basic elements .
- **Combinational control system :** It is a control system that is a combination of natural and man-made control systems, i.e. driving a car etc.
- **Time-variant control system :** It is a control system where any one or more variable of the control system vary with respect to time i.e. driving a vehicle.
- **Time-invariant control system** : It is a control system where none of its parameters or variable vary with respect to time i.e. control system made up of inductors, capacitors and resistors only.
- **Linear control system:** It is a control system that satisfies properties of homogeneity.
- **Non-linear control system :** It is a control system that does not satisfy properties of homogeneity .
- **Continuous -Time control system** : It is a control system where performances of control variable depends upon time , i.e. armature type speed control of motor.
- **Discrete -Time control system :** It is a control system where performances of control parameters depends upon discrete time i.e. microprocessor type speed control of motor.
- **Deterministic control system :** It is a control system where its output is predictable .
- **Stochastic control system :** It is a control system where its output is unpredictable .
- **Lumped-parameter control system :** if any control system is represented by mathematical ordinary differential equations then it is called Lumped-parameter control system .
- **Distributed-parameter control system :** if any control system is represented by electrical analogy components like resistors , inductors and capacitors and established a mathematically equation that it is called Distributed-parameter control system .
- **Single-input-single-output (SISO) control system :** It is a control system that has only one input and one output.
- **Multi-input-multi-output (MIMO) control system :** It is a control system that has only more than one input and more than one output.

But From the subject point of view , we will only study about two types of control system. These are described under here ?

- **Open loop control system**
- **Closed loop control system**

5.9. Open Loop Control System

An open-loop control system consists of controller or actuator takes input under the consideration and the control action is independent of output of the system or system feedback . In other word we can say that "**An open-loop control system consists of actuating device in order to control the process directly without considering the feedback** ".Fig -shows the block diagram of open loop control system in which process output is totally independent of controller action . Hence this system works on fixed operating condition having no disturbance & variation.

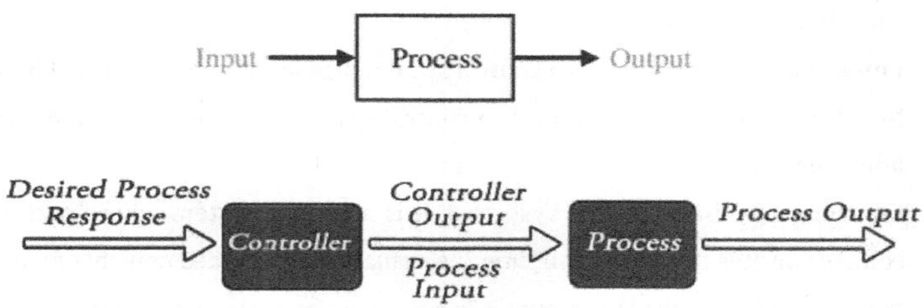

Diagram 5.15: Open Loop Control System (without feedback)

Following are common & practical example of open loop control system

- **Automatic Washing Machine -Automatic Washing** machine runs as per the pre-set time in respect of washing is completed or not.

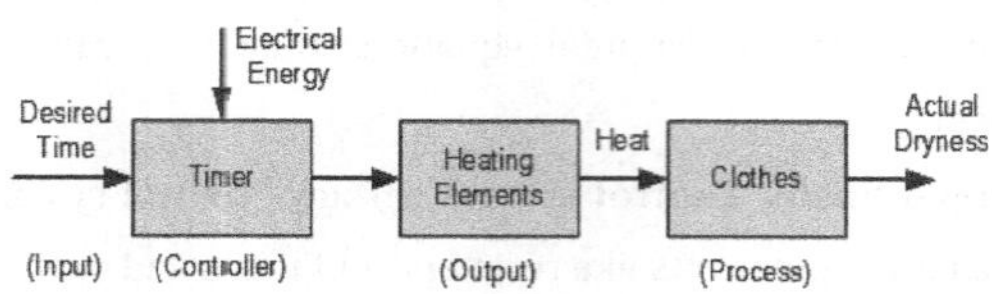

Diagram 5.16: Open Loop Control System

- **Bread Toaster : -** Bread Toaster operates as per adjusted time rather than toasting is completed or not.

Diagram 5.17: Open Loop Control System

- **Light Switch** - Lamps glow whenever light switch is on means independent of light is required or not.

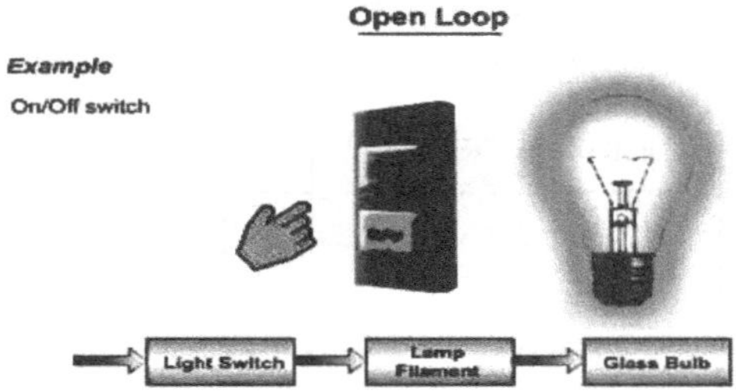

Diagram 5.18: Open Loop Control System

- **Volume on Stereo System :** Volume is adjusted manually irrespective of output volume level.
- **Electric Hand Drier** - Hot air (output) comes out as long as you keep your hand under the machine , irrespective of how much your hand is dried.
- **Automatic Tea / Coffee Maker** - These machines also function for pre adjusted time only.
- **Timer Based Clothes Drier** - This machine dries wet clothes for pre-adjusted time , it does not matter how much the clothes are dried.

Advantages of Open Loop Control System

- Simple in construction and design.
- Economical.
- Easy to maintain.
- Generally stable.
- Convenient to use as output is difficult to measure.

5.10. Closed Loop Control System

close loop control system is referred as feedback system. This system record output instead input and modified according to the requirement. **Open loop control system** can be converted in to closed loop control system by providing a feedback. This feedback automatically makes the suitable changes in the output due to external disturbance. In this way closed loop control system is called automatic control system. A feedback control system often uses a function of a prescribed relationship between the output and reference input to control the process. Often the difference between the output of the process under control and the reference input is amplified and used to control the process so that the difference is continually reduced.

For example-Air conditioning unit, A person steering an automobile. The schematic diagram of such a system is given below.

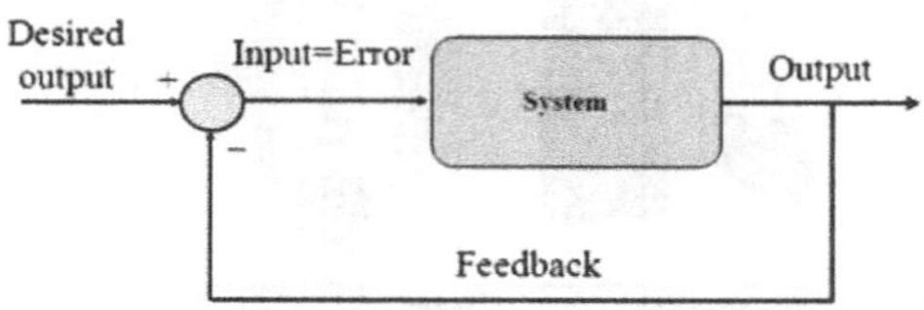

.... (a)

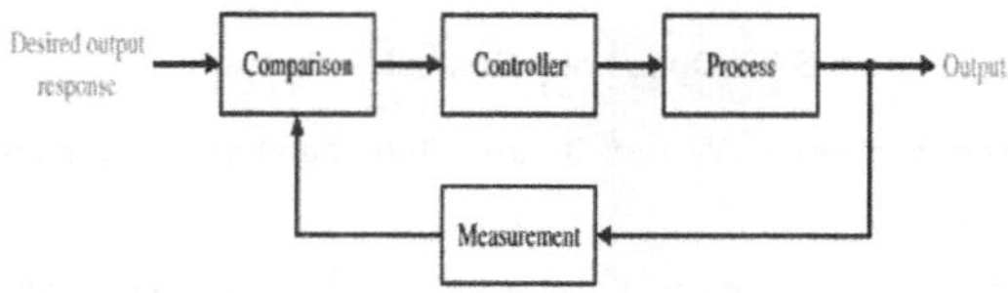

..........,(b)

Diagram 5.19: Closed Loop Control System

- A simple block diagram of **Closed loop control system** is given below in which an automobile steering control system is described ,

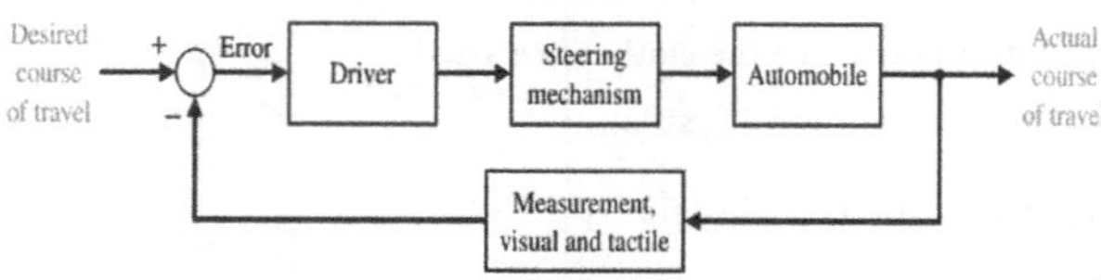

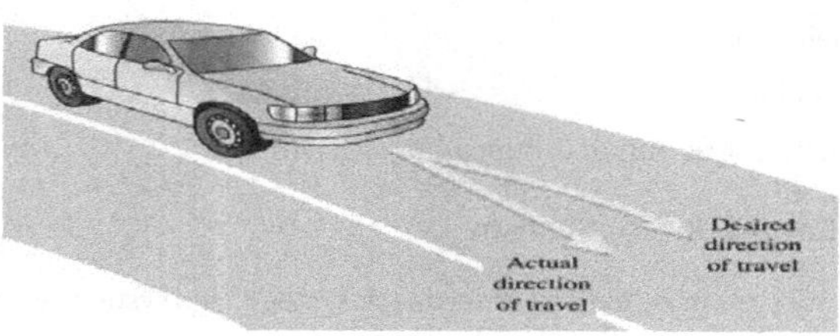

Diagram 5.20: Clcsed Loop Control System (Automobile Steering)

The closed-loop control systəm is given in which a person steering an automobile by looking at the car location on the road and making the appropriate adjustments. The introduction of feedback enables ɹs to control a desired output and can improve accuracy.

- An instrument that is used strip chart paper for plotting the record of a voltage. A critical component was a small potentiometer used to control the pen of the recorder through an actuator. The close lop control diagram is given below

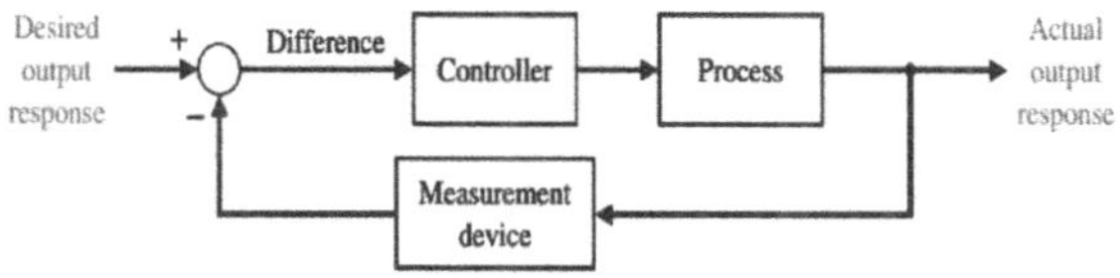

Diagram 5.21: Closed-loop Control System

- In electric clothes dryer, a sensor/transducer is input device which monitoring the temperature or dryness of the clothes and feed a signal relating to the dryness back to the controller as shown below.

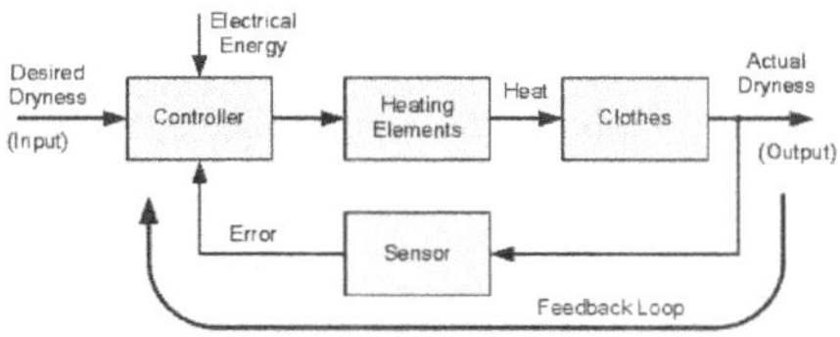

Diagram 5.22: Closed-loop Control System

Then calculate the difference between actual dryness & desired dryness of the clothes. Which representing the error value signal **(Error = Required Dryness – Actual Dryness)** is amplified by the controller, and the controller output makes the correction to the heating system for reducing the error.

5.11. Negative Feedback Control System

Control systems operate in a closed-loop having an accurate sensor, if the measured output & actual output of the system is equal. Then it is called as positive **feedback control system.** Means the difference between the desired output and the actual output is equal to the error, which is then adjusted by the control device (such as an amplifier). On the other hand if the output of the control device causes the actuator to modulate the process in order to reduce the error. Then it is called a **negative feedback** control system, because the output is subtracted from the input and the difference is used as the input signal to the power amplifier

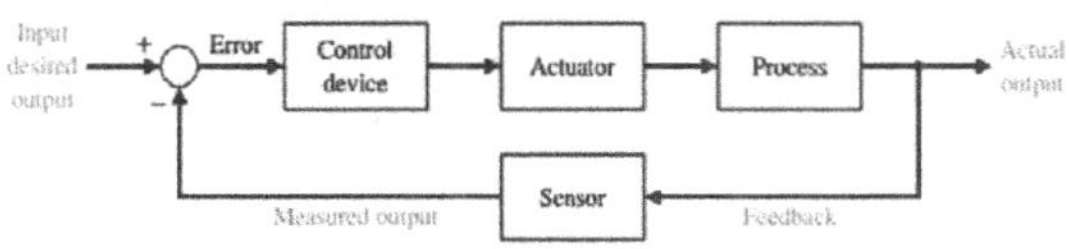

Diagram 5.23: Negative Feedback Control System

For example if a ship is heading to the right but rudder is actuated to direct the ship to the left.

Some other example of closed loop control system

- **Automatic Electric Iron -** Heating elements are controlled by output temperature of the iron.

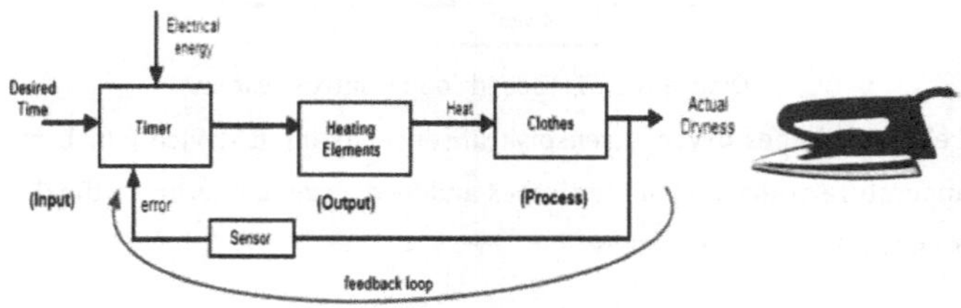

- **Water Level Controller -** Input water is controlled by water level of the reservoir.

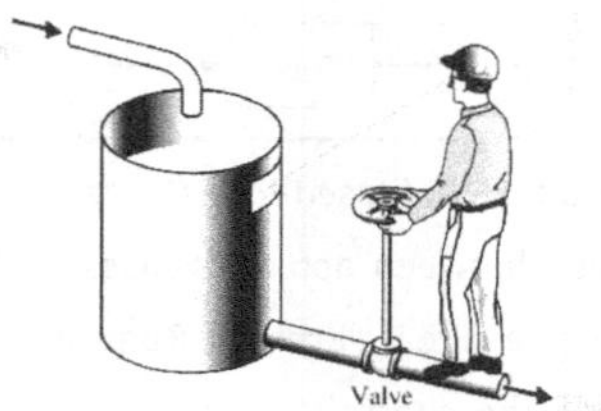

- **An Air Conditioner -** An air conditioner functions depending upon the temperature of the room.

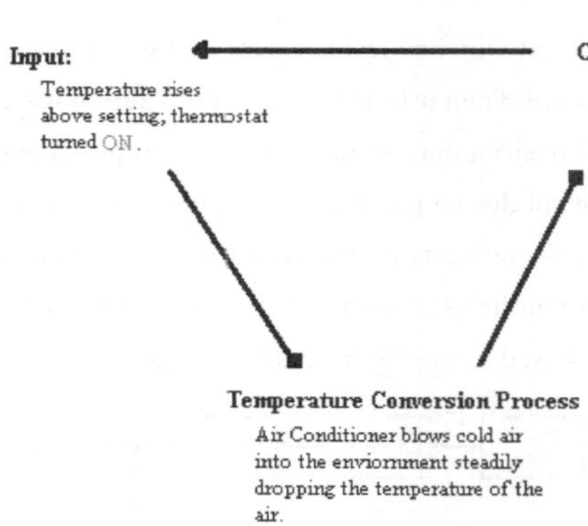

- **Cooling System in Car** - It operates depending upon the temperature which it controls.
- **Servo Voltage Stabilizer** - Voltage controller operates depending upon output voltage of the system.
- **Missile Launched and Auto Tracked by Radar** - The direction of missile is controlled by comparing the target and position of the missile.

Characteristics of Closed-loop Control

- To reduce errors by automatically adjusting the systems input.
- To improve stability of an unstable system.
- To increase or reduce the systems sensitivity.
- To enhance robustness against external disturbances to the process.
- To produce a reliable and repeatable performance.

Advantages of Closed Loop Control System

- Closed loop control systems are more accurate even in the presence of non-linearity.
- Highly accurate as any error arising is corrected due to presence of feedback signal.
- Bandwidth range is large.
- Facilitates automation.
- The sensitivity of system may be made small to make system more stable.
- This system is less affected by noise.

Disadvantages of Closed Loop Control System

- They are costlier.
- They are complicated in design.
- Required more maintenance.
- Feedback leads to oscillatory response.
- Overall gain is reduced due to presence of feedback.
- Stability is the major problem and more care is needed to design a stable closed loop system.

5.12. Characteristics of Good Control System

- **Accuracy :** Accuracy is the measurement tolerance of the instrument and defines the limits of the errors made when the instrument is used in normal operating conditions. Accuracy can be improved by using feedback elements. To increase accuracy of any control system error detector should be present in control system.

- **Sensitivity :** The parameters of control system are always changing with change in surrounding conditions, internal disturbance or any other parameters. This change can be expressed in terms of sensitivity. Any control system should be insensitive to such parameters but sensitive to input signals only.
- **Noise :** An undesired input signal is known as noise. A good control system should be able to reduce the noise effect for better performance.
- **Stability :** It is an important characteristic of control system. For the bounded input signal, the output must be bounded and if input is zero then output must be zero then such a control system is said to be stable system.
- **Bandwidth :** An operating frequency range decides the bandwidth of control system. Bandwidth should be large as possible for frequency response of good control system.
- **Speed :** It is the time taken by control system to achieve its stable output. A good control system possesses high speed. The transient period for such system is very small.
- **Oscillation :** A small numbers of oscillation or constant oscillation of output tend to system to be stable.

5.13. Block Diagrams of Control System

In control system , any kind of operation or process performed by system like physical , mechanical , electrical , chemical is represented with the help of pictorial block to simplify the control system process that's depends upon the cause-and-response relationship between input and output of the system and it is suitable way to derive transfer function of control element connected to the system .

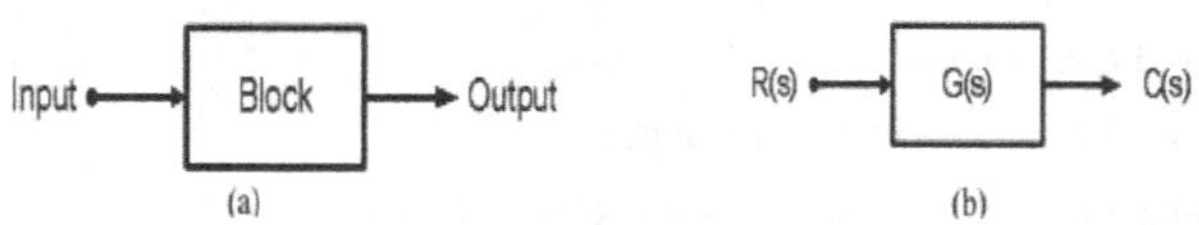

Diagram 5.24: Block Diagrams Representation

- **Error signal :** Consider the close loop control system in which following data is achieved as shown in figure ,

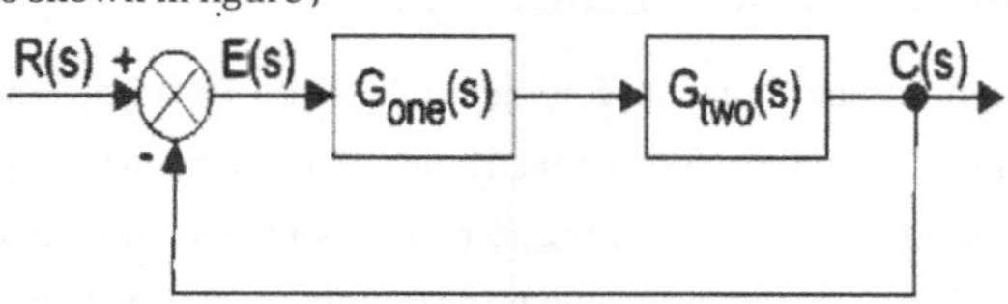

Where ,

C(s) is output signal or fed back of the system ,

R(s) is input of the system

Then **error signal** is defined as the difference between input and output

$$E(s) = R(s) - C(s)$$

- **Transfer Function :-** It is defined as the ratio of Laplace transform of output signal to Laplace transform of input signal assumption that all the initial conditions to be zero . as shown in figure ,

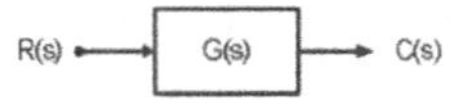

Where, **G(s)** transfer function ,**C(s)** is the output and **R(s)** is input

If the control system having several block than , the value of transfer function will be same i.e. ratio between **output / input** but we will consider the value of all the individual transfer function for each & every block after that summing up for calculating the final transfer function . such kind of operation is called block diagram reduction technique .

$$G(s) = \frac{C(s)}{R(s)} \quad \text{... 5.3}$$

- **Open-Loop Transfer Function :** It is defined as the ration between the ratio of the feedback signal to the actuating error signal .

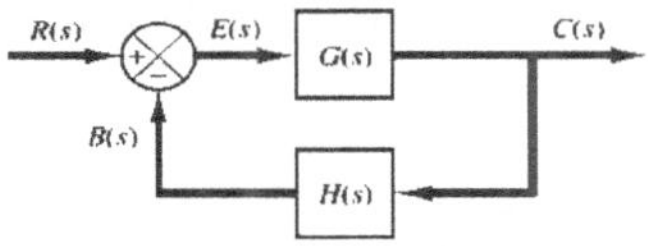

$$\textit{Open} - \textit{loop transfer function} = \frac{B(s)}{E(s)} = G(s)H(s)$$

- **Feed forward Transfer Function :-** It is defined as the ratio between output to the actuating error signal .

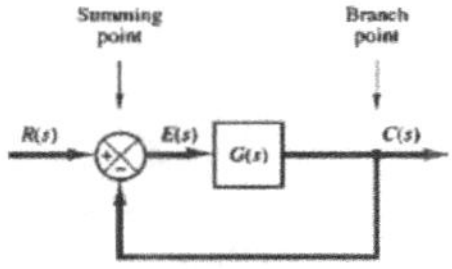

$$\textit{Feedforward transfer function} = \frac{C(s)}{E(s)} = G(s)$$

Closed-Loop Transfer Function :- The value of close loop transfer function for above problem is given as ,

$$\frac{C(s)}{R(s)} = \frac{G(s)}{1 + H(s)C(s)} \qquad \text{.. 5.4}$$

- **Closed-Loop System with Disturbance :** consider the block diagram of close loop control system in which disturbance is considered as shown in figure ,

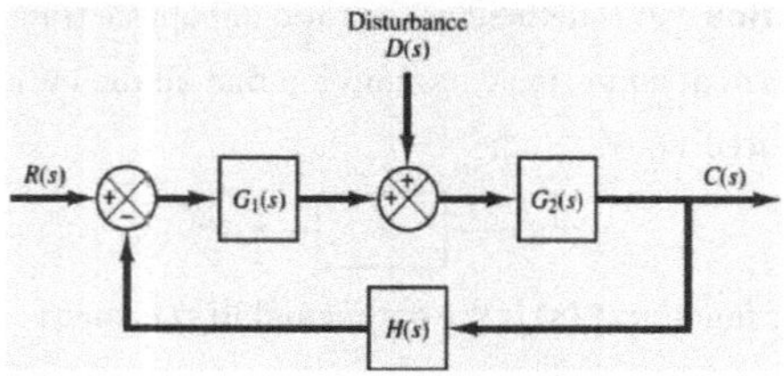

considering the response to the reference input R(s) and assume that disturbance is zero. Then the response C (s) to the reference input R(s) can be calculate as ,

$$\frac{C_R(s)}{R(s)} = \frac{G_1(s)G_2(s)}{1 + G_1(s)G_2(s)H(s)}$$

$$C(s) = C_D(s) + C_R(s)$$

$$= \frac{G_2(s)}{1 + G_1(s)G_2(s)H(s)}[G_1R(s) + D(s)] \qquad \text{...................... 5.5}$$

For calculating the effect of the disturbance , consider that reference input is zero , then

Disturbance D(s) =

$$\frac{C_D(s)}{D(s)} = \frac{G_2(s)}{1 + G_1(s)G_2(s)H(s)} \qquad \text{.................................... 5.6}$$

- **Take off Point of Block Diagram :** The Point which permit to supply same signal or input to more than one block or at summing point is known as Take off Point . Take off Point is shown in the figure below.

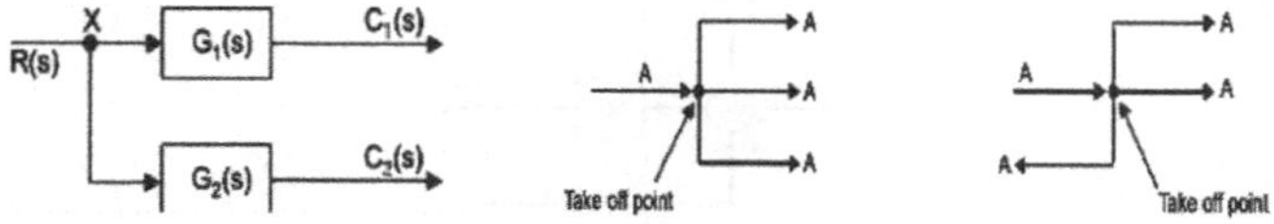

Take off Point

5.14. Laplace Transformation of System Model

Consider the input for any operational process whose input variable is analyze . Let be the real variable of system (r) be the function of time then it can be expressed as ,

$$F(s) = L\{f(t)\} = \int_{0}^{\infty} f(t)\ e^{-rt}\, dt \qquad \text{.................................. 5.7}$$

Where ,

F(s) = Laplace transform of f(t)

S = variable of system

Since r is variable of the system , having two parts real & imaginary parts of the system hence

$$r = \phi + j\,\sigma$$

ϕ = Real parts of complex variable

σ = Imaginary parts of complex variable

the value of f(t) will be inverse of Laplace transformation

$$f(t) = L^{-1}\ [F(s)] = L^{-1}\{L(f(t))\} \qquad \text{...................................... 5.8}$$

5.15. Calculation of Transfer Function of Control System

The following steps must be taken into consideration for calculating the transfer function .

- Write down the equation of all the variable as a function of time.
- Derive a Laplace transform equation for system variable with assumption that all the initial condition to be zero.
- Identify system input and output for the system,
- Expressed all the variable into equation, in terms of input variable and output variable.
- Take the ratio between Laplace transform of output to Laplace transform of input variable which will be the desired transfer function of the system.

(i) Cascade Blocks :- Whenever several control blocks are connected in cascaded manner, the transfer function of the entire system will be the product of transfer function of all individual blocks.

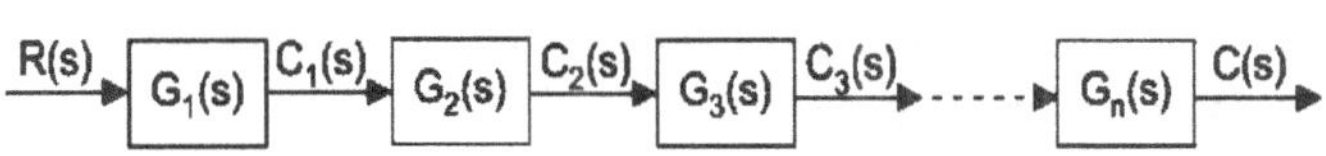

Diagram 5.25: Cascade Blocks

$$G(s) = \frac{C(s)}{R(s)} = \frac{C_1(s)}{R(s)} \cdot \frac{C_2(s)}{C_1(s)} \cdot \frac{C_3(s)}{C_2(s)} \cdot \cdot$$

Hence final block diagram and resultant transfer function can be represented as ,

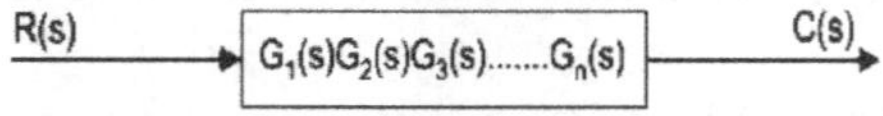

tant transfer function = $G_1(s)G_2(s)G_3(s) \cdots\cdot G_n(s)$ **5.9**

(ii) Parallel Blocks : whenever control block are arranged in parallel to each other and same input signal is provided then the resultant or over all transfer function will be equal to the algebraic sum of all individual transfer function of blocks. Such kind of system is as shown in figure .

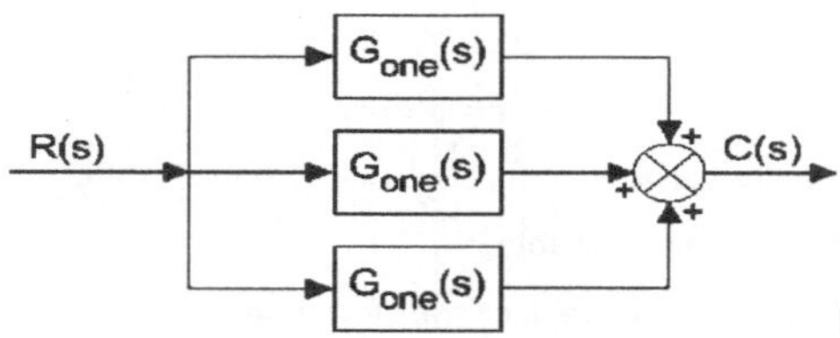

Diagram 5.26: Parallel Blocks Control System

$$G(S) = [C_1(S) / R_1(S)] + [C_2(S) / R_2(S)] + [C_3(S) / R_3(S)]$$
$$= G_1(S) + G_2(S) + G_3(S) \quad \text{...... } \mathbf{5.10}$$

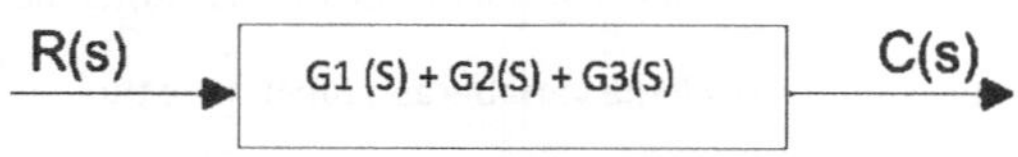

Diagram 5.27: Equivalent Transfer Function

5.16. Modeling of Control System

The control systems models is defined as a set of various variable of the process which is used to analysis and design in control systems for understanding how the process would behave in various conditions . The control system consists of actuators, sensors, or controllers. Hence the overall structure of process is to be used for analyses called control system model. Analysis is to investigate how the process responding at extreme conditions or at unstable conditions. It also gives insights into how stable the process could be under given set of conditions. This helps to understand how to control the process. The control system model can be classified in following categories.

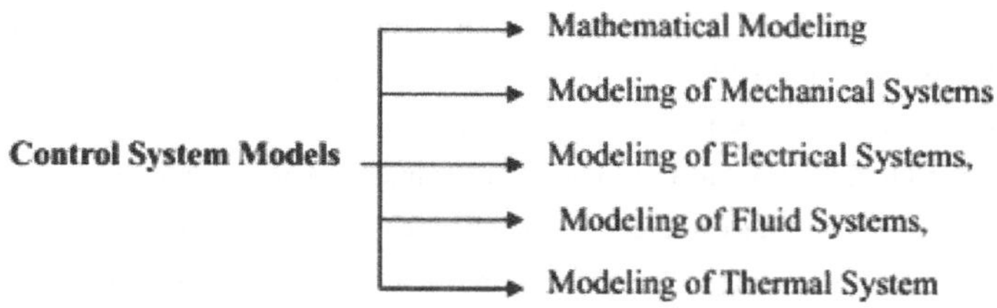

Diagram 5.28: Control System Model

5.17. Modeling of Mechanical Systems

The mechanical modeling system consists of three constituent elements , these are Inertia Elements , Spring Elements & Damping Elements . but before going through the study we will described the basic steps , usually applied in mechanical modeling .

Procedure for mechanical system modeling : Whenever mechanical system modeling is to be proceed , we will have to consider following steps .

- Select the coordinate system for modeling with appropriate sign convention , for linear motion Up side is taken to be (+) and toward left to right (+) as that of conventional coordinate system ,in case of rotational motion we will consider (+) for anticlockwise on the other hand for clock wise it must be considered as (-) .
- Define all the disturbing force associated with the system (input)
- Achieve output for the system as displacement and velocity etc. .
- Draw a suitable free body diagram for each mass which indicating forces and moments acting on it.
- Apply governing laws (Newton second law of motion) for each free body diagram ,

Force = Mass x Acceleration

- Deduced the related equation for appropriate solution.

(i) Inertia Elements :- In mechanical modeling system , the inertia elements is based on the phenomenon of kinetic energy storage /conservation of kinetic energy in the system .

- In case of translatory motion systems, the mass is usually denoted by "**m**" with unit as **kg** on the other hand in the systems, stiffness is represented by spring element '***k***'.
- In case of rotational system, the moments of inertia for rotation is usually represented by "**J**"with unit as **kg - m²** . On the other hand for rotational systems torsional spring element is represented by ' $\mathbf{k}_t$' in the system .

- The function of Inertia elements in the system is not only store energy but also dissipated it from the mechanical system for example Flywheels which is energy storage devices , dashpots is applied to releasing it .

(ii) Spring Elements : A linear spring is a mechanical elements whose function is to create deformation due to the effect of externally applied force . The deformation may be comes on the front line as a shortening or lengthening the elements. Hence the deformation is directly proportional to the applied force or sometimes applied torque. The schematic diagram is given as,

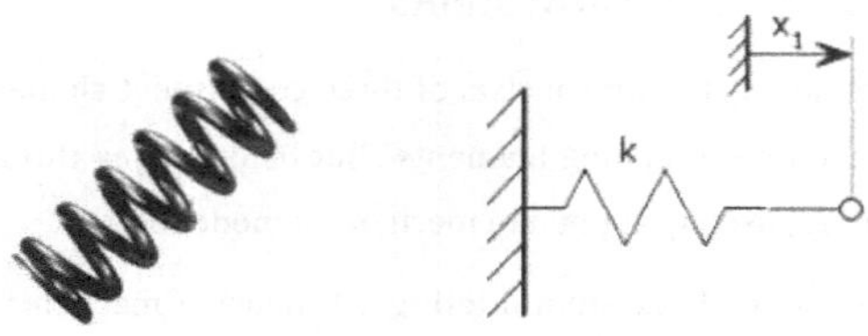

Diagram 5.29: Spring Elements

Since

Deformation **α** applied force

When no force is applied then,

$$F = kx_1 \qquad \text{.......................... 5.11}$$

When force is applied then,

$$F = k\Delta x \qquad \text{.......................... 5.12}$$

Where ,

Δx = Deformation in Spring

F = Force Applied

k = Spring Constant**(N/m).**

- In case of translational spring , the relationship between the applied force F and the achieved displacement x is given as ,

$$\mathbf{F = k\,x = k\,(x_1 - x_2)} \qquad \text{.................................. 5.13}$$

Where k is a constant of proportionality called a spring constant, the unit of **k is N/m**

- In case of torsional spring , when torque is applied then intensity of applied torque **T** is directly proportional to the angular displacement **θ** is

$$T = k\,\theta = k\,(\theta_1 - \theta_2) \qquad \text{..............................5.14}$$

Where **k** is a constant of proportionality called a torsional spring constant

The Spring Elements has two basic configurations which help to accumulate the resultant in mechanical modelling system. These are described as,

- **Series configurations**
- **Parallel configurations**

(i) Series configurations : In such type of configurations we have two or more springs in series combination as shown in figure . When springs are connected in series, they experience the same force but undergoes in different deflections. Such kind of combination having same equivalency as that of electrical resistance combination in parallel.

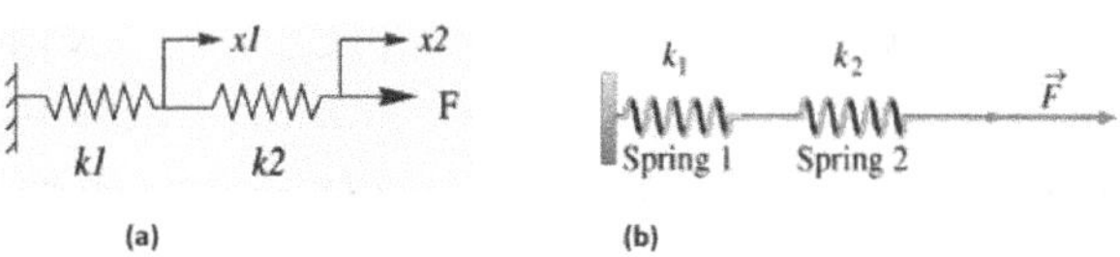

Diagram 5.30: Series Configurations

Let be assume, any mechanical system in which two springs are arrange in series combination having two spring constant $\mathbf{k_1}$and $\mathbf{k_2}$.suppose force " **F** " is applied according to the diagram but their deflection will be different. Let them be x_1 and x_2. Then

$$k_1 x_1 = k_2 x_2 = F$$

$$x_1 = \frac{F}{k_1} \qquad x_2 = \frac{F}{k_2} \qquad \text{.....................................5.15}$$

Since the total deflection is $x = x_1 + x_2$, and we have $F = k_{eq}\,x$, then equivalent deflection can be obtain as ,

$$\frac{F}{k_{eq}} = \frac{F}{k_1} + \frac{F}{k_2} \qquad \text{......................................5.16}$$

$$k_{eq} = \frac{1}{\frac{1}{k_1} + \frac{1}{k_2}} = \frac{k_1 k_2}{k_1 + k_2}. \qquad \text{.....................................5.17}$$

(ii) Parallel configurations : If the springs are in parallel combination and loaded with certain force intensity then the springs will be experienced same deflection .Therefore the equivalent value of spring constant will be the sum of individual . Such kind of combination having same equivalency as that of electrical resistance combination in series . such kind of system is as shown in figure ,

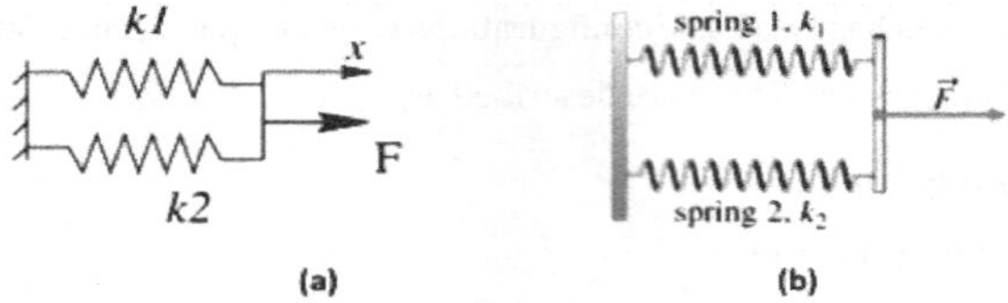

Diagram 5.31: Parallel Configurations

if k_1 and k_2 are spring constant and k_{eq} is equivalent spring constant , then the value of **keq** can be calculate as , Since two springs have same deflection , Therefore,

$$F = k_1 x + k_2 x = k_{eq} x \quad \text{.................................5.18}$$

$$K_{eq} = k_1 + k_2 \quad \text{.................................5.19}$$

Note – The spring will always follow the law of linear force-displacement or linear torque angular displacement.

5.18. Calculation of Transfer Function

Consider a mechanical spring system , as shown in figure in which , The spring force will be proportional to change in length .

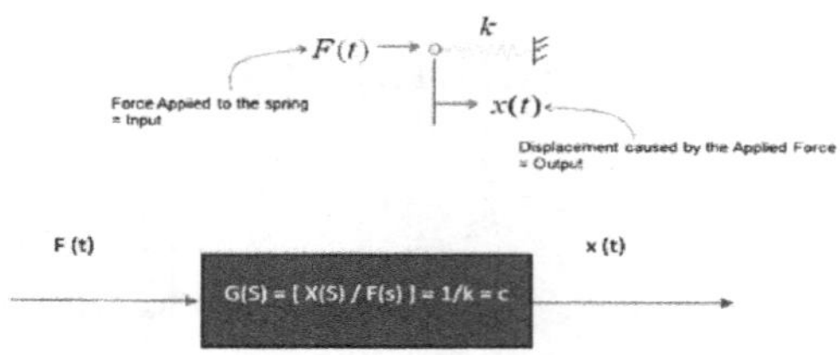

Diagram 5.32: Spring System and Building Block

Then ,

$$F(t) = k \cdot x(t) \quad \text{.. 5.20}$$

$$\mathbf{F(s) = k\,X(S)} \quad \text{..5.21}$$

the value of transfer function will be given as ,

$$[\,X(S) / F\,] = 1 / k = c \quad \text{...................... 5.22}$$

Where **C** is called mechanical capacitance , which is reciprocal of stiffness and having a analogy to electrical capacitance.

(i) Damping elements : Damping elements are non-conservative elements of the model which dissipate energy from the system. They convert the energy into another form of energy i.e., converts the mechanical energy to thermal energy. In general we have three types of damping elements in the system , A dashpot , A sliding viscous friction , A journal bearing.

In general, the dampers are of two types

- Translational damper
- Rotational damper

If viscosity is considered in mechanical modelling then the system modelling is associated with damping force which depends upon the relative velocity between the fluid and the surface. This damping force is taken to be proportional to the velocity across the damper. In this case we will only consider the linear damping force, which is a linear function of the relative velocity. These Energy dissipation elements are known as dampers or dashpots.

- A dashpot and a sliding viscous friction is to be considered as a Translational damper that's relate the relative velocity & the intensity of drag force. the drag force can be given as ,

$$f = c\dot{x} = c(\dot{x}_1 - \dot{x}_2) \quad \text{.............................. 5.23}$$

Where c is a **damping coefficient ,**

- In case of journal bearing is consider as a rotational damper , the drag torque can be given as

$$T = c\ \omega = c(\omega_1 - \omega_2) \quad \text{..................................5.24}$$

Where $\omega = \theta$ is the relative angular velocity.

As similar to the spring , damper also has two type of combination these are

Parallel combination – The parallel combination is given as ,

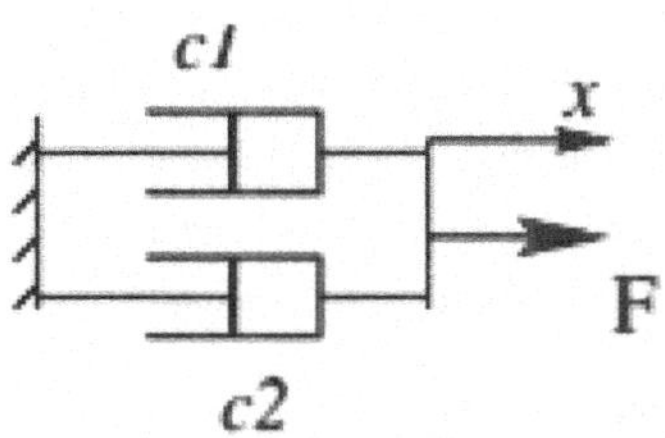

Diagram 5.33: Parallel Combination of Damper

$$c_{eq} = c_1 + c_2 \quad \text{.. 5.25}$$

Where ,

c_1 and c_2 are damping constant

c_{eq} is an equivalent damping effect

Series combination : The Series combination is given as ,

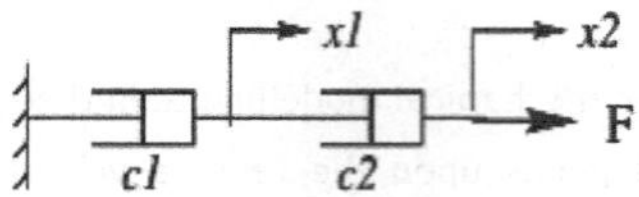

Diagram 5.34: Series Combination of Damper

$$c_{eq} = \frac{c_1 c_2}{c_1 + c_2} \quad \text{.. 5.26}$$

Where ,

c_1 and c_2 are damping constant

c_{eq} is an equivalent damping constant

5.19. Modeling of Simple Translational System

The model of Mechanical Translational Systems can be obtained by using three basic elements Mass, Spring and Dashpot. The weight of the mechanical system is represented by the Element Mass and it is assumed to be concentrated at the center of the body. The elastic deformation of the body can be represented by a spring. The friction existing in a mechanical system can be represented by a dashpot. basically modeling of Simple Translational System is based on the Newton's law of motion, according to it

- Newton's First Law states that if any particle will remain at rest or moving with uniform velocity in a straight line unless acted upon by an external force.
- Newton's second law state that the rate of change of momentum of an object is directly proportional to the resultant force applied and is in the direction of the resultant force. The resultant force is equal to the rate of change of momentum.

$$f = \frac{d(mv)}{dt} = m\frac{dv}{dt} = ma \quad \text{.................................. 5.27}$$

- Newton's third law state that for every action there is an equal and opposite reaction.

Modeling of translation system based on **straight line movement of particle or object. It** consists of three elements in which force behave as an input and displacement as an output. The brief description of all those elements is given below.

- Mass
- spring
- Dashpot or damper.

(i) Mass : If a force " **F** " is applied on a body of mass **M** , it experience a force in opposition due to its mass , this force of opposition is proportional to the acceleration thus according to Newton's second law Acceleration is the rate of change of velocity .In this context elasticity and friction force is considered as negligible value.

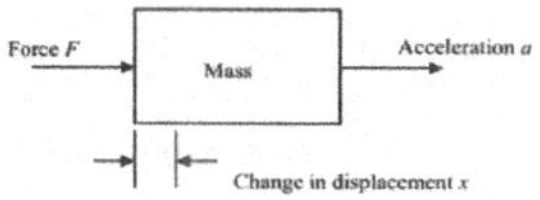

Mass

Mathematically is written as

$$F = ma = m\frac{dv}{dt} = m\frac{d(dx/dt)}{dt} = m\frac{d^2x}{dt^2} \qquad \text{.. 5.28}$$

Where,

F = applied force

M = mass

a = acceleration

x = displacement

in case of mass-damper system force is given by the following equation ,

$$\mathbf{F} = -kx - cv \qquad \text{.. 5.29}$$

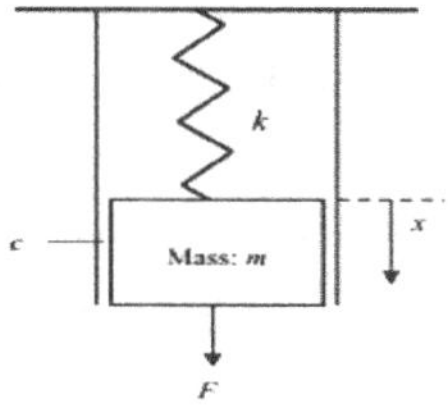

Diagram 5.35: Mass–Damper System

Where ,

F = applied force

c = coefficient of damping

V = is the relative velocity between the two points connected by the spring.

x = displacement

- **Calculation of transfer function :-** Whenever mass is in motion , the value of inertia force is in decreasing order and follow the newtons second law of motion , according to it

F = **Mass x Acceleration**

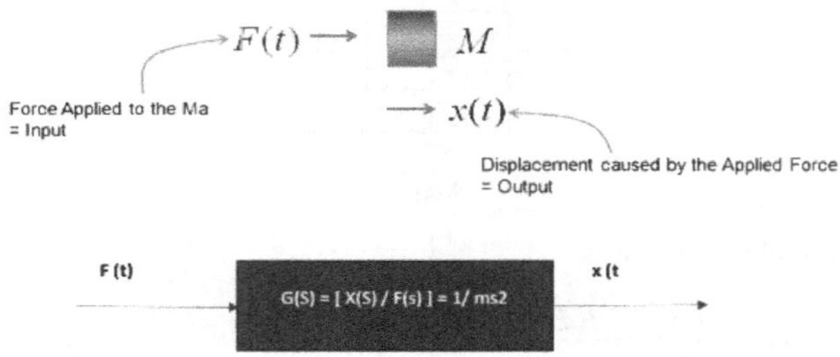

Diagram 5.36: Mass Damper System Transfer Function

Governing law for such a system will be =

$$F(t) = M \cdot \frac{d^2x(t)}{dt^2} \quad \text{...} \mathbf{5.30}$$

Laplace transformation will be =

$$F = m\, s^2 x \quad \text{...} \mathbf{5.31}$$

Transfer function for such a system will be given as ,

$$[\, X(S) / F \,] = 1 / ms^2 \quad \text{.......................................} \mathbf{5.32}$$

(i) Spring :- Spring is an element whose function is to stores potential energy. If a force F is applied on spring having a spring constant **K** , then an opposing force is acted on spring with respect to elasticity of spring. hence this opposing force will be proportional to the displacement of the spring or in other words the extension or compression will be proportional to the applied forces . In this context mass and friction force is considered as negligible value.

$$\mathbf{F = k\, x} \quad \text{...} \mathbf{5.33}$$

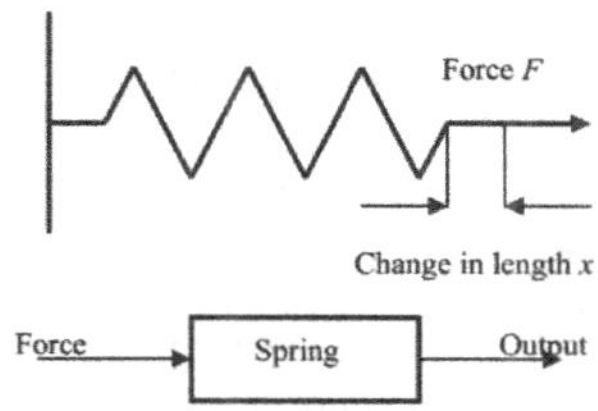

Diagram 5.37: Spring Element with Translation System

Where ,

F = applied force

M = mass

k = spring constant

x = displacement

(ii) Dashpot : If a force is applied on dashpot **B**, then it experiencing an opposing force due to friction of the dashpot. This opposing force is proportional to the velocity of the body. Assume mass and elasticity is negligible. As shown in figure,

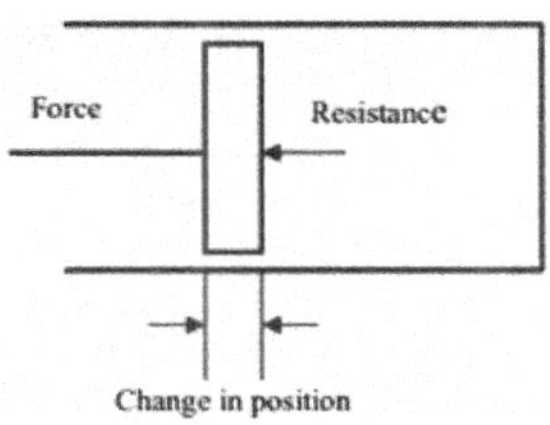

Diagram 5.38: Dashpot or Damper

Where ,

F = opposing force due to friction of dashpot

B = frictional coefficient

v = velocity

x = displacement

From the diagram the movement of the piston requires the fluid on one side of the piston to flow through or past the piston producing a damping or resistive force. The mass experiences a damping force in the direction opposite to the instantaneous motion. The relationship between the displacement x of the piston and the force as the input is depending upon on the rate of change of the output. The damping **F** can be calculated as,

$$\mathbf{F = c\,v} \quad \ldots\ldots\ldots\ldots\ldots\ldots\ldots\ldots \mathbf{5.34}$$

$$F = c\frac{dx}{dt} \quad \ldots\ldots\ldots\ldots\ldots\ldots\ldots\ldots \mathbf{5.35}$$

hence, the final modeling equation for simple translational systems is tabulated as ,

Model elements	Translational equation	Energy
Spring	F = kx	$E = \frac{1}{2}\frac{F^2}{k}$
Dashpot	$F = c\frac{dx}{dt}$	$P = cv^2$
Mass	$F = m\frac{d^2x}{dt^2}$	$E = \frac{1}{2}mv^2$

- **Calculation of transfer function ;-** Consider a damping system as shown in figure , the concern quantities is given as

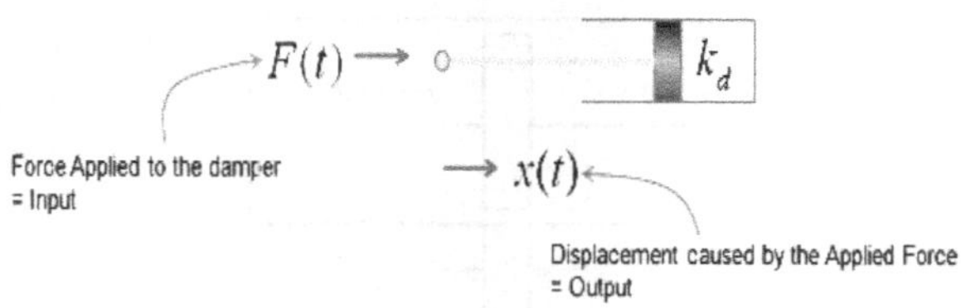

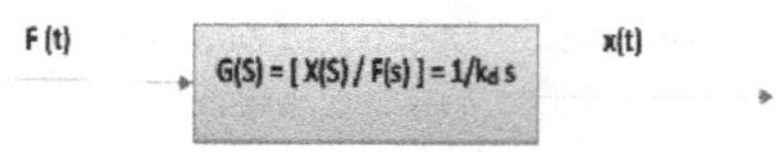

Diagram 5.39: Damper System and Building Block

Govrning law for dashpot or damper

$$F(t) = k_d \cdot \frac{dx(t)}{dt} \quad \ldots\ldots\ldots\ldots\ldots\ldots\ldots\ldots \mathbf{5.36}$$

Laplase equation for dashpot –

$$\mathbf{F = k_d\, s\, x} \quad \ldots\ldots\ldots\ldots\ldots\ldots\ldots\ldots \mathbf{5.37}$$

The transfer function =

$$\mathbf{\{X(s) / F\} = s / k_d} \quad \ldots\ldots\ldots\ldots\ldots\ldots\ldots\ldots \mathbf{5.38}$$

Where $\mathbf{k_d}$ = damping cofficient

Here we will described some other arrangement of mechanical system with suitable mechanical building block , which is very important from subject point of view .

(i) Spring-Mass mechanical system : Spring mass balance system consists of spring & a desired mass is attached to it. Since the motion of mass is one directional thus system indicating a single degree of freedom. the schematic diagram of spring mass balance system and Building Block is given as,

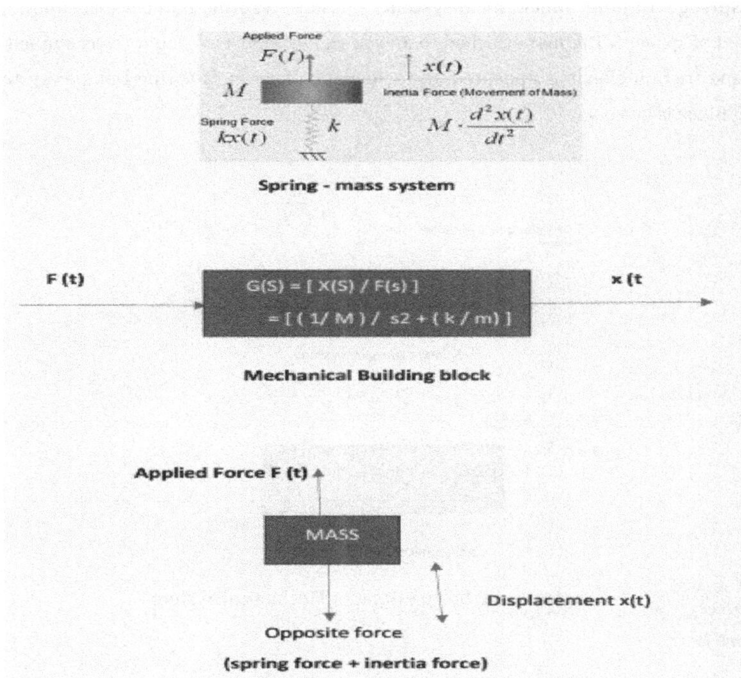

Diagram 5.40: Spring-Mass Mechanical System

Where ,

F (t) = Disturbing force (function of time t)

X(t) = Displacement, positive (upward motion) and function of time t

Whenever force is applied on such kind of system, it experienced two opposite force hence obeys Newton third law of motion.

Force by spring = **kx** , and Inertia force = **M (d^2x/dt^2)**

Total applied force = **k x + M (d^2x/dt^2)**

Applied D 'Almberts Principle , according to it

= Σ forces + Σ moments = 0

F - k x - M $(d^2x / dt^2) = 0$ **5.39**

Transfer function will be given as the ratio between **(Output / Input)**

G(s) = [x(s) / F (s)] = [(1/M) / (s^2 + K/M)] **5.40**

(ii) Spring – Damper mechanical system : - Assume a spring Damper mechanical system composed of spring & Damper attached to a fixed rigid support . When force is applied spring effects and friction effects is appeared. the schematic diagram of spring Damper system and Building Block is given as ,

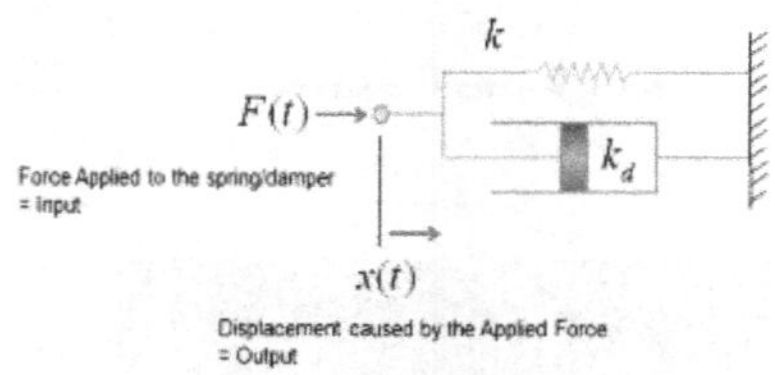

Spring - Damper system

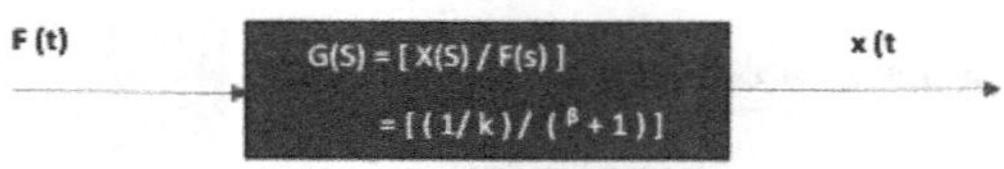

Mechanical Building block

Diagram 5.41: Spring-Damper Mechanical System

Where ,

F (t) = Disturbing force (function of time t)

X(t) = Displacement, positive (upward motion) and function of time t

$$F(t) = kx(t) + k_d \cdot \frac{dx(t)}{dt}$$

.... Governing equation

Whenever force is applied on such kind of system , it experienced two opposite force,

spring force = k x and **Friction force = $\mathbf{k_d}$ (dx/dt)**

Total applied force = **k x + $\mathbf{k_d}$ (dx/dt)**

In term of **' s '** it can be written as ,

Total applied force = **k x + $\mathbf{k_d}$ s x**

Thus Transfer function will be given as the ratio between **(Output / Input)**

G(s) = [x(s) / F (s)] = [(1/ K) / (k_d / k) s + 1)] **5.41**

(iii) Spring – Mass - Damper mechanical system : Consider a mechanical system in which spring , mass and Damper are attached as shown in figure .When tensile force is applied upon the system it experienced spring effects and friction effects and inertia effects .

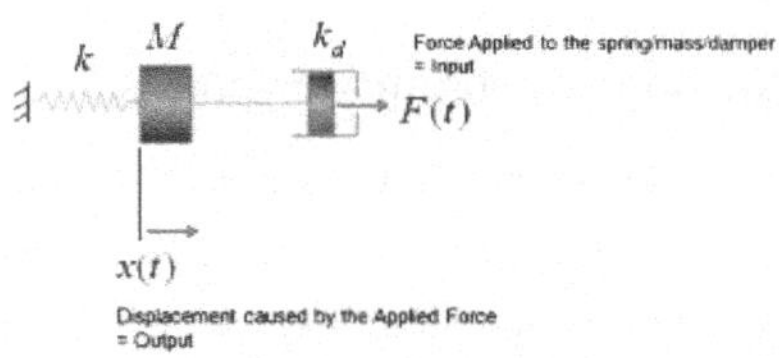

Spring – Mass -Damper system

Mechanical Building block

Diagram 5.42: Spring–Mass-Damper Mechanical System

Governing equation of spring – mass – damping system –

Spring Force   Damping Fource

Applied Force   Movement of Mass

$$F(t) = kx(t) + M\,\frac{d^2x(t)}{dt^2} + k_d \cdot \frac{dx(t)}{dt}$$ **5.42**

Where ,

F (t) = Disturbing force (function of time t)

X(t) = Displacement, positive (upward motion) and function of time t

Whenever force is applied on such kind of system, segmental force is given below,

Spring force = k x

Damping force or Friction force = k_d (dx/dt)

Inertia force = M (d^2x / dt^2)

Total Applied tensile force is accumulated as , = k x + k_d (dx/dt) + M (d^2x / dt^2)

In term of ' s ' it can be written as ,

$$F = (k x + k_d s x + M s^2 x)$$

Thus Transfer function will be given as the ratio between **(output / Input)**

$$G(s) = [x(s) / F (s)] = [(1/ K) / s^2(M/k) + (k_d / k) s + 1)]$$

$$\mathbf{G(S) = [X(S) / F(s)]= [(1/ k) / s^2(M/k) + (k_d / k) s + 1]} \quad \text{.............. 5.43}$$

5.20. Modeling of Rotational Mechanical Systems

The modeling of Mechanical Rotational Systems can be obtained by using three basic elements moment of inertia of Mass, Dashpot with rotational frictional coefficient and Torsion Spring with stiffness. The weight of the mechanical system is represented by the Moment Of Inertia of the mass and it is assumed to be concentrated at the center of the body. The elastic deformation of the body can be represented by a Torsion Spring. The friction existing in a mechanical system can be represented by a dashpot, In the Modeling of Rotational Mechanical Systems , the rotational elements rotate about fixed axis , then torque is given as ,

$$\mathbf{M} = \frac{d\mathbf{H}}{dt} = I\frac{d\omega}{dt} \quad \text{...5.44}$$

Where ,

M = External Applied Moments or torque

H = Angular Momentum

I = The body's moment of inertia about the axis

ω = The vector angular velocity

The Modeling of Rotational Systems consists of following three elements,

- Moment of inertia
- Torsional spring
- Dashpot.

Whenever any rotational system is subjected to torque , then applied torque will resists by opposing torque that is generated due to moment of inertia, elasticity and friction of the system. Since the intensity of applied torque and the opposing torques are equal but opposite in directions, then the resultant torque on the system will be zero.

(i) Moment of Inertia : In the rotational system, the function of moment of inertia is to stores kinetic energy. In this case it is considered that elasticity and friction will be negligible. When torque is applied then opposite torque generated due to moment of inertia will be proportional to angular acceleration. The schematic diagram is as shown in figure,

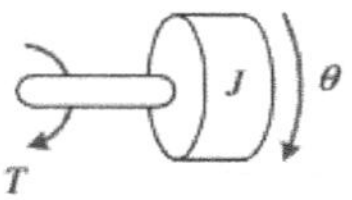

Diagram 5.43: Inertia of Rotational System

Then

$$\sum T = J\,\alpha \quad \text{.. } \mathbf{5.45}$$

Where ,

T = the applied torque

$\mathbf{T_j}$ = the opposing torque due to moment of inertia

J = moment of inertia

α = angular acceleration

θ = angular displacement

(ii) Torsional Spring : In this condition we will consider that moment of inertia and friction are negligible. When torque is applied on the torsional spring having a spring constant **K** then, opposing torque will be proportional to the angular displacement of the torsional spring.

Case – I : A torsional spring subjected to equal & opposites torque

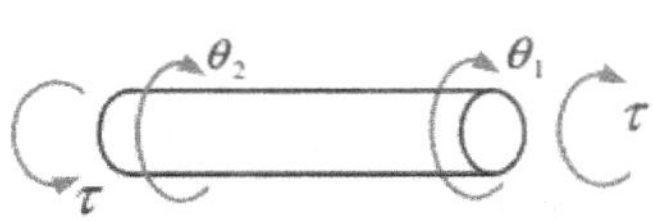

Diagram 5.44: A Torsional Spring Subjected to Equal & Opposites Torque

Consider the torsional spring as shown in Figure, since torques is equal in magnitude but opposite in direction, are applied to the ends of the spring. In this case, the torque T acting in the torsional spring is

$$T = k_t(\theta_1 - \theta_2) \quad \text{.. 5.46}$$

At each end, the spring acts in the direction opposite to that of the applied torque at that end.

Case – II :- A torsional spring one end fixed other is subjected to torque

Consider the torsional spring as shown in Figure , where one end is fixed and a torque τ is applied to the other

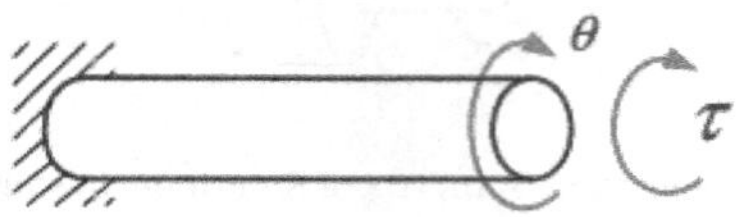

Diagram 5.45: A Torsional Spring One End Fixed Other is Subjected to Torque

The angular displacement of the free end is θ, then torque **T** in the torsional spring is given as ,

$$T = k_t \theta \quad \text{.................................... 5.47}$$

where θ is the angular displacement and k_t is the spring constant for torsional spring

(iii) Dashpot : In this condition we will consider that moment of inertia and elasticity are negligible. When a torque is applied on dashpot , then applied torque is opposed by a torque that is generated due to the rotational friction of the dashpot. thus opposing torque will be proportional to the angular velocity .

Consider the torsional damper as shown in Figure in which the torques τ is applied to the ends of the damper having equal magnitude, but opposite in direction.

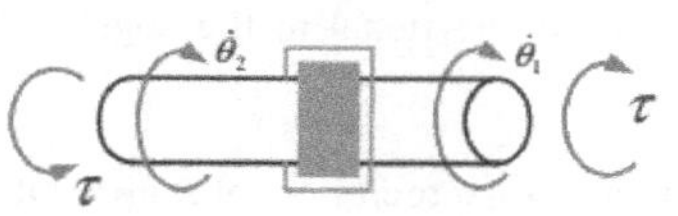

Diagram 5.46: Torsional or Rotational Damper

The angular velocities of the ends of the damper are $\dot{\theta}_1$ and $\dot{\theta}_2$. The damping torque T that arises in the damper will be proportional to the angular velocity differences $\dot{\theta}_1$ and $\dot{\theta}_2$ of the ends, or

$$T = b_T\left(\dot{\theta}_1 - \dot{\theta}_2\right) = b_T\dot{\theta} \quad \text{.......................................} 5.48$$

Where ,

b_T = the proportionality constant

T = Damping torque

$\dot{\theta}$ = The angular velocity difference or viscous friction coefficient

Summary of elements involved in linear mechanical systems

Element	Translation	Rotation
Inertia	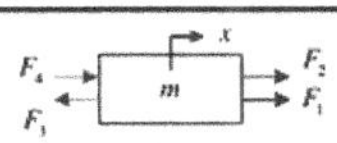 $\sum F = m\,a$	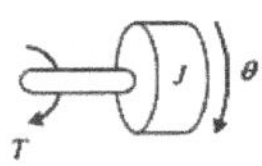 $\sum T = J\,\alpha$
Spring	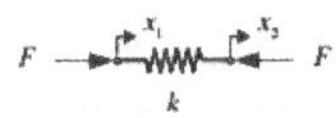 $F = k(x_1 - x_2) = kx$	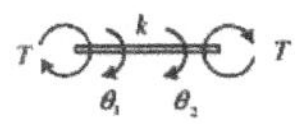 $T = k(\theta_1 - \theta_2) = k\theta$
Damper	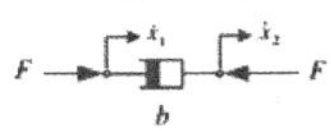 $F = b(\dot{x}_1 - \dot{x}_2) = b\dot{x}$	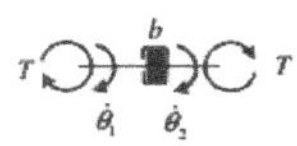 $T = b(\dot{\theta}_1 - \dot{\theta}_2) = b\dot{\theta}$

5.21. Calculation of Work, Energy, and Power

In mechanical modeling system, the amount of Work, Energy, and Power can be calculated as

(i) Work done : It is defined as the product of a force and a distance or a torque and the angular displacement . the schematic diagram is given as , In SI system the unit of work is joule (J) where 1-J = 1-N · m .

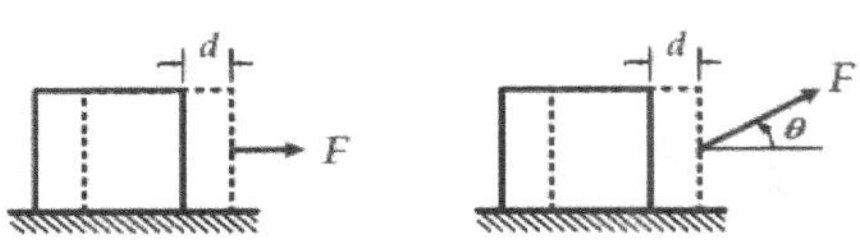

(a) (b)

Diagram 5.47: Work Done

Work done (w) = F d ....(a)

Work done (w) = $F\,d\cos\theta$ ....(b)

Work done by a spring may be calculated as,

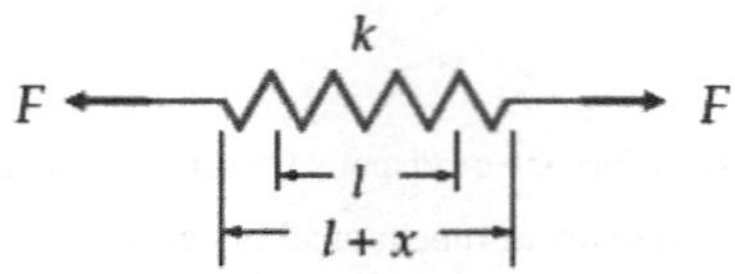

Then work done will be

$$W = \frac{1}{2} k\, x^2 \qquad \text{.................................... 5.49}$$

(ii) Energy : Energy is defined as the capacity to do the work , it can be available in chemical energy, electrical energy, or mechanical energy and so forth which can be converted from one form of the energy into another useful form because According to the law of conservation of energy, energy can be neither created nor destroyed. but in mechanical modeling system we will only consider two form of energy that is potential energy or kinetic energy . Thus potential energy due to gravity (when body is in static condition) is given as ,

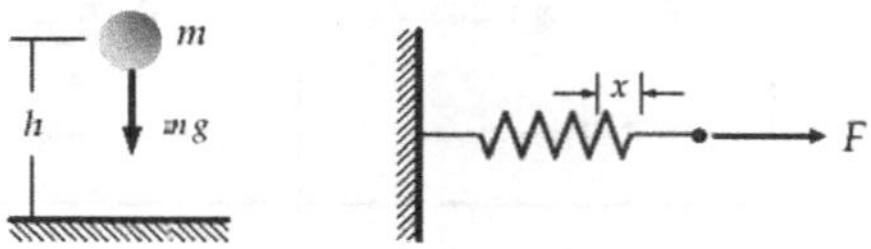

Diagram 5.48: Potential Energy

$$U = \int_0^h mg\,dx = mgh \qquad \text{................................ 5.50}$$

Where **h** is the altitude and ***g*** is a gravity

For a translational spring system, the potential energy or strain energy is formulated as

$$U = \int_0^x kx\,dx = \frac{1}{2}kx^2 \qquad \text{.................................... 5.51}$$

For a Rotational spring system , the potential energy is formulated as

$$U = \int_0^\theta k\theta\,d\theta = \frac{1}{2}k\theta^2 \qquad \text{..5.52}$$

- **Kinetic energy :-** The inertia elements in mechanical systems can store a kinetic energy.

For translation elements

$$T = \frac{1}{2}m\dot{x}^2 \qquad \text{...5.53}$$

Where x is velocity of the mass.

For Rotational elements

$$T = \frac{1}{2}J\dot{\theta}^2 \qquad \text{..5.54}$$

Where θ is the angular velocity of the moment of inertia (j).

- **Dissipated Energy.** Consider the damper in which one end is fixed and the other end is moved from $\mathbf{x_1}$to $\mathbf{x_2}$. The dissipated energy $\mathbf{\Delta W}$ in the damper is equal to the net work done on it:

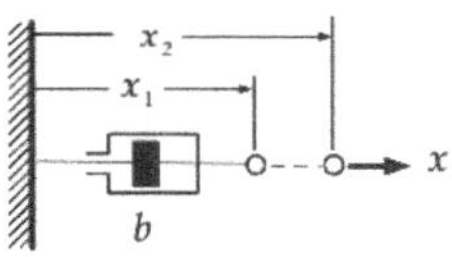

Damper

$$\Delta W = \int_{x_1}^{x_2} F\,dx = \int_{x_1}^{x_2} \underbrace{b\dot{x}}_{F}\,dx = b\int_{t_1}^{t_2} \dot{x}\frac{dx}{dt}\,dt = b\int_{t_1}^{t_2} \dot{x}^2\,dt \qquad \text{...............................5.55}$$

Power : - It is defined as the work done per unit time

Conservation of energy : When the mechanical system does not having any frictional element such as a damper then total energy will be constant , that is

$$T_1 + U_1 = T_2 + U_2 \qquad \text{... 5.56}$$

Here we will described some other arrangement of rotational mechanical system with suitable mechanical building block , which are very important from subject point of view .

(i) Torsional bar or Rotational Elements : consider a torsional bar or rotational elements whose one end is clamped and other end is subjected to twisting movement . when twisting force is applied an opposite torque is generated which is directly proportional to angle of twist, the overall mechanism is just behave like a mass and spring arrangement. the schematic diagram and mechanical building block is given as ,

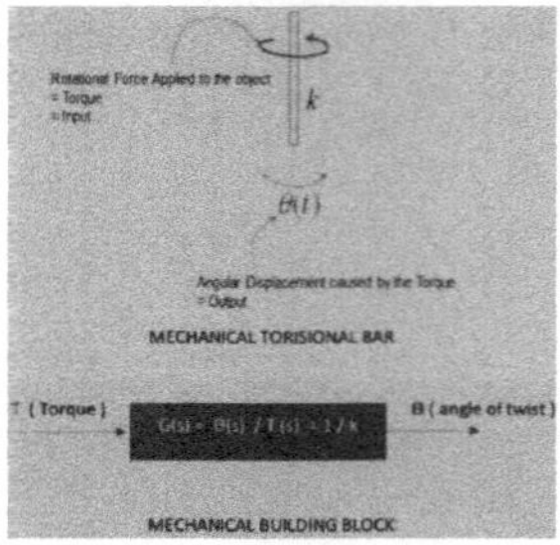

Diagram 5.49: Mechanical Torsional Bar

Where ,

T – Applied unbalanced torque (N /M)

Q – Angle of twist in radian

K – Torsional stiffness (NM / RAD)

Governing equation for Torsional bar

$$T(t) = k \cdot \theta(t)$$

Then the value of unbalanced torque = T (t) = k . Θ (t)

Laplace equation = T (s) = k . Θ (s)

Transfer function = G(s) = Θ (s) / T (s) = 1 / k 5.57

(ii) Torsional Damper : Consider a torsional damper which is subjected to rotational torque . when twisting force is applied then intensity of torque will be directly proportional to angular displacement, the overall mechanism and mechanical building block is given as ,

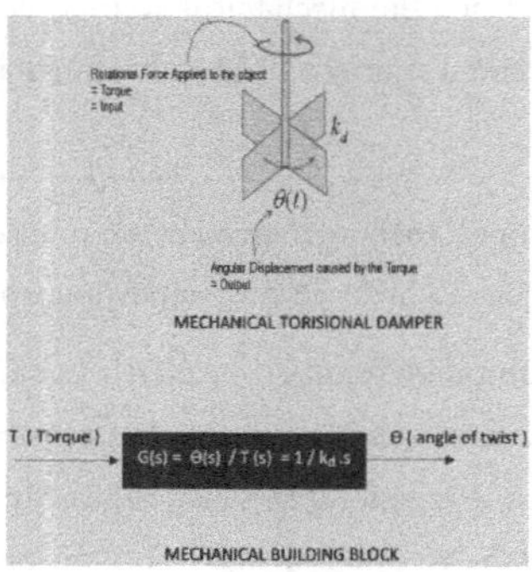

Diagram 5.50: Torsional Damper

Where ,

T – Applied unbalanced torque **(N /M)**

Θ – Angle of twist in radian

Governing equation for Torsional bar ;

$$T(t) = k_d \cdot \frac{d\theta(t)}{dt}$$

Then the value of unbalanced torque = **T (t) = K $_d$. dΘ / dt**

Laplace equation = **T (s) = k $_{d.}$ Θ (s)**

Transfer function **= G(s) = Θ (s) / T (s) = 1 / K$_d$ .s 5.58**

(iii) Rotational Inertia :- Consider a rotational mass which is subjected to rotational torque . when twisting force is applied then mass create resistance due to mass moment of inertia . the overall mechanism and mechanical building block is given below ,

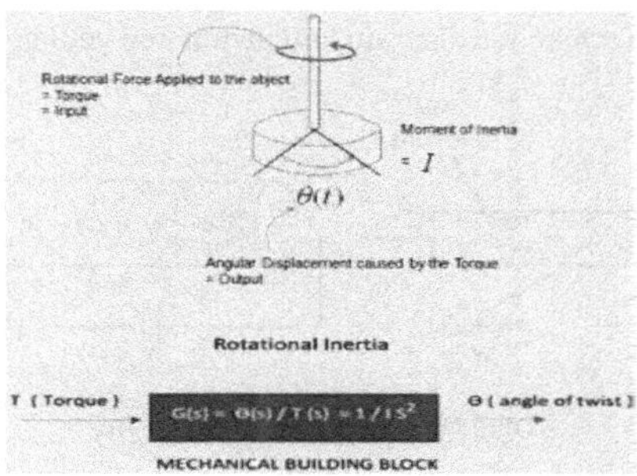

Diagram 5.51: Rotational Inertia System

Where ,

T – Applied unbalanced torque (N /M)

Θ – Angle of twist in radian

Governing equation for rotational inertia is based on newton second law of motion ;

$$T(t) = I \cdot \frac{d^2\theta(t)}{dt^2} \quad \text{...} \ \mathbf{5.59}$$

Laplace equation = **T (s) = I s ^{2}Θ**

Transfer function **= G(s) = Θ (s) / T (s) = 1 / I s ^{2} 5.60**

5.22. Modeling of Electrical Systems

In mechanical modeling system , we had been considered the three elements which was associated with the rotational system and translation system similarly in case of electrical modeling , we will also considered three elements & try to analyzed the effect of each variable upon the system . these are

- Resistors (R)
- Capacitors (C)
- Inductors (L)

Variables : Electrical modeling system consists of two basic variable **i.e.** Current & Voltage , which is described as ,

(i) Electric Current : It is defined as , the continuous flow of electrons or negative charge in an electric circuit is called an electric current or in other words we can say that it is the rate of flow of charge particles. the flow of charge depends upon the drift of free electrons in the conducting metal. The schematic diagram for current and voltage is given below,

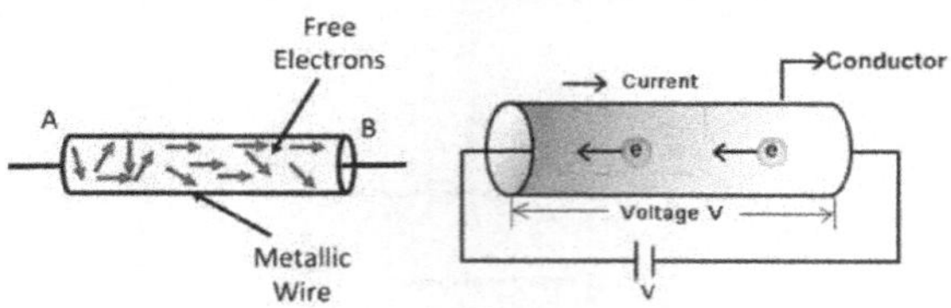

Diagram 5.52: Electric Current

Charge is measured in coulombs or coulombs/second and is measured with the help of an ammeter, abbreviated **"A"** . The **" I "** is the symbolic representation of the current.

Positive current :- Positive current is defined as which flows from the positive voltage end to the negative voltage end. But we always follow the actual current and voltage which is positive one. The sign convention for voltage and current is given as,

Correct sign convention for current and voltage

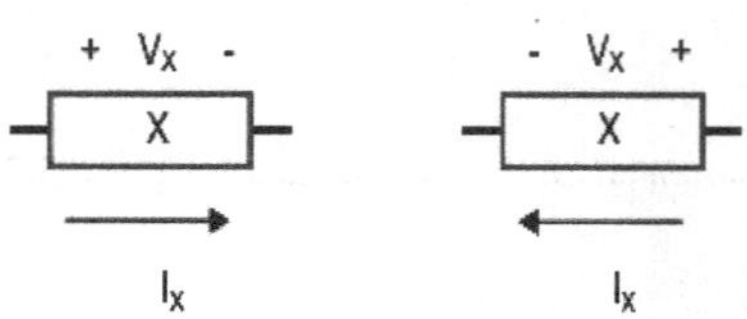

.... **For current**

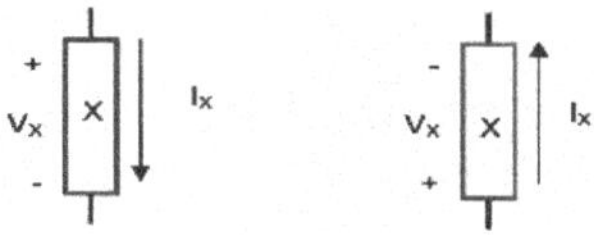

.... **For voltage**

Sign Conventions for Electrical Elements

(ii) Voltage : The voltage is a type of the electromagnetic force whose effect causes the flow of electrical current in the circuit which is a quantitative expression of the potential difference in charge between two points in an electrical circuit .

(iii) Resistors (R) : A resistor is an electrical device having a specific resistance whose function is to limits or regulates the flow of electrical current in a circuit. In order to fulfill the desired purpose like to reduce current flow, lowering the voltage levels, dissipation of heat and for protection & safe operation. the current through a resistor is inversely proportional to its resistance, and directly proportional to the voltage across it. Thus the current (I) is equal to the voltage (V) divided by the resistance (R) .

I=V/R 5.61

For a wire of length ' l ',

cross-sectional area –' A '

and specific resistivity – ' ρ '

$$R = \frac{\rho l}{A} \quad \text{.................. .... } \mathbf{5.62}$$

The circuit model for resistor is given as ,

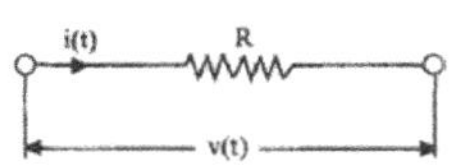

Circuit diagram of Resistor

Type of resistor : the classification of resistor is tabulated below .

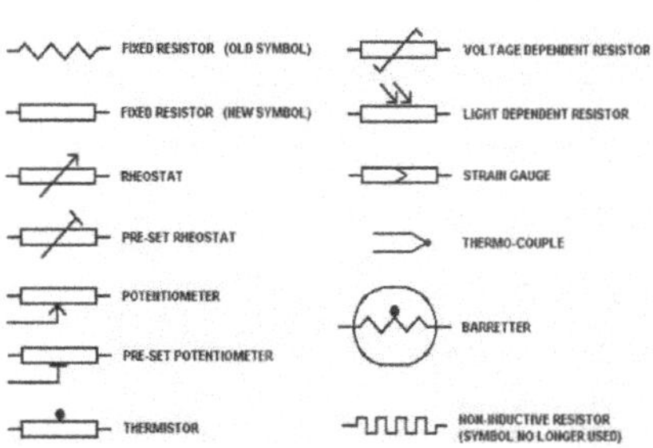

Fixed resistor : The function of fixed type of resistor is to resists the current .

Potentiometer : The function of variable resistor type potentiometer is to control the voltage .

variable resistor pre-set : The function of variable resistor pre-set is applied for altering the value of resistance

Note : The register is analogous to dampers in mechanical systems.

(iv) Energy dissipation : The calculation of heat/energy dissipation is obtained by ' Joule Effect ' according to ' Joule Effect ' The amount of heat produced in a current carrying resistor , is proportional to the square of the amount of current that is flowing through the resistor , for the given electrical resistance and the time hence

$$H \propto i^2 . R . t \quad \textbf{OR}$$

$$H = I^2 Rt$$

$$H = \frac{1}{J} . i^2 . R . t$$

$$= i^2 R t = \frac{V^2}{R} t \quad \ldots\ldots\ldots\ldots\ldots\ldots\ldots \textbf{5.63}$$

Where ,

' H ' is the heat generated in Joules,

' I ' is the current flowing through the resistor

' t ' is the time in seconds and ,

" V ' Applied Voltage

'J ' is a constant , known as Joule's constant .

Type of resistor :- the classification of resistor is tabulated below .

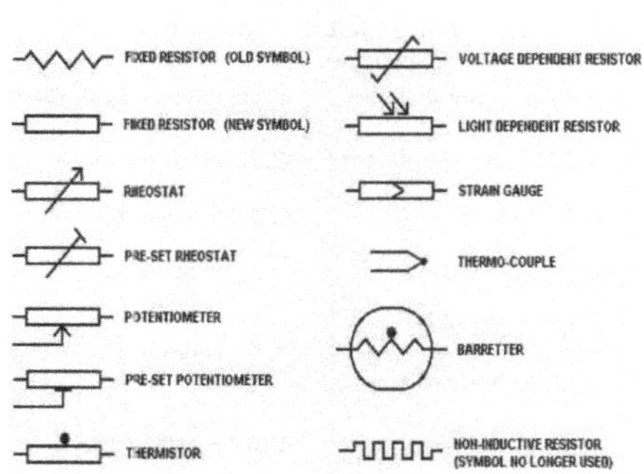

Fixed resistor : The function of fixed type of resistor is to resists the current .

Potentiometer : The function of variable resistor type potentiometer is to control the voltage .

variable resistor pre-set : The function of variable resistor pre-set is applied for altering the value of resistance

Note : The register is analogous to dampers in mechanical systems.

(v) Capacitors (C) : A capacitor is an electrical device which is used for accumulating electric charge or potential energy applied in electric the circuit. it has two conducting plates between which dielectric medium is provided. dielectric medium may be Air, Paper, Glass, Water. Vacuum whenever a voltage is applied across the plates we have two types of electrical charge, positive charge in the form of Protons and negative charge in the form of Electrons. When a DC voltage is placed across a capacitor, the positive (+ ve) charge quickly accumulates on one plate on the other hands opposite negative (- ve) charge accumulates on the other plate. For every particle of + ve charge that arrives at one plate a charge of the same sign will depart from the - ve plate. The flow of electrons onto the plates is known as the capacitors Charging Current which continues to flow until the voltage across both plates (and hence the capacitor) is equal to the applied voltage Vc. At this point the capacitor is said to be "fully charged" with electrons . as shown in figure .

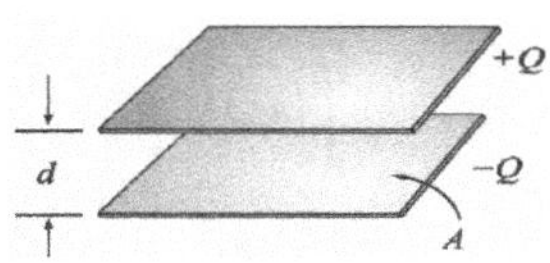

Capacitors

The capacitance in the circuit is directly proportional to the surface areas of the plates, and is inversely proportional to the separation between the plates. The capacity of a capacitor is measured in farads or amps / second.

Capacitance (c) = **(Q / V)**

Since in electrical circuits, the capacitor is the storage element, and therefore the voltage drop across the capacitor is proportional to the integral of the current , hence the voltage & energy equation can be written as,

$$V(t) = \frac{1}{C}\int_0^t i\,dt \qquad \text{.......... } \mathbf{5.64}$$

Energy equation

$$E = \frac{1}{2} C V^2 \qquad \text{.. 5.65}$$

Where ,

V is voltage across a capacitor,

C is a change in capacitance

Note :- The capacitor is analogous to the spring in mechanical systems.

The value of Capacitance for various Configuration

System	Capacitance
Isolated charged sphere of radius R	$C = 4\pi\varepsilon_0 R$
Parallel-plate capacitor of plate area A and plate separation d	$C = \varepsilon_0 \frac{A}{d}$
Cylindrical capacitor of length L, inner radius a and outer radius b	$C = \frac{2\pi\varepsilon_0 L}{\ln(b/a)}$
Spherical capacitor with inner radius a and outer radius b	$C = 4\pi\varepsilon_0 \frac{ab}{(b-a)}$

(v) Inductors (L) : The inductor is an electrical component has a conductor coil whose function is to store the electrical energy in the magnetic field when the electric current passes through it. The inductance of wire increases by increasing the number of turns. The figure below shows the symbolic representation of inductor.

(a)

The standard unit of inductance is the **Henry**, abbreviated H.

Inductor is classified as two types :-

- **Iron core inductors :** the inductor in which either the core is completely absent or ceramic material is used , such type of inductor is applied for high-frequency applications work where low inductance is required .

(b)

- **Iron Core Inductor (wound on ferrite core) :-** It is a fixed value inductor in which the iron core is kept between the coil.

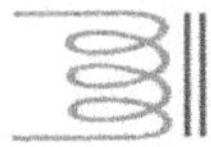

(c)

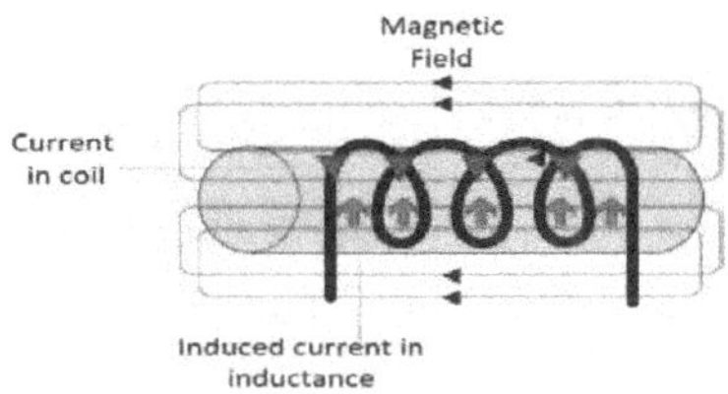

(d)

When current is flowing through the coil, magnetic field is generated & emf induced in the current carrying coil. thus, inductance will be proportional to the current rate. consider the inductor coil of length" L" the intensity of induced voltage (V) is given by the following equation,

$$V(t)=L\frac{d\,i(t)}{dt} \quad \text{.. 5.66}$$

For any cylindrical coil of diameter d length l, no. of turns n , and magnetic permeability μ , the inductance is given by :

$$L=\frac{\pi\,\mu\,n^2\,d^2}{4l} \quad \text{.. 5.67}$$

The magnetic energy stored may by calculated as :

$$E=\frac{1}{2}L\,i^2(t) \quad \text{.. 5.68}$$

Inductance for solenoids : Considers a solenoids of length " **L**" , Area " **A** " , having a turn " **N** " . as shown in figure .

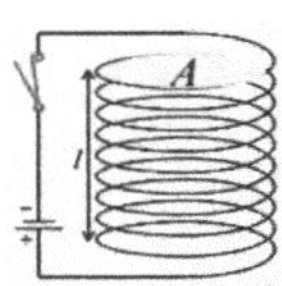

Induced emf	$E = -L\frac{dI}{dt}$
Length of Solenoids	$L = \mu_0 N^2 A / l$

Solenoids

Note : The inductor is analogous to the inertia due to the mass in mechanical systems

All the three elements, resistor, capacitors and inductors are arranged in series and parallel configurations. the table given below give the arrangement & their resultant.

Elements	Resultants
Resistors in series	$R_{total} = R_1 + R_2 + R_3$
Resistors in parallel	$\frac{1}{R_{total}} = \frac{1}{R_1} + \frac{1}{R_2} + \dots$
Capacitors in series	$\frac{1}{C_{total}} = \frac{1}{C_1} + \frac{1}{C_2} + \dots$
Capacitors in parallel	$C_{total} = C1 + C2 + \dots\dots$
Inductors in series	$L_{total} = L1 + L2 + \dots\dots$
Inductors in parallel	

5.23. Governing Law for Electrical Modeling

When the electrical elements get connected into a circuit and modelling of such a system is done then the value of unknown input quantity like current and voltage is determined in simple modeling system , complex circuits and networks , at the Nodal and Mesh and transfer of energy can be evaluated with the help of two laws . these law are basic law or governing law of electrical modeling , known as kirchoff "s law based on conservation of Energy , it has two type of configuration

- **Kirchhoff's Current Law (KCL)**
- **Kirchhoff's Voltage Law (KVL)**

(i) Kirchhoff's Current Law (KCL) : It is a Kirchhoff's first law and also recognized as a Kirchhoff 's point rule or Kirchhoff 's junction rule . it is a conceptual laws that deals with the conservation of charge entering and leaving a junction, used for electrical circuit analysis which is acronym as KCL.

According to Kirchhoff's first law " At any node or junction in an electrical circuit , the total current entering in to a circuits junction must be equal to the total current leaving the same junction with assumption that there is no loss of energy takes place throughout the circuit "

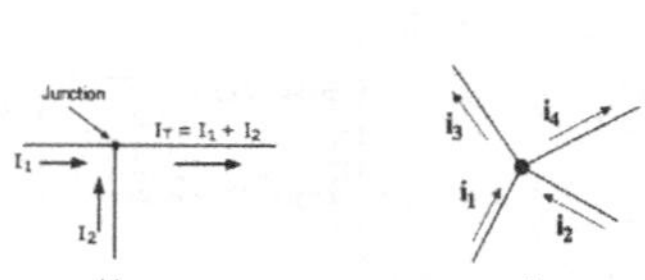

Diagram 5.53: Kirchhoff's First Law

Mathematically we can say that "The Algebraic sum of all the currents entering at any junction and after leaving that junction, current must be equal "

$\Sigma I_{IN} = \Sigma I_{OUT}$. **5.69**

Consider a diagram (b) indicating the Kirchhoff current law in which " The sum of the current entering a node has to equal the sum of the current leaving the node i.e.

$$\sum i_{in} - \sum i_{out} = 0 \quad \text{.......... 5.70}$$

$i_1 + i_2 - i_3 - i_4 = 0$ **5.71**

where ,

$(i_1 + i_2)$ is Entering currents .

$(-i_3 - i_4)$ is Leaving currents .

This law will be only valid if ,

- Charge is conserved,
- No charge builds up at the nodes ,
- In case of Lumped - element we considered the output will be zero.

(ii) Kirchhoff's Voltage Law (KVL) : - It is Kirchhoff 's second fundamental law that deals with the conservation of energy applied for a closed electrical circuit path . the basic purpose is to analyze and establish a suitable matematical equation for any electrical circuits.

According to Kirchhoff's second law " The algebraic sum of all voltages taken around a closed path in a circuit is zero " with assumption that there is no loss of energy takes place throughout the circuit "

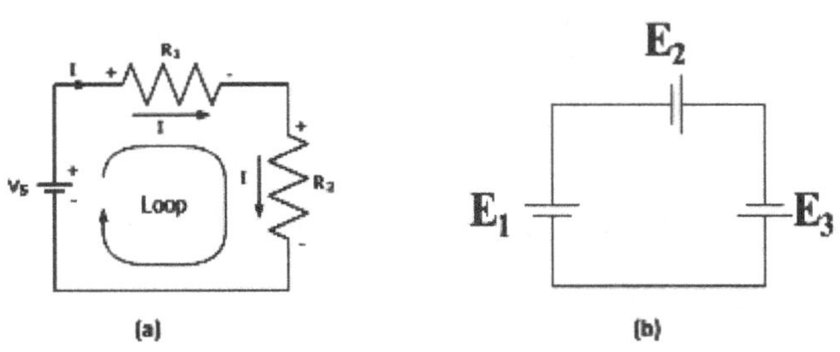

Diagram 5.54: Kirchhoff's Voltage Law (KVL)

Mathematically we can say that, The algebraic sum of all the potential differences around the closed loop circuit must be equal to zero .

$$\sum_{\substack{closed\\loop}} v_n = 0 \quad \text{.......... 5.72}$$

where , v_n is the voltage across circuit .

Consider a diagram (b) indicating the Kirchoff's voltage law in which " **The total voltage drop around a closed loop is zero i.e.**

$$E1 + E2 + E3 = 0 \quad \text{.......... 5.73}$$

This law will be only valid if ,

- Energy is conserved in the form of voltage ,
- Polarities , Signs of the sources and voltage drops across a resistor must be taken into account throughout the calculation .

5.24. Electrical Circuits Modeling for RLC Circuit

RLC circuit is very important circuit & generally applied for analyse of the linear systems . the model of the RLC circuit helps to understand the behaviors of an electrical control system . it consists of two type of modeling of series and parallel RLC circuits .

Case I : RLC series circuit

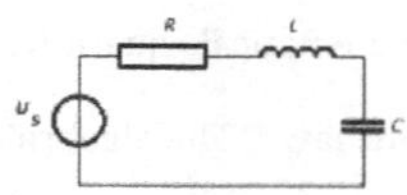

RLC Series Circuit

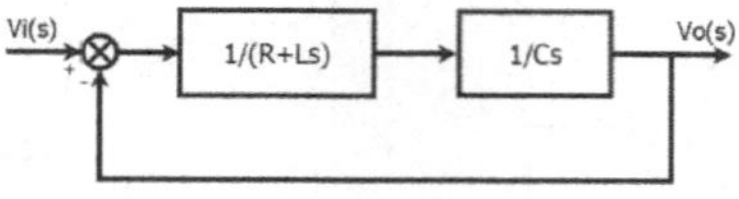

Diagram 5.55: Block Diagram for the RLC Series Circuit

The RLC series circuit is shown , the inputequation is given as

$$v_i(t) = Ri(t) + L\,\frac{di(t)}{dt} + \frac{1}{C}\int_0^t i(\tau)d\tau. \quad \text{.......... 5.74}$$

Let be $v_o(t)$ is output of capacitor C. Then, the output equation is represented by

$$v_o(t) = \frac{1}{C}\int_0^t i(\tau)d\tau. \quad \text{.......... 5.75}$$

In order to obtain transfer function, the Laplace transformation is applied then we have got .

$$V_i(s)=RI(s)+LsI(s)+\frac{1}{Cs}I(s), \quad \text{.. 5.76}$$

And

$$V_o(s)=\frac{1}{Cs}I(s). \quad \text{.. 5.77}$$

The transfer function will be written as ,

$$\frac{V_o(s)}{V_i(s)}=\frac{1}{LCs^2+RCs+1}. \quad \text{...5.78}$$

Hence output will be

$$v_o(t)=V_i\left(1-e^{-\frac{R}{2L}t}\cos\sqrt{\frac{4L+R^2C}{4L^2C}}t-\sqrt{\frac{R^2C}{4L+R^2C}}e^{-\frac{R}{2L}t}\sin\sqrt{\frac{4L+R^2C}{4L^2C}}t\right) \quad \text{............5.79}$$

Power in the RLC circuit :- The power calculation of RLC circuit can be given as ,

- Electrical voltage (U)
- Electrical current (i)

Then Power in the system : P = u i

$$u_R=iR$$
$$u_c=\frac{1}{c}\int idt$$
$$u_L=L\frac{di}{dt} \text{ or } i_L=\frac{1}{L}\int udt \quad \text{... 5.80}$$

The voltage over the elements are different

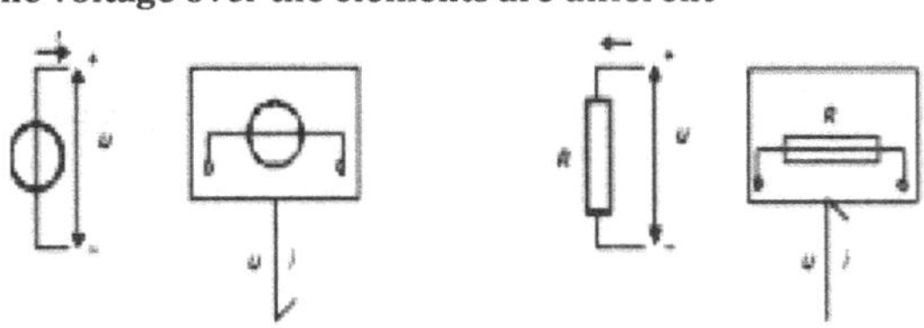

The current in the elements are same

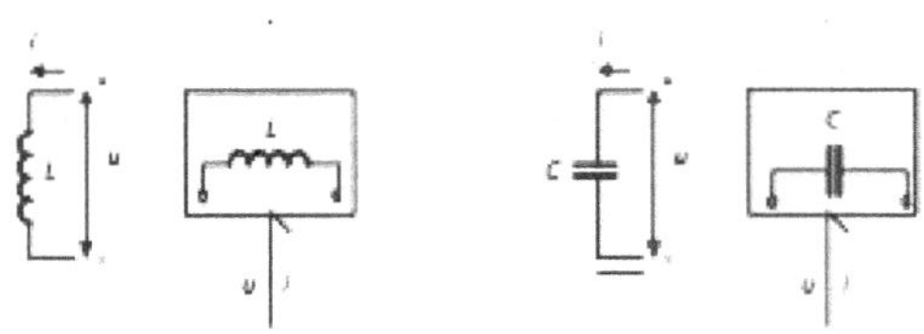

Case Ř: R-L-C Parallel circuit

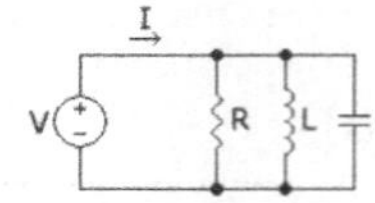

RLC parallel circuit.

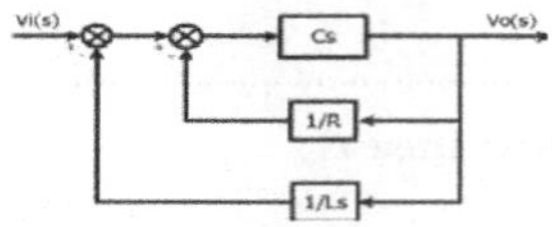

Diagram 5.56: Block Diagram for the R-L-C parallel Circuit

The equation of current will be ,

$$i(t)=\frac{v_R(t)}{R}+\frac{1}{L}\int_0^t v_L\ (\tau)d\tau+C\frac{dv_C(t)}{dt}. \quad \text{.........} \ \mathbf{5.81}$$

The output voltage through capacitor C

$$v_o(t)=v_R(t)=v_L\ (t)=v_C(t), \quad \text{.........} \ \mathbf{5.82}$$

Such that the mathematical model can be written in only one equation

$$i(t)=\frac{v_o(t)}{R}+\frac{1}{L}\int_0^t v_o(\tau)d\tau+C\frac{dv_o(t)}{dt}. \quad \text{.........} \ \mathbf{5.83}$$

For transfer function function operate laplace transformation then

$$I(s)=\frac{V_o(s)}{R}+\frac{1}{Ls}V_o(s)+CsV_o(s). \quad \text{.........} \ \mathbf{5.84}$$

Transfer function

$$\frac{V_o(s)}{I(s)}=\frac{RLs}{RLCs^2+Ls+R}. \quad \text{.........} \ \mathbf{5.85}$$

The output is given as ,

$$V_o(s)=\frac{RL}{RLCs^2+Ls+R}I, \quad \text{.........} \ \mathbf{5.86}$$

And solution will be ,

$$v_o(t)=2IR\sqrt{\frac{L}{4R^2C-L}}e^{-\frac{1}{2RC}t}\sin\sqrt{\frac{4R^2C-L}{4R^2L\ C^2}}t. \quad \text{.........} \ \mathbf{5.87}$$

5.25. Analogous Parameters for Electrical Modeling

Since mechanical modeling system always associated to the inertia elements , spring elements and damping elements whose effect could not be measured easily or when system having a complex nature so therefore it could be made corresponding analogy with electrical elements like resistor , capacitor and inductor in order to measuring the intensity of variable .There are two important types of analogous systems,

- Force Voltage Analogy (f-v)
- Force Current Analogy (f -i)

The fllowing table representing such kind of mechanical to electrical modeling system analogy .

For Force Voltage Analogy (f-v)

Translational mechanical system	Rotational control system	Electrical system analogy
Force (F)	Torque (T)	Voltage (V)
Mass (M)	Inertia (J)	Inductance (L)
Damper (D)	Damper (D)	Resistance (R)
Spring (K)	Spring (K)	Elastance (1/C)
Displacement (x)	Displacement (Θ)	Charge (q)
Velocity (u)$\dot{x}$	Velocity (u) $\dot{\theta}$	Current (i) = $\dot{q}$

For Force Current Analogy (f -i)

Translational mechanical system	Rotational control system	Electrical system analogy
Force (F)	Torque (T)	Current (i)
Mass (M)	Inertia (J)	capacitance (C)
Damper (D)	Damper (D)	conductance (1 / R)
Spring (K)	Spring (K)	Inductance (1/ L)
Displacement (x)	Displacement (Θ)	Flux linkage (ψ)
Velocity (u) = $\dot{x}$	Velocity (u) = $\dot{\theta}$	Voltage (v) = $\dot{\psi}$

Some other analogy

Translation mechanical system	Rotational mechanical system	Electrical system analogy
Static storage of Energy	Spring	Capacitor
Dynamic storage of Energy	Inertia	Inductor
Energy Dissipation	Dashpot/Damper	Resistor
Excitation	Force	Voltage
Response	Displacement	Charge

Impedance based representation

Element	Voltage-current	Current-voltage	Voltage-charge	Impedance, $Z(s)=V(s)/I(s)$
Capacitor	$v(t)=\frac{1}{C}\int_0^t i(\tau)d\tau$	$i(t)=C\frac{dv(t)}{dt}$	$v(t)=\frac{1}{C}q(t)$	$\frac{1}{Cs}$
Resistor	$v(t)=R\,i(t)$	$i(t)=\frac{1}{R}v(t)$	$v(t)=R\frac{dq(t)}{dt}$	R
Inductor	$v(t)=L\frac{di(t)}{dt}$	$i(t)=\frac{1}{L}\int_0^t v(\tau)d\tau$	$v(t)=L\frac{d^2q(t)}{dt^2}$	Ls

The following set of symbols and units are used:

v(t) = V (Volts) , i(t) = A (Amps) , q(t) = Q(Coulombs) , C = F (Farads) , R = Ω (Ohms) ,

L = H (Henries).

Component	Force-velocity	Force-displacement	Impedence $Z_M(s) = F(s)/X(s)$
Spring (K)	$f(t) = K\int_0^t v(\tau)d\tau$	$f(t) = Kx(t)$	K
Viscous damper (f_v)	$f(t) = f_v v(t)$	$f(t) = f_v\frac{dx(t)}{dt}$	$f_v s$
Mass (M)	$f(t) = M\frac{dv(t)}{dt}$	$f(t) = M\frac{d^2x(t)}{dt^2}$	Ms^2

5.26. Force Voltage Analogy

Consider the mechanical system in which both ends are rigidly fixed, the mass (m) is suspended between spring elements and damper, the electrical analogeous system is given in which spring elements having equivalency with inductance, damper is represented as resistance and force P will have output voltage of the system .

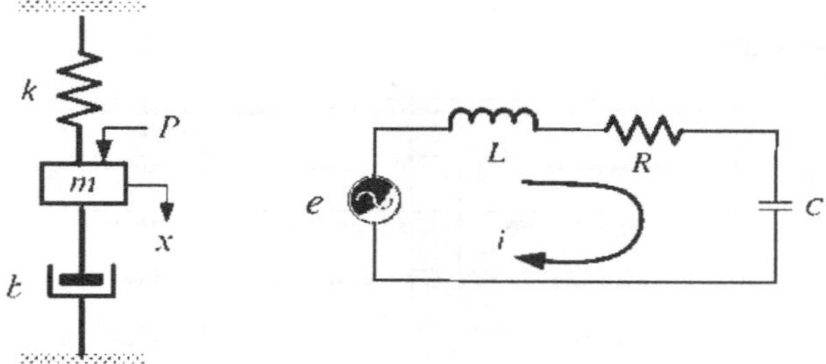

Diagram 5.57: Analogy between Mechanical and Electrical Systems

The differential equations for the mechanical system can be written as ,

$$m\frac{d^2x}{dt^2}+b\frac{dx}{dt}+kx=p \qquad \text{.......... 5.88}$$

Where

m = Suspended value of mass

x = Displacement of mass m

b = viscous – friction cofficient

k = Effect of spring elements

p = Applied force

The analogus differential equations for the electrical system can be written as ,

$$L\frac{di}{dt}+Ri+\frac{1}{C}\int idt=e \qquad \text{.......... 5.89}$$

Or

$$L\frac{d^2q}{dt^2}+R\frac{dq}{dt}+\frac{1}{C}q=e \qquad \text{.......... 5.90}$$

Where ,

L = Inductance, R = Resistance, i = current flow, e = voltage induced, q = value of charge Equation **(a)** and **(b)** representing the analogy between mechanical & electrial system

(i) For a rotational control system : Consider a simple mechanical rotational system in which angular velocity **(ϴ)** , torque (T) , inertia (J) is corressponds to electrical analogy voltage (v) , current (i) and capacitance (c) , the schematic diagram of the system is given below .

For a rotational control system

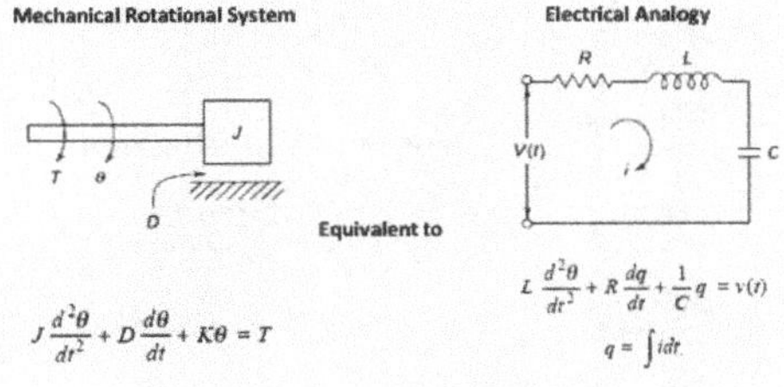

(a)

For a translational mechanical system

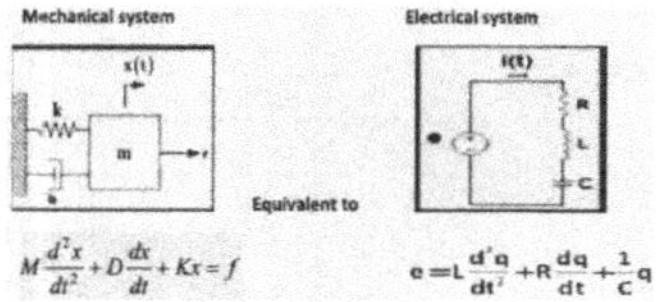

(b)

Diagram 5.58: Analogy for Rotational System (a) &Translational System (b)

Power in the system : - P = F v

$$F_d = cv$$

$$F_s = k_s \int v dt = \frac{1}{C_s} \int v dt \qquad \text{.........} \; \mathbf{5.91}$$

Applied consecutive law :

$$F_m = m\frac{dv}{dt} \; or \; v = \frac{1}{m}\int F_m dt$$

$$F_a = Force \qquad \text{.........} \; \mathbf{5.92}$$

Where ,

J = Inertia elements , D = Damper

k = Spring elements , θ= rotational angle

T = Torque , M = mass

x = displacement , f = Force

q = charge , C = Capacitance

(ii) Modeling of multiloop system : In multiloop modeling system , two or more than two elements of mechanical system will be involved in any particular mechanism , in such a system the analogy between mechanical translation system & electrical system can be given as ,

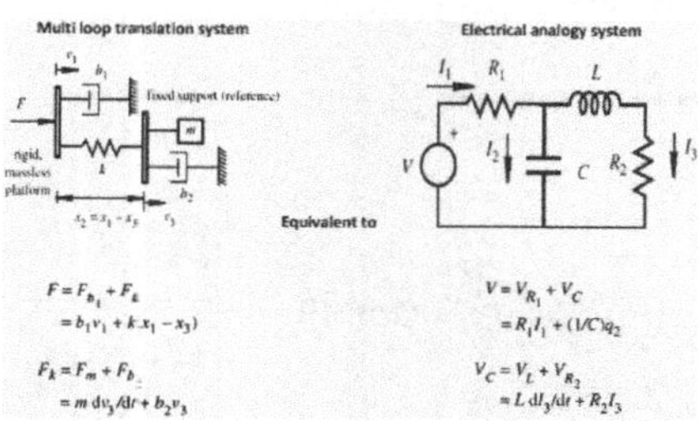

Diagram 5.59: Multiloop Modeling

Where ,

I_1 = current through resistor R_1 , I_2 = current through capacitorc

I_3 = current through resistor R_3 , L = Inductor

x_1 , x_2 , x_3 = Displacements, F_{b1} , F_{b2} = damper friction force

F_k = springs force

5.27. Liquid Level Systems

The basic concept of liquid level system is to be perform the analyses whether the flow is laminar or turbulent which can be identified with the help of Reynolds number ,

- If the value of Reynolds number range lies between 3000 to 4000 then representing the flow is turbulent which is characterized by the irregular movement of particles or sometimes, zig-zag motion, recirculation, eddies of fluid particles takes place. In turbulent flow changes in pressure and flow velocity always occurs.
- If the value of Reynolds number is less than 2000 then flow becomes laminar. it is representing the streamline flow in which fluid always flows in parallel layers, without occurring any kind of interruption between flow layers.

In any industrial application or process flow through the connecting pipes and tanks always subjected to turbulent flow. A simple liquid level system consists of a tank with an inlet valve and outlet valve and represents a single input single output (SISO) control system. In the system, while the out let flow liquid from the tank is manually controlled through the load vane, the inflow liquid into the tank is adjusted & controlled by a proportional valve. Normally, the outflow liquid from the tank is a load which is needed by process and continuously changes due to reasons beyond control. Therefore, the inflow liquid into the tank represents a manipulated variable depending on the liquid level. The out let flow liquid from the tank refers to a load or a disturbance. The top of the liquid tank is open and it has a cylindrical structure. The liquid level control system analyses is performed to calculate the difference between the inlet flow rate and outlet flow rate of the liquid in the tank. The schematic diagram of such a system is given below.

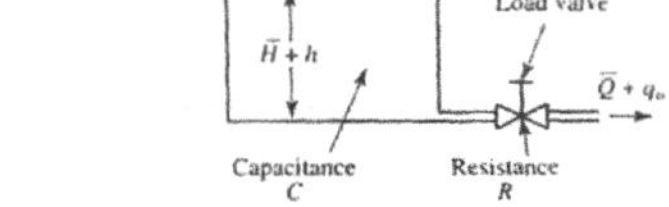

Diagram 5.60: A Schematic Diagram of the Liquid Level Control System

The dynamic characteristics of such a liquid level control system can be described in simple form by introducing the basic concept of resistance & capacitance in the liquid level system, which is described as,

5.27.1. Hydraulic Resistance of Liquid Level Systems

Consider a pipe which connects the two tanks , a valve is provided which create a hydraulic restriction or resistance for flowing fluid between both tanks . As shown in figure,

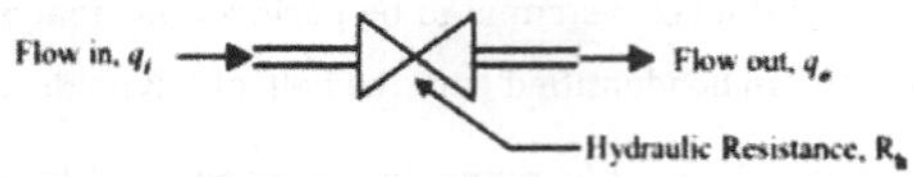

The hydrulic resistance is defined as the ratio between change in difference of the liquid levels between two tanks and unit change in flow rate .

$$\text{Resistance } (R) = \frac{\text{Change in liquid level}}{\text{Change in flow rate , } m^3/sec} = \frac{H}{Q}$$

Case – I :- If the flow is laminar then , flow rate is given as

$$Q = K H$$

Where ,

Q = steady state liquid flow rate in m3/sec

K = flow cofficient

H = Pressure head of liquid in meter

Resistance for laminar flow will be given as ,

$$R_{laminar} = (H / Q) \quad \text{.......................................} \quad 5.93$$

In this case the laminar flow is bsed on the coloumbs law that is " current is directly proportional to the potential difference " thus laminar flow resistance will be analogous to the the electrical resistance .

Case – II :- If the flow is turbulent through the restriction then , flow rate is given as ,

$$Q = K\sqrt{H}$$

$$= KA\sqrt{2gh} \quad \text{...} \quad 5.94$$

Where ,

Q = Steady state liquid flow rate in m^3/sec, K = Flow cofficient

H = Pressure head of liquid in meter, A = Area of discharge orifice

g = Gravitational costant , h = Pressure head of liquid

Resistance for turbulent flow will be given as ,

$$R_{turbulent} = (dH / dQ)$$

Then from equation (5.94)

$$dQ = \frac{K}{2\sqrt{H}} dH$$

On further calculation

$$\frac{dH}{dQ} = \frac{2\sqrt{H}}{K} = \frac{2\sqrt{H}\sqrt{H}}{Q} = \frac{2H}{Q}$$

Hence the above equation reduced as ,

$$R_t = \frac{2H}{Q} \quad \text{...} \ 5.95$$

In case of turbulent flow, the value of resistance will be two times with respect to laminar flow resistance. further if the value of head and flow rate is less, then the value of turbulent resistance will be considered as constant or negligible. From the above equation it was also concluded that resistance having analogy to electrical resistance. the resistance is directly proportional to two times of pressure head (H) but inversely proportional to flow rate (Q) .

5.27.2. Hydraulic Capacitance of Liquid Level Systems

In Liquid level systems the quantity of liquid in tank or variation in quantity is represented by electrical analogy elements by capacitance **'c '** . it is defined as the ration of quantity of stored liquid in tank to change in head of liquid .

$$C = \frac{(\text{Change in quantity of liquid stored , } m^3)}{(\text{change in head in meter})}$$

Suppose a tank is connected with supply of liquid pipes , having an surface area **' A '**, height of liquid level **' H '** and volume **' V '** then volume may be given as ,

$$V = A\ H \quad \text{.........................} \ 5.96$$

Here the volume and height of liquid level is depends upon the time and capacitance is depends upon the cross sectional area of tank. Then

$$A = V / H \quad \text{..........................} \ 5.97$$

$$= (\text{Quantity of liquid}) / (\text{height of liquid})$$

Since in the electrical capacitance **C = (Q / V)** having analogy with surface area of liquid in the tank , hence **C = A** , Differentiate the **equation** with respect to **time t** , then

$$(dV / dt) = A (H / dt)$$

(i) Transfer function : When we calculated the resistance of the liquid level system **R** and the capacitance **C** . Then the value of transfer function is the ratio between quantity of output to quantity of input . mathematically it is defined as,

$$G(s) = \frac{Q_o(s)}{Q_i(s)} \quad \text{..........................} \ 5.98$$

Demonstration of simple liquid level system :- Consider a liquid level system a shown in figure ,

Control valve

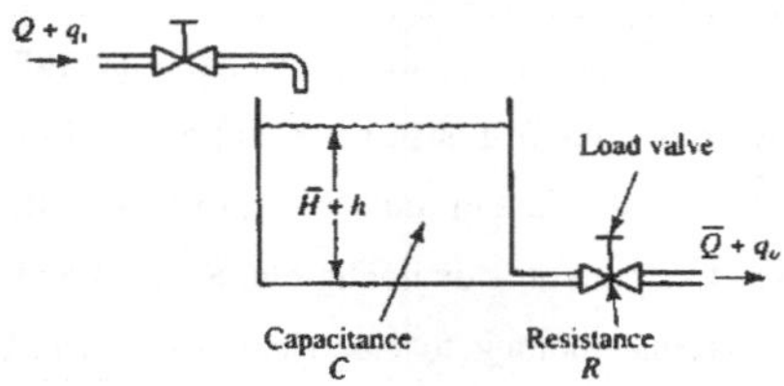

Diagram 5.61: Simple Liquid Level System

Where ,

Q = steady-state flow rate

q_i = small deviation of inflow rate from its steady-state value

q_o = small deviation of outflow rate from its steady-state value, m3/sec

H = steady-state head in m

h = small deviation of head from its steady-state value, m

then , the value of the resistance will be given as ,

$$q_o = \frac{h}{R}$$

The differential form of equation becomes

$$RC\frac{dh}{dt} + h = Rq_i$$

Apply laplace transformation

$$(RCs + 1)H(s) = RQ_i(s)$$

Transfer function will be **(output / input) = h / q_i** , then

$$\frac{H(s)}{Q_i(s)} = \frac{R}{RCs + 1}$$

If **Q_o** is output &**Q_i** is input then , transfer function will be

$$\frac{Q_o(s)}{Q_i(s)} = \frac{1}{RCs + 1} \quad \text{and} \quad Q_o(s) = \frac{1}{R}H(s) \qquad \text{.............. 5.99}$$

5.27.3. Liquid Level System for Complex System

Consider the liquid level system having two tanks in interaction , the value of related variable is given as , the schematic diagram of such a system is given as ,

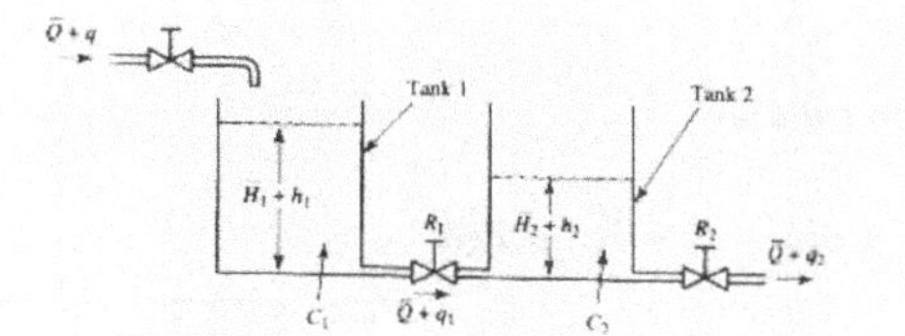

$$\frac{h_1 - h_2}{R_1} = q_1 , \quad C_1\frac{dh_1}{dt} = q - q_1 , \quad \frac{h_2}{R_2} = q_2 , \quad C_2\frac{dh_2}{dt} = q_1 - q_2$$

Diagram 5.62: Complex Liquid Level System

Where ,

$\bar{Q}$: Steady-state flow rate
$\bar{H}_1$: Steady-state liquid level of tank 1
$\bar{H}_2$: Steady-state liquid level of tank 2

Then the value f transfer function is given as ,

$$\frac{Q_2(s)}{Q(s)} = \frac{1}{R_1C_1R_2C_2s^2 + (R_1C_1 + R_2C_2 + R_2C_1)s + 1} \qquad \text{.............................. 5.100}$$

5.28. Modeling of Fluid System

In fluid system modeling , we will modeled elemets of fluid flow as we have been modeled to those for mechanical and electrical systems . the basic elements of such a system is given as

- **Resistance**
- **Capacitance**
- **Inertance**
- **Fluid energy source**

In fluid systems , the energy transmission takes place through the liquid so therefore the fluid system may be classified as incompressible system or compressible system . Since in this modeling we will consider the modeling of incompressible system thus we will be focused on few basic fluid flow governing laws which is given as ,

- **Bernoulli's Law and**
- **Law of Mass Conservation or continuity equation**

Before going through the discussion of modeling parameter, here a short description of the governing law is necessary ,

(i) Bernoulli 's Laws : Bernoulli 's equation is basic for modeling of fluid system for any particular flow . in case of incompressible flow if there is no energy losses or gains and mass density is constant then

According to Bernoulli's law ,

$$p_2 + \frac{\rho v_2^2}{2} + \rho g h_2 = p_1 + \frac{\rho v_1^2}{2} + \rho g h_1 \quad \text{.......... 5.27 a}$$

where **P_1 & P_2** are value of pressure at inlet and outlet , **h** head or vertical distance , and **V_1 &V_2**are the value of velocity at inlet and outlet .

(ii) Law of continuity : When a fluid is in motion , it must move in such a way that there is no occurrence of loss of fluid between input and output . or The algebraic sum of the flow rates at any junction in the loop must be zero.hencemass is conserved . and volume flow rate remain constant .The law of continuity having a same analogy as that of Kirchhoff's current law.

$$q_1 = q_2 \quad \text{.......... 5.27 b}$$

$$\sum_{\substack{\text{Any}\\\text{Node}}} q_j = 0$$

or

$$\sum q_{IN} = \sum q_{OUT}$$

$$\boxed{\rho A_1 V_1 = \rho A_2 V_2} \quad \text{.......... 5.27 c}$$

Where ,

q = volumetric flow (m^3 /s)

v = volume (m^3)

h = height of fluid in a vessel (m)

p = pressure (N/m2)

(iii) Compatibility Law : According to the law of Compatibility the sum of the pressure drops around a closed loop must be zero. The law of compatibility having a same analogy as that of Kirchhoff's voltage law.

$$\sum_{\substack{\text{Closed} \\ \text{Loop}}} \Delta p_j = \sum_{\substack{\text{Closed} \\ \text{Loop}}} p_{ij} = 0 \qquad \text{.......... 5.27 d}$$

5.28.1. Fluid Elements

In case of mechanical and electrical system, modeling is performed to inertia elements , spring elements, damper, storage capacity, losses , and energy input etc. in similar manner the fluid system modeling is also depends upon few elements which are associated with called primary elements . these elements are Inertance element based on inertia effects, fluid storage or volume storage and loss having a analogy to inductance, capacitances and resistance. which can be demonstrate as,

(i) Resistance : Resistance of any electrical conductor is based on ohms law , according to it **V = IR** , indicating a linear relationship between the variables . but in case of fluid systems the resistance imposed by the system does not linear. because the value of pressure drops indicating resistance depend upon the restriction created by Orifices , valves , nozzles and friction in pipes , change in area , are considered and modeled as a fluid resistors.

Consider a fluid system, in which the input is the volumetric rate of flow **q** and the output is the pressure difference **(p_1 – p_2)**. As shown in figure ,

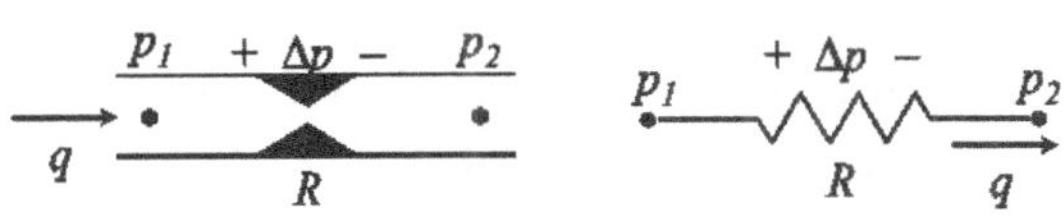

Diagram 5.63: Fluid Pressure with Electrical Analogy

Further,

$$\Delta p = p_1 - p_2 \qquad \textbf{(pressure drop)}$$

$$= R \cdot q \quad \textbf{and}$$

$$q = \frac{1}{R} \Delta p$$

Hence, the resistance imposed by restricted area is defined as the ratio between pressure difference to flow rate of liquid,

$$R = \frac{\text{Pressure difference}}{\text{Change in flow rate, m}^3/\text{s}} = \frac{p_1 - p_2}{q} \quad \text{.......................} \ 5.27\ e$$

This relationship is also called hydraulic resistance. But in case of sharp-edge orifices or if the flow is turbulent , it will represents the non linear relation ship amongs the variables and follow to law

$$q = K\sqrt{p}$$

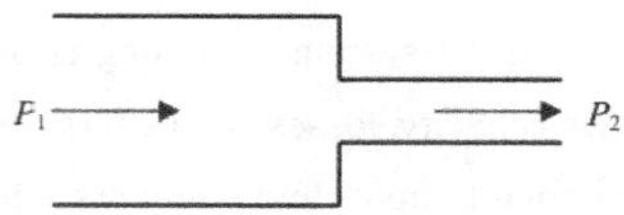

(ii) Capacitance : We know that the capacitance of any electrical capacitor is indicating the charge storage capacity of any capacitor with increasing the voltage but in case of fluid system , when the fluid is stored in storage tank , this fluid energy is stored as a potential energy and creating a pressure variation . the height between bottom of the tank to top surface of water level is called head for such a system so therefore ' The capacitance can be defined as in the ratio between the volume flow rate and the rate of pressure variation .

$$C = \frac{\text{Volume Flow Rate (m}^{3)}}{\text{Pressure Variation}}$$

$$C_{l,p} = \frac{q_v(t)}{\frac{dp(t)}{dt}} = \frac{q_v(t)\,dt}{dp(t)} = \frac{\frac{dV(t)}{dt}dt}{dp(t)} = \frac{dV}{dp} \quad \text{.......................} \ 5.27\ f$$

Some times it is also defined as the ratio between the change in quantity of stored liquid necessary to cause a unit change in the potential (head).

$$C = \frac{\text{Change in liquid stored, m}^3}{\text{Change in head, m}}$$

$$C_{l,h} = \frac{q_v(t)}{\frac{dh(t)}{dt}} = \frac{q_v(t)\,dt}{dh(t)} = \frac{\frac{dV(t)}{dt}dt}{dh(t)} = \frac{dV}{dh} \quad \text{.......................} \ 5.27\ g$$

(i) Force flow vs Current flow (a)

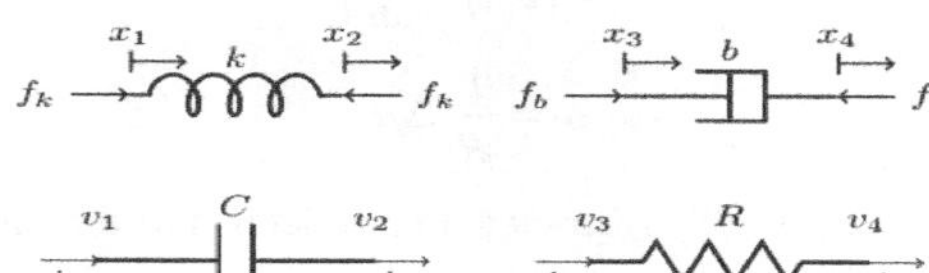

(a)

(ii) Electrical Analogy for Fluid Elements (b): Tank

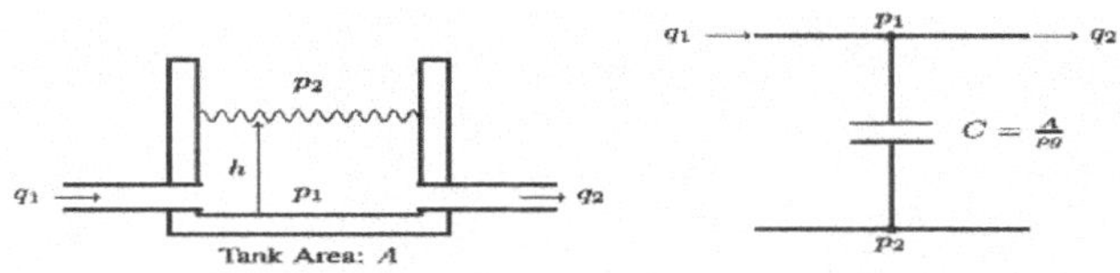

$\frac{A}{\rho g}\frac{d(p_1 - p_2)}{dt} = q_1 - q_2$ $\qquad$ $C\frac{d(p_1 - p_2)}{dt} = q_1 - q_2$

(b)

(iii) Electrical Analogy for Fluid Elements(c) : Valve

$p_1 - p_2 = Rq$ $\qquad$ $p_1 - p_2 = Rq$

(c)

Example of fluid capacitor :- Consider a reservoir containing fluid , as shown in figure

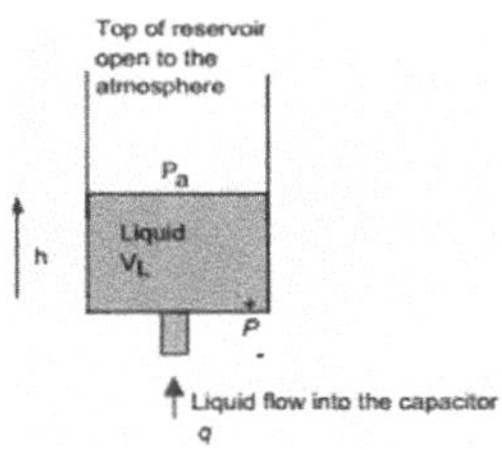

Diagram 5.64: Fluid Capacitor

Force due to weight of fluid = mg = $\rho V_L g$ = ρV_L

Where ,

ρ is the density of the fluid,

V_L is the volume of fluid in the tank,

Pressure difference = $p_{in} - p_a$

$$P = p_{in} - p_a$$

Then ,

$$p = \frac{\rho g V_L}{A} = \rho g h \qquad \text{5.27 h}$$

$$V_L = \left(\frac{A}{\rho g}\right)p \qquad \text{.............. 5.27 i}$$

Where ,

$$C = \left(\frac{A}{\rho g}\right).$$ **is representing the fluid capacitance ,**

Example – We can also understand it by another example for hydraulic capacitance as **(a) & (b)**

- **Spring-loaded chamber**
- **An air spring**

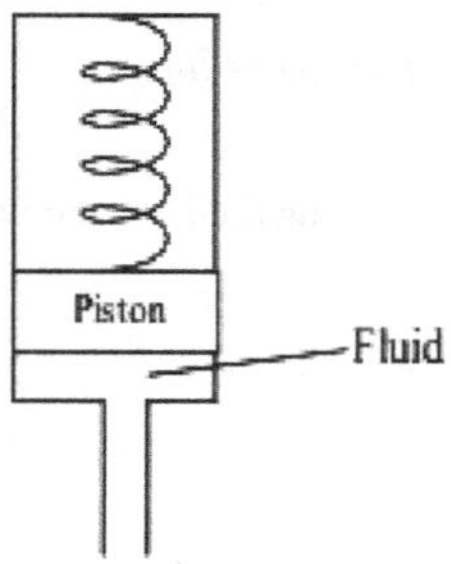

Spring-loaded Chamber

(a)

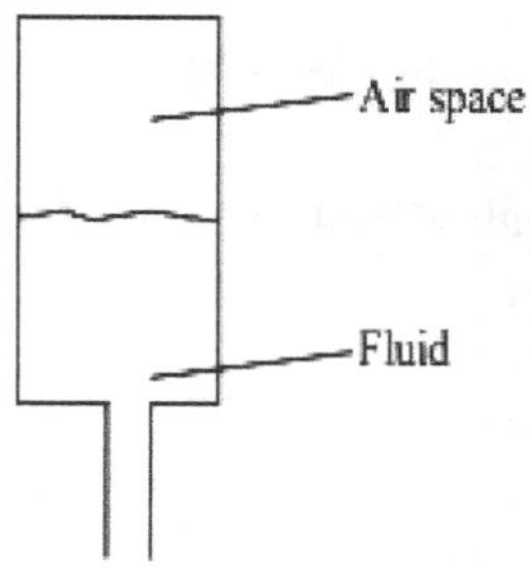

Air Spring

(b)

Energy stored in liquid capacitance

$$U_{l,p} = \frac{1}{2}C_{l,p}(\Delta p)^2 = \frac{1}{2} \times \frac{(\Delta V)^2}{C_{l,p}} = \frac{1}{2}(\Delta V)(\Delta p) \qquad \text{.............. 5.27 j}$$

Example – Discuss the modeling process for given hydraulic system.

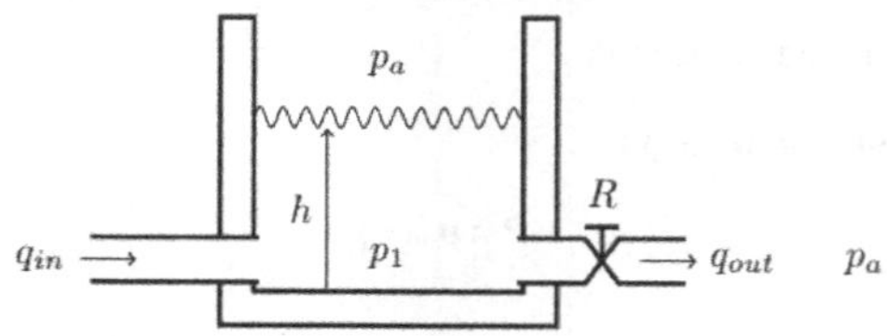

Tank Area: A

Step 1 Identify all node variables. For a fluid problem, the node variables are pressure. There are two unique pressure variables, P_1 , the pressure at the bottom of the tank, and P_a atmospheric pressure

Step 2 Identify one node as ground, or add a ground node. Most node variables are relative - for example, consider the ground on a circuit diagram. All voltages are measured with respect to ground, so the ground voltage is simply what we will consider to be 0 V. Negative voltages are lower than ground, positive voltages are higher than ground. If we identify a node as the ground, then we will measure all node variables relative to this node. For fluid systems, there are two typical choices: Either choose absolute zero pressure (i.e. a vacuum) as ground, or choose atmospheric pressure as ground. Pressure measured with respect to atmospheric pressure is called gaugepressure, and the translation is

Absolute Pressure = Gauge Pressure + Atmospheric Pressure

Step 3 Connect components between nodes

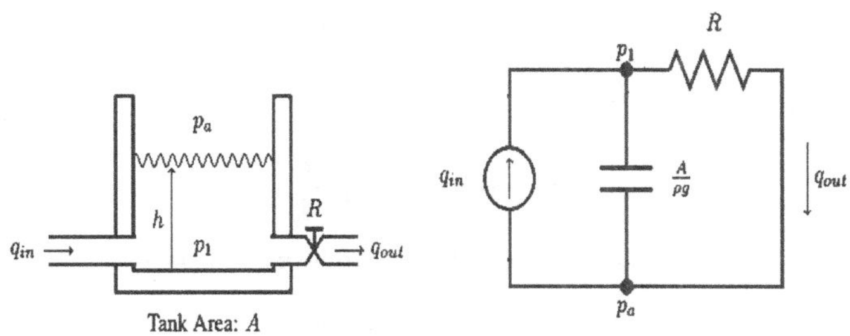

Analogy with Fluid System

(iii) Inertance or Inductance : Inertance in fluid flow system is subjected to inertia effects of flowing fluid in pipe which have the same equivalency of inductance as that of electrical systems or in mechanical systems . the significance of Inertance is to increasing the acceleration of flowing fluid by increasing the intensity of force finally thereby we will achieved the velocity of flowing fluid .

Thus inertance can be defined as the pressure difference required to produce a unit change in the rate of change of volume flow rate. which is denoted by **(I)** . Consider the diagram which shows the pressure drop Δp , across the element is proportional to the rate of change of the flow rate (q)

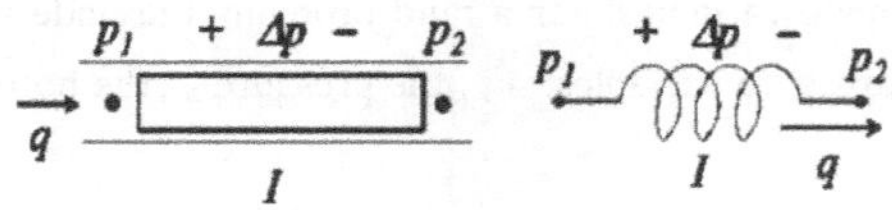

Diagram 5.65: Fluid Pressure with Electrical Analogy

Thus

$$\Delta P = (P_1 - P_2) A$$

Where ,

$(P_1 - P_2)$ is the pressure difference , **A** is the cross-sectional area **then ,**

Net required force for creating acceleration

$$F = (P_1 - P_2) = ma$$

Since a is acceleration so therefore

$$(p_1 - p_2)A = m\frac{dv}{dt}$$

But we know that $m = AL\rho$, then

$$(p_1 - p_2)A = AL\rho\frac{dv}{dt}$$

We also know that the volume flow rate of flowing fluid **q = A v** , hence

$$(p_1 - p_2)A = L\rho\frac{dq}{dt}$$

$$p_1 - p_2 = I\frac{dq}{dt}$$

Hence hydraulic inertance can be expressed as ,

$$I = \frac{L\rho}{A} \quad \text{...} \mathbf{5.27\ k}$$

The basic governing equation for inertance, capacitance and resistance in fluid flow modelling is tabulated as ,

Building block	Describing equation	Energy
Resistance	$q = \frac{p_1 - p_2}{R}$	$P = \frac{1}{R}(p_1 - p_2)^2$
Capacitance	$q = C\frac{d(p_1 - p_2)}{dt}$	$E = \frac{1}{2}C(p_1 - p_2)^2$
Inertance	$q = \frac{I}{L}\int(p_1 - p_2)dt$	$E = \frac{1}{2}Iq^2$

Example 1: Consider the case of fluid flow through a pipe having a diameter of 0.150 meter , length 17 meter, density 1000 kg / m3 , friction factor is 0.035 , viscocity is 1 then calculate Reynold's number, flow resistance and pressure drop in pipe .

Given that ,

Diameter (d) = 0.150 meter

Length (L) = 17 meter

density = 1000 kg / m3

friction factor = 0.035

viscocity (μ) = 1

Solution

(i) Calculate average velocity :

$V = 4Q / \prod d^2$

= (4 x 45) / (3.14 x 0.15 x 0.15) = 25.47

(ii) Calculate Reynolds number :

$Re = (\varrho V d) / \mu$

= (1000 x 25.4 x 0.15) / 1 = 3810

Since the value lies between 2000 and 4000 , then the flow is turbulent .

(iii) Resistance of flow :

$R_t = (16QfL\varrho) / (\prod^2 d^5)$

= (16 x 45 x 0.03209 x 17 x 1000) / (3.14) ^{2} x (0.15)^5 = 6.73 x 10^2 pas-m^3

(iv) Pressure Drop :

$P = k_t Q^2$

= 7.48 x 10^4 x 45 x 45

Example 2 : Calculate the value of inertance of water flowing through the pipe having a diameter of 1.5cm , length of pipe is 47 meter , density is 1000 .

Solution

(i) Area of pipe = ($\prod d^2 / 4$)

= (3.14 x 0.015 x 0.015) / 4 = 7 x 10^3 m^2

(ii) Inertance (I) = $L\varrho / A$

= (1000 x 47) / 7000 = 6.7142

Example 3 : A tank has a diameter of 1.83 meter , height 10 feet , calculate the capacitance of the tank .

Solution :

C = A / ϱg

= 2.63 / (1000 x 9.8) = 2.68x10^{-4} m^{3} / pa

5.29. Fluid Energy Source

In fluid level system the head difference is achieved with the help of hydraulic actuators whose function is to convert input fluid pressure into kinetic energy at the output in fluid storage device like tank and fluid flow medium like pipes .the pumps act as sources, analogous to voltage or current sources in electrical systems. hence such kind of process is called as the fluid energy source for any particular fluid system .

For example – Mechanical hydrulic pump is applied for improving the equivalent head . when it operate transfoming the input electric energy into out put liquid work or improve the energy of fluid . thus the head generated by the pump is given by following equation

$$h = h_g - K_h q_v^2 \qquad \text{.................................. 5.28 a}$$

where ,

h_g = Geometric head,

K_h = Coefficient related to the energy losses.

h = Head difference .

q_v = Pressure head - Flow Rate

in the form of pressure it can be written as ,

$$p = p_0 - K_p q_v^2 \qquad \text{.................................. 5.28 b}$$

Where ,

p_o is the maximum flow rate ,

K_p is a constant

5.30. Modeling of Thermal System

Before going through the modeling of such a system, here we will take the short description of the modes of heat transfer, which has great importance for importance for modeling , there are three mechanisms of heat transfer by which heat can flow from one substance to another substance and the heat transfer only possible if temperature difference exists between the bodies . these are

- Condustion
- Convection
- Radiation

(i) Conduction : Thermal conduction is mechanism of heat transfer in which heat transferred takes place from higher temperature to lower temperature such kind of heat transfer takes place through adjoining bodies or different medium in direct physical contact . heat transfer via conduction is achieved by movement of free electrons , vibrational energy or random molecular motion . then heat flow rate may be given as ,

$$q = (k A \Delta\theta)/\Delta L \quad \text{..} \quad \mathbf{5.28\ c}$$

Where ,

q = Heat flow rate , Kcal / sec

$\Delta\theta$ = Tempetrature difference in ^{o}C

K = Thermal conductivity , Kcal / sec ^{o}C

Δ L = Thickness of conductor in meter

For example : The flow of heat that occurs through the wall of a building if the temperature inside is higher (or lower) than the temperatures outside.

(ii) Convection : Heat transferred by thermal convection is carried out by the mixing of fluid medium and only applicable for fluid and gases . the heat transfer between surface and fluid is define by newtons law of cooling . according to it ,

$$q = (H A \Delta\theta) \quad \text{..} \quad \mathbf{5.28\ d}$$

Where ,

q = Heat flow rate , **Kcal / sec**

$\Delta\theta$ = Tempetrature difference in^{o}**C**

A = Area normal to heat flow in $\mathbf{m^2}$

H = Convection heat transfer cofficient in Kcal / m^2sec ^{o}**C**

(iii) Radiation : Heat transfer due to thermal radiation is only possible if heat is available in the form of radient energy or in wave form . heat transferred with the radiation are based on stefen - boltzman law . Convective heat transfer involves the flow of heat in a liquid or gas, as when a fan blows cool air across a hot object; the air carries away some of the heat of the object. according to it .

$$q = (H A \Delta\theta) \quad \text{..........} \quad 5.28\ e$$

Where ,

E_b = Energy radiate per unit time

T = Absolute temperature of surfaces

σ_b = Stefen – boltzman constant

Radiative heat transfer , like conductive transfer , is caused by a temperature difference between objects , does not require a physical medium for heat flow it **operate via vacuum**

5.31. Governing Laws

In any process industries , the basic thermal process may be mixing of hot and cold fluid , exchange heat through solid body contact , heat transferred between solid and liquid , heat generation due to chemical process etc. all these processes are implemented on particular governing law . these are given below .

- The first law of thermodynamics applied to a regular interface and implies that the heat loss by a system must pass integrally to another system . It tells about the way of heat generation process as well as how much amount of heat is being generated throughout that process .
- The second law of thermodynamics implies that the hotter system gives off heat while the colder one takes it , and described about the rate of heat flow . For example – In any body if heat is supplied then internal energy is increased thus change in body temperature will be proportional to the heat flow rate .

5.32. Analogy for Thermal Modeling

Thermal systems are those that involve the storage and transfer of heat. The thermal system have analogy and analized in terms of resistance and capacitance . in the thermal modeling we will only consider the conduction and convection , except radiation because radiation may be considered if and only if the temperature of emitter is high enough that of the receiver . Following are important elements of modeling process .

- **The Energy Balance**
- **Thermal Resistance**
- **Thermal Capacitance**

(i) Energy balance : For developing a mathematical model of a thermal system we will consider the concept of an energy balance , According to it at any given condition or in a junction in a thermal System , the heat at that junction must be equal to the heat out of the junction plus heat stored in that condition .

Heat in = Heat out + Heat stored

(ii) Thermal resistance (R) : The thermal resistance in thermal modeling is represented by ' **R** ' which is similar to electrical resistance . The significance of thermal resistance , indicating the rate of heat flow through the system . Thus thermal resistance is defined as the ratio between temperature difference and heat flow rate .

$$R = \frac{\text{change in temperature difference, °C}}{\text{change in heat flow rate, kcal/sec}}$$

Consider the the circular cylinder having an uniform cross section , with heat flow in one direction (perpendicular to the cross section) . as shown in diagram ,

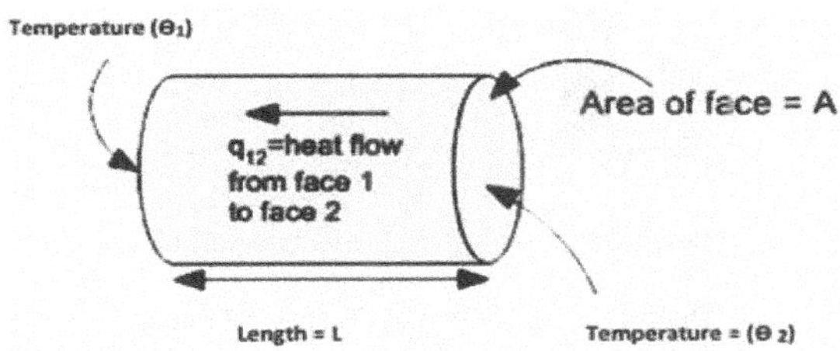

Diagram 5.66: Heat Flow Through Cylinders

Where ,

A is the area of the cylinder,

L is the length of the cylinder,

k is the thermal conductivity of the cylinder.

θ_1 = One end of the cylinder temperature ,

θ_2 = One end of the cylinder temperature ,

Then heat flow will be given as ,

$$q = (kA\Delta\theta)/\Delta L \quad \text{or} \quad \ldots\ldots\ldots\ldots\ldots \; 5.29\,f$$

For conduction or convection,

Heat flow rate, $q = K\,\Delta\theta$

On differentiating we get,

$dq = K\, d(\Delta\theta)$

$d(\Delta\theta) / dq = 1/K$

But thermal resistance , $R = d(\Delta\theta) / dq$

Thermal resistance , $R = 1 / K$ for conduction

$= 1/K = 1/HA =$ For convection

Note: For radiation it will be -- $R = 1/4Kr\,\theta^3$

Example 4 : Consider the problem in which there is a wall , one side of wall having interior temperature, Denoted by $\mathbf{T_i}$ and other side at ambient temperature $\mathbf{T_a}$. The wall has a thermal resistance of $\mathbf{R_{ia}}$. Then electrical analogy of thermal system will be as ,

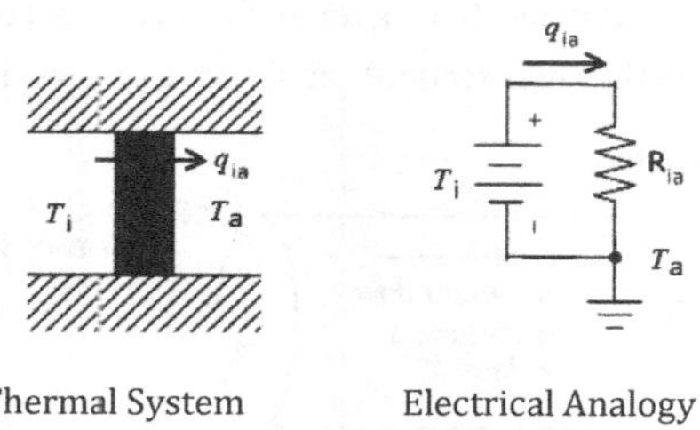

Diagram 5.67: Thermal Electrical Analogy

Where ,

T_a = Ambient temperature or ground temperature considered as zero

T_i = Interior temperature

R_{ia} = Resistance of wall

q_{ia} =Heat flow rate

(i) Thermal conductance (C) : The thermal conductance in thermal modeling is represented by **' c '** which is similar to electrical capacitance . In addition to thermal resistance, objects can also have thermal capacitance . The thermal capacitance of an object or body is a measure of how much heat it can store. Thus thermal capacitance is defined as the ratio between change in heat stored to the temperature difference .

$$C = \frac{\text{change in heat stored, kcal}}{\text{change in temperature, °C}}$$

$$C\frac{dT}{dt} = q \quad \text{.......................................} \ 5.39\ g$$

where

C = thermal capacitance (cal/°C)

dT/dt = the rate of change of temperature (°C/s)

q = heat flow (cal/s)

Thermal capacitance (C) = Mass x Specific heat of liquid

c = M c

where ,

M = mass of substance considered in , kg

c = specific heat of liquids Kcal / kg °c

Example 5 : consider a diagram in which thermal anology is represent with the help of electrical analogy .

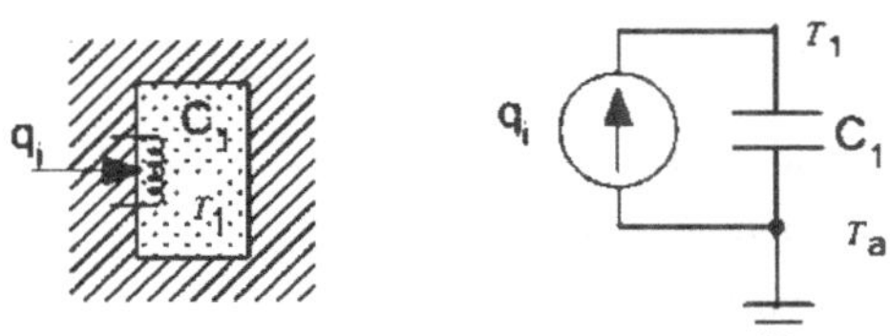

Thermal Model Electrical Analogy

Diagram 5.68: Thermal–Electrical Analogy

$$C_1\frac{dT_1}{dt} = q_1$$

Where ,

q_i = Heat flow Equivalent to voltage supply ,

c_1 = Thermal capacitace equvalent to electrical capacitance ,

T_1 = Increased fluid temperature ,

T_a = Ground temperature or ambient temperature .

Example 6 : Consider a problem in which we have an internal temperature θi , and an ambient temperature θa , with two resistances between them . consider θa = o , the schematic diagram is as shown in figure .

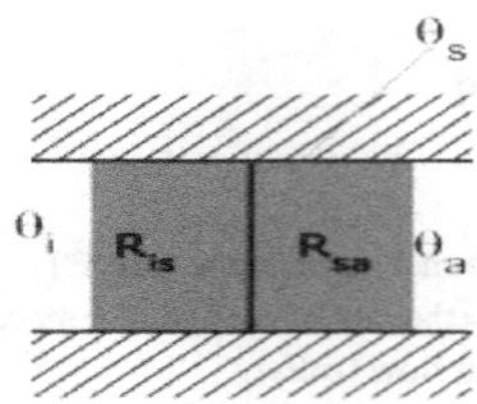

(a) Draw a Electrical Analogy Circuit for Thermal System

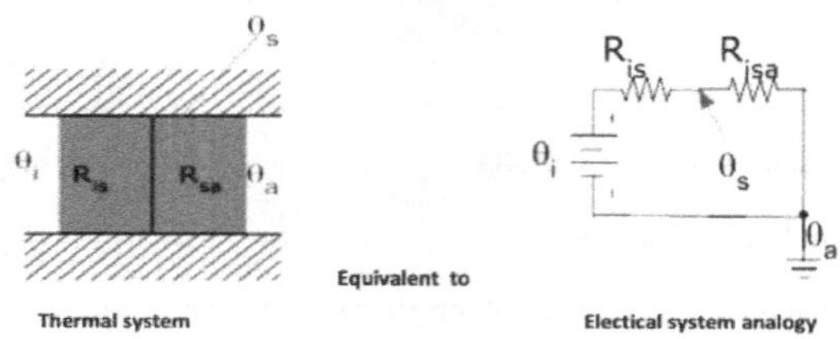

Diagram 5.69: Thermal–Electrical Analogy

Temperature input (θi) is equivalent to voltage sources .

- **θa** is ambient temperature .
- **θs** unknown temperature .
- Two resistances ($\mathbf{R}_{is}$and $\mathbf{R}_{sa}$)

(b) developed a Mathematical Model

Since problem is showing only thermal resistance and there is no heat storing so therefore we have to calculate only the value of (θs) .

According to energy balance equation

Heat in = Heat out + Heat stored

Since heat store is zero hence , Heat in=Heat out

$$\frac{\theta_i - \theta_s}{R_{is}} = \frac{\theta_s - \theta_a}{R_{sa}}$$

After solving we have got

$$= \frac{\theta_s}{R_{sa}}$$

a) Calculate the value of **θs = ?** , If **θi** , = 37°C , **θa** = 9°C , $\mathbf{R}_{is}$= 0.75°/W and $\mathbf{R}_{sa}$= 2.25°/W .

Solution

$$\frac{\theta_i - \theta_s}{R_{is}} = \frac{\theta_s}{R_{sa}}$$

$$(\theta_i - \theta_s)R_{sa} = \theta_s R_{is}$$

$$\theta_s(R_{is} + R_{sa}) = \theta_i R_{sa}$$

$$\theta_s = \theta_i \frac{R_{sa}}{R_{is} + R_{sa}}$$

$$= 28^\circ \frac{2.25}{0.75 + 2.25} = 28(0.75)$$

$$= 21^\circ$$

Example 7 : Consider a building with a single room. The resistance of the walls between the room and the ambient is R_{ra}, and the thermal capacitance of the room is C_r, the heat into the room is q_i, the temperature of the room is θr, and the external temperature is a constant, θa. Consider the ambient temperature = 0 , The schematic diagram of thermal system is given below ,

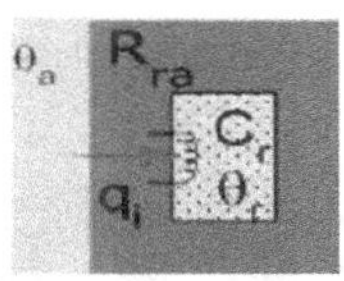

Solution

a) develop a electrical analogy models for thermal system .

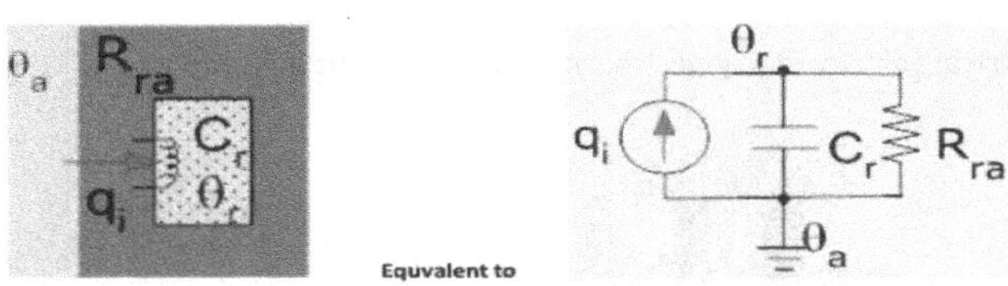

Thermal system

Electrical analogy

Diagram 5.70: Thermal–Electrical Analogy Model

- A current source q_I representing equivalancy to heat input .
- Temperature increment of room expressing the storage of energy having an equivalancy to thermal capacitance
- Heat flow from the room to atmosphere expressing equivalancy to resistance .

b) Develop a mathematical model for given problem .

Consider energy balanced equation

Heat in = Heat out + Heat stored

$$q_i = \frac{\theta_r - \theta_a}{R_{ra}} + C\frac{d\theta_r}{dt}$$

$$q_i = \frac{\theta_r}{R_{ra}} + C\frac{d\theta_r}{dt}$$

consider $\theta_a = 0$

Example 8 : Consider the problem of two adjacent room , the quantities of both room is given as ,

- The resistance of the walls room 1 and ambient is R_{1a} ,
- The resistance of the walls room 2 and ambient is R_{2a} ,
- The resistance between room 1 and room 2 is R_{12} ,
- The capacitance of rooms 1 and 2 are C_1 and C_2 and temperature θ1 and θ2 ,
- A heater in room 1 generates a heat q_{in}. The temperatur external temperature is a constant, θa.

Draw an equvalent electrical analogy for thermal system which is given below ,

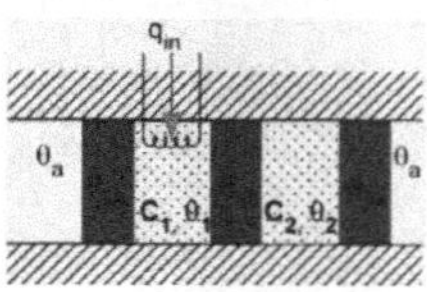

Solution

The electrical analogy for a given thermal system will be given as ,

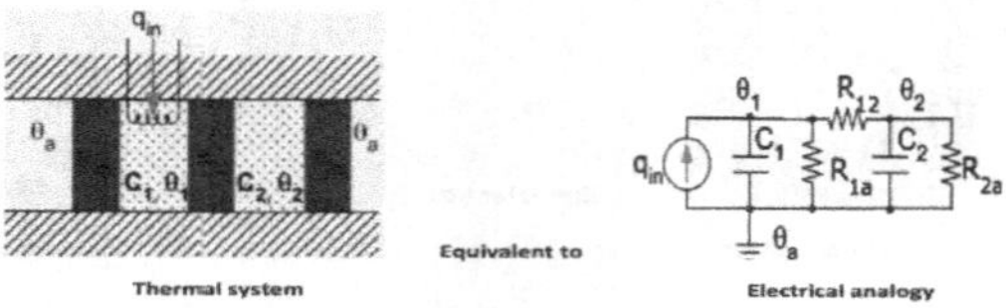

Diagram 5.71: Thermal-Electrical Analogy Model

5.33. Transfer Function of Thermal System

Consider a simple thermal system as shown in figure. assume that the tank is insulated means there is no heat transfer takes place to surroundings and no heat storage in the insulation. the liquid is kept at constant temperature with the help of stirrer, derive the transfer function of thermal system.

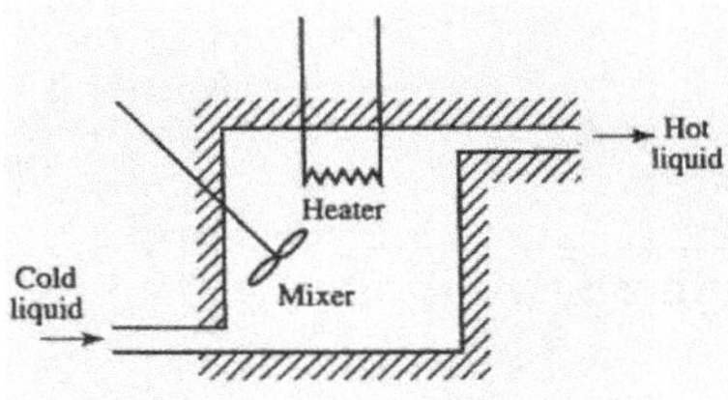

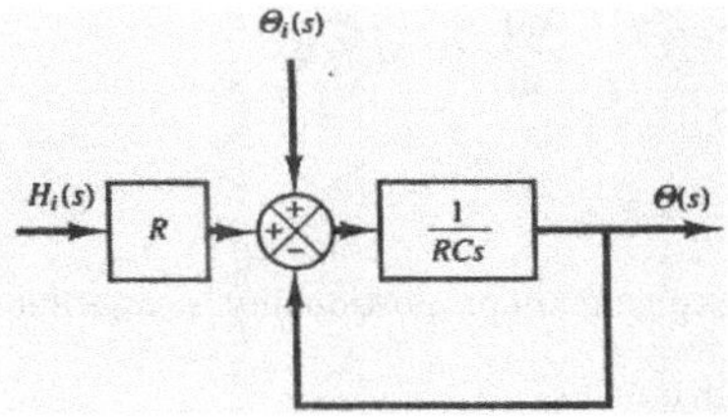

Thermal System Block Diagram

Let θ_1 = Steady state temperature of inflowing liquid, °C

θ_0 = Sandy state temperature of outflowing liquid, °C

G = Steady state liquid flow rate, Kg/sec

M = Mass of liquid in tank, Kg

c = Specific heat of liquid, Kcal/Kg °C

R = Thermal resistance, °C - sec/Kcal

C = Thermal capacitance, Kcal/°C

Q = Steady state heat input rate, Kcal/Sec

Let us assume that the temperature of in flowing liquid is kept constant. Let the heat input rate to the thermal system supplied by the heater is suddenly changed from **Q to Q + q_1**.Due to this, the heat output flow rate will gradually change from **Q to Q + q_0**.The temperature of the outflowing liquid will also be changed from **θ_0 to $\theta_0 + \theta$.**

For this system the equation for q_0, C and R are obtained as follows,

$$\left.\begin{matrix}\text{Change in output}\\ \text{heat flow rate,}\end{matrix}\right\} q_0 = \begin{matrix}\text{Liquid flow}\\ \text{rate, G}\end{matrix} \times \begin{matrix}\text{Specific heat of}\\ \text{liquid, c}\end{matrix} \times \begin{matrix}\text{Change in}\\ \text{temperature, } \theta\end{matrix} \quad = Gc\theta \qquad \ldots\ldots\ldots\ldots 1$$

Thermal capacitance (C) = Mass (M) x Specific heat of liquid (c) = **M c**

$$\text{Thermal resistance, R} = \frac{\text{Change in temperature, } \theta}{\text{Change in heat flow rate, } q_0} = \frac{\theta}{q_0} \qquad \ldots\ldots\ldots\ldots 2$$

On substituting for q_o from equation (1) in equation (2) we get,

$$R = \frac{\theta}{Gc\theta} = \frac{1}{Gc}$$

In this thermal system, rate of change of temperature is directly proportional to change in heat input rate.

$$\therefore \frac{d\theta}{dt} \;\alpha\; q_i - q_0 \qquad \text{.......................... 3}$$

$$\therefore C\frac{d\theta}{dt} = q_i - q_0 \qquad \text{.......................... 4}$$

The constant of proportionality is capacitance **C** of the system

From equation (2) ,

$R = \theta / q_0$

$q_0 = \theta / R$

On substituting for q_o from equation (3) in equation (4) we get,

$$C\frac{d\theta}{dt} = q_i - \frac{\theta}{R} \quad \Rightarrow \quad C\frac{d\theta}{dt} = \frac{Rq_i - \theta}{R} \quad \Rightarrow \quad RC\frac{d\theta}{dt} = Rq_i - \theta$$

$$\therefore RC\frac{d\theta}{dt} + \theta = Rq_i \qquad \text{.................................. 5}$$

Let , L {θ} = θ(s); L{dθ/dt} = sθ(s) ; L{q_1} = Q_1(s), On taking Laplace transform of equation (5)

$$RC\, s\, \theta(s) + \theta(s) = R\, Q_1(s)$$

$$\theta(s)\, [sRC + 1] = R\, Q_1(s)$$

Since according to definition of the transfer function it is a ratio of **output to input** .if model is represented by the "**s**" then the transfer function is given as ,

G(s) = (Output / Input)

θ (s) / Q_1 (s) is the required transfer function of thermal system.

$$\therefore \frac{\theta(s)}{Q_i(s)} = \frac{R}{sRC+1} = \frac{R}{RC\left(s+\frac{1}{RC}\right)} = \frac{\frac{1}{C}}{s+\frac{1}{RC}} \qquad \text{.......................... 5.29 h}$$

5.34. Laplace Transform

Laplace transforms convert differential equations into algebraic equations. They are related to frequency response.

$$L\{x(t)\} = X(s) = \int_0^\infty x(t)e^{-st}dt$$

$$L\{x(t)\} = X(s) = \int_0^\infty x(t)e^{-st}dt \qquad \text{.. 5.32 a}$$

No.	Function	Time-domain $x(t)= \mathcal{L}^{-1}\{X(s)\}$	Laplace domain $X(s)= \mathcal{L}\{x(t)\}$
1	Delay	$\delta(t-\tau)$	$e^{-\tau s}$
2	Unit impulse	$\delta(t)$	1
3	Unit step	u(t)	$\frac{1}{s}$
4	Ramp	t	$\frac{1}{s^2}$
5	Exponential decay	$e^{-\alpha t}$	$\frac{1}{s+\alpha}$
6	Exponential approach	$\left(1-e^{-\alpha t}\right)$	$\frac{\alpha}{s(s+\alpha)}$
7	Sine	$\sin \omega t$	$\frac{\omega}{s^2+\omega^2}$
8	Cosine	$\cos \omega t$	$\frac{s}{s^2+\omega^2}$
9	Hyperbolic sine	$\sinh \alpha t$	$\frac{\alpha}{s^2-\alpha^2}$
10	Hyperbolic cosine	$\cosh \alpha t$	$\frac{s}{s^2-\alpha^2}$
11	Exponentially decaying sine wave	$e^{-\alpha t} \sin \omega t$	$\frac{\omega}{(s+\alpha)^2+\omega^2}$
12	Exponentially decaying cosine wave	$e^{-\alpha t} \cos \omega t$	$\frac{s+\alpha}{(s+\alpha)^2+\omega^2}$

5.35. Signal Flow Graph for Control System

The representation of Signal Flow Graph is invented by S. J. Mason for simplifying a complex control system . In Control System any kind of problem is represented with the help of transfer function, summing symbol and take off point . But Signal Flow Graph of Control System replaced all these elements by nodes which relates to concern branches. Each node represents a system variable, and each branch acts as a signal multiplier. The direction of flow indicated by the arrow and transfer function is considered as transmittance. Then the related elements are to be converted into linear algebraic equations.

Basic equation of signal flow graph :

$$y_j = \sum_{k=1}^{N} a_{kj} y_k \quad j = 1, 2, \ldots, N \qquad \text{.......... 5.33 a}$$

$$j\text{th effect} = \sum_{k=1}^{N} (\text{gain from } k \text{ to } j) \times (k\text{th cause}) \qquad \text{.......... 5.33 b}$$

$$\text{Output} = \sum (\text{gain}) \times (\text{input})$$

$$Y_j(s) = \sum_{k=1}^{N} G_{kj}(s) Y_k(s) \quad j = 1, 2, \ldots, N \qquad \text{.......... 5.33 b}$$

Where ,

y_j = Output of control system

a_k = Gain of control system

y_k = Input to control system

the above equation can easily understand by the following example

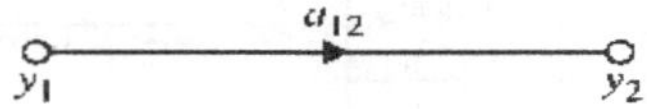

Signal flow graph of $y_2 = a_{12}y_1$.

$$y_2 = a_{12}y_1$$

,

$$y_1 = \frac{1}{a_{12}} y_2$$

Block diagram and equivalent signal flow diagram

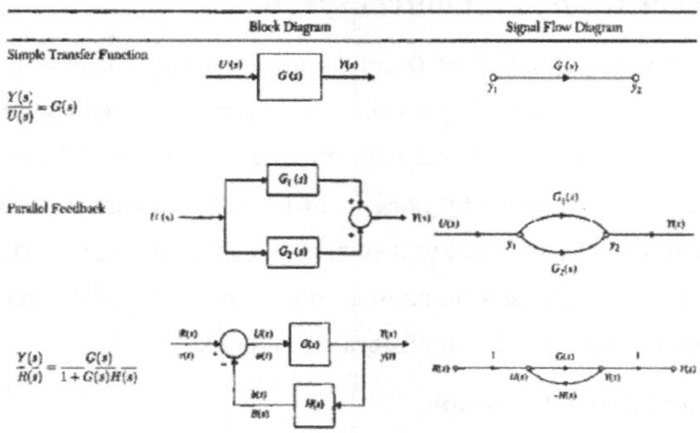

Here the short definition of the each elements will be as

(i) Transfer function :- The transfer function is defined as , the ratio of Laplace transform of the output (response function) to the Laplace transform of the input (drive function) with consideration that the value of all the initial variable are zero.

Transferfunction = $G(s)$ = (Output / Input)

$G(s) = Y(s) / X(s)$

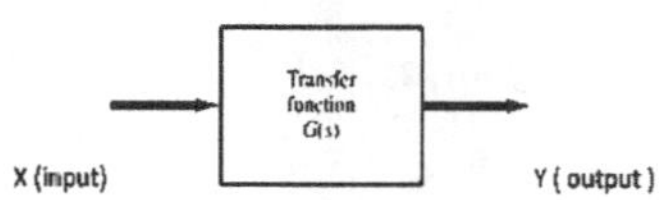

Where ,

X is the input

Y is the output

Example : Find system transfer function between capacitance and voltage for the following RLC circuit as shown in Figure ,

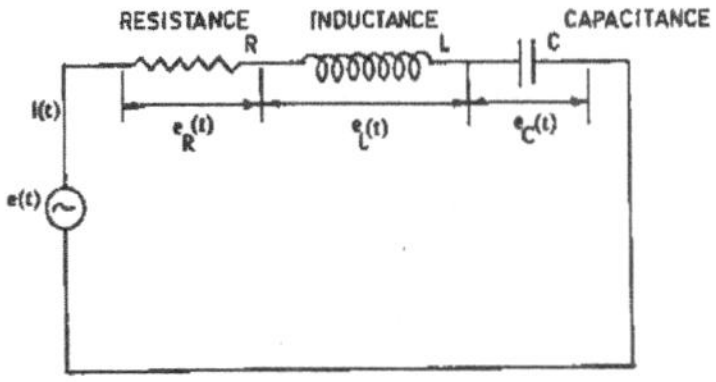

Diagram 5.72: R-L-C Circuit

In this problem we have three elements whose value can be calculated as ,

Voltage drop across the Resistance = e_R (t) = I R

Voltage across the Inductance = e_L (t) = L (di / dt)

Voltage across the capacitance = e_C (t) = (1/c) ∫ i dt

Hence ,

$$\text{Total voltage, } e(t) = iR + L\frac{di}{dt} + \frac{1}{C}\int idt \quad \text{.......................................} \ \mathbf{5.33\ d}$$

Laplace transformation will be given as

$$E(s) = I(s)\left(R + Ls + \frac{1}{Cs}\right)$$

Transfer function between capacitance voltage and source voltage

$$\frac{E_C(s)}{E(s)} = \frac{1}{Cs\left(R + Ls + \frac{1}{Cs}\right)} = \frac{\omega_n^2}{\left(s^2 + 2\zeta\omega_n s + \omega_n^2\right)}$$

$$\text{where } \omega_n = \frac{1}{\sqrt{LC}} \text{ and } \zeta = \frac{R}{2\sqrt{\frac{L}{C}}} \quad \text{...} \ \mathbf{5.33\ d}$$

ω_n & ζ Indicating Natural Frequency & Damping ratio

(ii) Error in control system :- In open loop control system , the compensation of error could not be possible but in case of Feedback or closed-loop control system it can be minimized .This can be seen by using the final value theorem to estimate the steady state error of the open and closed loop control systems as shown in figure ,

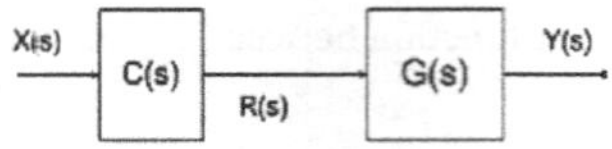

Consider the open loop control system, transfer function will be ,

$$H_{open} = C(s)G(s)$$

The steady state error is the values of the error signal

$$e(t) = x(t) - y(t)$$

Apply final value theorem we have got ,

$$e_{ss} = \lim_{s\to 0} sE(s)$$

$$E(s) = X(s) - Y(s) = (1 - C(s)G(s))X(s)$$

Then for a unit step $x(t)$

$$e_{ss} = \lim_{s\to 0} s(1 - C(s)G(s))\frac{1}{s} = 1 - C(0)G(0)$$

Consider the close loop control system , the transfer function will be ,

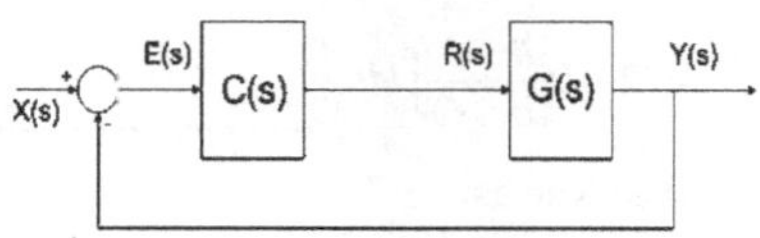

$$H_{closed} = \frac{C(s)G(s)}{1 + C(s)G(s)}$$

$$E(s) = X(s) - \frac{C(s)G(s)}{1 + C(s)G(s)}X(s) = \frac{1}{1 + C(s)G(s)}X(s)$$

Then , error

$$e_{ss} = \lim_{s\to 0} s\frac{1}{1 + C(s)G(s)}\frac{1}{s} = \frac{1}{1 + C(0)G(0)} \quad \text{.......... 5.33 e}$$

(iii) System Sensitivity :- The term system Sensitivity is defined as the ratio of the relative change in system transfer function **H(s)** to the relative change in the plant transfer function **G(s)** for a small change. The system sensitivity is represented by **"s "** ,

$$S = \left|\frac{\partial H/H}{\partial G/G}\right| = \left|\frac{\partial H}{\partial G}\frac{G}{H}\right| \quad \text{.......... 5.33 f}$$

Sensitivity for open loop control system will be given as ,

$$S = C(s)\frac{G(s)}{C(s)G(s)} = 1 \quad \text{.......... 5.33 g}$$

Sensitivity for close loop control system will be given as ,

$$S = \frac{C(s)}{(1+C(s)G(s))^2} \frac{G(s)}{C(s)G(s)/(1+C(s)G(s))} = \frac{1}{1+C(s)G(s)} \quad \text{...................... .... 5.33 h}$$

(iv) Block Diagrams : A block diagram of a system is a pictorial representation of the functions performed by each component and of the flow signals.

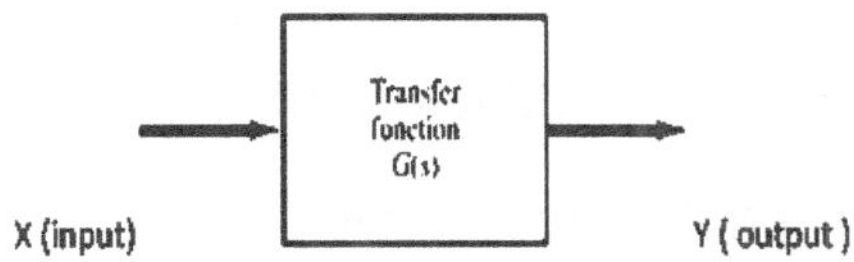

- **Summing Point :** A circle with a cross is representing to the summing operation symbol . The plus or minus sign at each arrowhead indicates whether that signal is to be added or subtracted.

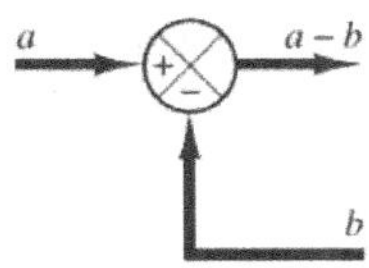

- **Branch Point :** A branch point is a point from which the signal from a block relates to next blocks or summing points.

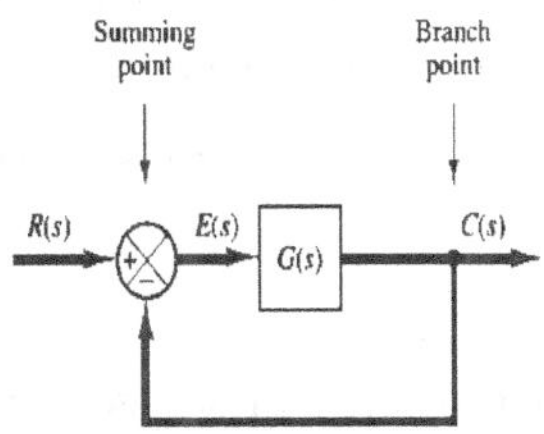

(v) Definition of basic terms associated with signal flow graph

- **Node :** The point at which branches meet is known as node. If a node has only outgoing branches, then it is known as input node. While if the same has only incoming branches, it is known as output node .
- **Forward Path :** The path from input to output without repeating any node.
- **Loop :** The path which originates and terminates on the same node with no repetition of other nodes.
- **Path gain :** The product of the branch gains along the path .
- **Loop gain :** The product of the branch gain of the branches coming in the loop.
- **Non Touching Loop :** The loops which have no common nodes, branches and paths

For example :- Consider the following diagram

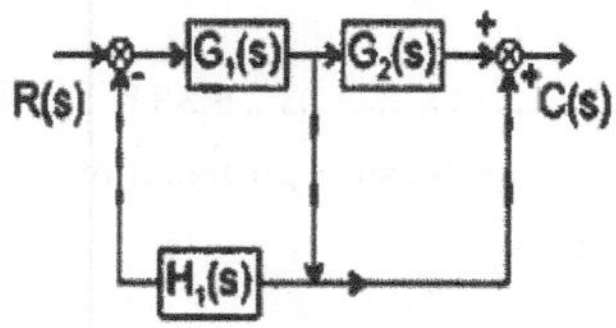

Block Diagram

Equivalent to

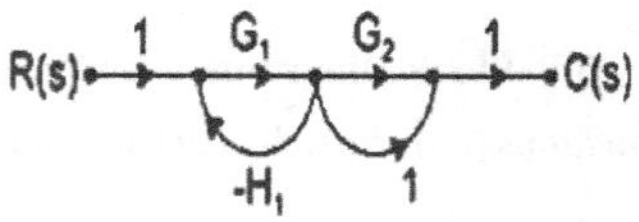

Signal Flow Graph

5.36. Construction of Signal Flow Graph

For construction of single flow graph we will have to implements , the following procedure in which variables are behave like a node on the other hand and each block considered as a branch.

- Construct the input to output nodes from left to right.
- Connect the all nodes by appropriate branches and construct a linear flow graph parallel to the block diagram.
- If the desired output node has outgoing branches, add a dummy node and a unity gain branch
- Applied mason's gain formula for finding transfer function of the system.

5.36.1. Governing Rules for Signal Flow Graph

Following rules are very important that has been generally applied for evaluating the signal flow graph .

- **Addition rules : -** According to it ,

  $$\Sigma(\textbf{All variables})_{\textbf{nodes}} = \Sigma\,(\textbf{ All signal entering })_{\textbf{nodes}}$$

- **Transmission rule :-** According to it , The value of the variable designated by a node is transmitted on every branch leaving the node.
- **Multiplication rule :-** A cascaded connection of **n-1** branches with transmission functions can be replaced by a single branch with new transmission function equal to the product of the old ones.

5.36.2. Mason's Gain Formula

The basic concept of this formula is to be used for calculating the transfer function of the a signal flow graph and is defined as , the net gain achieved between input and output nodes indicated by overall gain of the system or signal flow graphs . which can be given as ,

$$M = \frac{Y}{X} = \frac{\sum_{j=1}^{N} M_j \Delta_j}{\Delta} \quad \text{.. 5.34 a}$$

Where ,

M = Transfer Function or Gain of the System

X = Input Nodes

Y = Output nodes

N = Total number of forward paths between ' X ' and ' Y '

Δ = 1 – (Sum of all individual loop gains) + (Sum of product of gains of two non-touching loops) – (Sum of product of gains of combinations of three non-touching loops) + ...

Δ_j = 1 – (Sum of all individual loop gains not touching find determinant j_{th} forward path) + (Sum of product of gains of two non-touching loops not touching the j_{th} forward path) – ....

For example : Consider the following signal flow graphs , find , Forward paths and their corresponding gains , Find the loops and their corresponding gains , Find determinant Δ , Find determinant Δ_j and Find out transfer function .

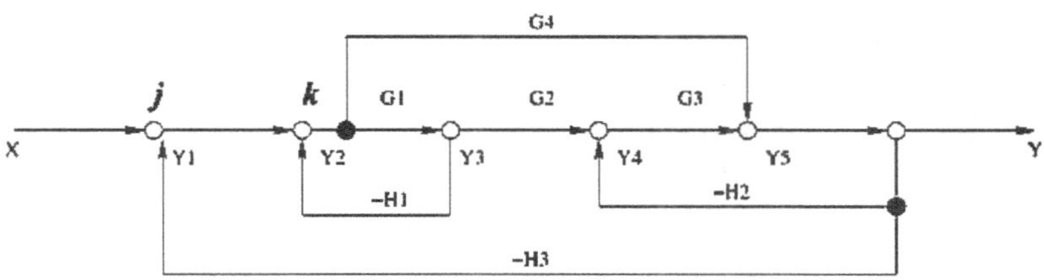

Signal Flow Graphs

Solution

- **There are two forward paths is representing in the diagram**

$$M_1 = G_1 G_2 G_3 \text{ and } M_2 = G_4$$

- **Find the loops and their corresponding gains**

Since there are four loops exists, the corresponding gain will be calculated as ,

(i) $Loop_1 = - G_1 H_1$

(ii) $Loop_2 = -G_3 H_2$

(iii) $Loop_3 = - G_1 G_2 G_3 H_3$

(iv) $Loop_4 = - G_4 H_3$

- **Find determinant Δ j**
  (i) When we eliminate the path $M_1 = G_1 G_2 G_3$ then, there is no complete loops is remain, hence $\Delta_1 = 1$.
  (ii) When we eliminate the path $M_2 = G_4$ then, there is no complete loops is remaining, hence $\Delta_2 = 1$.
- **Find determinant Δ**
  (i) We have two non-touching loops these are p_1 & p_2

Σ non-touching loops gains = $(-G_1H_1)(-G_3H_2)$ then ,

Δ = 1 – Σ (all individual loop gains) + Σ (product of gains of two non-touching loops)

$$= 1 - (-G_1H_1 - G_3H_2 - G_1G_2G_3H_3 - G_4H_3) + (-G_1H_1)(-G_3H_2)$$
$$= 1 + G_1H_1 + G_3H_2 + G_1G_2G_3H_3 + G_4H_3 + G_1G_3H_1H_2$$

- **Find out transfer function**

Apply meason's formula for calculating the transfer function of signal flow graph.

$$M = \frac{Y}{X} = \frac{\sum_{j=1}^{N} M_j\Delta_j}{\Delta} = \frac{G_1G_2G_3 + G_4}{1 + G_1H_1 + G_3H_2 + G_1G_2G_3H_3 + G_4H_3 + G_1G_3H_1H_2}$$

Example 1: Constructing a signal flow Graph for the following set of linear algebraic equations.

$$\begin{aligned} y_2 &= a_{12}y_1 + a_{32}y_3 \\ y_3 &= a_{23}y_2 + a_{43}y_4 \\ y_4 &= a_{24}y_2 + a_{34}y_3 + a_{44}y_4 \\ y_5 &= a_{25}y_2 + a_{45}y_4 \end{aligned} \qquad \text{.............................. a}$$

Solution

the solution of following equation is given below .

- **Step – 1**

Signal flow diagram for equation $y_2 = a_{12} y_1 + a_{32} y_3$

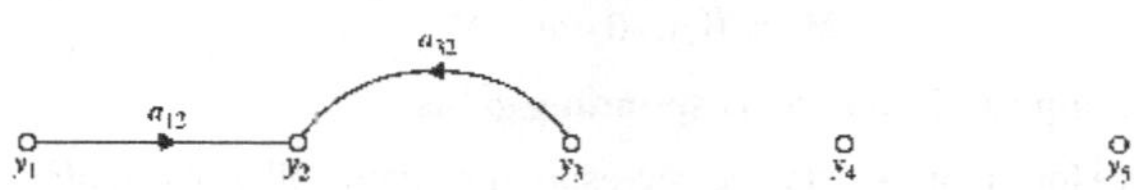

- **Step – 2**

Signal flow diagram for equation

$y_2 = a_{12} y_1 + a_{32} y_3$,

$y_3 = a_{23} y_2 + a_{43} y_4$

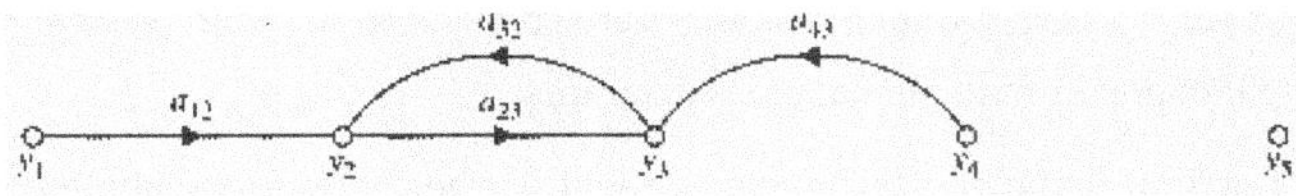

- **Step -3**

Signal flow diagram for equation

$y_2 = a_{12} y_1 + a_{32} y_3$,

$y_3 = a_{23} y_2 + a_{43} y_4$,

$y_4 = a_{24} y_2 + a_{34} y_3 + a_{44} y_4$

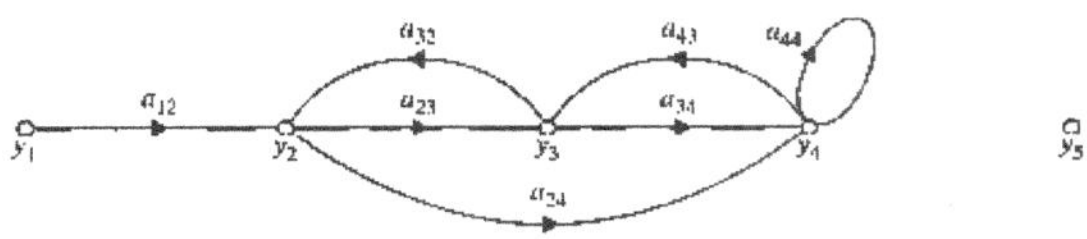

- **Step - 4 Final signal Flow Diagram**

Signal flow diagram for equation

$y_2 = a_{12} y_1 + a_{32} y_3$,

$y_3 = a_{23} y_2 + a_{43} y_4$,

$y_4 = a_{24} y_2 + a_{34} y_3 + a_{44} y_4$,

$y_5 = a_{25} y_2 + a_{45} y_4$.

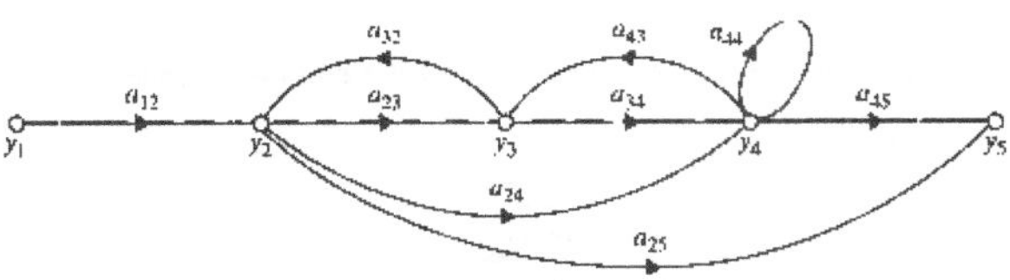

5.37. Block Diagram Reduction technique

In general we know that the control system consists of so many components and these components are connected together to perform a their specific function and represented with the help of individual block, hence " **A block diagram of a system is a pictorial representation of the functions performed by each component of the system** "

A control system has not been a single block with a definite input, an output and a transfer function but also it has composed of multiple subsystems. when multiple subsystems are enclosed in control system then it becomes very complex & complicated. such kind of condition will always be unfavorable to determine the overall transfer function of control system. Hence evaluation of such a complex system becomes very complex so therefore system performance required simplification or reduction of block diagrams.

In Block Diagram Reduction technique, we will consider how to combine the individual blocks so that we can represent a whole system as a single block , and therefore a single transfer function can be calculated. Here we will consider following elements that always be combined with the sub system, these are **signals , system , summing junctions** and **pickoff points.**

(i) **Signals** .... signals

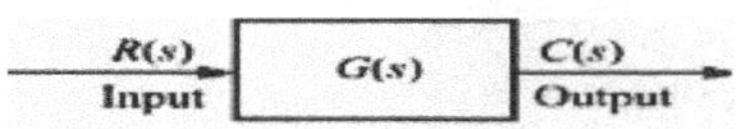

(ii) .... system

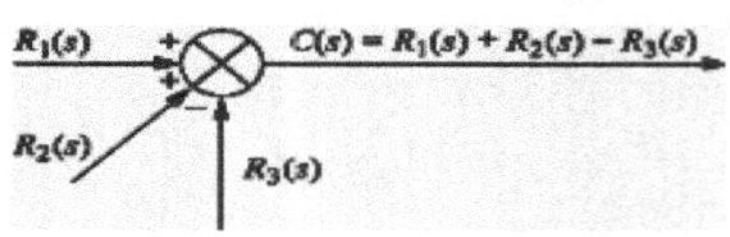

(iii) .... summing junctions

Output signals = Algebraic sum of the all Input Signals

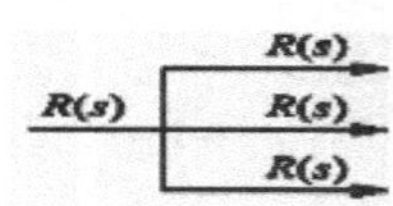

(iv) .... pickoff points

Distribution of input signal on multiple path

5.37.1. Reduction Process of Block Diagram

Following steps must be taken into consideration for Reduction process

Step. 1 : Reduce the Cascade Blocks of the system ,

Step. 2 : Reduce the parallel Blocks of the system ,

Step. 3 : Reduce the internal Feedback loops of control system ,

Step. 4 : If necessary shift take off points towards right and summing points towards left.

Step. 5 : Repeat steps the process up to single block achievements .

Step. 6 : Find transfer function of the overall system using the formula [**C(S) / R (S)**]

(i) Reduction of cascade block : When multiple sub systems are connected in series , then output of one subsystem becomes the input to the next , similarly for further block . thus, these sub systems are said to be cascade block. the schematic diagram of cascade block and resultant transfer function will be given as,

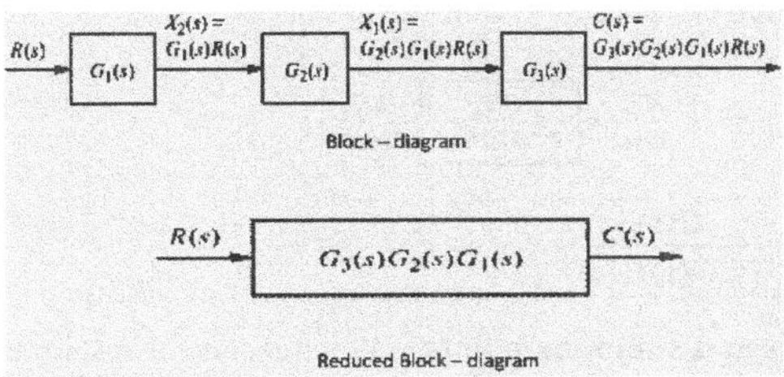

.... **5.3 a**

Where **R (s)** = signal input , **C (s)** = signal output

(ii) Reduction of parallel block : A parallel block system have a common input and the equivalent transfer function Ge(s) is the algebraic sum of all the transfer function value of individual subsystems ' transfer functions . Ge (s) = G_1 (s) ± G_2(s) ± G_3(s)

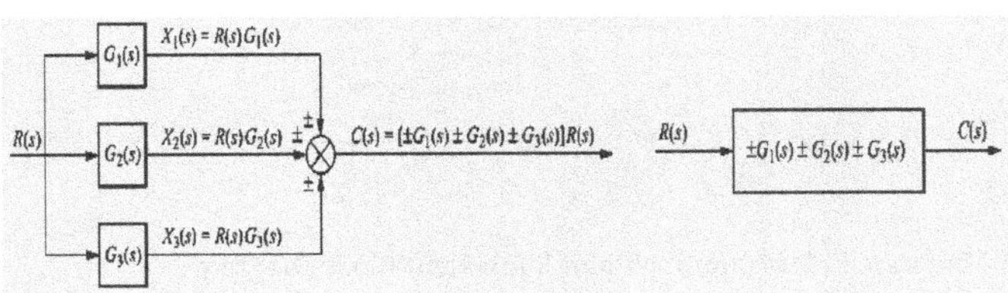

.... **5.3 b**

Diagram 5.73: Parallel Block–Reduction

Feedback loop Form :- For example consider the feedback system as shown in figure

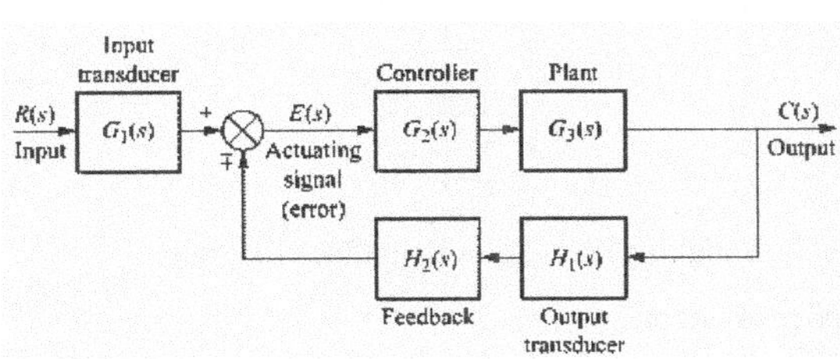

.... **5.3c**

The simplify cascade block model is reduced as ,

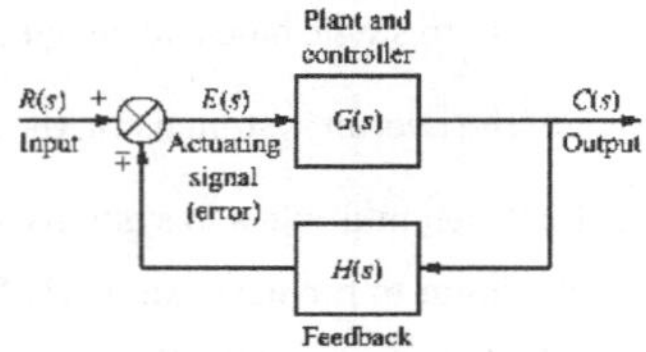

The equivalent transfer function of such a reduced system will be **E(s) = R(s) ± C(s) H(s)**.

The overall transfer function of closed loop control system is given as ,

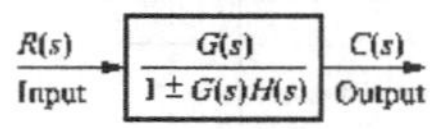

$$G_e(s) = \frac{G(s)}{1 \pm G(s)H(s)} \qquad \text{..................................... 5.3d}$$

(iii) Shift take off points and summing points :- When a control system is complex and complicated then shifting of take-off point and summing point or block helps in reducing the control block system . consider the following figure that will representing such a situation,

Case – I Summing junction

- Summing junction at left and equivalent block diagram to right

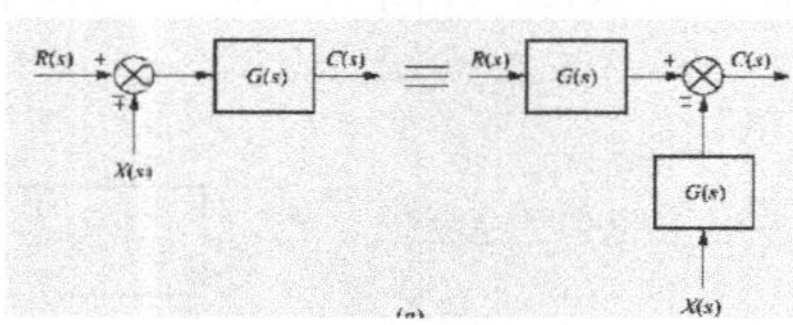

Diagram 5.74: Summing Point Equivalent Block Diagram

- Summing junction at right and equivalent block diagram to left

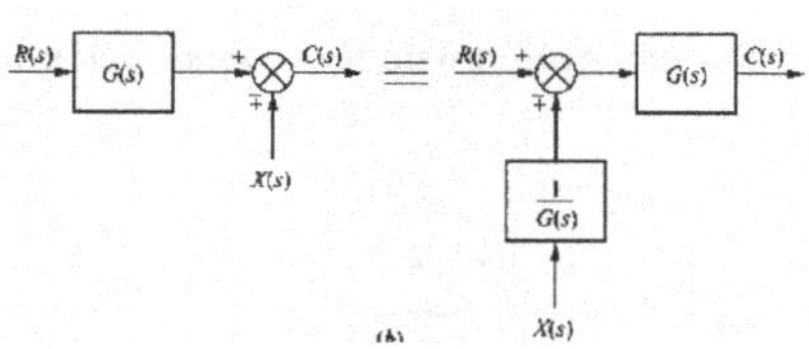

Diagram 5.75: Summing Point Equivalent Block Diagram

Case – II Pickoff point in control system

- Pickoff point at left and equivalent block diagram to right

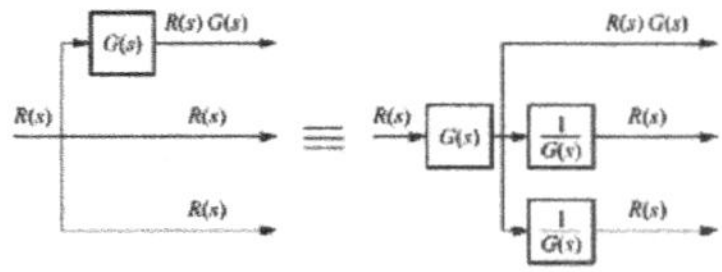

Diagram 5.76: Pickoff Point Equivalent Block Diagram

- Pickoff point at right and equivalent block diagram to left

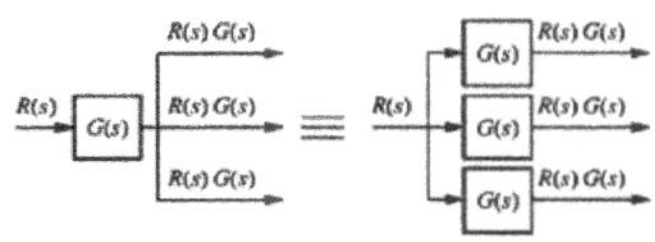

Diagram 5.77: Pickoff Point Equivalent Block Diagram

(iv) Signal flow graph : Signal-flow graphs are also a block diagram reduction technique in which Signal-flow graphs are an alternative to block diagrams. Unlike block diagrams, which blocks , signals , summing junctions , and pickoff points are replaced by branches and nodes , as shown in figure ,

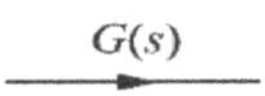

...A line with an arrow (direction of signal-flow through the system) and

G(s) transfer function

.... Nodes

5.38. Block – Diagram Reduction Equation

Block reduction is tabulated as

	Manipulation	Original Block Diagram	Equivalent Block Diagram	Equation
1	Combining Blocks in Cascade	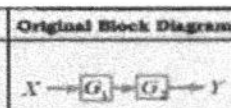	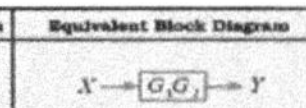	$Y=(G_1G_2).X$
2	Combining Blocks in Parallel; or Eliminating a Forward Loop	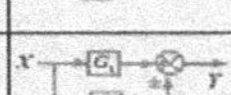	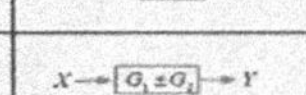	$Y=(G_1 \pm G_2).X$
3	Moving a pickoff point behind a block	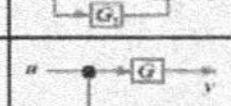	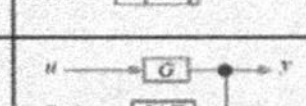	$y = Gu$ $u = \frac{1}{G}y$
4	Moving a pickoff point ahead of a block	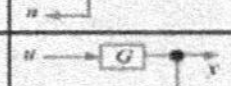	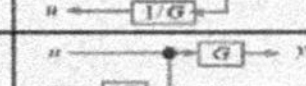	$y = Gu$
5	Moving a summing point behind a block	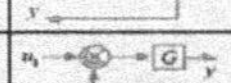	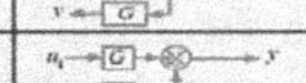	$e_2 = G(u_1 - u_2)$
6	Moving a summing point ahead of a block	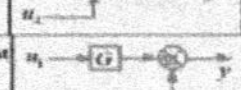	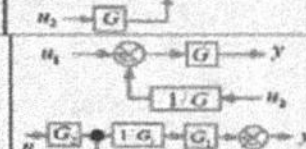	$y = Gu_1 - u_2$ $y=(G_1 - G_2)$

Example 1 : Using block diagram reduction technique find closed loop transfer function **C(s) / R(s)** shown in figure below .

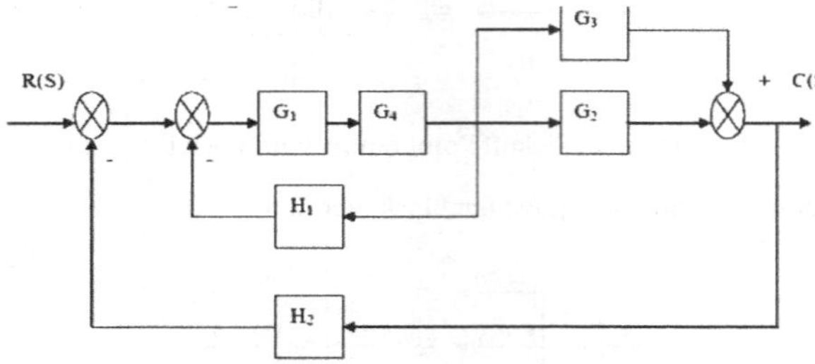

Step 1 : Combine the blocks G_1 &G_2 which are in cascade and combine the blocks G_2& G_3 which are in parallel .

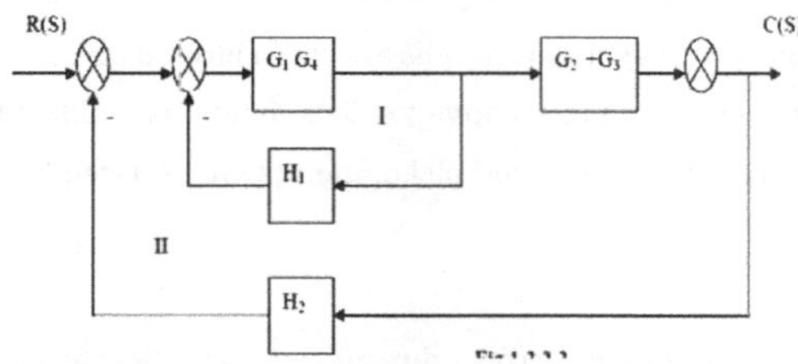

Step 2: Eliminate feedback loopI and combine the blocks (G_1G_4 / 1 + $G_1G_4H_1$) & (G_2 + G_3) which are in parallel as shown ,

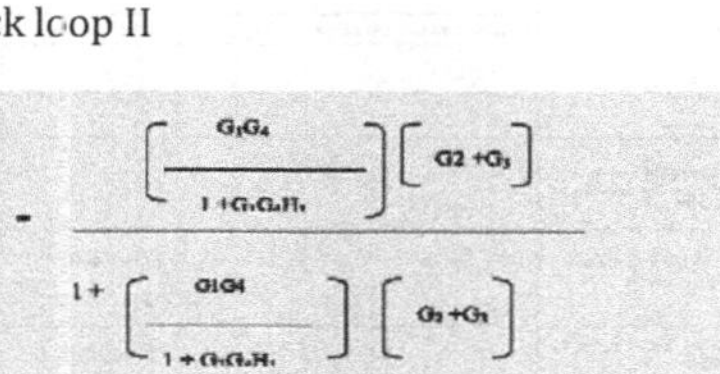

Step 3 : eliminate feedback loop II

$$\frac{C(s)}{R(s)} = \frac{\left[\dfrac{G_1G_4}{1+G_1G_4H_1}\right]\left[G_2+G_3\right]}{1+\left[\dfrac{G_1G_4}{1+G_1G_4H_1}\right]\left[G_2+G_3\right]}$$

$$= \frac{\left[G_1G_4\right]\left[G_2+G_3\right]}{\left[1+G_1G_4H_1\right]+\left[G_1G_4\left(G_2+G_3\right)\right]}$$

Example 2 : Determine the transfer function C(s) / R(s) of control system by block diagram reduction method.

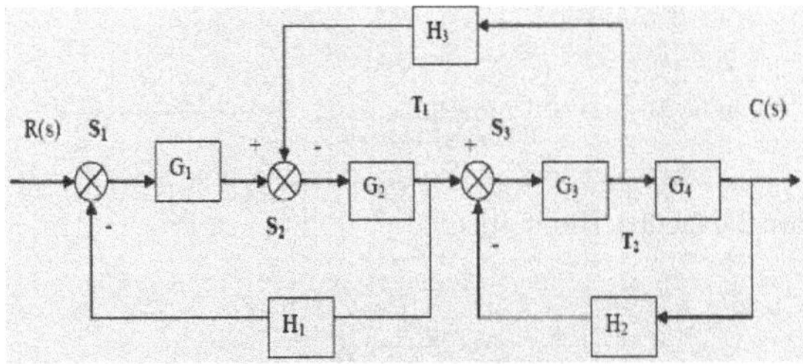

Step 1 : Shifting the summing point S_2 before the block G_1 and shifting the take off point T_2 after the block G_4.

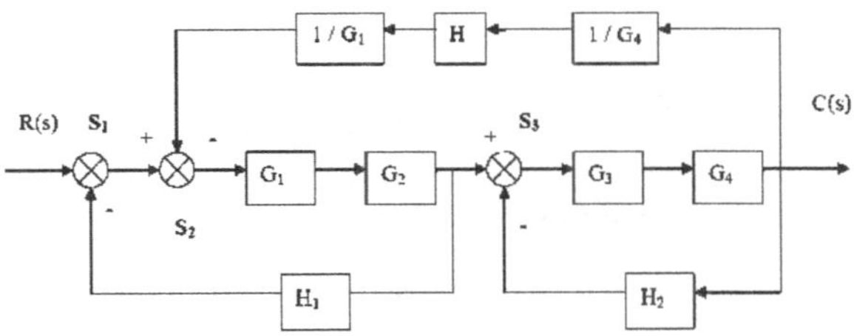

Step : 2 Exchange the summing points and take off points using associative law and combining the series blocks we get .

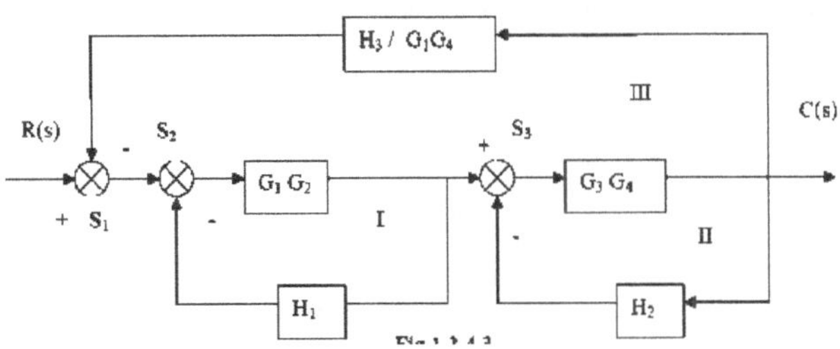

Step : 3 Eliminating inner feedback loops I, II

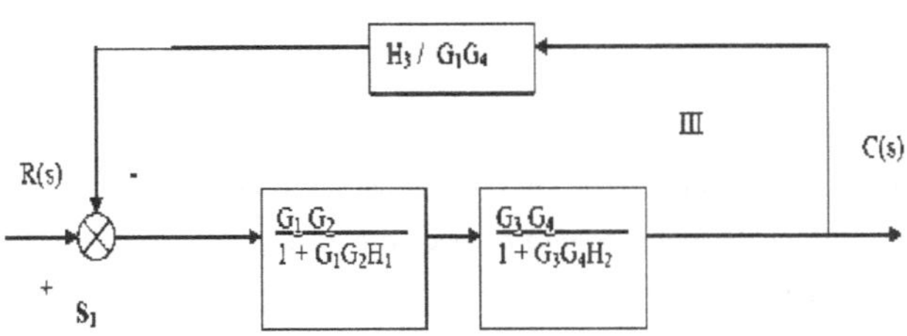

Step : 4 Combine the blocks in series

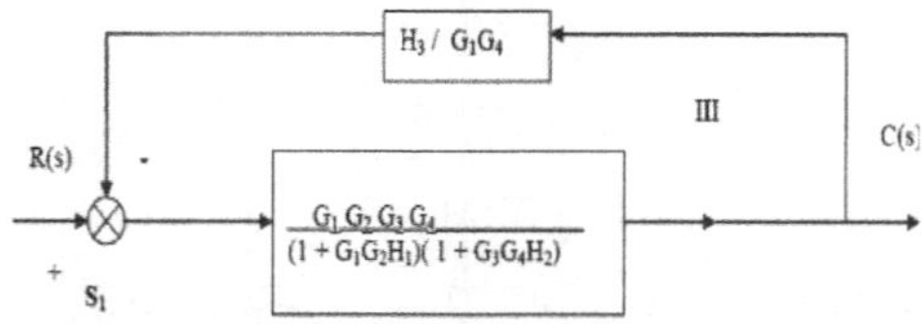

Step 5 : Eliminate the feedback loop III

$$\frac{C(s)}{R(s)} = \frac{\left\{\dfrac{G_1 G_2 G_3 G_4}{(1+G_1G_2H_1)(1+G_3G_4H_2)}\right\}}{1+\left\{\dfrac{G_1 G_2 G_3 G_4}{(1+G_1G_2H_1)(1+G_3G_4H_2)}\right\}\left\{\dfrac{H_3}{G_1G_4}\right\}}$$

$$\frac{C(s)}{R(s)} = \frac{G_1G_2G_3G_4}{(1+G_1G_2H_1)(1+G_3G_4H_2)+G_2G_3H_3}$$

5.39. Transient Response Specifications

Transient response specifications is very basic and important parts of control design and analysis because it 's not only define the efficiency of control system but also indicating how quickly and accurately a system responds to changing inputs suddenly .

Following are some important terms that must be considered when ever impulse response function analyses is performed .

(i) Rise time : The rise time is defined as the time that is taken by the transient response to move from10% to 90% of the steady state response.

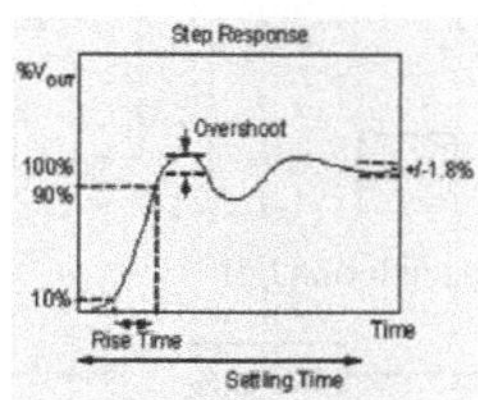

Diagram 5.78: Rise Time

(ii) Maximum Overshoot : The maximum overshoot is the percentage by which the maximum value ofthe transient response exceeds the steady state value , both the process is given below in the diagram.

$$M_p = \frac{|y_{max} - y_{ss}|}{|y_{ss}|} \cdot 100\ \% \qquad \text{.......... 5.3 a}$$

y_{max} = Upper peak value of overshoot

y_{ss} = Lower peak value of overshoot

(iii) Peak Time : The peak time is defined as the time at which the maximum overshoot occurs or it is a maximum peak value of control system .

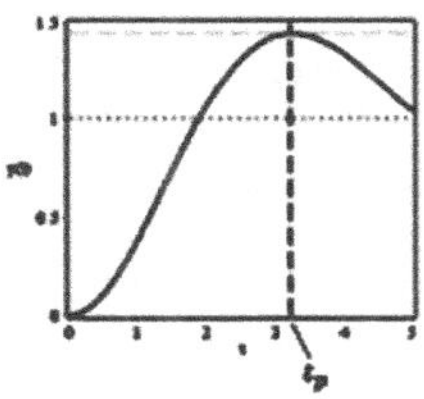

Diagram 5.79: Peak Time

(iv) Settling Time : The settling time is defined as the time after which the output is within a specified steady state value . it is represented by horizontal dotted line at reading 1 vertically , in the above diagram.

(v) Settling Min time : The minimum value represented by the dotted line at which the response has been risen.

(vi) Settling Min : Minimum value of the response from which it has been fallen .

Note – the value may be tolerated from steady state response will be ± 2 % to ± 5 %

5.40. Characteristics of First Order Systems

In many engineering application, we have so many example which can be represented by a simple model . like the fluid flow from a tank, and the cooling of a cup of coffee . for such kind of problem, the first - order differential equation, can be written in the standard form as follows ,

$$\tau \frac{dy}{dt} + y(t) = f(t) \qquad \text{.......... 5.3 b}$$

Where ,

y = the measured and controlled output variable

$f(t)$ = the input function or the forcing function for ordinary differential equation ,

T = time constant of the process .

Consider the equation of system indicating the first-order state equation, which is given as

$$\dot{x} = ax + bu$$
$$y = cx + du. \quad \text{.. 5.3 c}$$

$$\frac{dy}{dt} - ay = bu, \quad \text{.. 5.4 d}$$

Divided equation **5.4 d** by , **-a** then we have got ,

$$-\frac{1}{a}\frac{dy}{dt} + y(t) = -\frac{b}{a}u(t) \quad \text{.................................. 5.5 e}$$

Where T = - (1 / a) known as time constant .

Time Constant :- It is denoted by T = - (1 / a) whose function is to measure the response speed . if the value of time constant is large then system response takes a longer time to rise. Suppose for any system the Laplace domain characteristic of the system is represented by following equation ,

$$G(s) = \frac{Y(s)}{U(s)} = \frac{K \cdot a}{s + a} \quad \text{.............................. 5.5 f}$$

Where **K is constant = D C Gain**

DC Gain :- It is defined as , the ratio between the magnitude of the steady-state output of the system to the input.

For a unit-step input

$$= \mathcal{L}^{-1}[K] = K \quad \text{...................................... 5.6 g}$$

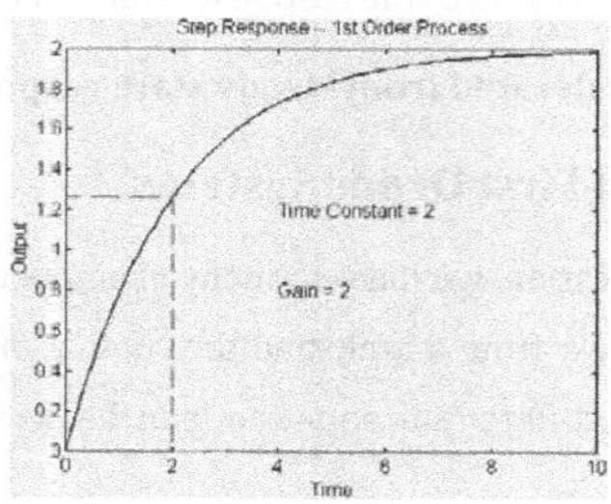

Step response of first order system

The above curve shows the "step response "of a 1st order system which has same shape equivalency and applicable for all type first order system. the above curve representing time constant 2.0 and "Unit step response" means that the forcing function has magnitude 1. When the system reach the steady state condition, the system response attain a constant value. Thus we have an equation,

C / (1 + k T) .... **5.7 h**

Where ,

C = magnitude of the forcing function

T = Time constant

K = D C Gain

Response Specifications for a First Order System :- The time response of a first order system to a step-input can be given as ,

$$y(t) = K\cdot\left(-e^{-\frac{t}{\tau}} + 1\right) \qquad \text{.. 5.8}$$

Then

$$t = -\tau\cdot\ln\left(1 - \frac{y(t)}{K}\right) \qquad \text{.. 5.10 j}$$

Rise time calculation :- the value of rise time is calculated by following relationship ,

$$t_d := -\tau\cdot\ln\left(1 - \frac{0.9\cdot K}{K}\right) - \left(-\tau\cdot\ln\left(1 - \frac{0.1\cdot K}{K}\right)\right)$$

OR

$$t_d = 2.197\ \tau \qquad \text{.. 5. 9 j}$$

Settling time calculation :- the value of Settling time is calculated by following relationship ,

$$t_s := -\tau\cdot\ln\left(1 - \frac{0.95\cdot K}{K}\right) \qquad \text{.................................... 5. 10 k}$$

Where ,

$f(t) = (-b/a)u(t).$ **is called forcing function**

If y(t) is output variable then , Input /output differential equation is given as ,

$$\frac{dy}{dt} - ay = d\frac{du}{dt} + (bc - ad)\,u.$$

Then standard form , will be given as

$$-\frac{1}{a}\frac{dy}{dt} + y(t) = -\frac{d}{a}\frac{du}{dt} + \frac{ad - bc}{a}u(t).$$

OR

$$f(t) = -\frac{d}{a}\frac{du}{dt} + \frac{ad - bc}{a}u(t). \qquad \text{.. 5. 11 L}$$

5.41. The Characteristic Response of First-Order Systems

In general the response function is divided into two types

- A transient response in which the system responding dynamically,
- A steady-state response , in which the system is gain to its final value .

Consider the standard form of input /output differential equation for any system can be given as ,

$$\tau\frac{dy}{dt}+y=f(t). \qquad \text{.......... 5.39 a}$$

If time constant **T** , initial condition **y(0) = 0** then the solution of **f(t)** is known as the characteristic first-order response. Then total response will be given as ,

$$\begin{aligned} y(t) &= y_h(t)+y_p(t) \\ &= Ce^{-t/\tau}+y_p(t) \end{aligned} \qquad \text{.......... 5.39 b}$$

Where , **C** is a constant at **y(0) = 0 ,**

y_p **(t)** is a particular solution with respect to forcing function **f(t).**

$y_h(t) = e^{-\lambda t}$ **= exponential function**

Where , $\lambda = 1/\tau$.

(i) The Characteristic Unit Step Response :- Whenever sudden changes in input is achieved then the unit step $u_s(t)$ is characterize for any system at time t = 0 , may be given as

$$f(t)=u_s(t)=\begin{cases} 0 & t<0. \\ 1 & t\geq 0. \end{cases} \qquad \text{.......... 5.39 c}$$

Solution of equation :

$$y_s(t)=Ce^{-t/\tau}+1 \qquad \text{.......... 5.40 d}$$

The value of characteristic response when t = 0, $y_s(0) = 0$ then above solution is reduced as

$$y_s(t)=1-e^{-t/\tau}.$$

y_s**(t)** = step response

5.42. Second Order Systems

The characteristic form of a second order system can be written as , A second order system behave as a spring-mass system with damper .

$$\ddot{y}(t)+2\cdot\zeta\cdot\omega_n\cdot s\cdot\dot{y}(t)+\omega_n^{\,2}\cdot y=K\cdot\omega_n^{\,2}\cdot u(t) \qquad \text{.......... 5.41 e}$$

In Laplace transformation , it can be written as , Then

$$G(s) = \frac{Y(s)}{U(s)} = \frac{K \cdot \omega_n^2}{s^2 + 2 \cdot \zeta \cdot \omega_n \cdot s + \omega_n^2} \qquad \text{.. 5.42 f}$$

Where ,

ζ = damping ratio ,

ω_n = natural frequency

K = DC gain.

(iii) Response Specifications for a Second Order System : Response specification consists of following elements .

- **Rise time :** Following governing equation can be used to calculating the rise time of Second Order System .

$$t_r \cong 1.8 \cdot \omega_n$$

$$t_r \cong \frac{0.8 + 2.5 \cdot \zeta}{\omega_n}$$

$$t_r \cong \frac{1 - 0.4167 \cdot \zeta + 2.917 \cdot \zeta^2}{\omega_n} \qquad \text{.. 5.43 g}$$

- **Settling time :** settling time can be calculated as ,

$$t_s \cong \begin{cases} \dfrac{3.2}{\zeta \cdot \omega_n} & 0 < \zeta \le 0.69 \\ \dfrac{4.5 \cdot \zeta}{\omega_n} & 0.69 < \zeta < 1 \end{cases} \qquad \text{.. 5.44 h}$$

Where ,

ζ = damping ratio ,

ω_n = natural frequency

- **Peak time and maximum overshoot :** The value of peak time can be calculated analytically , the mathematical formula is given as ,

For peak time :-

$$t_p = \frac{\pi}{\sqrt{1 - \zeta^2} \cdot \omega_n} \qquad \text{... 5.45 i}$$

- **Maximum overshoot :** The maximum overshoot for is represented with the % exponential value .

$$M_p = e^{-\frac{\zeta \cdot \pi}{\sqrt{1-\zeta^2}}} \% \quad \text{.. 5.46 j}$$

Where , ζ = damping ratio

Example 1: The transfer function of a plant is by the following equation,

$$G_p(s) = \frac{18}{s^2+3\cdot s+9}$$

Assume that the natural frequency of system is ω_n = 3and damping ratio is $\zeta = (3/2\omega_n)$, then Calculate the following elements of the system

- The rise time (t_r)
- The 5% settling time (t_s)
- The percent overshoot (M_P)
- The peak time (t_p)

Solution

(i) The rise time

$$t_r := \frac{1-0.4167\cdot\zeta+2.917\cdot\zeta^2}{\omega_n}$$

t_r = [(1- 0.4167 x 0.5 + 2.917 x (0.5) ^{2}] / 3 = **0.50 sec.**

(ii) The 5% settling time

$$t_s := \frac{3.2}{\zeta\cdot\omega_n}$$

t_s = (3.2 / 0.5 x 3) = **2 .15 sec.**

(iii) The percent overshoot

$$M_p := e^{-\frac{\zeta\cdot\pi}{\sqrt{1-\zeta^2}}}\cdot 100$$

After substituting the value , we have

$$M_P = 100\, e^{-\frac{1}{3}\pi\sqrt{3}}$$

= 16.35 %

The peak time :

$$t_p := \frac{\pi}{\sqrt{1-\zeta^2}\cdot\omega_n}$$

Put the value or damping ration and natural frequency then , we have

$$\frac{2}{9}\pi\sqrt{3}$$

= 1.25 sec.

Answer

The rise time = 0.50

The 5% settling time = 2.15

The percent overshoot = 16.35 %

Peak time = 1.25 sec.

5.43. Derivation for Second-Order System Transient Response

The second order transient response system consists of two state variable. consider the example of physical second-order system in which one elements is used to storing energy on the other hands another is applied for heat energy exchange or dissipation. In mechanical system, energy exchange takes places between mass and stiffness elements. In electrical it is between capacitors and inductors , in fluid or hydraulic systems it is between fluid inheritance and capacitance . But in some cases the study may also performed between two independent energy storage components, for example energy may be exchanged between a mechanical mass and a fluid capacitance (tank) through a piston , or between an electrical inductance and mechanical inertia as in case of electric motor .

Let be the state equations $\dot{x} = Ax + Bu$ for a linear second-order system with a single input are a pair of coupled first-order differential equations in the two state variables ,

$$\begin{bmatrix}\dot{x}_1\\ \dot{x}_2\end{bmatrix} = \begin{bmatrix}a_{11} & a_{12}\\ a_{21} & a_{22}\end{bmatrix}\begin{bmatrix}x_1\\ x_2\end{bmatrix} + \begin{bmatrix}b_1\\ b_2\end{bmatrix}u. \quad \text{.. 5.4 a}$$

$$\frac{dx_1}{dt} = a_{11}x_1 + a_{12}x_2 + b_1u$$

$$\frac{dx_2}{dt} = a_{21}x_1 + a_{22}x_1 + b_2u. \quad \text{.. 5.4 b}$$

Taking the laplace transformation of equation 2

$$\begin{aligned}(s\mathbf{I}-\mathbf{A})\mathbf{X}(s) &= \mathbf{BU}(s)\\ \mathbf{X}(s) &= (s\mathbf{I}-\mathbf{A})^{-1}\mathbf{BU}(s)\\ &= \frac{1}{\det[s\mathbf{I}-\mathbf{A}]}\begin{bmatrix} s-a_{22} & a_{12}\\ a_{21} & s-a_{11}\end{bmatrix}\begin{bmatrix} b_1\\ b_2\end{bmatrix}U(s)\end{aligned}$$

$$\det[s\mathbf{I}-\mathbf{A}]\mathbf{X}(s) = \begin{bmatrix} s-a_{22} & a_{12}\\ a_{21} & s-a_{11}\end{bmatrix}\begin{bmatrix} b_1\\ b_2\end{bmatrix}U(s)$$

$$\frac{d^2x_1}{dt^2}-(a_{11}+a_{22})\frac{dx_1}{dt}+(a_{11}a_{22}-a_{12}a_{21})x_1 = b_1\frac{du}{dt}+(a_{12}b_2-a_{22}b_1)u.$$

Then ,

$$\frac{d^2x_2}{dt^2}-(a_{11}+a_{22})\frac{dx_2}{dt}+(a_{11}a_{22}-a_{12}a_{21})x_2 = b_2\frac{du}{dt}+(a_{21}b_1-a_{11}b_2)u. \quad \text{......... 5.4 c}$$

This equation **can be written in the form of natural frequency (ω_n) and damping ratio is ζ**

$$\begin{aligned}\frac{d^2x_1}{dt^2}+2\zeta\omega_n\frac{dx_1}{dt}+\omega_n^2x_1 &= b_1\frac{du}{dt}+(a_{12}b_2-a_{22}b_1)u\\ \frac{d^2x_2}{dt^2}+2\zeta\omega_n\frac{dx_2}{dt}+\omega_n^2x_2 &= b_2\frac{du}{dt}+(a_{21}b_1-a_{11}b_2)u.\end{aligned}$$

Compare the equation 3 & 4 , we get

$$\omega_n = \sqrt{a_{11}a_{22}-a_{12}a_{21}} \quad \text{.. 5.4 d}$$

$$\begin{aligned}\zeta &= -\frac{1}{2\omega_n}(a_{11}+a_{22})\\ &= \frac{-(a_{11}+a_{22})}{2\sqrt{a_{11}a_{22}-a_{12}a_{21}}}\end{aligned} \quad \text{.. 5.4 e}$$

Where

ω_n = undamped natural frequency with units of radians/second ,

ζ = is a damping ratio of system which is dimensionless parameter.

Generation of a Differential Equation in an Output Variable :- let be the output equation for any system is given as ,

y = Cx + Du **9**

$$
\begin{aligned}
y(t) &= \begin{bmatrix} c_1 & c_2 \end{bmatrix} \begin{bmatrix} x_1 \\ x_2 \end{bmatrix} + [d]\, u(t) \\
&= c_1 x_1(t) + c_2 x_2(t) + d u(t).
\end{aligned}
$$

.... 5.5 f

Then the differential form of the equation will be given as ,

$$
\frac{d^2 y}{dt^2} - (a_{11} + a_{22}) \frac{dy}{dt} + (a_{11}a_{22} - a_{12}a_{21})\, y = q_2 \frac{d^2 u}{dt^2} + q_1 \frac{du}{dt} + q_0 u
$$

.... 5.8 g

This equation can be written in the form of natural frequency (ω_n) and damping ratio is ζ

$$
\frac{d^2 y}{dt^2} + 2\zeta\omega_n \frac{dy}{dt} + \omega_n^2 y = q_2 \frac{d^2 u}{dt^2} + q_1 \frac{du}{dt} + q_0 u
$$

Where ,

$$
\begin{aligned}
q_0 &= c_1(-a_{22}b_1 + a_{12}b_2) + c_2(-a_{11}b_2 + a_{21}b_1) + d(a_{11}a_{22} - a_{12}a_{21}) \\
q_1 &= c_1 b_1 + c_2 b_2 - d(a_{11} + a_{22}) \\
q_2 &= d.
\end{aligned}
$$

.... 5.8 g

The following tables gives the summary of transient response of the system

Damping ratio	Input $f(t)$	Characteristic Response $y(t)$
$0 \le \zeta < 1$	$f(t) = u_r(t)$	$y_r(t) = \frac{1}{\omega_n^2}\left[t + \frac{e^{-\zeta\omega_n t}}{\omega_n}\left(2\zeta\cos\omega_d t + \frac{2\zeta^2 - 1}{\sqrt{1-\zeta^2}}\sin\omega_d t\right) - \frac{2\zeta}{\omega_n}\right]$
	$f(t) = u_s(t)$	$y_s(t) = \frac{1}{\omega_n^2}\left[1 - \frac{e^{-\zeta\omega_n t}}{\sqrt{1-\zeta^2}}\cos(\omega_d t - \psi)\right]$
	$f(t) = \delta(t)$	$y_\delta(t) = \frac{e^{-\zeta\omega_n t}}{\omega_n\sqrt{1-\zeta^2}}\sin(\omega_d t)$
$\zeta = 1$	$f(t) = u_r(t)$	$y_r(t) = \frac{1}{\omega_n^2}\left[t + \frac{2}{\omega_n}e^{-\omega_n t} + te^{-\omega_n t} - \frac{2}{\omega_n}\right]$
	$f(t) = u_s(t)$	$y_s(t) = \frac{1}{\omega_n^2}\left[1 - e^{-\omega_n t} - \omega_n t e^{-\omega_n t}\right]$
	$f(t) = \delta(t)$	$y_\delta(t) = te^{-\omega_n t}$
$\zeta > 1$	$f(t) = u_r(t)$	$y_r(t) = \frac{1}{\omega_n^2}\left[t + \frac{\omega_n}{2\sqrt{1-\zeta^2}}\left(\tau_1^2 e^{-t/\tau_1} - \tau_2^2 e^{-t/\tau_2}\right) - \frac{2\zeta}{\omega_n}\right]$
	$f(t) = u_s(t)$	$y_s(t) = \frac{1}{\omega_n^2}\left[1 - \frac{\omega_n}{2\sqrt{\zeta^2-1}}\left(\tau_1 e^{-t/\tau_1} - \tau_2 e^{-t/\tau_2}\right)\right]$
	$f(t) = \delta(t)$	$y_\delta(t) = \frac{1}{2\omega_n\sqrt{\zeta^2-1}}\left(e^{-t/\tau_1} - e^{-t/\tau_2}\right)$

The above table applied for transient response of the system ,

$$
\ddot{y} + 2\zeta\omega_n\dot{y} + \omega_n^2 y = f(t)
$$

Where ,

U_r (t) = unit ramp , U_s (t) = unit step , δ(t) = impulse response

The damped natural frequency $\omega_d = \sqrt{1-\zeta^2}\omega_n$ for $0 \le \zeta < 1$.

The phase angle $\psi = \tan^{-1}\left(\zeta/\sqrt{1-\zeta^2}\right)$ for $0 \le \zeta < 1$.

For over-damped systems ($\zeta > 1$) the time constants are

$\tau_1 = 1/\left(\zeta\omega_n - \sqrt{\zeta^2-1}\omega_n\right)$, and $\tau_2 = 1/\left(\zeta\omega_n + \sqrt{\zeta^2-1}\omega_n\right)$.

Example :- Determine the response of a physical system with differential equation

$\frac{d^2y}{dt^2} + 8\frac{dy}{dt} + 16y = 3\frac{du}{dt} + 2u$ to a step input u(t) = 2 for t > 0 or t = 1

Solution

The characteristic equation can be given as ,

$$\lambda^2 + 10\lambda + 16 = 0 \qquad \text{.......... 1}$$

Here $\lambda_1 = -2$, $\lambda_2 = -8$ then $\omega_n = 4$ rad /sec and $\zeta = 1.25$.

Since system is over damped then response to equation will be

$$y_s(t) = \frac{1}{\omega_n^2}\left[1 - \frac{\omega_n}{2\sqrt{\zeta^2-1}}\left(\tau_1 e^{-t/\tau_1} - \tau_2 e^{-t/\tau_2}\right)\right] \qquad \text{.......... 2}$$

For above equation

$$\tau_1 = 1/2, \text{ and } \tau_2 = 1/8,$$

Then equation will reduced as ,

$$\begin{aligned} y_s(t) &= \frac{1}{16}\left[1 - \frac{8}{3}\left(\frac{1}{2}e^{-2t} - \frac{1}{8}e^{-8t}\right)\right] \\ &= \frac{1}{16} - \frac{1}{12}e^{-2t} + \frac{1}{48}e^{-8t} \end{aligned} \qquad \text{.......... 3}$$

For step unit u(t) ,

$$\begin{aligned} y(t) &= 2\left[3\frac{dy_s}{dt} + 2y_s\right] \\ &= 2\left[3\left(\frac{1}{6}e^{-2t} - \frac{1}{6}e^{-8t}\right) + 2\left(\frac{1}{16} - \frac{1}{12}e^{-2t} + \frac{1}{48}e^{-8t}\right)\right] \\ &= \frac{1}{4} - \frac{2}{3}e^{-2t} + \frac{11}{12}e^{-8t} \end{aligned}$$

.... **Answer**

5.44. The Step Response of a Second-Order System

Consider the step of unit amplitude , The forced differential equation will be

$$\frac{d^2y_s}{dt^2} + 2\zeta\omega_n\frac{dy_s}{dt} + \omega_n^2 y_s = u_s(t). \qquad \text{5.41 a}$$

Where , $u_s(t)$ is the unit step function.

Since the Total solution = Homogeneous response + Particular solution

Where λ_1 and λ_2 is root of the equation

$$\begin{aligned} y_s(t) &= y_h(t) + y_p(t) \\ &= C_1e^{\lambda_1 t} + C_2e^{\lambda_2 t} + y_p(t). \end{aligned}$$

Putt y_p (t) = K in differential equation , we have

$$\omega_n^2 K = 1$$

OR

$$y_s(t) = C_1e^{\lambda_1 t} + C_2e^{\lambda_2 t} + \frac{1}{\omega_n^2}.$$

Then the value of C_1 and C_2 will be

$$C_1 = \frac{\lambda_2}{\omega_n^2(\lambda_1 - \lambda_2)}, \qquad C_2 = \frac{\lambda_1}{\omega_n^2(\lambda_2 - \lambda_1)}.$$

If λ_1 and λ_2 is root well defined then unit step response ,

$$\begin{aligned} y_s(t) &= \frac{1}{\omega_n^2}\left[1 - \left(\frac{\lambda_2}{\lambda_2 - \lambda_1}e^{\lambda_1 t} + \frac{\lambda_1}{\lambda_1 - \lambda_2}e^{\lambda_2 t}\right)\right] \\ &= \frac{1}{\omega_n^2}\left[1 - \frac{\lambda_2\lambda_1}{\lambda_2 - \lambda_1}\left(\frac{1}{\lambda_1}e^{\lambda_1 t} - \frac{1}{\lambda_2}e^{\lambda_2 t}\right)\right] \end{aligned} \qquad \textbf{5.41 b}$$

If root are identical then ,

$$y_s(t) = \frac{1}{\omega_n^2}\left[1 - \frac{1}{\tau_2 - \tau_1}\left(\tau_2 e^{-t/\tau_2} - \tau_1 e^{-t/\tau_1}\right)\right] \quad \text{for } \zeta > 1, \qquad \textbf{5.41 c}$$

If system is under damped then ,

$$y_s(t) = \frac{1}{\omega_n^2}\left[1 - \frac{e^{-\zeta\omega_n t}}{\sqrt{1-\zeta^2}}\cos(\omega_d t - \psi)\right] \quad \text{for } 0 < \zeta < 1. \qquad \textbf{5.41 d}$$

5.45. The Impulse Response

The impulse response of the system is nothing but it is a transfer function of the system . to understand the impulse , consider the following diagram .

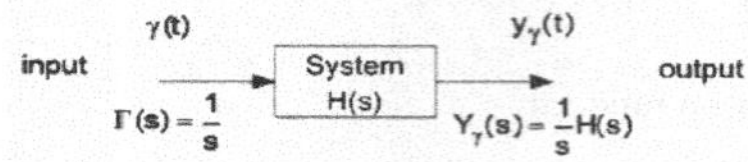

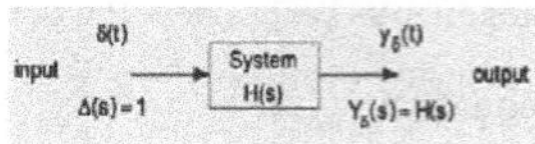

Where ,

H(s) is transfer function of the system

h(t) is impulse response of a system or inverse Laplace Transform of **H(s)**

γ(t) = unit step function

δ(t) = unit impulse response

then , impulse response of the system can be given as ,

$$h(t) = \mathcal{L}^{-1}\{H(s)\}$$

5.46. The Characteristic Impulse Response

The impulse function **δ(t)**is defined as the limit of a pulse of duration T and amplitude **1 / T as T** approaches zero , and it is considered as the derivative of the unit step function.

Suppose that , in a linear systems impulse response is **δ(t)** , characteristic step response is **y** $_s$**(t)** and forcing function is given by **f(t) = δ(t)**. then Impulse Response is given as ,

$$\begin{aligned} y_\delta(t) &= \frac{dy_s}{dt} = \frac{d}{dt}\left(1 - e^{-t/\tau}\right) \\ &= \frac{1}{\tau}e^{-t/\tau} \quad \text{for } t \geq 0. \end{aligned}$$

.... **5.43 a**

First order differential equation

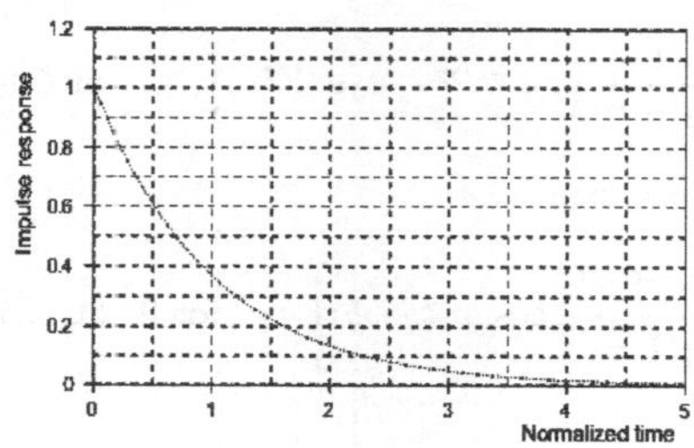

The impulse response curve for

$$\tau \dot{y} + y = \delta(t).$$

From the above curve we have seen that, the impulse response representing the exponential form having declining nature , which is discontinuous at time t = 0 with an initial value y (0^+) = 1/T , where the superscript 0^+ indicates a time increment > zero .

The Characteristic Ramp Response : The unit ramp u_r (t) = t for t > 0 or t = 0 , then the integral of the unit step function $u_s(t)$, called the unit ramp response function .

$$u_r(t) = \int_0^t u_s(t)dt.$$

$$\begin{aligned} y_r(t) &= \int_0^t y_s(t)dt = \int_0^t \left(1 - e^{-t/\tau}\right) dt \\ &= t - \tau\left(1 - e^{-t/\tau}\right) \end{aligned}$$

Where ,

y_r (t) = characteristic response when ramp forcing function f(t) = u_r(t) ,

y_s (t) = step response

The Input /Output Impulse Response: we know that from impulse response function ,

$$\begin{aligned} y_\delta(t) &= \frac{dy_s}{dt} = \frac{d}{dt}\left(1 - e^{-t/\tau}\right) \\ &= \frac{1}{\tau} e^{-t/\tau} \qquad \text{for } t \geq 0. \end{aligned}$$

.... **5.43 b**

Input $u(t)$	Characteristic Response	Input/Output Response $y(t)$ for $t \geq 0$
$u(t) = 0$		$y(t) = y(0)e^{-t/\tau}$
$u(t) = u_r(t)$	$y_r(t) = t - \tau\left(1 - e^{-t/\tau}\right)$	$y(t) = \left[q_0 t + (q_1 - q_0\tau)\left(1 - e^{-t/\tau}\right)\right]$
$u(t) = u_s(t)$	$y_s(t) = y_s(t) = 1 - e^{-t/\tau}$	$y(t) = \left[q_0 - \left(q_0 - \frac{q_1}{\tau}\right)e^{-t/\tau}\right]$
$u(t) = \delta(t)$	$y_\delta(t) = \frac{1}{\tau}e^{-t/\tau}$	$y(t) = \frac{q_1}{\tau}\delta(t) + \left(\frac{q_0}{\tau} - \frac{q_1}{\tau^2}\right)e^{-t/\tau}$

When t = 0 , then the derivative of characteristic impulse response y_δ(t) is given as ,

$$\frac{d}{dt}y_\delta(t) = \frac{1}{\tau}\delta(t) - \frac{1}{\tau^2}e^{-t/\tau}$$ and

The response of the first-order linear system $\tau \dot{y} + y = q_1 \dot{u} + q_0 u$ for the singularity inputs.

May be given as ,

$$\begin{aligned} y(t) &= q_1 \frac{dy_\delta}{dt} + q_0 y_\delta(t) \\ &= \frac{q_1}{\tau}\delta(t) + \left(\frac{q_0}{\tau} - \frac{q_1}{\tau^2}\right)e^{-t/\tau}, \end{aligned}$$

.... **5.43 c**

Impulse response of a Second-Order System : For any linear system , impulse response $y_\delta(t)$ can be achieved by step response ys (t) .

$$y_\delta(t) = \frac{dy_s}{dt} \quad \text{because} \quad \delta(t) = \frac{d}{dt}u_s(t)$$

Where ,

u_s (t) = The unit step function and , f(t) = δ(t)

Governing standard form of the equation is same as that of transient response system which can be given as ,

$$\frac{d^2y}{dt^2} + 2\zeta\omega_n\frac{dy}{dt} + \omega_n^2 y = f(t).$$

.. 5.43 d

Then differentiating above equation , we have impulse response equation

$$\frac{d^2y_\delta}{dt^2} + 2\zeta\omega_n\frac{dy_\delta}{dt} + \omega_n^2 y_\delta = \delta(t).$$

If the root λ_1 and λ_2 are well defined of characteristic equation , then governing equation will be given as ,

$$= \frac{1}{\omega_n^2}\left[1 - \frac{\lambda_2\lambda_1}{\lambda_2 - \lambda_1}\left(\frac{1}{\lambda_1}e^{\lambda_1 t} - \frac{1}{\lambda_2}e^{\lambda_2 t}\right)\right]$$

For obtaining the impulse response, Differentiate it ,

$$= \frac{1}{\omega_n^2}\frac{\lambda_1\lambda_2}{\lambda_1 - \lambda_2}\left(e^{\lambda_1 t} - e^{\lambda_2 t}\right)$$

$$= \frac{1}{\lambda_1 - \lambda_2}\left(e^{\lambda_1 t} - e^{\lambda_2 t}\right).$$

.. 5.43 e

Since $\omega^2_n = \lambda_1\lambda_2$ and for real root $\zeta > 1$ then ,

$$\lambda_1 = -\zeta\omega_n + \sqrt{\zeta^2 - 1}\omega_n\ ,$$

$$\lambda_2 = -\zeta\omega_n - \sqrt{\zeta^2 - 1}\omega_n\ .$$

Hence the equation reduced as ,

$$y_\delta(t) = \frac{1}{2\omega_n\sqrt{\zeta^2 - 1}}\left(e^{(-\zeta+\sqrt{\zeta^2-1})\omega_n t} - e^{(-\zeta-\sqrt{\zeta^2-1})\omega_n t}\right)$$

$$= \frac{1}{2\omega_n\sqrt{\zeta^2 - 1}}\left(e^{-t/\tau_1} - e^{-t/\tau_2}\right)$$

.. 5.43 f

Where

T_1 = -1 / λ_1 and T_2 = -1 / λ_2

Example : Find the response of a physical system with the differential equation ,

$$\frac{d^2y}{dt^2} + 8\frac{dy}{dt} + 4y = \frac{d^2u}{dt^2} + 2\frac{du}{dt} + u ,$$

to a step input u(t) = 2 for t > 0 or t = 0

Solution

The characteristic equation is

$$\lambda^2 + 4\lambda + 4 = 0$$

Here the system is critically damped because **$\lambda_1 = \lambda_2 = -2$** , so therefore **$\omega_n = 2$ rad / sec** .

Hence the characteristic equation for impulse response can be given as ,

$$y_\delta(t) = te^{-\omega_n t} = te^{-2t}.$$

Since q_2 is not equals to zero , then equation can be written

$$\begin{aligned} y(t) &= q_2\delta(t) + (q_1 - 2b_2\zeta\omega_n)\frac{dy_\delta}{dt} + \left(q_0 - b_2\omega_n^2\right) y_\delta(t) \\ &= \delta(t) - 4\frac{dy_\delta}{dt} - 2y_\delta \\ &= \delta(t) - te^{-2t} - 2e^{-2t} \end{aligned}$$

Example 1 : A mass *m* = 10 kg is at rest on a horizontal plane with viscous friction coefficient *B* = 20 N-s/m, as shown in Fig. 9. A short impulsive force of amplitude 200 N and duration 0.01 s is applied. Determine how far the mass travels before coming to rest, and how long it takes for the velocity to decay to less than 1% of its initial value.

Solution

The differential equation relating the velocity of the mass to the applied

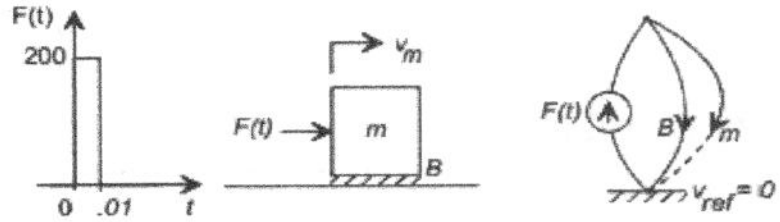

Then force will be represented as ,

$$\frac{m}{B}\frac{dv_m}{dt} + v_m = \frac{1}{B}F_{in}(t) \qquad \text{...} 1$$

The value of time constant will be as

T = m / B = 10 / 20 = 0.5 seconds.

The duration of the force pulse is much less than the time constant hence the system impulse response may given as ,

$$v_m(t) = \frac{1}{m}e^{-Bt/m} \qquad \text{.......... 2}$$

Further if u(t) = 2δ(t) , then response will be ,

$$v_m(t) = 0.2e^{-2t}.$$

The distance x travelled will be ,

$$x = \int_0^{\infty} 0.2e^{-2t}dt = 0.1 \text{ m.} \qquad \text{.......... 3}$$

The time T for the velocity to decay to less than 1% of its original value is obtained by the following relation

$$v_m(T)/v_m(0) = 0.01 = e^{-2T}, \text{ or } T = 2.303 \text{ seconds.}$$

Example 2 : Determine the differential equations in the state variables x_1 (t) and x_2 (t) for the system

$$\begin{bmatrix} \dot{x}_1 \\ \dot{x}_2 \end{bmatrix} = \begin{bmatrix} -1 & -2 \\ 2 & -3 \end{bmatrix} \begin{bmatrix} x_1 \\ x_2 \end{bmatrix} + \begin{bmatrix} 1 \\ 0 \end{bmatrix} u. \qquad \text{.......... 1}$$

Calculate the undammed natural frequency ω_n and damping ratio ζ for this system .

Solution

For this system

$$[sI - A] = \begin{bmatrix} s+1 & 2 \\ -2 & s+3 \end{bmatrix}$$

and

$$\det[sI - A] = s^2 + 4s + 7 \qquad \text{.......... 2}$$

Thus state variable $x_1(t)$ can be given as ,

$$\frac{d^2x_1}{dt^2} + 4\frac{dx_1}{dt} + 7x_1 = \frac{du}{dt} + 3u. \qquad \text{.......... 3}$$

$$\frac{d^2x_1}{dt^2} + 2\zeta\omega_n\frac{dx_1}{dt} + \omega_n^2 x_1 = b_1\frac{du}{dt} + (a_{12}b_2 - a_{22}b_1)u \qquad \text{.........................} 4$$

Similarly for $x_2(t)$

$$\frac{d^2x_2}{dt^2} + 4\frac{dx_2}{dt} + 7x_2 = 2u. \qquad \text{...} 5$$

$$\frac{d^2x_2}{dt^2} + 2\zeta\omega_n\frac{dx_2}{dt} + \omega_n^2 x_2 = b_2\frac{du}{dt} + (a_{21}b_1 - a_{11}b_2)u. \qquad \text{.........................} 6$$

Comparing equation (3) & (4) and (5) & (6) , we get

$$\omega_n = \sqrt{7} \text{ rad / sec}$$

$$\zeta = 2/\sqrt{7} = 0.755.$$

Example 3 : A rotational system consists of an inertial load *J* mounted in viscous bearings *B*, and driven by an angular velocity source **Ω_{in} (t)** through a long light shaft with significant torsional stiffness *K* Derive a pair of second-order differential equations for the variables **Ω_j** and **Ω_K**.

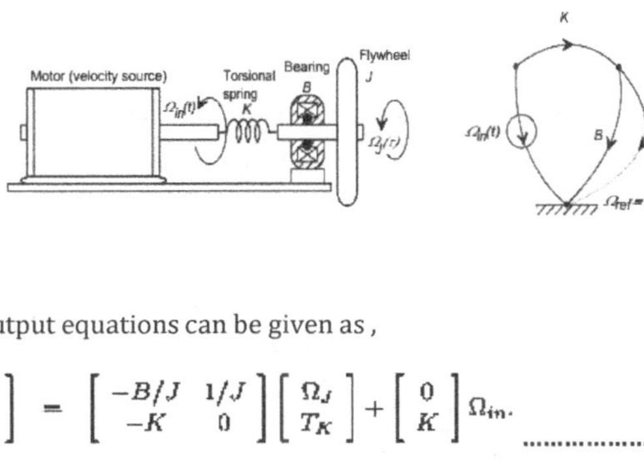

Solution

The state and output equations can be given as ,

$$\begin{bmatrix} \dot{\Omega}_j \\ \dot{T}_k \end{bmatrix} = \begin{bmatrix} -B/J & 1/J \\ -K & 0 \end{bmatrix} \begin{bmatrix} \Omega_J \\ T_K \end{bmatrix} + \begin{bmatrix} 0 \\ K \end{bmatrix} \Omega_{in}. \qquad \text{...........................} 1$$

$$\begin{bmatrix} \Omega_J \\ \Omega_K \end{bmatrix} = \begin{bmatrix} 1 & 0 \\ -1 & 0 \end{bmatrix} \begin{bmatrix} \Omega_J \\ T_K \end{bmatrix} + \begin{bmatrix} 0 \\ 1 \end{bmatrix} \qquad \text{...........................} 2$$

In this case there are two outputs and the transfer function matrix is

$$\begin{aligned} \mathbf{H}(s) &= \mathbf{C}[s\mathbf{I} - \mathbf{A}]^{-1}\mathbf{B} + \mathbf{D} \\ &= \frac{\mathbf{C}\mathrm{adj}[s\mathbf{I} - \mathbf{A}]\mathbf{B} + \mathbf{D}}{\det[s\mathbf{I} - \mathbf{A}]} \\ &= \begin{bmatrix} \dfrac{K/J}{s^2 + (B/J)s + K/J} \\ \dfrac{s^2 + (B/J)s}{s^2 + (B/J)s + K/J} \end{bmatrix} \end{aligned} \qquad \text{.....................................} 3$$

The required differential equation is given as ,

$$\frac{d^2\Omega_J}{dt^2} + \frac{B}{J}\frac{d\Omega_J}{dt} + \frac{K}{J}\Omega_J = \frac{K}{J}\Omega_{in}. \quad \text{.................................} \; 4$$

And

$$\frac{d^2\Omega_K}{dt^2} + \frac{B}{J}\frac{d\Omega_K}{dt} + \frac{K}{J}\Omega_K = \frac{d^2\Omega_{in}}{dt^2} + \frac{B}{J}\frac{d\Omega_{in}}{dt} \quad \text{.................................} \; 5$$

Then un-damped natural frequency will be

$$\omega_n = \sqrt{\frac{K}{J}}$$

Damping ratio

$$\zeta = \frac{B/J}{2\sqrt{K/J}} = \frac{B}{2\sqrt{KJ}}.$$

Example 4 : Simplify the block diagram as given below

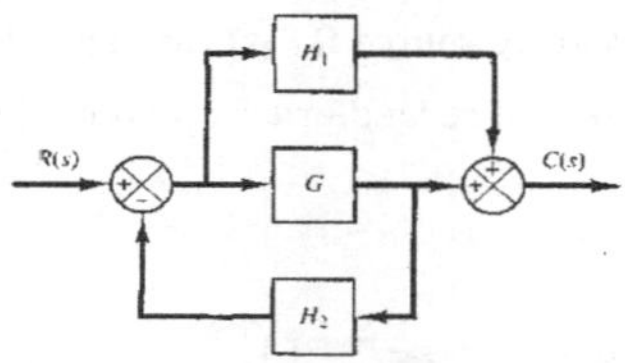

Solution : First, move the branch point of the path involving H_1 outside the loop involving H_2 as shown in Figure (a). Then eliminating two loops results in Figure (b). Combining two blocks into one gives Figure(c)

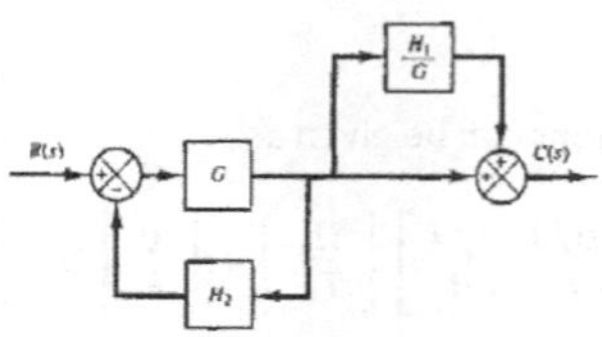

(a)

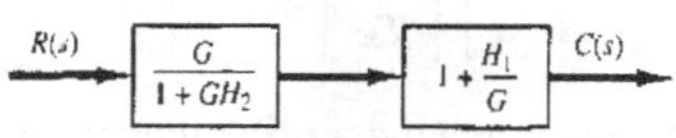

(b)

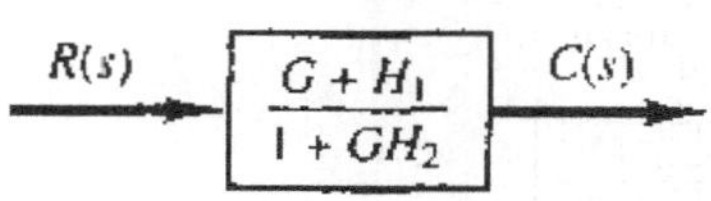

(c)

Example 5 : Calculate the transfer function of following diagram C (s) and R(s).

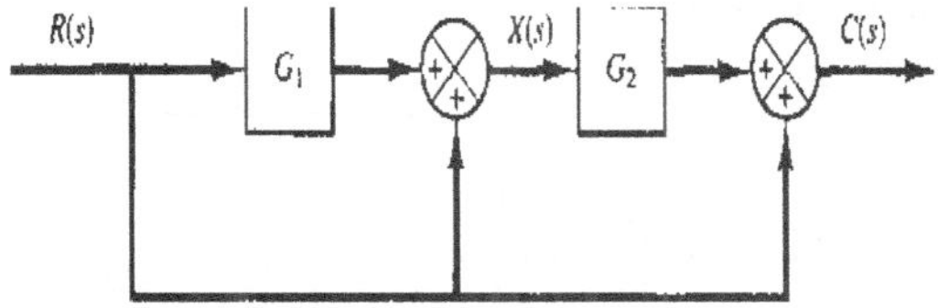

Solution

Reduction of block diagram is given as ,

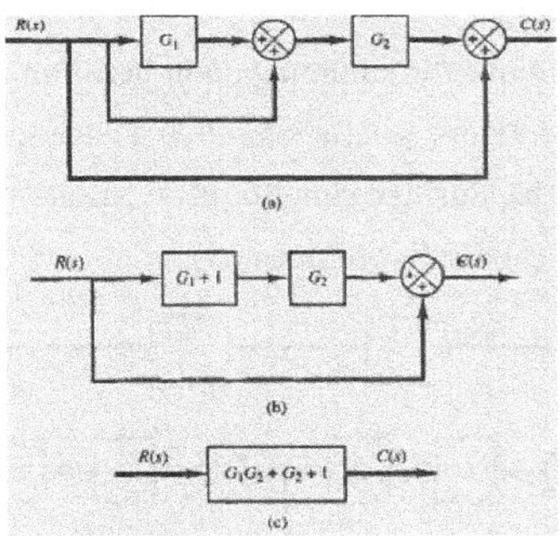

The block diagram of Figure can be modified to that of as shown in Figure(a). Eliminating the minor feed forward path, we obtain Figure (b), which can be simplified to that of as shown in Figure (c).The transfer function **C (s) / R (s)** is thus given by

$$\frac{C(s)}{R(s)} = G_1G_2 + G_2 + 1 \qquad \text{.............................. 1}$$

The same result can also be obtained by proceeding as follows: Since signal X (s) is the sum of two signals $G_1R(s)$and R(s),we have

$$X(s) = G_1R(s) + R(s) \qquad \text{...................... 2}$$

The output signal C (s)is the sum of $G_2X(s)$ and R(s).Hence

C (s)= G_2 X (s)+ R(s)

= $G_2[G_1R(s)$+ R (s)] + R(s) 3

Hence transfer function can be given as ,

$$\frac{C(s)}{R(s)} = G_1G_2 + G_2 + 1 \qquad \text{.. 4}$$

Example 6 : Simplify the block diagram shown in Figure , Then, obtain the closed-loop transfer function C(s) / R(s) ,

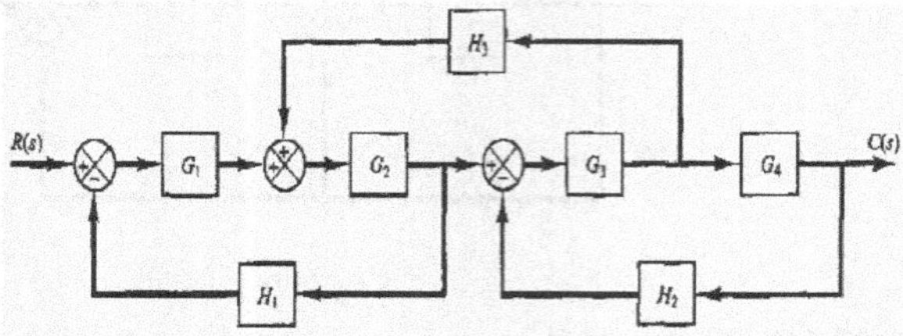

Solution : First move the branch point between G_3 and G_4 to the right-hand side of the loop containing G_3, G_4 and H_2. Then move the summing point between G_1 and G_2 to the left-hand side of the first summing point. See Figure (a). By simplifying each loop, the block diagram can be modified as shown in Figure (b). Further simplification results in Figure (c), from which the closed-loop transfer function C(s)/R(s)is obtained as ,

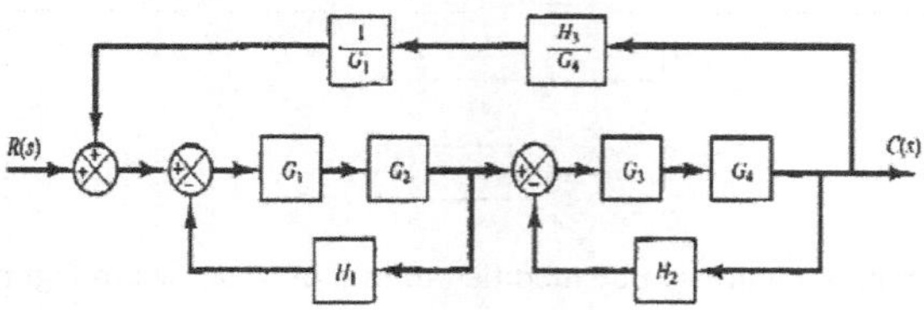

(a)

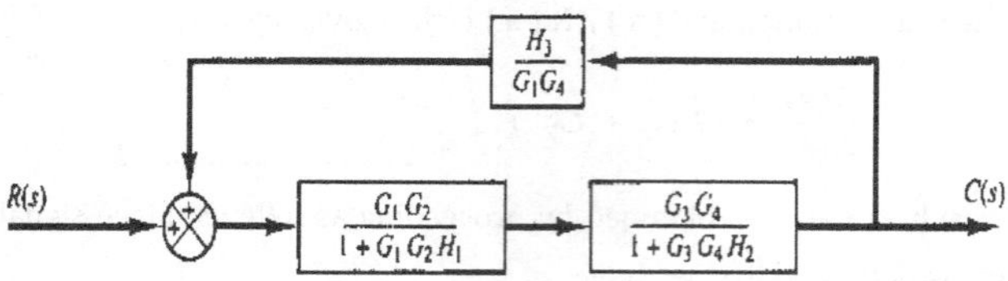

(b)

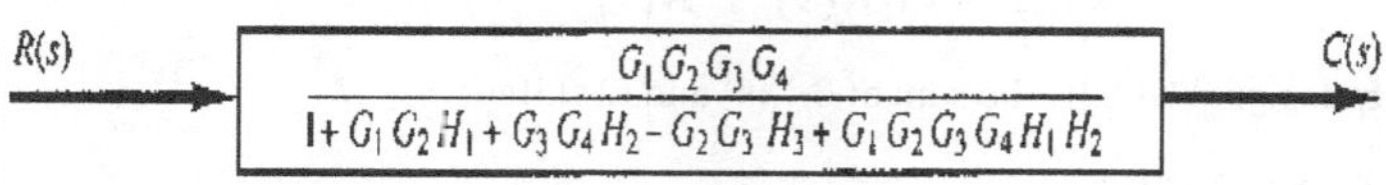

(c)

Then transfer function will be

$$\frac{C(s)}{R(s)} = \frac{G_1G_2G_3G_4}{1 + G_1G_2H_1 + G_3G_4H_2 - G_2G_3H_3 + G_1G_2G_3G_4H_1H_2}$$

Example 7 : Obtain transfer functions **C(s) / R(s)** and C (s) / D (s)of the system as shown in Figure ,

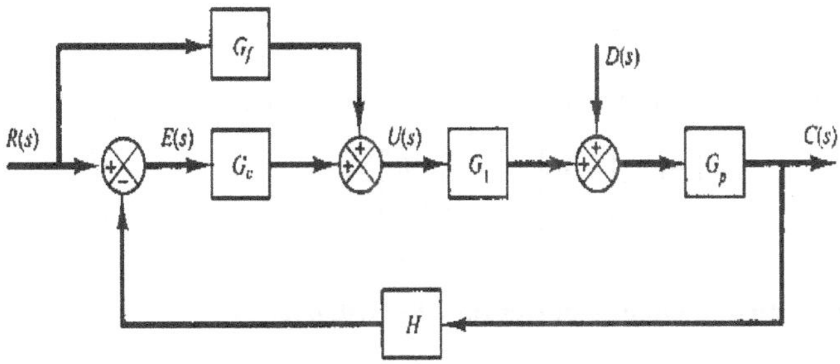

Solution

$$U(s) = G_f R(s) + G_c E(s) \quad \ldots\ldots 1$$

$$C(s) = G_p [D(s) + G_1 u(s)] \quad \ldots\ldots 2$$

$$E(s) = R(s) - H C(s) \quad \ldots\ldots 3$$

From equation **(1) & (2)** , we have got ,

$$C(s) = G_p D(s) + G_1 G_p [G_f R(s) + G_c E(s)] \quad \ldots\ldots 4$$

From equation **(3) & (4)** , we have got ,

$$C(s) = G_p D(s) + G_1 G_p \{G_f R(s) + G_c [R(s) - HC(s)]\}$$

transfer function **C (s) / R (s)** will be ,

$$\frac{C(s)}{R(s)} = \frac{G_1 G_p (G_f + G_c)}{1 + G_1 G_p G_c H}$$

transfer function **C (s) / D (s)** will be

$$\frac{C(s)}{D(s)} = \frac{G_p}{1 + G_1 G_p G_c H}$$

Example 8 : Obtain the transfer function of the system defined by

$$\begin{bmatrix} \dot{x}_1 \\ \dot{x}_2 \\ \dot{x}_3 \end{bmatrix} = \begin{bmatrix} -1 & 1 & 0 \\ 0 & -1 & 1 \\ 0 & 0 & -2 \end{bmatrix} \begin{bmatrix} x_1 \\ x_2 \\ x_3 \end{bmatrix} + \begin{bmatrix} 0 \\ 0 \\ 1 \end{bmatrix} u$$

$$y = [1 \quad 0 \quad 0] \begin{bmatrix} x_1 \\ x_2 \\ x_3 \end{bmatrix}$$

Solution

For above equation the transfer function G(s)is given as ,

$$G(s) = \mathbf{C}(s\mathbf{I} - \mathbf{A})^{-1}\mathbf{B} + D$$

Where , the value of transfer function elements by the matrices is given as

$$\mathbf{A} = \begin{bmatrix} -1 & 1 & 0 \\ 0 & -1 & 1 \\ 0 & 0 & -2 \end{bmatrix}, \quad \mathbf{B} = \begin{bmatrix} 0 \\ 0 \\ 1 \end{bmatrix}, \quad \mathbf{C} = [1 \;\; 0 \;\; 0], \quad D = 0$$

On further calculation the transfer function will become ,

$$G(s) = [1 \;\; 0 \;\; 0] \begin{bmatrix} s+1 & -1 & 0 \\ 0 & s+1 & -1 \\ 0 & 0 & s+2 \end{bmatrix}^{-1} \begin{bmatrix} 0 \\ 0 \\ 1 \end{bmatrix}$$

$$= [1 \;\; 0 \;\; 0] \begin{bmatrix} \frac{1}{s+1} & \frac{1}{(s+1)^2} & \frac{1}{(s+1)^2(s+2)} \\ 0 & \frac{1}{s+1} & \frac{1}{(s+1)(s+2)} \\ 0 & 0 & \frac{1}{s+2} \end{bmatrix} \begin{bmatrix} 0 \\ 0 \\ 1 \end{bmatrix}$$

Hence the value of transfer function will be as ,

$$= \frac{1}{(s+1)^2(s+2)} = \frac{1}{s^3 + 4s^2 + 5s + 2}$$

Example 9 : what will be the state-space representation of given mechanical system which is given below ,

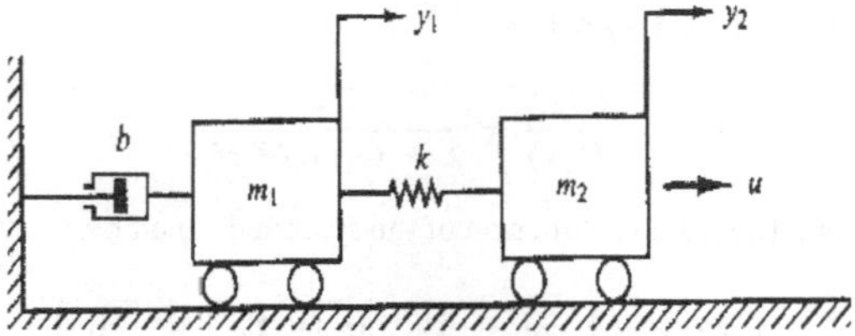

Solution

The system equations are

$$m_1\ddot{y}_1 + b\dot{y}_1 + k(y_1 - y_2) = 0$$
$$m_2\ddot{y}_2 + k(y_2 - y_1) = u$$

The output variables for this system are $\mathbf{y}_1$ and $\mathbf{y}_2$ Define state variables as

$$\begin{aligned} x_1 &= y_1 \\ x_2 &= \dot{y}_1 \\ x_3 &= y_2 \\ x_4 &= \dot{y}_2 \end{aligned}$$

Then equation of the system will be as ,

$$\begin{aligned} \dot{x}_1 &= x_2 \\ \dot{x}_2 &= \frac{1}{m_1}\{-b\dot{y}_1 - k(y_1 - y_2)\} = -\frac{k}{m_1}x_1 - \frac{b}{m_1}x_2 + \frac{k}{m_1}x_3 \\ \dot{x}_3 &= x_4 \\ \dot{x}_4 &= \frac{1}{m_2}\{-k(y_2 - y_1) + u\} = \frac{k}{m_2}x_1 - \frac{k}{m_2}x_3 + \frac{1}{m_2}u \end{aligned}$$

Thus the state equation and output equation of the system can be represented as ,

$$\begin{bmatrix} \dot{x}_1 \\ \dot{x}_2 \\ \dot{x}_3 \\ \dot{x}_4 \end{bmatrix} = \begin{bmatrix} 0 & 1 & 0 & 0 \\ -\frac{k}{m_1} & -\frac{b}{m_1} & \frac{k}{m_1} & 0 \\ 0 & 0 & 0 & 1 \\ \frac{k}{m_2} & 0 & -\frac{k}{m_2} & 0 \end{bmatrix} \begin{bmatrix} x_1 \\ x_2 \\ x_3 \\ x_4 \end{bmatrix} + \begin{bmatrix} 0 \\ 0 \\ 0 \\ \frac{1}{m_2} \end{bmatrix} u$$

...... State Equation

$$\begin{bmatrix} y_1 \\ y_2 \end{bmatrix} = \begin{bmatrix} 1 & 0 & 0 & 0 \\ 0 & 0 & 1 & 0 \end{bmatrix} \begin{bmatrix} x_1 \\ x_2 \\ x_3 \\ x_4 \end{bmatrix}$$

.... Output Equation

Example 10 : Calculate the equivalent spring constant for the following system .

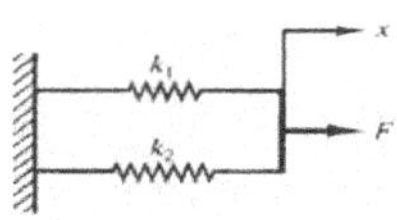

(a)

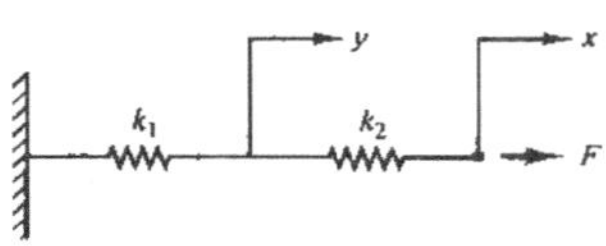

(b)

Solution : for figure (a) , since spring is in parallel the equivalent spring constant is given as

$$k_1 x + k_2 x = F = k_{eq} x \qquad \text{..........} \; 1$$

$$k_{eq} = k_1 + k_2 \qquad \text{..........} \; 2$$

For figure (b) , since spring is in series connection the equivalent spring constant is given as ,

$$k_1 y = F \quad \text{...} 3$$

$$k_2(x - y) = F \quad \text{...} 4$$

From equation (3) & (4) , we have

$$k_2 x = F + \frac{k_2}{k_1} F = \frac{k_1 + k_2}{k_1} F$$

The equivalent spring constant will be ,

$$k_{eq} = \frac{F}{x} = \frac{k_1 k_2}{k_1 + k_2} = \frac{1}{\frac{1}{k_1} + \frac{1}{k_2}}$$

Example 11 : Represent the electrical analogy of given mechanical system as shown in Figure below , also Calculate the transfer function of the system.

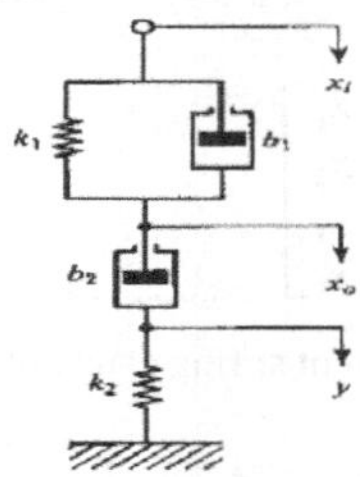

Solution: Since for making electrical analogous system , the damper friction is represented by capacitance . thus, equivalent analogy will be given as ,

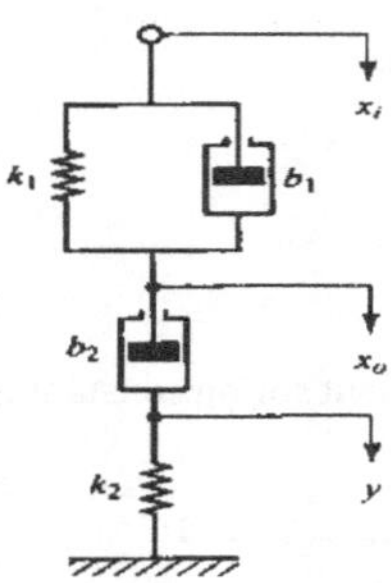

Equivalent to

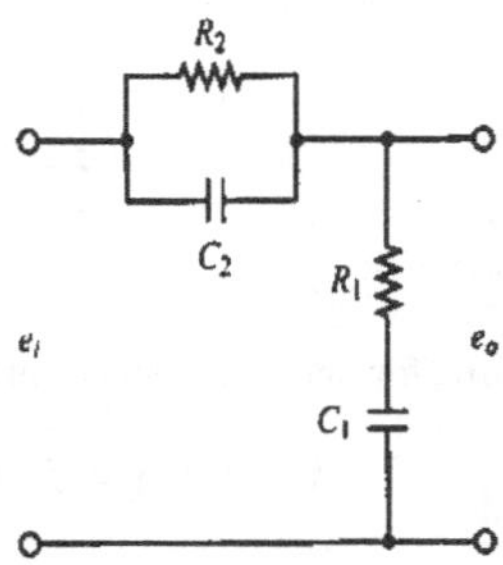

Mechanical System **Electrical Analogy**

Solution

The transfer function **E $_o$ (s) / E $_i$ (s)** is calculated as ,

$$\frac{E_o(s)}{E_i(s)} = \frac{R_1 + \dfrac{1}{C_1 s}}{\dfrac{1}{(1/R_2) + C_2 s} + R_1 + \dfrac{1}{C_1 s}}$$

$$= \frac{(R_1 C_1 s + 1)(R_2 C_2 s + 1)}{(R_1 C_1 s + 1)(R_2 C_2 s + 1) + R_2 C_1 s}$$

Example 12 : Calculate the transfer function for the closed-loop transfer function **C (s) / R (s)** for which the signal flow graph is given below .

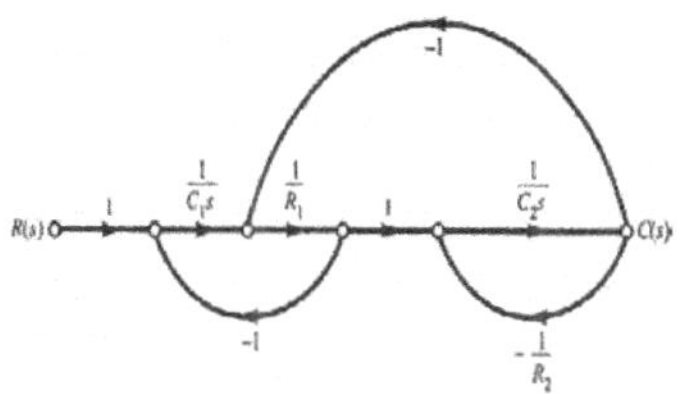

Signal Flow Graph

Solution

In this case we have single forward path which connects the input R (s) and the Output C (s) .Then ,

$$P_1 = \frac{1}{C_1 s} \frac{1}{R_1} \frac{1}{C_2 s}$$

The signal flow graph representing three individual loops , which is given as ,

$$L_1 = -\frac{1}{C_1 s} \frac{1}{R_1} \quad \text{..} \; 1$$

$$L_2 = -\frac{1}{C_2 s} \frac{1}{R_2} \quad \text{..} \; 2$$

$$L_3 = -\frac{1}{R_1} \frac{1}{C_2 s} \quad \text{..} \; 3$$

Since ,

Loop **L** $_1$ does not touch loop **L** $_2$

Loop **L** $_1$ touches loop **L** $_3$

Loop L $_2$ touches loop L $_3$

Hence

Determinant will be as ,

$$\Delta = 1 - (L_1 + L_2 + L_3) + (L_1 L_2)$$

$$= 1 + \frac{1}{R_1C_1s} + \frac{1}{R_2C_2s} + \frac{1}{R_1C_2s} + \frac{1}{R_1C_1R_2C_2s^2}$$

Since all three loops touch the forward path **P1**, we remove **L 1 , L 2 , L 3 ,** and **L 1 , L 2 .** then cofactor

Hence the closed-loop transfer function ,

$$\frac{C(s)}{R(s)} = \frac{P_1\Delta_1}{\Delta}$$

$$= \frac{\dfrac{1}{R_1C_1C_2s^2}}{1 + \dfrac{1}{R_1C_1s} + \dfrac{1}{R_2C_2s} + \dfrac{1}{R_1C_2s} + \dfrac{1}{R_1C_1R_2C_2s^2}} = \frac{R_2}{R_1C_1R_2C_2s^2 + (R_1C_1 + R_2C_2 + R_2C_1)s + 1}$$

Question for Discussion

1. List the advantages and disadvantages of an open loop system & closed loop system.
2. State the procedure for reduction of block diagram , convert the system in unity feed back system ,

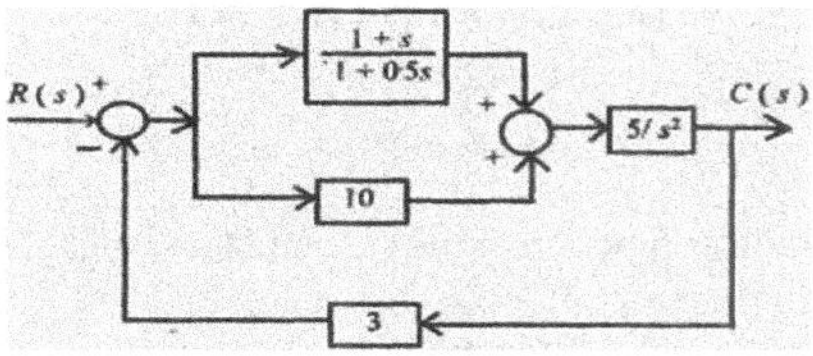

3. Calculate the transfer of the system ,

$$\dot{x}_1 = -4x_1 - x_2 + 3u$$
$$\dot{x}_2 = -2x_1 - 3x_2 + 5u$$
$$y = x_1 + 2x_2.$$

4. Explain the terms " open loop system & closed loop system " of an industrial process , with the help of suitable schematic or block diagram where necessary .
5. Explain the procedure of signal flow graph from a block diagram . how can you construct a signal flow graph from linear equations .
6. Determine the unit step response , y(t) , for the following system ,

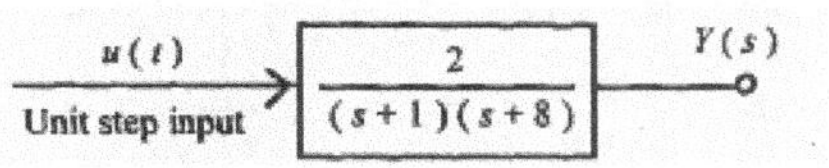

7. How can you select state variable for different system , write state variable system for mechanical system ,

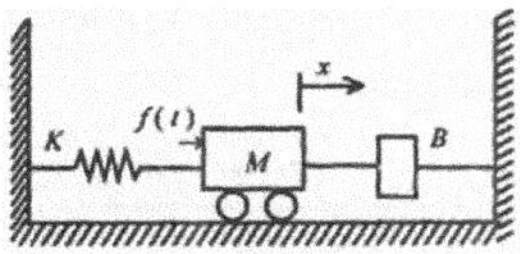

8. Explain, with an example, what do you mean by an open-loop control system. What are the limitations of open-loop control system ?
9. Explain transfer function for various type of control system . draw the electrical analogy of the mechanical system .

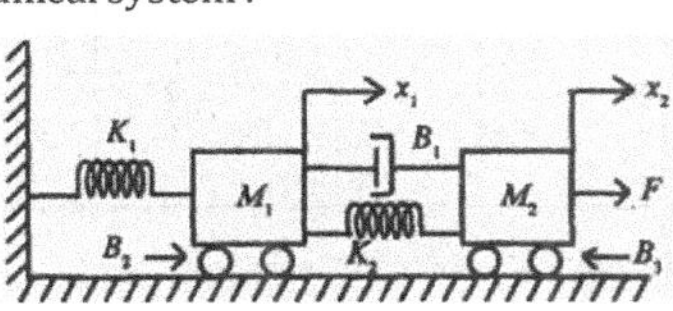

10. Draw signal flow graph for following set of algebraic equation .

$$
\begin{aligned}
(i)\quad x_1 &= -x_2 - 3x_3 + 3 \\
x_2 &= 5x_1 - 2x_2 + x_3 \\
x_3 &= 4x_1 + x_2 - 5x_3 + 5
\end{aligned}
$$

$$
\begin{aligned}
(ii)\quad 2x_1 + 3x_2 + x_3 &= -1 \\
x_1 - 2x_2 - x_3 &= 1 \\
3x_2 + x_3 &= 0
\end{aligned}
$$

11. Find the overall transfer function of the system whose signal flow graph is given as ,

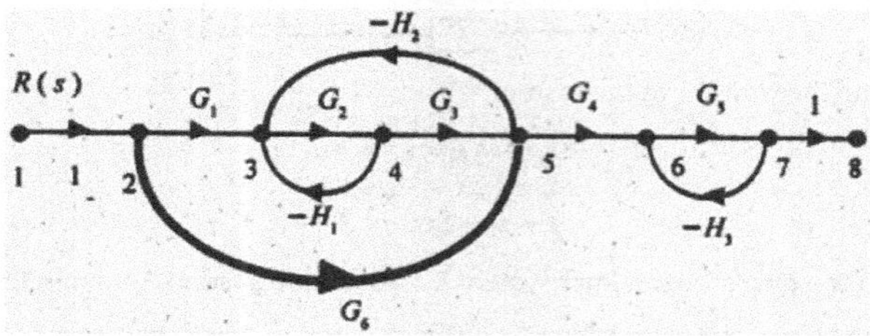

12. Explain the relationship between state equations & transfer function .
13. Determine the transfer function of following electrical system ,

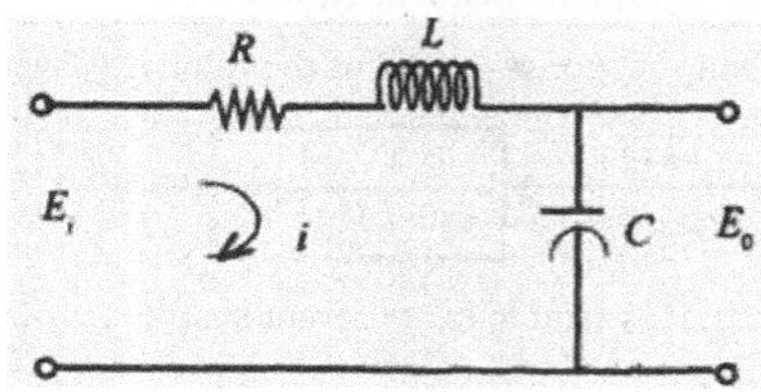

14. The close loop transfer function is given as , determine the response of the system when a unit step is applied at input ,

$$G(s) = \frac{s(s^2 + 9s + 19)}{s^3 + 7s^2 + 14s + 8}$$

15. Draw the block diagram of R-L-C circuits , also find out the transfer function of the circuit as shown in figure ,

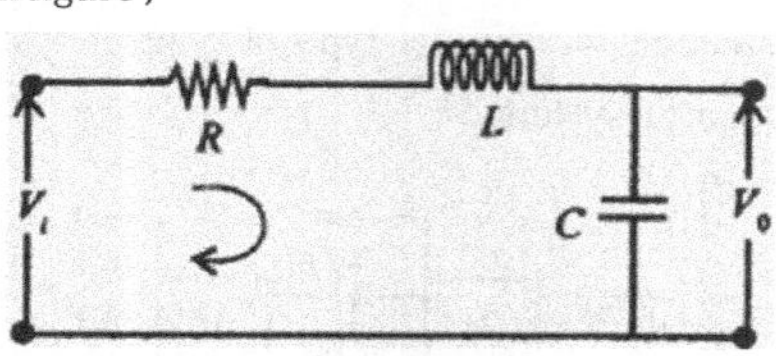

Where , $\mathbf{V_i}$ = input voltage & $\mathbf{V_o}$ = output voltage

16. Draw the signal flow graph & transfer function of following block diagram ,

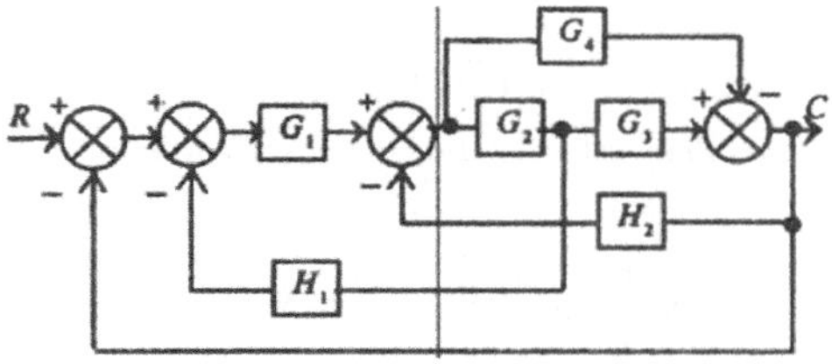

17. Determine the values of rise time and maximum overshoot Mp in its step response if the value of damping ratio is 0.5 in the following diagram .

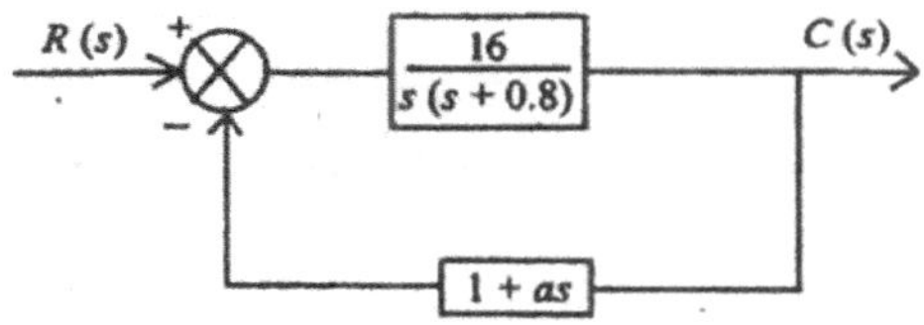

18. Determine the value of transfer function with the help of signal flow graph technique for the following block diagram .

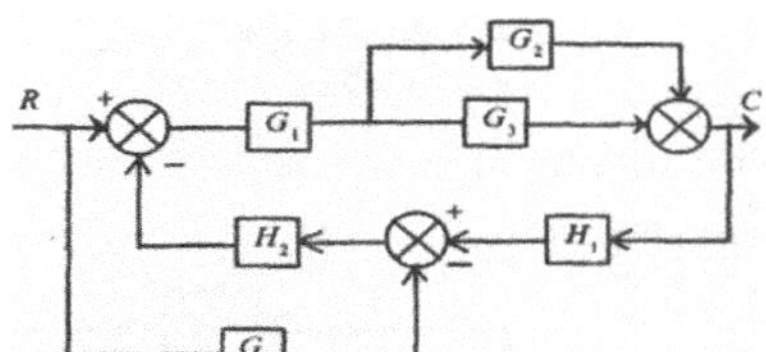

19. The block diagram of unity control feedback system id given below ,

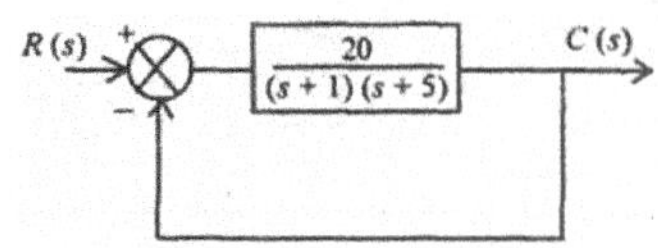

Calculate , Transfer function of the system and value of **ω_n , ζ , t_p , M_p .**

20. Using block diagram reduction technique, determine the overall transfer function for the system , as shown in figure .

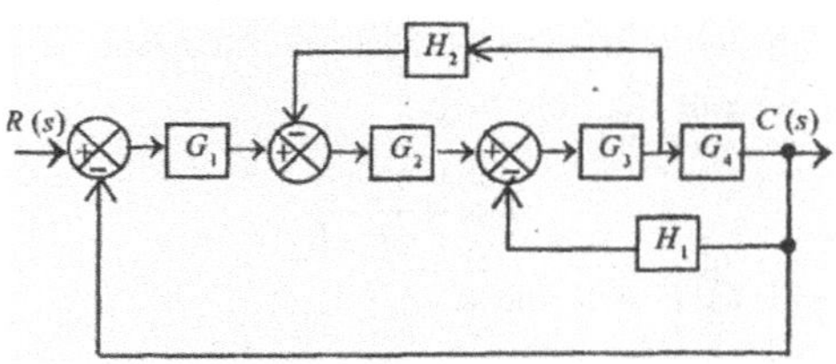

21. State Mason's gain formula. Give the procedure for converting block diagram to signal flow graph .

22. Use the block diagram reduction technique and signal flow graph method to determine the overall transfer function of the system shown in Fig.

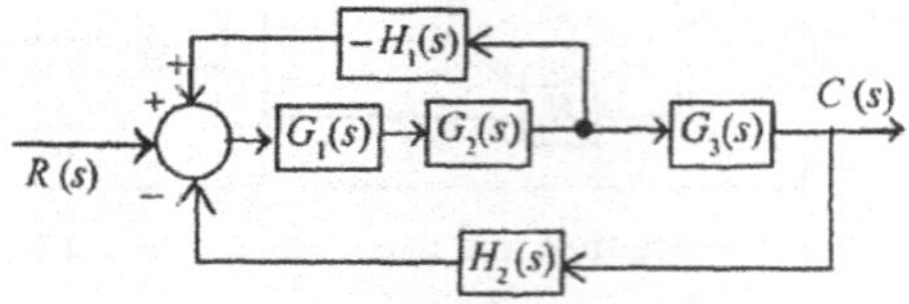

23. The forward path of a unity feedback system is described as ,

$$G(s) = \frac{36}{s(s+0.72)}$$

Determine its characteristic equation , damping ratio , peak overshoot , peak time , settling time for unit step input.

24. Obtain the transfer function **Y (s) / U(s)** of the system whose governing equations are given below

$$\dot{x}_1 = -4x_1 - x_2 + 3u$$
$$\dot{x}_2 = 2x_1 - 3x_2 + 5u$$
$$y = x_1 + 2x_2$$

25. Determine the overall transfer function relating C and R for the system whose block diagram is as Shown in Fig

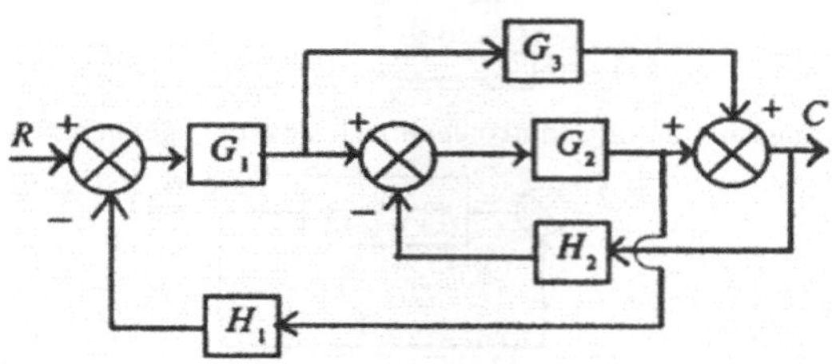

26. The forward path transfer function of a unity feedback control system is given by

$G(s) = \frac{2}{s(s+3)}$ Obtain an expression for unit step response of the system.

27. The transfer function of a unity-feedback control system is given by the following equation ,**if ω_n = 3 rad / sec and ζ = 0.2**

$$G(s) = \frac{K(s+2)}{s^3 + \beta s^2 + 4s + 1}$$

Then calculate the value of **K** and **β**

28. Obtain the transfer function **y(s) / x(s)** for following signal flow diagram as shown in figure ,

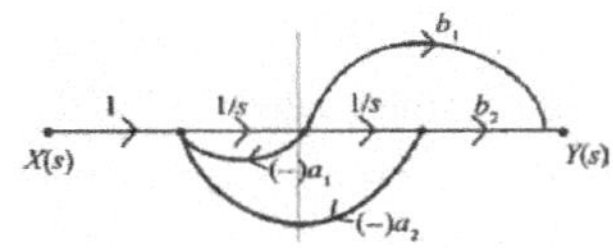

29. Obtain the transfer function **C(s) / R(s)** for following block diagram as shown in figure ,

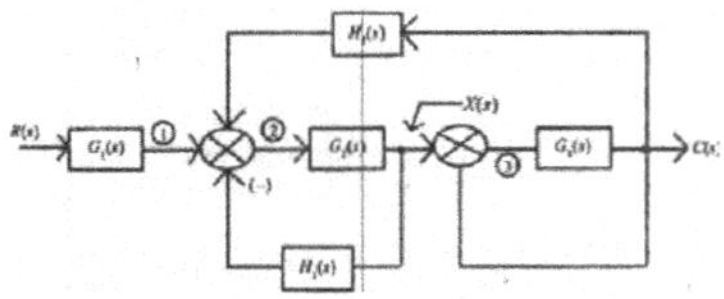

30. In closed loop control system , calculate sensitivity s $_{g1}$ & s $_{g2}$.

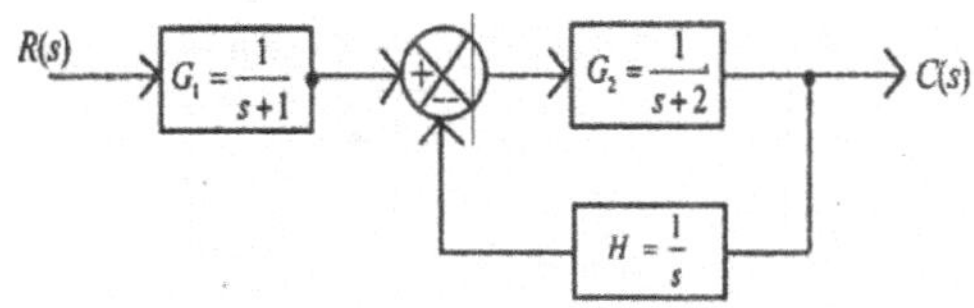

31. Obtain the transfer function of the given system as shown in Fig

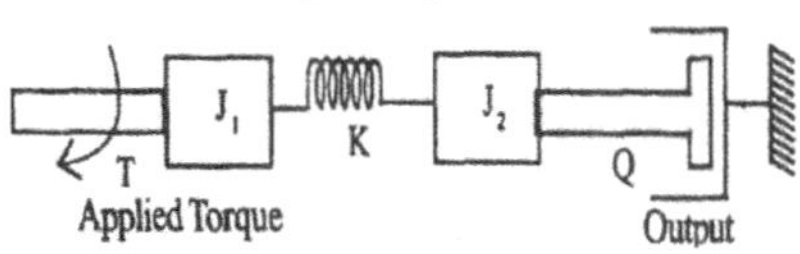

32. The unit impulse response of a system is given as ,

$e^{-t}(1 - \cos 3t)$, Determine its transfer function .

33. Draw signal flow graph in the Laplace domain for a system given by the equation

$\frac{d^2y}{dx^2} + \frac{2dy}{3dx} + \frac{11y}{2} = x$. Ignore initial conditions.

34. What are the three main factors or terms found in all process control systems? List examples of each type.

35. What is the primary requirement of any process control system? Define system error

36. A measuring system has a natural frequency of 0.5 rad/s, a damping ratio of 0.5, and a static sensitivity of 0.5 m/V. Estimate its 90% rise time and settling time if F(t) = 2 U(t) and the initial condition is zero. Plot the response y(t) and indicate its transient and steady responses

37. An instrument having a resonance frequency of 1414 rad/s with a damping ratio of 0.5 is used to measure a signal of 6000 Hz. Estimate the expected dynamic error and phase shift.

38. A force transducer having a damping ratio of 0.5 and a natural frequency of 4000 Hz is available for use to measure a periodic signal of 2000 Hz. Estimate its resonance frequency

39. Explain how system properties of static sensitivity, natural frequency, and damping ratio affect the output information from a measurement system.

40. Compute the transfer function of the depicted block diagram by reduction method.

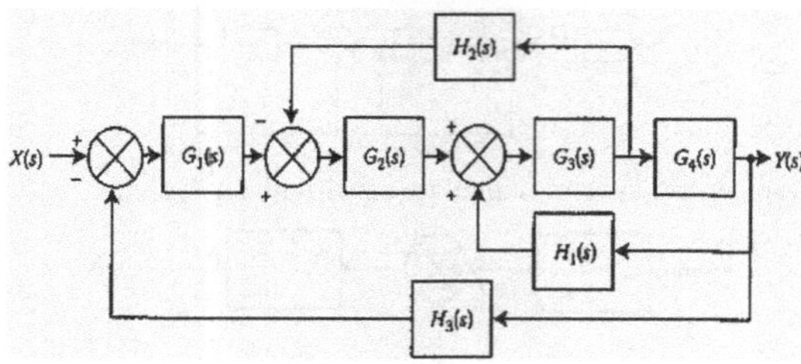

References

1. D.M. Considine, (ed.). Process Industrial Instruments and Controls Handbook, 4th ed., New York: McGraw-Hill, 1983.
2. The Foxboro Company. Introduction to Process Control. Foxboro Company, 1986.
3. C.D. Johnson, Process Control Instrumentation Technology, 2d ed., New York: John Wiley & Sons, 1982.
4. P.W. Murrill, Fundamentals of Process Control Theory, 3d ed., Research Triangle Park, NC: ISA, 2000.
5. Control system by U A Bakshi " technical publication pune year 2010 "
6. R.C. Weyrick, Fundamentals of Automatic Control, New York: McGraw-Hill, 1975.
7. G.F. Franklin et al. "Feedback Control of Dynamic Systems", 5th Edition. Upper Saddle River, NJ, 2006, Pearson Education, Inc.
8. T.E. Marlin, Process Control: Designing Processes and Control Systems for Dynamic Performance, McGraw-Hill, Pp. 61-65, 1995.
9. D.E. Seborg, T.F. Edgar, D.A. Mellichamp, Process Dynamics and Control, John Wiley, Pp. 105-110, 1989.
10. Katsuhiko Ogata, Yanjuan Yang, Modern control engineering, 1970.
11. R. Burns, Advanced control engineering. Butterworth-Heinemann, 2001.
12. Richard C. Dorf, Modern control systems, Addison-Wesley Longman Publishing Co., Inc., 1995.
13. S. Ghosh, Control Systems: Theory and Applications, 2nd Edition, Pearson.
14. D. Roy Chaudhury, Modern Control Engineering, 4th Edition, PHI.
15. J.J. Distefano, III, A.R. Stubberud, I.J. Williams, Feedback and Control Systems, 2nd Edition, TMH, Schaums Outlines.
16. K. Ogata, Modern Control Engineering, 4th Edition, PHI.
17. P.R. Stuart, Standards for the measurement of pressure', Measurement and Control, Vol.20, No.8, Pp.7–11, 1987.
18. D. Karnopp, D. Margolis, R. Rosenberg, System dynamics : Modeling and simulation of mechatronics systems, John Wiley & Sons,2000.
19. Control Systems Engineering, N S Ni J h Wil & Sons or man S Nise, John Wiley & Sons Systems Dynamics and Response, S. Graham Kelly, Thomson Publisher
20. H. Merritt, Hydraulic Control Systems, John Wiley, New York, 1967.
21. Nicolae Lobontiu, System Dynamics for Engineering Students, Academic Press.

Section–II

Lab Work

Practical 1

Object : To Study about the Resistance Temperature Detector.

Apparatus Required: Resistance temperature detector (R.T.D), Digital panel meter, Heater and container, Thermometer.

Theory : It consists of glass or metal tube which houses a ceramic mandrel on which a platinum resistance wire is wound. Such glass or metal bulb is evacuated or filled with inert gas to protect the resistance wire sensing element. The lead wires of the sensing element projected out of the ceramic mandrel. This arrangement becomes the resistance thermometer.

Basic Principle : This type of transducer is used for temperature measurement. It is based on the principle that whenever an electric conductor is subjected to temperature change, the resistance of the conductor changes. Thus, the change in resistance of the electric conductor becomes a measure of the change in temperature when calibrated. The unknown temperature is determined in terms of electrical resistance of the conductor, which senses the temperature. The change in resistance of this device is precisely determined either by bridge circuit or by ohmmeter. Resistance of a conductor changes with change in temperature.

Generally, the metals used are Platinum due to following features:

- Platinum provides good stability and accuracy.
- It can operate on wide range of temperature.
- It has good linearity over wide temperature range.
- Less errors during operation

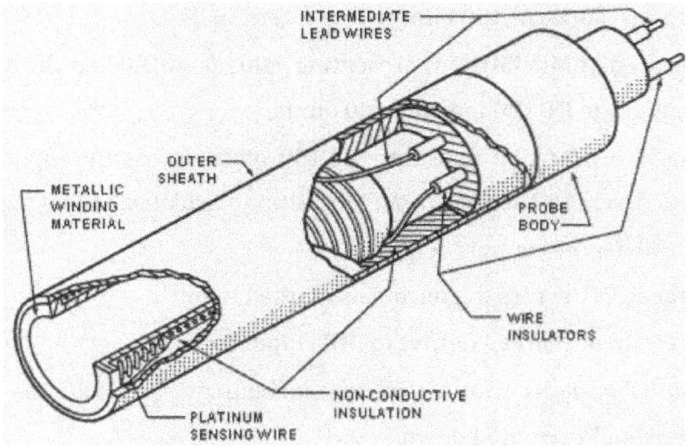

Diagram 1.1: Resistance Temperature Detector

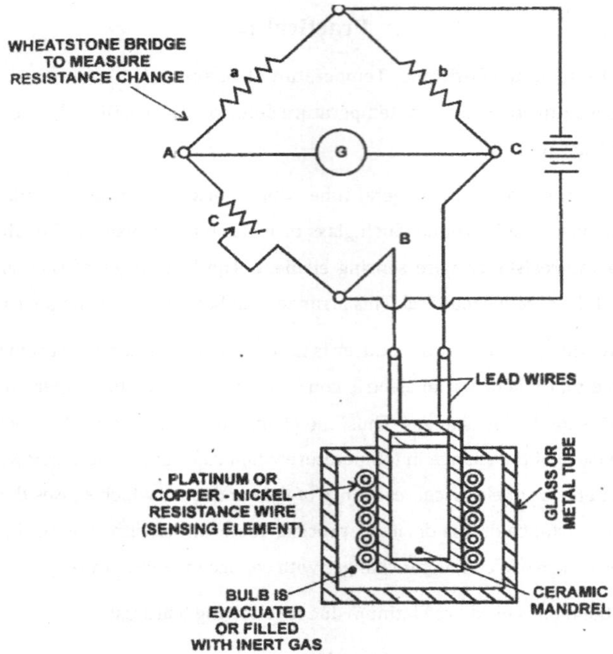

Diagram 1.2: Resistance Temperature Detector (R.T.D) Circuit Diagram

Procedure

- Set the switch $\mathbf{S_2}$ at marked – **TEMPERATURE**
- Connects a resistance of 100 ohms to input terminals
- Set pot marked at **MINIMUM** representing value **0** on DPM on the other hand at zero-degree centigrade PT 100 indicate 100 ohms.
- Now connect a precision resistance of 139 ohms across the input terminals. Adjust MAX pot (P_1) to read 100.0 on DPM. This action simulates boiling point temperature of water i.e.100 degree centigrade.
- Connect the R.T.D. across the terminals marked input.
- Connect terminals marked output to DPM input terminals.
- Note that RTD may be immersed in the boiling water and the resistance of the transducer should be noted down.
- Note down the readings of the thermometer and RTD.

Observations

S.No	Thermometer reading (°c)	Temperature on DPM (°c)	Resistance on DPM (ohms)	Percentage Error

Precautions

- Make the connections properly.
- Take the readings when they are steady.
- RTD should not touch the bottom of the vessel.
- Please handle RTD very carefully.

Graph Resistance versus Temperature

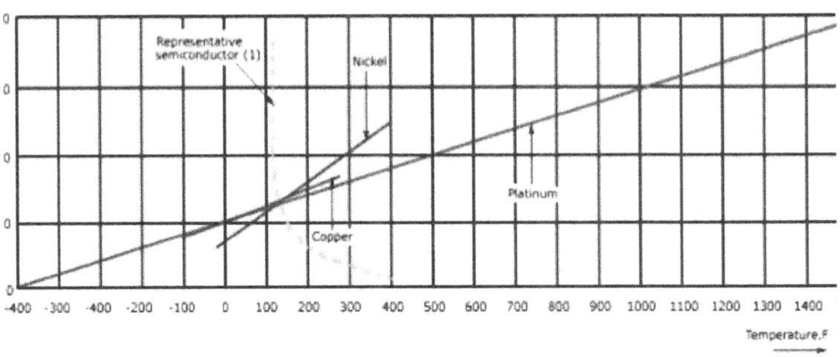

Result &discussion :- …………………

…………………

Practical 2

Objective: To study about various Temperature Measuring Instruments

- Mercury – in glass thermometer
- Thermocouple
- Electrical resistance thermometer
- Bi–metallic strip

Apparatus used: Mercury–in glass thermometer, Thermocouple, Electrical resistance thermometer, Bi–metallic strip.

Mercury–in glass thermometer: The liquid–in–glass thermometer is one of the most common temperature measuring devices. Both liquid and glass expand on heating and their differential expansion is used to indicate the temperature. The lower temperature limit is–37.8°Cfor mercury down to–130°C for pentane. The higher temperature range is 340°C but this range maybe extended to 560°C by filling the space above mercury with CO2 or N2 at high pressure thereby increasing its boiling point and range. Though the liquid–in–glass thermometer has certain laboratory applications, yet it is not much used industrially because of its fragility, and because of the inevitable proximity of the display to the measuring point. The schematic diagram is given below,

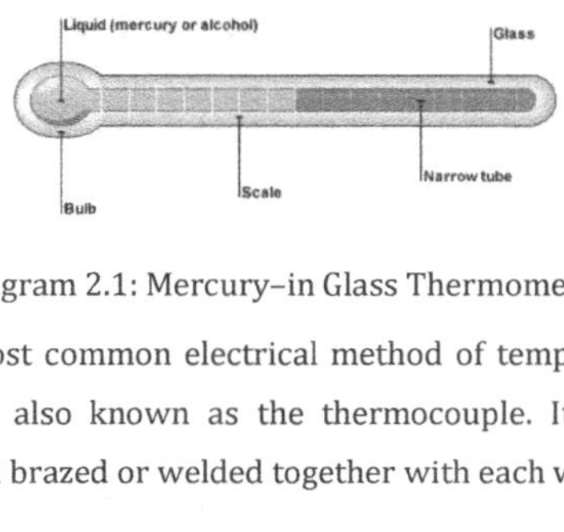

Diagram 2.1: Mercury–in Glass Thermometer

Thermocouple: The most common electrical method of temperature measurement uses the thermo-electric sensor also known as the thermocouple. It consists of two wires of different metals twisted and brazed or welded together with each wire covered with insulation which may be either:

- Mineral (**magnesium oxide**) insulation for normal duty,
- ceramic insulation for heavy duty.

When two conductors of dissimilar metals, say *A* and *B* are joined together to form a loop (thermocouple) and two unequal temperatures T_1 and $T_2$are interposed at two junctions J1 and J2respectively, then an infinite resistance voltmeter detects the electromotive force *E* or if a low resistance ammeter is connected, a current flow I is measured

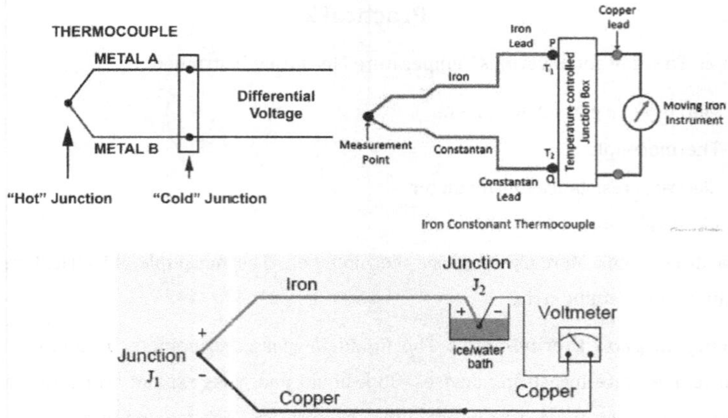

Diagram 2.2: Iron–copper Thermocouple

Electrical resistance thermometer: In resistance thermometers, the change in resistance of various materials. In general, the resistance of the highly conducting materials (metals) increases with increase in temperature and the coils of such materials are called metallic resistance thermometers. Whereas the resistance of semiconductor materials generally (not always) decreases with increases in temperature. Thermo-sensitive resistors having such negative temperature characteristics are commonly known as NTC thermistors.

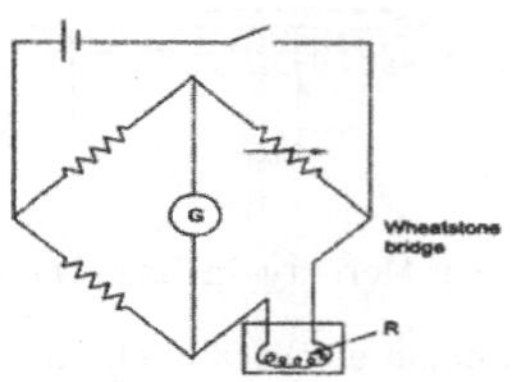

Diagram 2.3: Electrical Resistance Thermometer–Circuit Diagram

Bi-metallic strip : This type of thermometer also employs the principle of solid expansion and consists of a 'bimetal' strip usually in the form of a cantilever beam. This comprises strips of two metals, having different coefficients of thermal expansion, welded or riveted together so that relative motion between them is prevented. An increase in temperature causes the deflection of the free end of the strip, assuming that metal A has the higher coefficient of expansion. The deflection with the temperature is nearly linear, depending mainly on the coefficient of linear thermal expansion. Invar is commonly employed as the low expansion metal. This is an iron–nickel alloy containing 36% nickel. Its coefficient of thermal expansion is around 1/20th of the ordinary metals. Brass is used as high expansion material for the measurement of low temperatures whereas nickel alloys are used when higher temperatures have to be measured.

The sensitivity is improved by using a longer strip in a helical form. One end of the helix is anchored to the casing and the other end which is free is conveniently connected to the pointer which sweeps over a circular dial graduated in degrees of temperature. As shown in Figure,

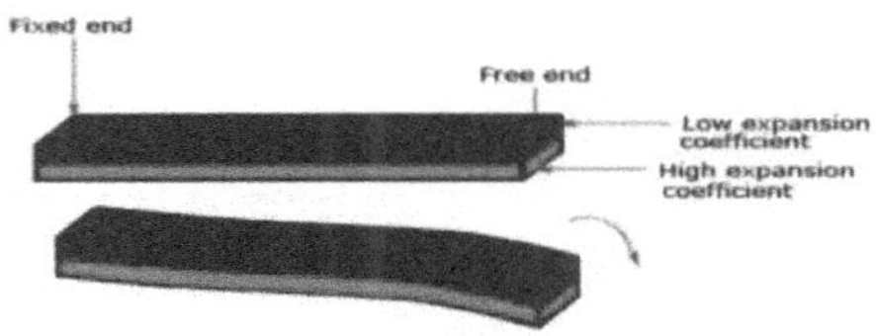

Diagram 2.4: Bi–metallic Strip

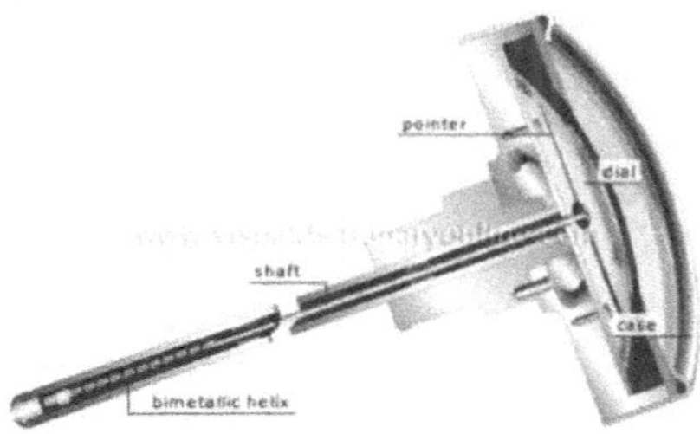

Diagram 2.5: Helical Bi–metallic Strip

Practical 3

Objective: To measure torque of a rotating shaft using torsion meter/strain gauge Torque transducer.

Apparatus required : Torsion meter/strain gauge torque transducer.

Introduction: Instruments for measuring torque and power in rotating system. The power transmitted can be calculated from the torque, using the equation,

$$P = \omega T$$

Where,

P is the power (in watts),

T is torque (N m)

ω is angular speed (rad / s).

Devices used for power measurement are also known as dynamometers,

- Transmission type dynamometers, in which the power being transmitted through the device is measured. The device is neither a power generator nor a power absorber and is used on the shaft transmitting power, between the prime mover and the load.
- Driving type dynamometer, in which drive is obtained from the dynamometer itself or the dynamometer is the power generator like an electric motor.
- Absorption type dynamometer, in which mechanical energy is absorbed after it is measured. The power generator may be an engine or a motor.

Theory &Working: Angular twist θ in a shaft due to torque is given by equation

$$\mathbf{T = G\,\theta\pi d^4 / 32\,L}$$

After measurement of θ, the torque 'T' can be easily calculated from the above relations. Either an electrical or optical arrangement can be used to measure θ.

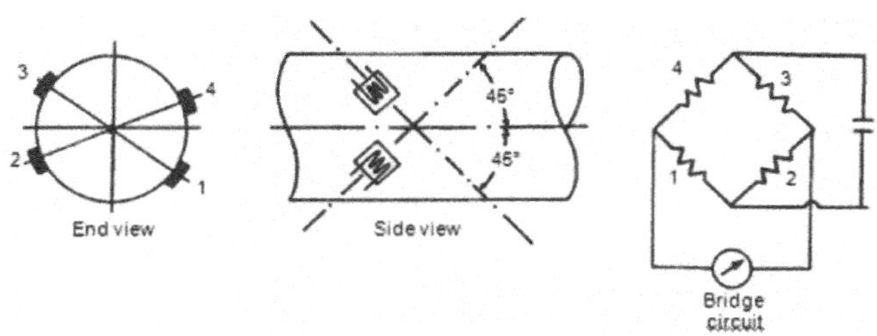

Diagram 3.1: Strain Gauge Torque Transducer

In the optical arrangement two discs A and B are mounted at a distance 'L' on the shaft, move relative to each other through an angle θ. This is recorded by the observer. Disc 'C' is only to help in viewing the relative angular displacement of A & B. Optical techniques for torque measurements have become available recently with the development of laser diodes and fibre-optic light transmissicn systems. One such system is shown in Figure given below. Two black-and-white striped wheels are mounted at either end of the rotating shaft and are in alignment when no torque is applied to the shaft. Light from a laser diode light source is directed by a pair of optic-fibre cables onto the wheels. The rotation of the wheels causes pulses of reflected light and these are transmitted back to a receiver by a second pair of fiber-optic cables. Under zero torque conditions, the two pulse trains of reflected light are in phase with each other. If torque is now applied to the shaft, the reflected light is modulated. Measurement by the receiver of the phase difference between the reflected pulse trains therefore allows the magnitude cf torque in the shaft to be calculated.

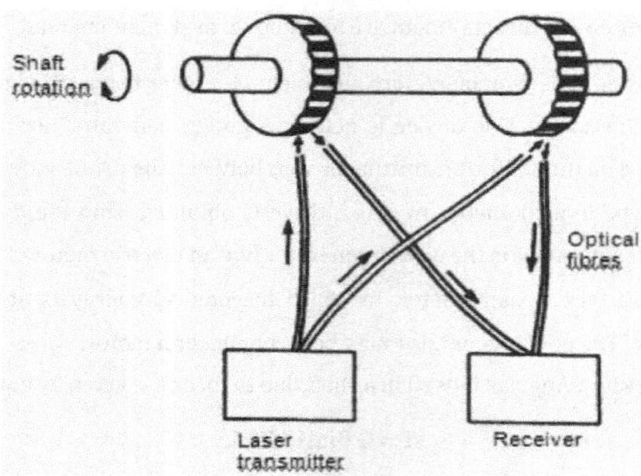

Diagram 3.2: Torsion Meter

Result &discussion :- ....

Practical 4

Objective : To measure the linear displacement by using L.V.D.T

Apparatus required : Digital instrumentation tutor , L.V.D.T

Theory: LVDT converts linear motion into an electrical signal. It is used for measuring displacement. The LVDT consists of a primary winding and two secondary windings, which are wound on cylindrical former. The two secondary windings S1 & S2 have equal number of turns. The secondary windings are placed identically on either side of the primary winding. The primary winding is connected to an a.c. source. A movable core is placed inside the cylindrical former.

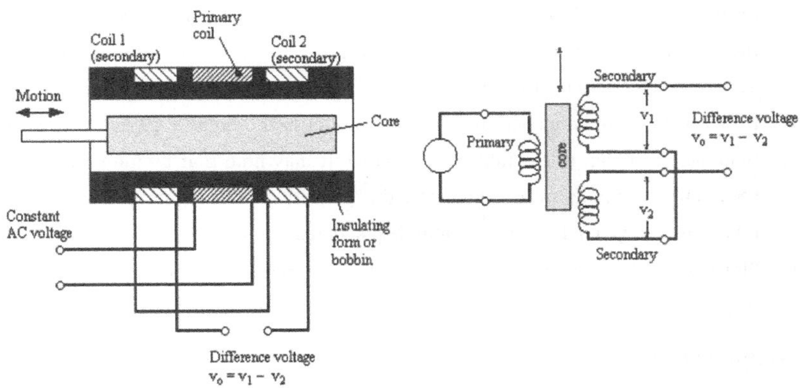

Diagram 4.1: L.V.D.T& Respective Circuit Diagram

Working Principle : As the primary winding is connected to the a.c. source. It is excited and hence a magnetic field is produced. Due to this magnetic field, a voltage is induced in the secondary windings. The differential output is $E_0=(E_{s1}-E_{s2})$ When the core is in the normal (null) position, the magnetic field linking with both the secondary windings S_1& S_2 are equal. Hence the e.m.f. induced in them are also equal Therefore null position E_0 = 0, (Es1 = Es2). When the core is moved to the right of the null position, more magnetic field links with the winding S_2 and less with winding S_1. Therefore $E_{s2}> E_{s1}$. The output voltage $E_0=E_{s2} - E_{s1}$ and is in the phase E_{s2}. When the core is moved to the left of the null position more magnetic field links with the winding S_1 and less with the winding S_2. Therefore $E_{s1}> E_{s2}$. The output voltage of the L.V.D.T. gives a measure of the physical position of the core and it's displacement.

Procedure

- Connect the terminals marked "PRIMARY" on the front panel of the instrument to the terminals marked "PRIMARY" on the transducer itself, with the help of flexible wires provided along with. Observe the colour code for the wires provided and the colour of the binding posts an d similarly connect the terminals marked secondary.
- Keep pot marked "MAX" in most anti clock-wise position.
- The magnetic core may be displaced and the pointer may be brought to zero position. If the DPM is not indicating zero, use potentiometer marked "MIN" to get a zero on DPM at zero mechanical position. If the core is displaced in both directions, the meter must show indications with appropriate polarity. Now displace the core to 19 mm positions in one of the directions. Adjust the "MAX" pot to get an indication of 19.00 on the DPM under this condition. Now the setup is ready for experimentation. You may again check for zero position also.
- Now the core can be moved by a known amount in the range ± 19 mm and the DPM readings are taken in the table given below. It may note that by inter changing the secondary terminals or the primary, the polarity of the meter indication can be reversed for a given direction of input displacement.
- Plot the graph of input displacement and the output indication on the x and y axis respectively

Observations Table

Sl.No.	Input Displacement (mm)	Output Indication (mm)	Difference (mm)	Percentage Error

Precautions

1. Move the core with a gentle fashion.
2. While connecting lead wire from panel to transducer make proper connections with respective colour code.
3. Avoid starting of the excitations source terminals.
4. Do not try to affect the core movement beyond 20 mm as per the given range.

Observations

- Study the linearity of input and output displacements.
- Note the effects of interchanging the secondary connections on the meter output polarity
- Note that when the core is mechanically at zero position, a small electrical output is obtained due to imperfections of the transducer.

Result &discussion : ....

Practical 5

Objective : Measuring torque and power generated by a prime mover by using pony brake dynamometer.

Apparatus required: Tachometer, Measuring tape, spring balance.

Theory: A rope brake dynamometer consists of one or more rope wrapped around the flywheel of engine and brake drum whose power is to be measured. The ropes are placed evenly across the width of the drum. The upward rings of ropes are connected to the spring balance on each side. The rotation of flywheel produces frictional force and ropes are tightened and consecutively a force is exerted on the drum. Due to this enormous amount of heat is produced. The heat is removed using water as coolant. The schematic diagram of device is given below,

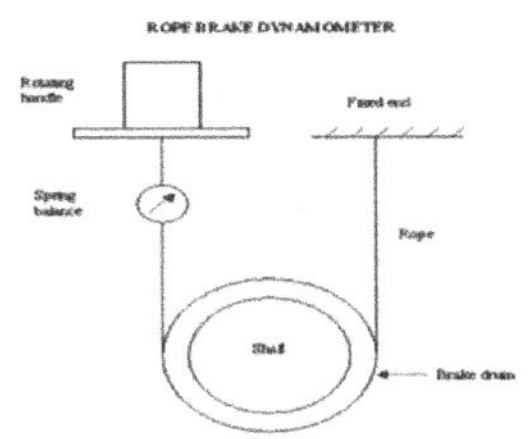

Rope Brake Dynamometer

Diagram 5.1: Rope Brake Dynamometer

Procedure

- Check the fuel supply to engine, lubrication oil level in the oil pump, water circulation in the cooling system, etc.
- Start the engine and ensure no load condition on the brake drum.
- Allow the engine to stabilize before loading.
- Now the load is applied gradually on the rope brake dynamometer.
- For every increase in load, the speed is measured using tachometer.
- Repeat the procedure for various load and power is measured.
- Calculate the torque and power using the given formula.

Torque T.

$$T = F \times R$$

Where

F = Applied Force

R = Radius of rope on Drum

Work done & power calculation

Work done / minute = Force x distance = F **X** R

= Average torque **x** Angular displacement = **(2 π N T / 60)**

Then power (p) =(2 π N T / 60) watts

Sample Data

- Speed, N = 1500 rpm
- Radius of the brake drum, R = 0.2 m
- Theoretical power applied = 1.5 kW
- Theoretical torque = P*60/(2*pi*N) N-m
- Actual Torque= W * 9.81 * R N-m

Result & discussion:- ....

Practical 6

Object : Measuring velocity of fluid flow with the help of Ventura meter and orifice meter .

Apparatus required: Ventura meter/orifice meter/Pitot-tube, U–tube manometer, Differential pressure gauge.

Theory: Venturi-meter and orifice meter are the commonly used flow meters for measuring mass/volumetric flow rate or velocity of the flowing fluid. These flow meters are also known as variable head meters.

Venturi-meter: The venture-meter has a converging conical inlet, a cylindrical throat and a iverging recovery cone. The following figure shows the venture-meter with uniform cylindrical section before converging entrance, a throat and divergent outlet.

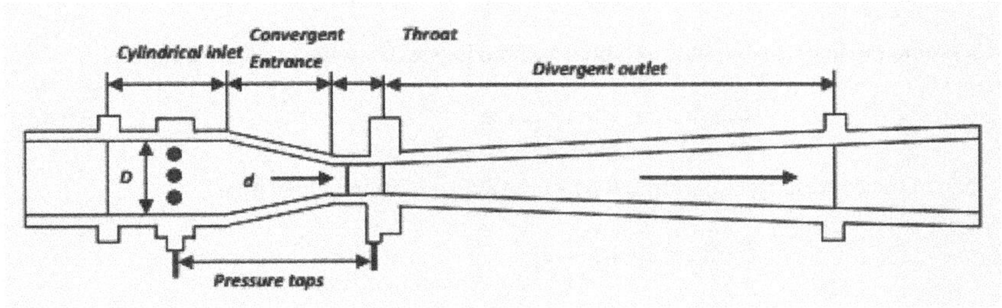

Diagram 6.1: Venturi-Meter

The converging inlet section decreases the area of the fluid stream, causing the velocity to increase and the pressure to decrease. The low pressure is measured in the center of the cylindrical throat as the pressure will be at its lowest value, where neither the pressure nor the velocity will be changing. As the fluid enters the diverging section the pressure is largely recovered lowering the velocity of the fluid. The major disadvantages of this type of flow detection are the high initial costs for installation and difficulty in installation and inspection. The Venturi effect is the reduction in fluid pressure that results when a fluid flows through a constricted section of pipe. The fluid velocity must increase through the constriction to satisfy the equation of continuity, while its pressure must decrease due to conservation of energy:

The gain in kinetic energy is balanced by a drop in pressure or a pressure gradient force. An equation for the drop in pressure due to Venturi effect may be derived from a combination of Bernoulli's principle and the equation of continuity.

Operation of Venture-meter

- The fluid whose flow rate is to be measured enters the entry section of the venture-meter with a pressure $\mathbf{P_1}$.
- As the fluid from the entry section of venture-meter flows into the converging section, its pressure keeps on reducing and attains a minimum value $\mathbf{P_2}$ when it enters the throat. That is, in the throat, the fluid pressure $\mathbf{P_2}$will be minimum.
- The differential pressure sensor attached between the entry and throat section of the venture-meter records the pressure difference $\mathbf{(P_1\text{-}P_2)}$ which becomes an indication of the flow rate of the fluid through the pipe when calibrated.
- The diverging section has been provided to enable the fluid to regain its pressure and hence its kinetic energy. Lesser the angle of the diverging section, greater is the recovery.

Assume incompressible flow and no frictional losses, from Bernoulli's Equation

$$\frac{P_1}{\gamma}+\frac{V_1^2}{2g}+Z_1=\frac{P_2}{\gamma}+\frac{V_2^2}{2g}+Z_2 \quad \text{.............................. 1}$$

Use of the continuity Equation Q = A1V1 = A2V2, equation (1) becomes

$$\frac{P_1-P_2}{\gamma}+Z_1-Z_2=\frac{V_2^2}{2g}\left[1-\left(\frac{A_2}{A_1}\right)^2\right] \text{.................................2}$$

$$V_2=\frac{1}{\sqrt{1-\left(\frac{A_2}{A_1}\right)^2}}\sqrt{2g\left(\frac{P-P_2}{\gamma}+(Z_1-Z_2)\right)} \quad \text{..............................3}$$

Theoretical

$$Q_{theo}=A_2\,V_2=\frac{A_2}{\sqrt{1-\left(\frac{A_2}{A_1}\right)^2}}\sqrt{2g\left(\frac{P_1-P_2}{\gamma}+(Z_1-Z_2)\right)} \quad \text{................................... 4}$$

$$Q_{theo}=A_2\,V_2=\frac{A_2}{\sqrt{1-\left(\frac{A_2}{A_1}\right)^2}}\sqrt{2g(\Delta h)} \quad \text{......................... 5}$$

The term $\frac{P_1-P_2}{\gamma}+(Z_1-Z_2)$ represents the difference in piezo-metric head (Δh) between the two sections 1 and 2. The above expression for V2 is obtained based on the assumption of one-dimensional frictionless flow. Hence the theoretical flow can be expressed as

$$Q_{theo}=\sqrt{\frac{2g\Delta h}{\left(\frac{1}{A_2^2}-\frac{1}{A_1^2}\right)}} \quad \text{.............................. 6}$$

Because of the above assumptions, the actual flow rate, Qact from Qtheoand the ratio between them is called the discharge coefficient, Cd which can be written as

$$C_d = \frac{Q_{act}}{Q_{theo}} \quad \text{.....................................} \; 7$$

Application

- It is used where high-pressure recovery is required.
- Can be used for measuring flow rates of water, gases, suspended solids, slurries and dirty liquids.
- Can be used to measure high flow rates in pipes having diameters in a few meters.

Limitations

- They are large in size and hence where space is limited, they cannot be used.
- Expensive initial cost, installation and maintenance.
- Require long laying length. That is, the venture-meter has t_i be proceeded by a straight pipe which is free from fittings and misalignments to avoid turbulence in flow, for satisfactory operation. Therefore, straightening vanes are a must.
- Cannot be used in pipes below 7.5 cm diameter.

Result & discussion :- ....

Orifice Meter

Apparatus required : Orifice Meter, U – tube manometer , Differential pressure gauge

working Principle : When an orifice plate is placed in a pipe carrying the fluid whose rate of flow is to be measured, the orifice plate causes a pressure drop which varies with the flow rate. This pressure drop is measured using a differential pressure sensor and when calibrated this pressure drop becomes a measure flow rate.

Theory of Orifice Meter : The main parts of an orifice flow meter are as follows:

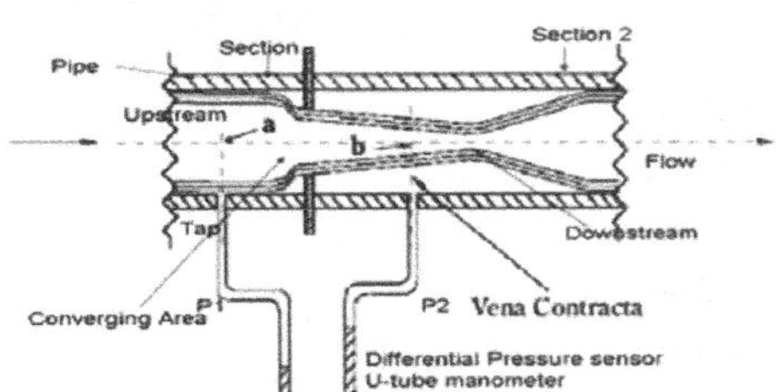

Diagram 6.2: Orifice Meter

- A stainless-steel orifice plate which is held between flanges of a pipe carrying the fluid whose flow rate is being measured.
- It should be noted that for a certain distance before and after the orifice plate fitted between the flanges, the pipe carrying the fluid should be straight in order to maintain laminar flow conditions.
- Openings are provided at two places 1 and 2 for attaching a differential pressure sensor (U-tube manometer, differential pressure gauge etc) as shown in the diagram.

Operation of Orifice Meter

- The detail of the fluid movement inside the pipe and orifice plate has to be understood.
- The fluid having uniform cross section of flow converges into the orifice plate's opening in its upstream. When the fluid comes out of the orifice plate's opening, its cross section is minimum and uniform for a particular distance and then the cross section of the fluid starts diverging in the downstream.
- At the upstream of the orifice, before the converging of the fluid takes place, the pressure of the fluid(P_1) is maximum. As the fluid starts converging, to enter the orifice opening its pressure drops. When the fluid comes out of the orifice opening, its pressure is minimum (P_2) and this minimum pressure remains constant in the minimum cross section area of fluid flow at the downstream.
- This minimum cross sectional area of the fluid obtained at downstream from the orifice edge is called **VENA-CONTRACTA.**
- The differential pressure sensor attached between points 1 and 2 records the pressure difference($P_1 - P_2$) between these two points which becomes an indication of the flow rate of the fluid through .
- The discharge may calculated by the following relationship ,

$$Qa = (Cd.A2/\sqrt{1-(A2/A1)}\hat{}2)*(\sqrt{2(P1-P2)/P})$$

Where, Q_a = flow rate

C_d = Discharge coefficient

A_1 = Cross sectional area of pipe

A_2 = Cross sectional area of orifice

P_1, P_2 = Static Pressures

Result &discussion :- ....

Practical 7

Object : Measuring velocity of fluid flow with the help of pitot-tube .

Apparatus required :pitot-tube

Theory of Pitot tube : The Pitot tube is a simple device that allows for the measurement of the flow pressure in a moving fluid. This device is a section of tube that measures the pressure at the tip and the pressure at the side of the tube. Reading this differential pressure and applying Bernoulli's equation will allow for the calculation of the fluid velocity.

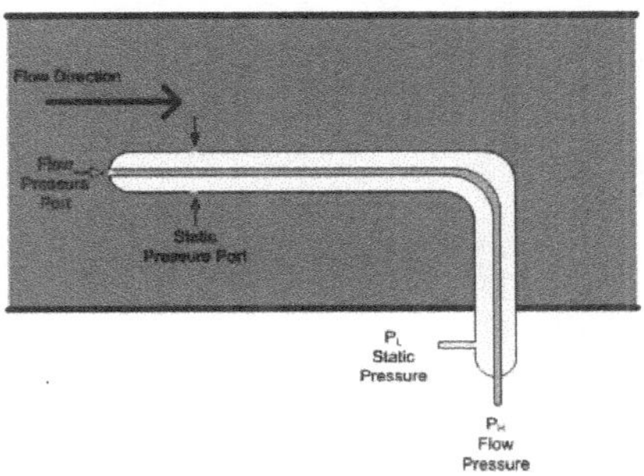

Diagram 7.1: Pitot Tube

The above diagram shows how the Pitot tube is constructed of two tubes, one inside the other, to create a static pressure port and a flow pressure port. Applying Bernoulli's equation we get:

$$P_S + r\left(\frac{V^2}{2}\right) = P_F \qquad \text{Where} \quad \begin{aligned} r &= \text{Density} \\ V &= \text{Velocity} \\ P_S &= \text{Static Pressure} \\ P_F &= \text{Flow Pressure} \end{aligned}$$

If we solve for the velocity we get the following equation:

$$V^2 = \frac{2(P_F - P_S)}{r}$$

With the velocity of the fluid now known, you can simply multiply it by the area of the duct to get the total volume flow.

Then discharge will be –

$$Q = C_d A \sqrt{\frac{2(P_f - P_s)}{r}}$$

P_s = Static Pressure

P_F = Flow Pressure

C_d = Coefficient of Discharge

A = Cross sectional area of pipe

Result & discussion :- ....

Practical 8

Object: Study of various pressure measuring devices like manometers, mercury in glass pressure gauge etc.

Apparatus required: U-tube manometer, inclined-tube manometer, Multiple liquid manometer, Float-type manometer, Mercury in glass pressure gauge, Bourdon tube pressure gauge, McLeod Vacuum Gauge

Theory

U-tube manometer: The simplest type of liquid column manometer is the U-tube manometer. When the pressures $\mathbf{P_1}$ and $\mathbf{P_2}$ are equal, the height difference $\mathbf{\Delta h}$-and therefore $\mathbf{\Delta p}$-is zero. With the same internal diameter, surface consistency and material, the capillary elevation has no effect, the schematic diagram is given below.

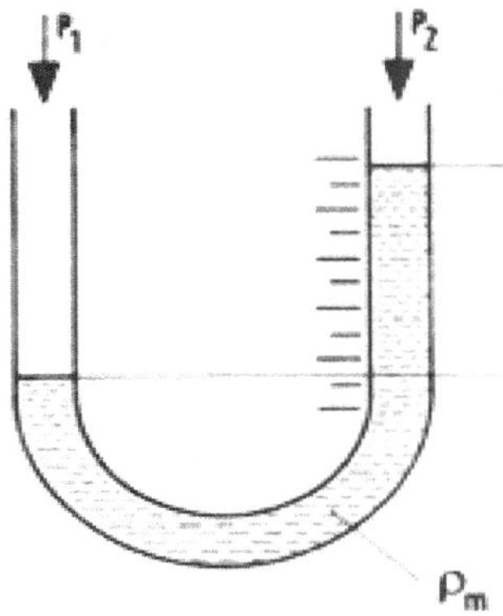

Diagram 8.1: U-tube Manometer

The maximum pressure differences $\mathbf{\Delta p}$ depend on the length of the tubes and on the density of the liquid.

Inclined-tube manometer: The inclined-tube manometer is used to measure very low pressure, The sloping design of the tube stretches the graduation by an amount proportional to the angle of inclination $\boldsymbol{\alpha}$. For this reason, the angle of inclination of many inclined tube manometers can be adjusted. With unequal areas $\mathbf{A_1}$ and $\mathbf{A_2}$, the graduation will need to be corrected accordingly due to the changing level of liquid at $\mathbf{A_1}$. For high precision, the measurement must be made very carefully. Generally, these instruments are equipped with a bubble leveler for precise horizontal adjustment. the schematic diagram is given below.

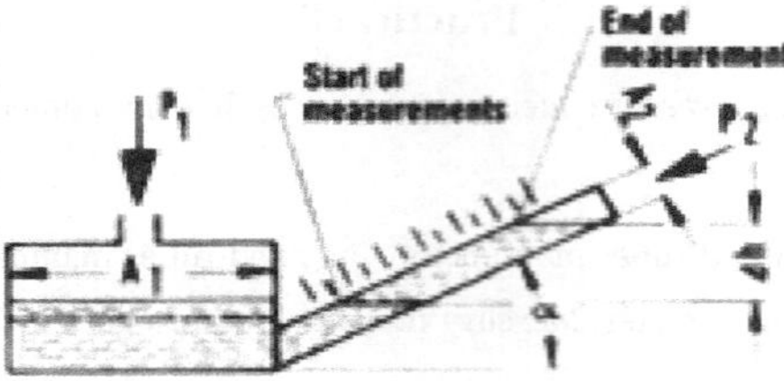

Diagram 8.2: Inclined-tube Manometer

Multiple liquid manometer: A multiple liquid manometer allows magnification of the measuring range by a factor of 8 to 10 because the measurement is based solely on the difference of the two densities. With multiple liquid manometers it is important that the separating liquids not mix with each other nor with the process fluid. If the process fluid density ρ_m and the separating liquid density ρ_{m1} differ, the change of height of the upper liquid level must be taken into account. This is particularly important for the measurement of gas pressures. The schematic diagram is given below.

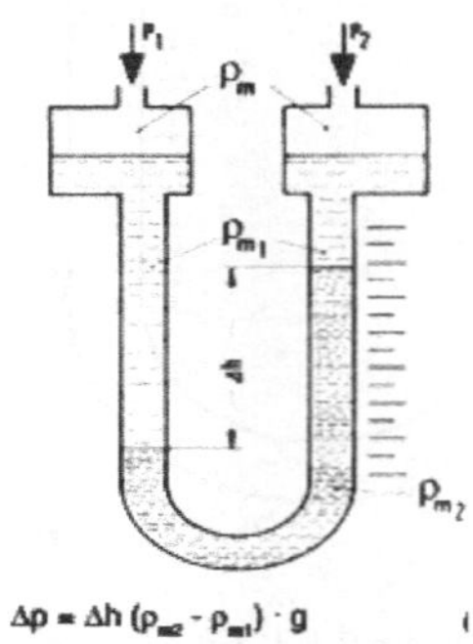

Diagram 8.3: Multiple Liquid Manometer

Float-type manometer: The float-type manometer tries to combine the advantages of easy reading on a graduated scale with the advantages of a liquid column. A float **S** follows the height of the liquid column and relays this height to the outside. The measuring range can be changed by reversing the ratio $\mathbf{A_1 : A_2}$. The main problem with this type of instrument is that the friction occurring from the pressure-tight transmission of the measurement results adds error to the reading. Additional equipment can be added to the float-type manometer that determines the position of the float from the outside which then transmits the results to the graduated scale. The schematic diagram is given below.

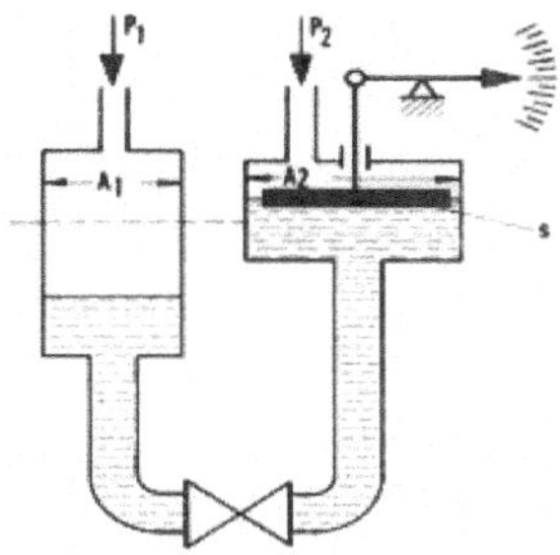

Diagram 8.4: Float-type Manometer

Mercury in glass pressure gauge : Mercury barometer consists essentially of a glass tube sealed at one end and mounted vertically in a bowl or cistern of mercury so that the open end of the tube is submerged below the surface of mercury in the cistern. In the Fig. 8.5 , the zero level of mercury in the cistern is shown under the influence of atmospheric pressure P_1. When $P_2 = P_1$ = Atmospheric pressure, the zero level in the cistern can be marked. If the tube is open and if different pressures act in the cistern and in the tube, then there will be difference in the levels of mercury. If P_1 is greater than P_2 as shown in Fig.8.5 (ii), then mercury will be forced down in the cistern and corresponding rise will be there in the tube.

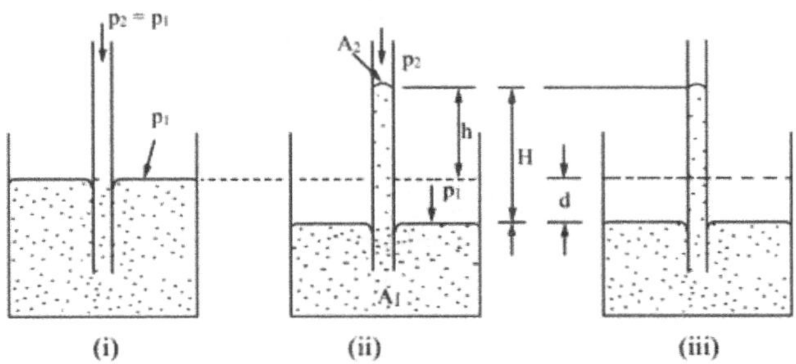

Diagram 8.5: Mercury in Glass Pressure Gauge

The balance of pressures will be given by $\mathbf{P_1 = P_2 + H\rho}$ (where **H** is the difference between levels of mercury in the cistern and in the tube and **ρ** the density of mercury). From the Figure (ii), $\mathbf{H = h + d}$. The quantity of mercury that has left the cistern is same as that has risen in the tube.

$$A_1 d = A_2 h$$

where $\mathbf{A_1}$ and $\mathbf{A_2}$ are areas of the cistern and the tube respectively.

$$d = \frac{A_2}{A_1}h$$

$$\therefore\ H = h + h\left(\frac{A_2}{A_1}\right)$$

$$= h\left(1 + \frac{A_2}{A_1}\right)$$

$$p_1 = p_2 + h\left(1 + \frac{A_2}{A_1}\right)\rho$$

$p_2 \rightarrow 0$ in practical cases as the tube will be evacuated and sealed.

$$p_1 = h\left(1 + \frac{A_2}{A_1}\right)\rho$$

Bourdon tube : Bourdon tube is the oldest pressure sensing element .It is a length of metal tube of elliptical cross section and shaped into letter " **C** ".One end is left free and the other end is fixed and is open for the pressure source to be applied. A tube of elliptical cross section has a smaller volume than a circular one of the same length and perimeter. As shown in figure ,

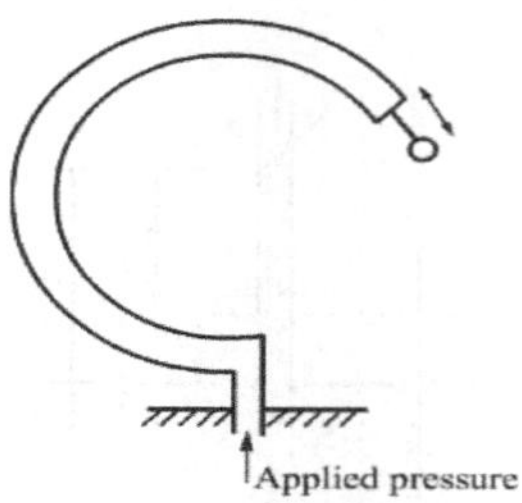

Diagram 8.5: Bourdon Tube Pressure Gauge

When connected to the pressure source it is made to accommodate more of the fluid. Resultant of all reactions will produce maximum displacement at the free end. Within close limits, the change in angle subtended at the center by a tube is proportional to the change of internal pressure and within the limits of proportionality of the material; the displacement of the free end is proportional to the applied pressure. The ratio between major and minor axes decides the sensitivity of the Bourdon tube. Materials of the Bourdon tube is Phosphor bronze, Beryllium bronze or Beryllium Copper

McLeod Vacuum Gauge : The main parts of McLeod gauge are as follows:

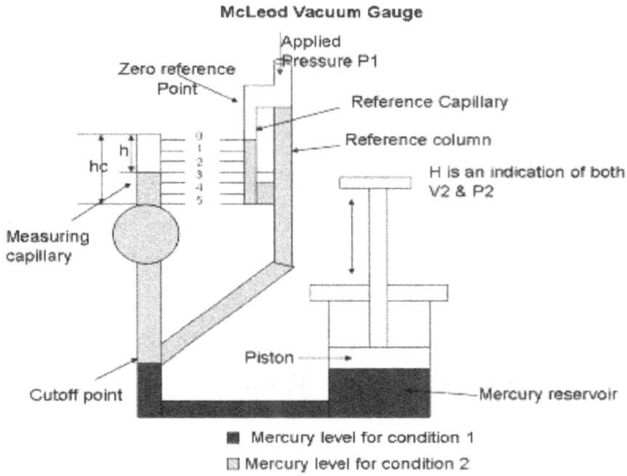

Diagram 8.7: McLeod Vacuum Gauge

A reference column with reference capillary tube. The reference capillary tube has a point called zero reference point. This reference column is connected to a bulb and measuring capillary and the place of connection of the bulb with reference column is called as cut off point. (It is called the cutoff point, since if the mercury level is raised above this point, it will cut off the entry of the applied pressure to the bulb and measuring capillary. Below the reference column and the bulb, there is a mercury reservoir operated by a piston.

Operation of McLeod Vacuum Gauge

The pressure to be measured **(P_1)** is applied to the top of the reference column of the McLeod Gauge as shown in diagram. The mercury level in the gauge is raised by operating the piston to fill the volume as shown by the dark shade in the diagram. When this is the case (condition – 1), the applied pressure fills the bulb and the capillary. Now again the piston is operated so that the mercury level in the gauge increases. When the mercury level reaches the cutoff point, a known volume of gas **(V_1)** is trapped in the bulb and measuring capillary tube. The mercury level is further raised by operating the piston so the trapped gas in the bulb and measuring capillary tube are compressed. This is done until the mercury level reaches the "Zero reference Point" marked on the reference capillary (condition – 2). In this condition, the volume of the gas in the measuring capillary tube is read directly by a scale besides it. That is, the difference in height **'H'** of the measuring capillary and the reference capillary becomes a measure of the volume **(V_2)**and pressure **(P_2)** of the trapped gas. Now as **V_1,V_2 and P_2** are known, the applied pressure **P_1** can be calculated using Boyle's Law given by;

$$P_1V_1 = P_2V_2$$

Let the volume of the bulb from the cutoff point up to the beginning of the measuring capillary tube = **V**

Let area of cross – section of the measuring capillary tube = a

Let height of measuring capillary tube = **h_c.**

Therefore, Initial Volume of gas entrapped in the bulb plus measuring capillary tube

$$V_1 = V + ah_c.$$

When the mercury has been forced upwards to reach the zero-reference point in the reference capillary, the final volume of the gas = **V_2 +ah.**

Where,

h = height of the compressed gas in the measuring capillary tube

P_1 = Applied pressure of the gas unknown.

P_2 = Pressure of gas at final condition, that is, after compression = **P_1+h**

We have

$$P_1V_1 = P_2V_2 \text{ (Boyle's Law)}$$

Therefore ,**$P_1V_1 = (P_1 + h)\ ah\ P_1 V_1 = P_1 ah + ah^2$**

$P_1V_1 - P_1ah = ah^2\ P_1 = ah^2 / (V_1 - ah)$

Since ah is very small when compared to **V_1**, it can be neglected. Therefore,

$P_1 = ah^2 / V_1$Thus the applied pressure is calculated using the McLeod Gauge.

Applications : The McLeod Gauge is used to measure vacuum pressure.

Advantages of the McLeod Gauge

- It serves as a reference standard to calibrate other low-pressure gauges.
- There is no need to apply corrections to the McLeod Gauge readings.

Limitations of McLeod Gauge

- The gas whose pressure is to be measured should obey the Boyle's law
- Moisture traps must be provided to avoid any considerable vapor into the gauge.
- It cannot give a continuous output.

Result &discussion: ....

Practical 9

Object : To Study about the measuring devices like Vernier Caliper & Micrometer

Apparatus: Vernier Caliper, Micrometer

Working principle: When two scales or division slightly different in size are used, the difference between them can be utilized to enhance the accuracy of measurement.

Type of Vernier Caliper: There are three type of Vernier Caliper generally used in measurement

- Type A has jaws on both side for internal and external measurement and also has a blade for depth measurement.
- Type B is provided with jaw on one side for internal and external measurement.
- Type C has jaws on both sides for making the and for making operations.

Working: Vernier caliper consists of two steel rules and these can slide along each other. The main scale is engraved on solid L-shaped frame. On this scale cm. graduation are divided in to 20 parts. So that one small division equals 0.05 cm. One end of the frame consist a fixed jaw which is shaped into a contact lip at it's extremely. The alignment of the distance boundaries with the corresponding graduation of the rule is ensured by means of the positive contact member. The sliding jaw, coupled to a Vernier scale contains another measuring at it's left. The final adjustment of movable jaws can be done by the adjusting screw.

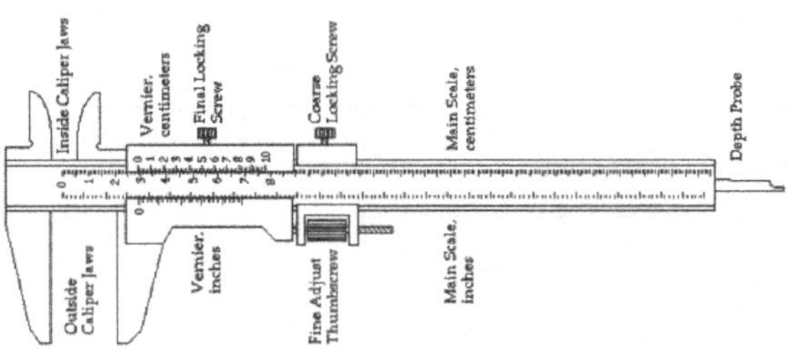

Diagram 9.1: Vernier Caliper

First the whole movable jaws assembly is adjusted so that the two measuring tips just touch the part to be measured. The lock nut B is tightened. Final adjustment depending upon the sense of correct feel is made by the adjusting screw. The movement of adjusting screw makes the part containing locking nut A and sliding jaws to move, q\as the adjusting screw rotates on a screw which is in way fixed to the movable jaw. After final adjustment has been done, the locking nut A is also tightened and the reading is noted down. The measuring tips are so designed as to measure inside as well outside dimensions

Calculation of Least Count

Suppose Vernier scale reading = 65 division. This division consider with exactly 63th on the main scale.

LEAST COUNT = (division coincide with main scale/division coincide with Vernier scale)= (63 /65) =0.96

Therefore, Difference between main scale and Vernier scale small division and one Vernier division is **1 - 0.96** = **0.04** mm.

Reading :suppose that small division of the main scale is 0.028 unit. Say the Vernier scale contain 28 division and these coincide exactly 27 division of main scale. So now one Vernier division is equal to 1/28 of 27 scale division i.e.(1/28 x 27 x 0.028) = 0.027 units. Therefore, the difference between one main scale small division and one Vernier division (Least count of instrument) equal 0.028 -0.027 i.e. 0.001 unit.

Observation Table

Sr.no	M.S.R. (A)	V.S.R x LC (B)	Total C=A+B	Average Dimension
1				
2				
3				
4				

Result &Discussion :- ....

....

....

Practical 10

Object : To Study about the measuring devices like Micrometer.

Apparatus : Micrometer

Working Principle : The micrometer screw gauge essentially consist of an accurate screw having about 10 or 20 thread per cm and revolves in a fixed nut. The end of the screw forms one measuring tip and the other measuring tip is constituted by a stationary anvil in the base of the frame the screw is threaded for certain length and is plain at inward. The plain portion is called sleeve and its end is the measuring surface. The spindle is advanced or retracted by turning a thimble connected to the spindle. The barrel is graduated in unit of 0.05 cm i.e. 20 divisions per cm, which is the lead of the screw for one complete revolution. The thimble has. 25 divisions around its periphery noncircular portion. Thus it sub-divides each revolution of the screw in 25 equal parts i.e. each division corresponds to 0.002 cm. the schematic diagram is given as,

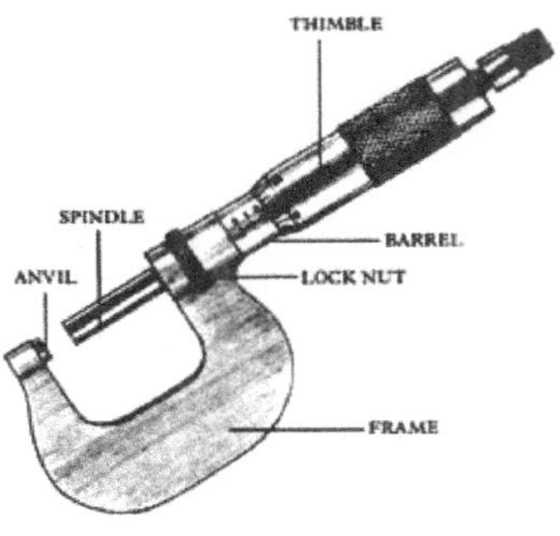

Diagram 10.1: Micrometer

A lock nut is provided for locking a dimension by preventing motion of the spindle. A ratchet stop is provided to maintain uniform measuring pressure. When spindle is brought in contact with the work at correct measuring pressure, the clutch, starts slipping and no further movement of the spindle takes place the rotation of ratchet. In the backward movement it is positive due to shape of ratchet.

Least Count

$$= \frac{\text{smallest division on main scale}}{\text{No. of division on vernier scale}}$$

$$= 0.5 / 50$$

$$= 0.01\text{mm.}$$

Reading : Before taking reading anvil and spindle must be brought together carefully and the initial reading noted down. In metric micrometers the pitch of the screw thread is0.5mm so that one revolution of the screw moves it exactly by 0.5mm. Main scale on barrel has least division of 0.5mm. The thimble has 50 division on its circumference therefore one division on thimble is = 0.5 / 50 =0.01mm.

Observation Table

Sr.no	M.S.R. (A)	V.S.R x LC (B)	Total C=A+B	Average Dimension
1				
2				
3				
4				

Result & discussion :- ....

....

....

Practical 11

Object : To measure the speed of a motor by 'Photo Pickup' and Magnetic Pickup methods.

Apparatus : Tachometer , DPM

Theory

Magnetic pickup : In this type of device, a small toothed wheel is to be attached to the shaft whose speed is to be determined. A permanent magnet with a coil wound around it is placed near the rotating toothed wheel. As the wheel rotates, the magnetic flux linking the magnet and coil changes. As a result, voltage is induced in the coil. The frequency of the pulses depends upon the number of teeth on the wheel.

$$\text{Speed of the shaft} = \frac{\text{Pulses per Second}}{\text{Number of teeth on the wheel}}$$

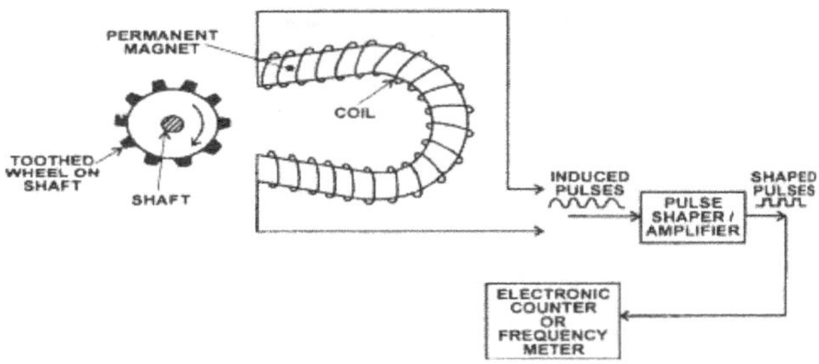

Diagram 11.1: Magnetic Pickup

Photo electric pickup: This is another method of speed measurement based on photo electric effect. An opaque disc with evenly spaced holes on its periphery is to be attached to the shaft whose speed to be measured. A light source is placed on one side and a light sensitive transducer on the other side of the disc. Both are in alignment with the holes in disc. As the disc rotates the intermittent light falling on the photocell produces voltage pulses whose frequency is a measure of the speed of the shaft. Remaining is same as inductive pickup.

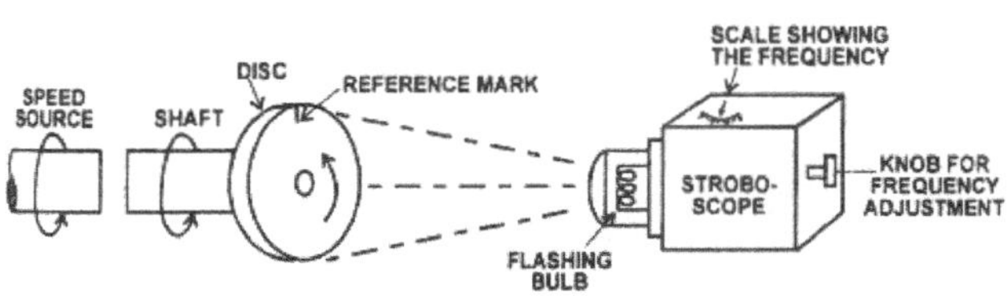

Diagram 11.2: Photo Electric Pickup

Procedure

- Connect the calibration source to the input socket by the cable provided and connect the O/P terminals to the DPM observing polarity.
- Adjust the pot marked "max" to get 1500 as the meter (DPM). Now the set up is calibrated from 1500rpm, Neglect decimal point.
- Connect the photo electric and magnetic pickup in proper fashion.
- Connect the motor terminal cable to the O/P terminals of D.C. supply ensure that dimmer knob is in zero position. Switch on the power supply and control slowly so on changing the speed of motor.
- Motor speed will be indicated by the DPM for each pickup
- Compare the DPM reading with digital tachometer reading

OBSERVATIONS:

S.No	Tachometer (RPM)	Photo pickup (RPM)	Percentage error	Tachometer (RPM)	Magnetic pickup (RPM)	Percentage error

Precautions

- Always start the motor with zero speed and operate the knob slowly.
- As soon as the experiment is over disconnecting the photo-electric and magnetic pick-ups.

Result & discussion :- ....

....

....

Practical 12

Object : To measure the pressure or force applied with the help of piezoelectric load cells.

Apparatus used : Power supply , piezoelectric crystal.

Theory : This transducer works on the principle of piezoelectric effect ,When a mechanical force is applied to piezoelectric material the dimensions of these material gets changed resulting the generation of electric charge or electric potential across the surface of crystal. In the reverse manner, if electric charge or electric potential is applied across the surface of crystal, it results the change in dimension of crystal due to deformation. This principle is used in piezoelectric transducers for the measurement of pressure, force and acceleration i.e. mechanical input and electrical output. The schematic diagram is given as,

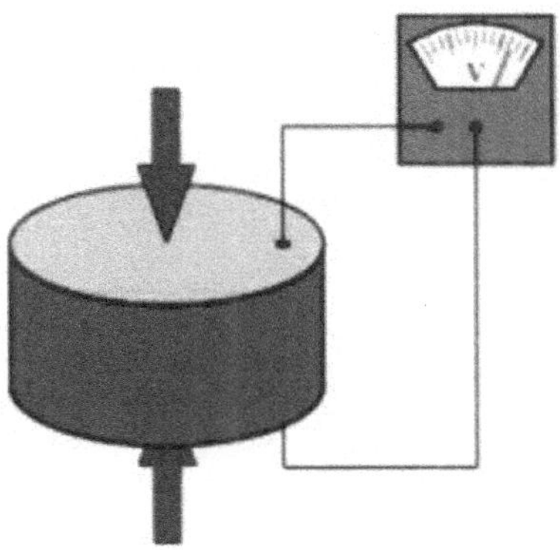

Diagram 12.1: Piezoelectric Load Cells

Piezoelectric materials: Quartz crystal, Rochelle salt, Lithium sulphate, Barium Titanate , Thin Polymer films .

Operating Steps

- Connect the circuit as shown in diagram – 12.1.
- Switch on the power supply.
- Plot graph according to readings taken

Result & discussion :- ....

....

Practical 13

Object : To study the measurement of strain by strain gauge device.

Apparatus used : Digital panel meter (DPM) , cantilever beam, weights

Theory : Strain gauges are devices used to measure the dimensional change of components under test. These are used in many applications like force measuring devices, measurement of vibration, measurement of pressure etc., In this experiment bonded strain gauges are used. These gauges are directly bonded or pasted on the surface to study. In this fine wire strain gauges are used. A fine resistance wire of diameter of 0.025mm, which is bent again and again as shown in figure. This is due to increase the length of the wire so that it permits a uniform distribution of stress. This resistance wire is placed between the two covering like paper, Bakelite or Teflon, which are cemented to each other. The carrier base protects the gauge from damages. Loads are provided for electrically connecting then wheat stone bridge are applied to measuring the change in strain.

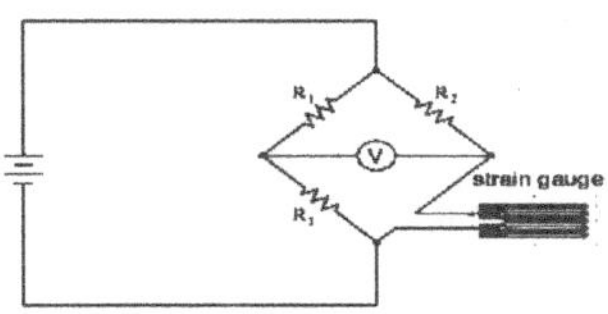

Diagram 13.1: Quarter–Bridge Strain Gauge Circuit

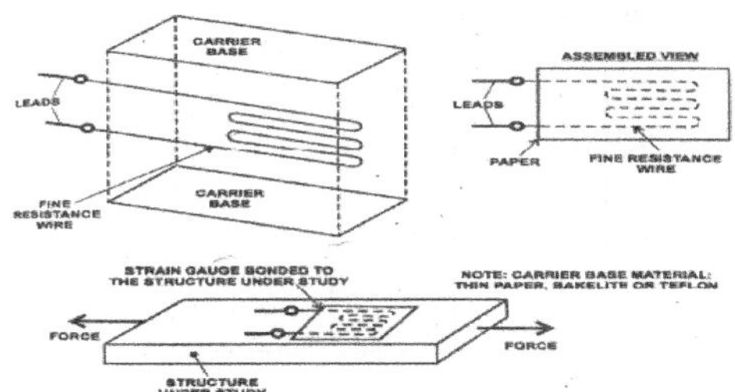

Diagram 13.2: Type of Strain Gauge

Operation

- Connect the flexible wires provided with the strain gauge cantilever beam as shown in figure.
- Turn 'ON' the main supply, note down the initial reading say P_1 and after a predefined time take reading P_2 on the meter.

- Now apply a gentle pressure with hand on the end of the cantilever beam, the Digital Panel Meter (DPM) should indicate some change in reading. This indicates the strain gauge setup is ready for experiment dial.
- Now keep P_3 pot in minimum clock-wise position corresponding to position of gain = 100. Check for null balance again.
- Now apply weight of 1kg, 2kg etc...... so on and note down the DPM reading.

Observation & Calculations

V_O = V_I (ΔR / R)& Gauge Factor = (ΔR / R) / (ΔL / L)

Stress = WL / (1 / 6) BT^2

W = Applied load

L = Length of cantilever beam

BT = Breadth of cantilever beam

T = Thickness of cantilever beam

Theoretical strain = Stress / E

Strain = Change in Length / Original Length

Observation Table

SL.NO.	LOAD	DPM INDICATION	STRAIN(N/mm²		Percentage error	Strain		Percentage error
			Theoretical	Practical		Theoretical	Practical	

Precautions

- Make the connection to the binding posts and terminals very carefully.
- Provide a warm up time of about 10 to 15 minutes before taking readings.
- Ensure that the cantilever beam arrangement is fixed to the table.

Result & discussion :- ....

....

....

Practical 14

Objective : To Study about the displacement by variable capacitance transducers.

Apparatus : Capacitance transducer.

Theory: A variable capacitance transducer works on the principal of change of capacitance may be due to its a displacement sensitive transducer. Due to motion there is a change in the capacitance between two plates. Suitably circulating is used to generate a voltage corresponding to capacitance change. The capacitance between two plates is given by the equation

$$\mathbf{C} = 1/3.\,(\,6\pi\,\epsilon\,A\,/\,d\,)$$

Where,

C= Capacitance in μf

A= Area of plate cm^2

D = Distance between plates

Є = Dielectric constant of the medium between the plates

Advantages

- They require very small force
- They are extremely sensitive
- Good frequency response
- Loading effect is minimum

They are used where inductive transducer are ineffective due to magnetic field.

Disadvantages

- They require insulation as its metallic parts are required to be earth.
- The capacitance transducer shows nonlinear behavior which may be an account of edge current.

Result & discussion :- ....

....

....

Practical 15

Objective: To Study about the calibration of the given pressure gauge.

Apparatus: Pressure cell/Sensor/Gauge, Dial type pressure cell indicator, Digital pressure Indicator .

Theory: Instruments used to measure pressure are called **pressure gauges** or **vacuum gauges**. A **'manometer'** is an instrument that uses a column of liquid to measure pressure, although the term is often used nowadays to mean any pressure measuring instrument. A **pressure sensor/gauges** measures pressure, typically of gases or liquids. Pressure is an expression of the force required to stop a fluid from expanding and is usually stated in terms of force per unit area. A pressure sensor usually acts as a transducer; it generates a signal as a function of the pressure imposed. Pressure sensors are used for control and monitoring in thousands of everyday applications. Pressure sensors can also be used to indirectly measure other variables such as fluid/gas flow, speed, water level, and altitude.

Procedure

- Used a dead weight tester for pressure measurement and connect it to pressure indicator.
- Connect the instrument to mains i.e., 230 volts power supply and switch on the instrument.
- Check up the dead weight pressure tester plunger is to the extreme end so that there should not be any load or pressure on the piston.
- Now adjust the zero point of the indicator, to indicate zero.
- Apply the load of 10kg on the piston.
- Move the plunger to apply pressure on the piston. When applied pressure reaches **10 kg/cm^2**, piston will start moving up.
- Now read the pressure gauge reading and then release the pressure.
- Repeat the experiment for different loads on the piston step by step, and note down the readings of dial gauge and pressure indicator.

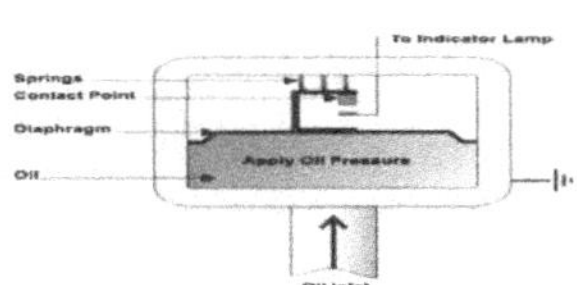

Diagram 15.1: Dead Weight Tester

Observation Table

SL No.	Actual Pressure (P_a) (kg/cm^2)	Pressure shown in digital indicator (P_i) (kg/cm^2)	Error $P_i - P_a$	% Error $\frac{Pi - Pa}{Pa} \times 100$
1				
2				
3				
4				
5				

Calculation: $$\% \text{ Error} = \frac{\text{Indicated Pressure} - \text{Actual Pressure}}{\text{Actual Pressure}} \times 100$$

Applications

- Measurement of steam pressure in power plants & boilers.
- Measurement of pressure in large pumping stations/water works/or minor/major irrigations.

Result & discussion :- ....

....

....

References

1. Mechanical Measurements, Beckwith Marangoni and Lienhard, Pearson Education, 6th Ed., 2006.
2. Dieter, G.E., Mechanical metallurgy, 1988, SI metric edition, McGraw-Hill, ISBN 0-07-100406-8.
3. Unit Operations of Chemical Engineering' by McCabe and Smith, 5th Edition, McGraw Hill, Inc. pages 214-221.
4. Engineering Metrology, I.C. Gupta, Dhanpat Rai Publications, Delhi.
5. Mechanical Measurements, R.K. Jain Khanna Publishers, 1994.
6. Industrial Instrumentation, Alsutko, Jerry. D. Faulk, Cengage Asia Pvt. Ltd. 2002.
7. Measurement Systems Applications and Design, Ernest O. Doebelin, 5th Ed., McGraw Hill Book Co.
8. Metrology & Measurement, Anand K. Bewoor & Vinay A. Kulkarni, Tata McGraw Hill Pvt. Ltd., New Delhi.

Section–III

Formula for Numerical Problem

Formula for Numerical Problem

(i) Average Velocity For Flowing Fluid **$V = 4Q / \Pi d^2$**

(ii) Reynolds number **$Re = (\varrho V d) / \mu$**

(iii) Resistance of flow **$R_t = (16QfL\varrho) / (\Pi^2 d^5)$**

(iv) Pressure Drop **$P = k_t Q^2$**

(v) **Hydrulic inertance**

$$I = \frac{L\rho}{A}$$

(vi) Tank capacitance **$C = A / \varrho g$**

(vii) Convection $q = (H A \Delta\Theta)$

(viii) Thermal resistance $q = (H A \Delta\Theta)$

(ix) Heat flow $q = (k A \Delta\Theta) / \Delta L$

(x) **Thermal capacitance (C) = Mass x Specific heat of liquid**

$c = M c$

(xi) **System Sensitivity**

$$S = \left|\frac{\partial H/H}{\partial G/G}\right| = \left|\frac{\partial H}{\partial G}\frac{G}{H}\right|$$

(xii) **Maximum Overshoot**

$$M_p = \frac{|y_{max} - y_{ss}|}{|y_{ss}|} \cdot 100\ \%$$

(xiii) **Rise time**

$$t_d = 2.197\ \tau$$

(xiv) Settling time

$$t_s := -\tau \cdot \ln\left(1 - \frac{0.95 \cdot K}{K}\right)$$

(xv) Rise time second order

$$t_r \cong \frac{1 - 0.4167 \cdot \zeta + 2.917 \cdot \zeta^2}{\omega_n}$$

(xvi) peak time second order

$$t_p = \frac{\pi}{\sqrt{1 - \zeta^2} \cdot \omega_n}$$

(xvii) Maximum overshoot second order

$$M_p = e^{-\frac{\zeta \cdot \pi}{\sqrt{1 - \zeta^2}}}\ \%$$

(xviii) Response equation fcr **over damped system**

$$y_s(t) = \frac{1}{\omega_n^2}\left[1 - \frac{\omega_n}{2\sqrt{\zeta^2 - 1}}\left(\tau_1 e^{-t/\tau_1} - \tau_2 e^{-t/\tau_2}\right)\right]$$

(xix) Response equation fcr under damped

$$y_s(t) = \frac{1}{\omega_n^2}\left[1 - \frac{e^{-\zeta\omega_n t}}{\sqrt{1-\zeta^2}}\cos(\omega_d t - \psi)\right] \quad \text{for } 0 < \zeta < 1.$$

(xx) Undamped natural frequency

$$\omega_n = \sqrt{\frac{K}{J}}$$

(xxi) Damping ratio

$$\zeta = \frac{B/J}{2\sqrt{K/J}} = \frac{B}{2\sqrt{KJ}}.$$

(xxii) **Air density**

$$\rho = P / R_g T$$

(xxiii) **kinematic viscosity**

$$(\mu / \rho)$$

(xxiv) **Capacity of bellows**

$$C = \frac{A^2 \rho}{K} + \frac{V}{nR_g T}$$

(xxv) **Air Resistance**

$$R = \frac{8 v L}{\pi r^4}$$

(xxvi) **Time Constant**

$$\tau = RC$$

(xxvii) **Pressure Gauge Capacitance**

$$C = \frac{A^2 \rho}{K}$$

(xxviii) Thickness of the diaphragm

$$t = a\sqrt{\frac{0.75(p_1 - p_2)(1-v^2)\times 10^3}{E}}\,\frac{1}{\sqrt{\frac{V_o}{V_s}}}$$

(xxix) Maximum of the diaphragm

$$y_{max} = \frac{3(p_1 - p_2)\left(1 - \upsilon_p^2\right) r^4}{16 E_m t^3}$$

(xxx) Natural frequency of the diaphragm

$$\omega_n = 10.21\sqrt{\frac{E_m t^2}{12\left(1 - \upsilon_p^2\right)\rho r^4}}$$

(xxxi) **Absolute pressure**

$$P_{ab} = \text{Atm pressure} + (\rho g H / W)$$

(xxxii) **Gauge pressure** $P_{Gauge} = Pabs - P_o$

(xxxiii) buoyancy effects (e_b)

$$e_2 = -\gamma_{air}/\gamma_{masses}$$

(xxxiv) Gravity effects (e_g)

$$e_t = -(2.637 \times 10^{-3} \cos 2\phi + 9.6 \times 10^{-8} z + 5 \times 10^{-5})$$

(xxxv) sensitivity of capacitive sensor

$$Ks = (\frac{\partial C}{\partial t})$$

(xxxvi) **velocity approach factor**

$$E = 1 / (1 - \beta^4)^{1/2}$$

(xxxvii) **Flow rate (Q)**

$$K_0 A_0 Y \sqrt{2gH[(S_m/S) - 1]}$$

(xxxviii) Pressure difference

$$\mathbf{P_a - P_b = (\rho_m - \rho) g\, h_m}$$

(xxxix) Velocity at the venture – throat

$$V_b = \frac{C_v}{\sqrt{1-\beta^4}} \sqrt{\frac{2(P_a - P_b)}{\rho}}$$

(xl) venture velocity & flow rate

Velocity $$V_2 = \frac{C_d A_1}{\sqrt{A_1^2 - A_2^2}} \sqrt{2gh_m\left(\frac{\rho_m}{\rho} - 1\right)} \text{ and } Q = V_2 \times A_2$$

Flow rate $$Q = \frac{C_d A_1 A_2}{\sqrt{A_1^2 - A_2^2}} \sqrt{2gh_m\left(\frac{\rho_m}{\rho} - 1\right)}$$

(xli) Flow velocity of orifice meter

$$V = C_v \sqrt{2gh}$$

(xlii) Coefficient of contraction =

$$C_c = \frac{C_d}{C_v}$$

(xliii) Velocity approach factor =

$$M_{v.a} = \frac{1}{\sqrt{1 - (d/D)^4}}$$

(xliv) Velocity in electromagnetic field = v = (e / B l)

(xlv) Induced Emf (e) = $B v l$

(xlvi) Volume flow rate (Q) in rota meter

$$Q = Cd \cdot \frac{D_p^2 - D_f^2}{D_f} \sqrt{\frac{\pi g V_f (\rho_f - \rho)}{2\rho}}$$

INDEX